Methods in Enzymology

Volume 281

VITAMINS AND COENZYMES

Part K

METHODS IN ENZYMOLOGY

EDITORS-IN-CHIEF

John N. Abelson Melvin I. Simon

DIVISION OF BIOLOGY
CALIFORNIA INSTITUTE OF TECHNOLOGY
PASADENA, CALIFORNIA

FOUNDING EDITORS

Sidney P. Colowick and Nathan O. Kaplan

Methods in Enzymology

Volume 281

Vitamins and Coenzymes

Part K

EDITED BY

Donald B. McCormick

DEPARTMENT OF BIOCHEMISTRY
EMORY UNIVERSITY
ATLANTA, GEORGIA

John W. Suttie

DEPARTMENTS OF BIOCHEMISTRY AND NUTRITIONAL SCIENCES
UNIVERSITY OF WISCONSIN–MADISON
MADISON, WISCONSIN

Conrad Wagner

DEPARTMENT OF VETERANS AFFAIRS MEDICAL CENTER
AND DEPARTMENT OF BIOCHEMISTRY
VANDERBILT UNIVERSITY SCHOOL OF MEDICINE
NASHVILLE, TENNESSEE

ACADEMIC PRESS
San Diego London Boston New York Sydney Tokyo Toronto

This book is printed on acid-free paper. ♾

Academic Press
15 East 26th Street, 15th floor, New York, New York 10010
http://www.apnet.com

Academic Press Limited
24-28 Oval Road, London NW1 7DX, UK
http://www.hbuk.co.uk/ap/

International Standard Book Number: 0-12-182182-X

PRINTED IN THE UNITED STATES OF AMERICA
97 98 99 00 01 02 MM 9 8 7 6 5 4 3 2 1

Table of Contents

Section I. Folic Acid

Section II. Vitamin B_{12} and Cobalamins

Section III. Heme

Section IV. Miscellaneous Vitamins and Coenzymes

Section V. Methodologies Broadly Applicable to Vitamins and Coenzymes

Contributors to Volume 281

Article numbers are in parentheses following the names of contributors.
Affiliations listed are current.

RHONE K. AKEE (11), *SAIC Frederick, National Cancer Institute—Frederick Research and Development Center, Frederick, Maryland 21702-1201*

ROBERT H. ALLEN (28), *University of Colorado Health Sciences Center, Hematology Division, Denver, Colorado 80220*

DAVID H. ALPERS (29), *Washington University School of Medicine, St. Louis, Missouri 63110*

NOBUYUKI AMANO (46), *Molecular and Clinical Nutrition Section, National Institute of Diabetes and Digestive and Kidney Diseases, National Institutes of Health, Bethesda, Maryland 20892-1372*

DEAN R. APPLING (22, 26), *Department of Chemistry and Biochemistry and The Biochemical Institute, The University of Texas, Austin, Texas 78712*

IAN ATKINSON (17), *Department of Nutritional Sciences, University of California, Berkeley, California 94720-3104*

SARAH J. AWAN (36), *Department of Molecular Genetics, Institute of Ophthalmology, University College London, London EC1V 9EL, United Kingdom*

JUNE E. AYLING (1), *Department of Pharmacology, University of South Alabama, College of Medicine, Mobile, Alabama 36688*

PAMELA J. BAGLEY (2), *Jean Mayer USDA Human Nutrition Research Center on Aging, Tufts University, Boston, Massachusetts 02111*

STEVEN W. BAILEY (1), *Department of Pharmacology, University of South Alabama, College of Medicine, Mobile, Alabama 36688*

RUMA BANERJEE (23), *Department of Biochemistry, University of Nebraska, Lincoln, Nebraska 68588-0664*

SANTANU BOSE (32), *Department of Biochemistry, Medical College of Wisconsin, Milwaukee, Wisconsin 53226*

ALBERT BRENNER (17), *Department of Nutritional Sciences, University of California, Berkeley, California 94720-3104*

ZHIQIANG CHEN (23), *Department of Biochemistry, University of Nebraska, Lincoln, Nebraska 68588-0664*

ROBERT J. COOK (16), *Department of Biochemistry, Vanderbilt University School of Medicine, Nashville, Tennessee 37232-0146*

EDWIN A. COSSINS (18), *Department of Biological Sciences, University of Alberta, Edmonton, Alberta T6G 2E9, Canada*

HARRY A. DAILEY (38, 39, 43), *Department of Microbiology and Department of Biochemistry and Molecular Biology, University of Georgia, Athens, Georgia 30602-7229*

TAMARA A. DAILEY (38, 39), *Department of Biochemistry and Molecular Biology, University of Georgia, Athens, Georgia 30602-7229*

GEORGE H. ELDER (40), *Department of Medical Biochemistry, University of Wales College of Medicine, Cardiff CF4 4XN, United Kingdom*

JIANGUO FAN (12), *Molecular Pharmacology and Therapeutics, Memorial Sloan-Kettering Cancer Center, New York, New York 10021*

ANTHONY L. FITZHUGH (11), *SAIC Frederick, National Cancer Institute—Frederick Research and Development Center, Frederick, Maryland 21702-1201*

KERRY FLUHR (24), *Biophysics Research Division and Department of Biological Chemistry, University of Michigan, Ann Arbor, Michigan 48109-1055*

TIMOTHY GARROW (17), *Division of Foods and Nutrition, University of Illinois, Urbana, Illinois*

MARILYN M. GORDON (29), *Washington University School of Medicine, St. Louis, Missouri 63110*

CELIA W. GOULDING (24), *Biophysics Research Division, University of Michigan, Ann Arbor, Michigan 48109-1055*

JESSE F. GREGORY III (13), *Food Science and Human Nutrition Department, University of Florida, Gainesville, Florida 32611-0370*

CHARLES B. GRISSOM (27), *Department of Chemistry, University of Utah, Salt Lake City, Utah 84112*

SUMEDHA GULATI (23), *Department of Biochemistry, University of Nebraska, Lincoln, Nebraska 68588-0664*

KAZUYUKI HATAKEYAMA (15), *University of Pittsburgh, Department of Surgery, Pittsburgh, Pennsylvania 15213*

DONALD W. HORNE (4), *VA Medical Center (151), Department of Biochemistry, Vanderbilt University, Nashville, Tennessee 37212-2637*

MASAAKI HOSHIGA (15), *Osaka Medical College, Department of Internal Medicine, Osaka, Japan*

SHA HUANG (24), *Biophysics Research Division, University of Michigan, Ann Arbor, Michigan 48109-1055*

FRANK HUENNEKENS (12), *Division of Biochemistry, Department of Molecular and Experimental Medicine, The Scripps Research Institute, La Jolla, California 92037*

HELENA C. IMESON (18), *Department of Biology, University of Lethbridge, Lethbridge, Canada*

JOSEPH T. JARRETT (24), *Biophysics Research Division, University of Michigan, Ann Arbor, Michigan 48109-1055*

JACQUES JOLIVET (20), *Centre de Recherche, Centre Hospitalier de L'Université de Montréal, Pavillon Hôtel-Dieu, Montreal, Quebec H2W 1T8, Canada*

EVDOKIA KASTANOS (26), *Department of Chemistry and Biochemistry, The University of Texas, Austin, Texas 78712*

J. FRED KOLHOUSE (3), *Department of Medicine, Division of Hematology, University of Colorado Health Sciences Center, Denver, Colorado 80220*

ORAN KWON (46), *Molecular and Clinical Nutrition Section, National Institute of Diabetes and Digestive and Kidney Diseases, National Institutes of Health, Bethesda, Maryland 20892-1372*

PIERRE LABBE (42), *Laboratoire de Biochimie des Porphyrines, Institut Jacques Monod, Université Paris VII, 75 251 Paris-Cedex 05, France*

ROBERT LEADBEATER (37), *Department of Biochemistry, School of Biological Science, University of Southampton, Southampton S016 7PX, United Kingdom*

ROBERT J. LEEMING (7), *Clinical Chemistry Department, Birmingham Children's Hospital, Birmingham B16 8ET, United Kingdom*

JEREMY E. LELEAN (35), *Department of Biochemistry, School of Biological Science, University of Southampton, Southampton S016 7PX, United Kingdom*

MARK LEVINE (46), *Molecular and Clinical Nutrition Section, National Institute of Diabetes and Digestive and Kidney Diseases, National Institutes of Health, Bethesda, Maryland 20892-1372*

ADRIAN J. LLOYD (35), *Department of Biochemistry, School of Biological Science, University of Southampton, Southampton S016 7PX, United Kingdom*

ROBERT E. MACKENZIE (21), *Department of Biochemistry, McGill University, Montreal, Quebec H3G 1Y6, Canada*

ROWENA G. MATTHEWS (24), *Biophysics Research Division and Department of Biological Chemistry, University of Michigan, Ann Arbor, Michigan 48109-1055*

JOE MCPARTLIN (8), *Vitamin Research, Sir Patrick Duns Trinity College Laboratory, Central Pathology, St. James Hospital, Dublin 8, Ireland*

SHELDON MILSTIEN (14), *Laboratory of Cell and Molecular Regulation, National Institute of Mental Health, National Institutes of Health, Bethesda, Maryland 20892*

ANNE M. MOLLOY (5), *Department of Clinical Medicine, Trinity College Dublin, Dublin 2, Ireland*

YOSHIHISA NAKANO (33, 34), *Department of Applied Biological Chemistry, Osaka Prefecture University, Sakai, Osaka 593, Japan*

ETTAYA NATARAJAN (27), *Department of Chemistry, University of Utah, Salt Lake City, Utah 84112*

JAE PARK (46), *Molecular and Clinical Nutrition Section, National Institute of Diabetes and Digestive and Kidney Diseases, National Institutes of Health, Bethesda, Maryland 20892-1372*

LAURA B. PASTERNACK (26), *Department of Chemistry and Biochemistry, The University of Texas, Austin, Texas 78712*

CHRISTINE M. PFEIFFER (13), *Food Science and Human Nutrition Department, University of Florida, Gainesville, Florida 32611-0370*

LIAN QIAN (31), *State University of New York, Health Science Center at Brooklyn, Brooklyn, New York 11203*

EDWARD V. QUADROS (30, 31), *State University of New York, Health Science Center at Brooklyn, Brooklyn, New York 11203*

ANDREW G. ROBERTS (40), *Department of Medical Biochemistry, University of Wales College of Medicine, Cardiff CF4 4XN, United Kingdom*

SHELDON P. ROTHENBERG (30, 31), *State University of New York, Health Science Center at Brooklyn, Brooklyn, New York 11203*

STEVEN C. RUMSEY (46), *Molecular and Clinical Nutrition Section, National Institute of Diabetes and Digestive and Kidney Diseases, National Institutes of Health, Bethesda, Maryland 20892-1372*

GREG RUSSELL-JONES (29), *Biotech Australia, Roseville, New South Wales, Australia*

CHERUPPOLIL R. SANTHOSH-KUMAR (3), *Regional Cancer Centre, Medical College Campus, Trivandrum, Kerala State, India*

VERNE SCHIRCH (9, 10, 19), *Department of Biochemistry, Virginia Commonwealth University, Richmond, Virginia 23298-0614*

JOHN M. SCOTT (5, 8), *Department of Biochemistry, Trinity College Dublin, Dublin 2, Ireland*

BELLUR SEETHARAM (32), *Division of Gastroenterology, Department of Medicine and Biochemistry, Medical College of Wisconsin, Milwaukee, Wisconsin 53226*

JACOB SELHUB (2), *Jean Mayer USDA Human Nutrition Research Center on Aging, Tufts University, Boston, Massachusetts 02111*

VERA M. SELLERS (43), *Department of Microbiology, University of Georgia, Athens, Georgia 30602-2605*

BARRY SHANE (17), *Department of Nutritional Sciences, University of California, Berkeley, California 94720-3104*

PETER M. SHOOLINGIN-JORDAN (35, 36, 37), *Department of Biochemistry, School of Biological Science, University of Southampton, Southampton S016 7PX, United Kingdom*

DAVID P. SUNDIN (28), *Department of Medicine/Nephrology, Indiana University School of Medicine, Indianapolis, Indiana 46202*

HELMUT WACHTER (6), *Institut für Medizinische Chemie und Biochemie der Universität Innsbruck, A-6020 Innsbruck, Austria*

YAOHUI WANG (46), *Molecular and Clinical Nutrition Section, National Institute of Diabetes and Digestive and Kidney Diseases, National Institutes of Health, Bethesda, Maryland 20892-1372*

MARTIN J. WARREN (36), *Department of Biochemistry, School of Biological Science, University of Southampton, Southampton S016 7PX, United Kingdom*

FUMIO WATANABE (33, 34), *Department of Food and Nutrition, Kochi Women's University, Eikokuji-cho 5-15, Kochi 780, Japan*

ERNST R. WERNER (6), *Institut für Medizinische Chemie und Biochemie der Universität Innsbruck, A-6020 Innsbruck, Austria*

GABRIELE WERNER-FELMAYER (6), *Institut für Medizinische Chemie und Biochemie der Universität Innsbruck, A-6020 Innsbruck, Austria*

MARY G. WEST (22), *Department of Chemistry and Biochemistry, The University of Texas, Austin, Texas 78712*

ROBERT H. WHITE (44), *Department of Biochemistry, Virginia Polytechnic Institute and State University, Blacksburg, Virginia 24061-0308*

YAKOV Y. WOLDMAN (26), *Department of Chemistry and Biochemistry, The University of Texas, Austin, Texas 78712*

TAKEO YOSHINAGA (41), *Department of Public Health, Faculty of Medicine, Kyoto University, Kyoto 606-01, Japan*

HOWARD ZALKIN (25), *Department of Biochemistry, Purdue University, West Lafayette, Indiana 47907-1153*

JANOS ZEMPLENI (45), *Department of Pediatrics (Gastroenterology), Arkansas Children's Hospital Research Institute, Little Rock, Arkansas 72202*

Preface

From 1970 to 1986, eight "Vitamins and Coenzymes" volumes were published in the *Methods in Enzymology* series. Volumes XVIII A, B, and C appeared in 1970–1971 and Volumes 62 (D), 66 (E), and 67 (F) in 1979–1980. These volumes were edited by D. B. McCormick and L. D. Wright. Volumes 122 (G) and 123 (H), published in 1986, were edited by F. Chytil and D. B. McCormick. In the decade that has elapsed since the last volume was published, considerable progress has been made, so it was reasonable to update the subject of "Vitamins and Coenzymes."

In this current set of volumes (279, 280, 281, and 282) we have attempted to collect and collate many of the newer techniques and methodologies attendant to assays, isolations, and characterizations of vitamins, coenzymes, and those systems responsible for their biosynthesis, transport, and metabolism. There are examples of procedures that are modifications of earlier ones as well as of those that have newly evolved. As before, there has been an attempt to allow such overlap as would offer flexibility in the choice of methods, rather than presume any one is best for all laboratories. Where there is no inclusion of a particular subject covered in earlier volumes, we believe the subject was adequately treated and the reader should refer to those volumes.

The information provided reflects the efforts of our numerous contributors to whom we express our gratitude. We are also grateful to our secretaries at our academic home bases and to Shirley Light and the staff of Academic Press. Finally, one of us (D. B. M.) recalls fondly the encouragement proffered years ago by Drs. Nathan O. Kaplan and Sidney P. Colowick who saw the need for "Vitamins and Coenzymes" within the *Methods in Enzymology* series, which they initiated.

DONALD B. MCCORMICK
JOHN W. SUTTIE
CONRAD WAGNER

METHODS IN ENZYMOLOGY

VOLUME I. Preparation and Assay of Enzymes
Edited by SIDNEY P. COLOWICK AND NATHAN O. KAPLAN

VOLUME II. Preparation and Assay of Enzymes
Edited by SIDNEY P. COLOWICK AND NATHAN O. KAPLAN

VOLUME III. Preparation and Assay of Substrates
Edited by SIDNEY P. COLOWICK AND NATHAN O. KAPLAN

VOLUME IV. Special Techniques for the Enzymologist
Edited by SIDNEY P. COLOWICK AND NATHAN O. KAPLAN

VOLUME V. Preparation and Assay of Enzymes
Edited by SIDNEY P. COLOWICK AND NATHAN O. KAPLAN

VOLUME VI. Preparation and Assay of Enzymes (*Continued*)
Preparation and Assay of Substrates
Special Techniques
Edited by SIDNEY P. COLOWICK AND NATHAN O. KAPLAN

VOLUME VII. Cumulative Subject Index
Edited by SIDNEY P. COLOWICK AND NATHAN O. KAPLAN

VOLUME VIII. Complex Carbohydrates
Edited by ELIZABETH F. NEUFELD AND VICTOR GINSBURG

VOLUME IX. Carbohydrate Metabolism
Edited by WILLIS A. WOOD

VOLUME X. Oxidation and Phosphorylation
Edited by RONALD W. ESTABROOK AND MAYNARD E. PULLMAN

VOLUME XI. Enzyme Structure
Edited by C. H. W. HIRS

VOLUME XII. Nucleic Acids (Parts A and B)
Edited by LAWRENCE GROSSMAN AND KIVIE MOLDAVE

VOLUME XIII. Citric Acid Cycle
Edited by J. M. LOWENSTEIN

VOLUME XIV. Lipids
Edited by J. M. LOWENSTEIN

VOLUME XV. Steroids and Terpenoids
Edited by RAYMOND B. CLAYTON

VOLUME XVI. Fast Reactions
Edited by KENNETH KUSTIN

VOLUME LV. Biomembranes (Part F: Bioenergetics)
Edited by SIDNEY FLEISCHER AND LESTER PACKER

VOLUME LVI. Biomembranes (Part G: Bioenergetics)
Edited by SIDNEY FLEISCHER AND LESTER PACKER

VOLUME LVII. Bioluminescence and Chemiluminescence
Edited by MARLENE A. DELUCA

VOLUME LVIII. Cell Culture
Edited by WILLIAM B. JAKOBY AND IRA PASTAN

VOLUME LIX. Nucleic Acids and Protein Synthesis (Part G)
Edited by KIVIE MOLDAVE AND LAWRENCE GROSSMAN

VOLUME LX. Nucleic Acids and Protein Synthesis (Part H)
Edited by KIVIE MOLDAVE AND LAWRENCE GROSSMAN

VOLUME 61. Enzyme Structure (Part H)
Edited by C. H. W. HIRS AND SERGE N. TIMASHEFF

VOLUME 62. Vitamins and Coenzymes (Part D)
Edited by DONALD B. MCCORMICK AND LEMUEL D. WRIGHT

VOLUME 63. Enzyme Kinetics and Mechanism (Part A: Initial Rate and Inhibitor Methods)
Edited by DANIEL L. PURICH

VOLUME 64. Enzyme Kinetics and Mechanism (Part B: Isotopic Probes and Complex Enzyme Systems)
Edited by DANIEL L. PURICH

VOLUME 65. Nucleic Acids (Part I)
Edited by LAWRENCE GROSSMAN AND KIVIE MOLDAVE

VOLUME 66. Vitamins and Coenzymes (Part E)
Edited by DONALD B. MCCORMICK AND LEMUEL D. WRIGHT

VOLUME 67. Vitamins and Coenzymes (Part F)
Edited by DONALD B. MCCORMICK AND LEMUEL D. WRIGHT

VOLUME 68. Recombinant DNA
Edited by RAY WU

VOLUME 69. Photosynthesis and Nitrogen Fixation (Part C)
Edited by ANTHONY SAN PIETRO

VOLUME 70. Immunochemical Techniques (Part A)
Edited by HELEN VAN VUNAKIS AND JOHN J. LANGONE

VOLUME 71. Lipids (Part C)
Edited by JOHN M. LOWENSTEIN

VOLUME 72. Lipids (Part D)
Edited by JOHN M. LOWENSTEIN

VOLUME 73. Immunochemical Techniques (Part B)
Edited by JOHN J. LANGONE AND HELEN VAN VUNAKIS

VOLUME 74. Immunochemical Techniques (Part C)
Edited by JOHN J. LANGONE AND HELEN VAN VUNAKIS

VOLUME 75. Cumulative Subject Index Volumes XXXI, XXXII, XXXIV–LX
Edited by EDWARD A. DENNIS AND MARTHA G. DENNIS

VOLUME 76. Hemoglobins
Edited by ERALDO ANTONINI, LUIGI ROSSI-BERNARDI, AND EMILIA CHIANCONE

VOLUME 77. Detoxication and Drug Metabolism
Edited by WILLIAM B. JAKOBY

VOLUME 78. Interferons (Part A)
Edited by SIDNEY PESTKA

VOLUME 79. Interferons (Part B)
Edited by SIDNEY PESTKA

VOLUME 80. Proteolytic Enzymes (Part C)
Edited by LASZLO LORAND

VOLUME 81. Biomembranes (Part H: Visual Pigments and Purple Membranes, I)
Edited by LESTER PACKER

VOLUME 82. Structural and Contractile Proteins (Part A: Extracellular Matrix)
Edited by LEON W. CUNNINGHAM AND DIXIE W. FREDERIKSEN

VOLUME 83. Complex Carbohydrates (Part D)
Edited by VICTOR GINSBURG

VOLUME 84. Immunochemical Techniques (Part D: Selected Immunoassays)
Edited by JOHN J. LANGONE AND HELEN VAN VUNAKIS

VOLUME 85. Structural and Contractile Proteins (Part B: The Contractile Apparatus and the Cytoskeleton)
Edited by DIXIE W. FREDERIKSEN AND LEON W. CUNNINGHAM

VOLUME 86. Prostaglandins and Arachidonate Metabolites
Edited by WILLIAM E. M. LANDS AND WILLIAM L. SMITH

VOLUME 87. Enzyme Kinetics and Mechanism (Part C: Intermediates, Stereochemistry, and Rate Studies)
Edited by DANIEL L. PURICH

VOLUME 88. Biomembranes (Part I: Visual Pigments and Purple Membranes, II)
Edited by LESTER PACKER

VOLUME 89. Carbohydrate Metabolism (Part D)
Edited by WILLIS A. WOOD

VOLUME 90. Carbohydrate Metabolism (Part E)
Edited by WILLIS A. WOOD

VOLUME 125. Biomembranes (Part M: Transport in Bacteria, Mitochondria, and Chloroplasts: General Approaches and Transport Systems)
Edited by SIDNEY FLEISCHER AND BECCA FLEISCHER

VOLUME 126. Biomembranes (Part N: Transport in Bacteria, Mitochondria, and Chloroplasts: Protonmotive Force)
Edited by SIDNEY FLEISCHER AND BECCA FLEISCHER

VOLUME 127. Biomembranes (Part O: Protons and Water: Structure and Translocation)
Edited by LESTER PACKER

VOLUME 128. Plasma Lipoproteins (Part A: Preparation, Structure, and Molecular Biology)
Edited by JERE P. SEGREST AND JOHN J. ALBERS

VOLUME 129. Plasma Lipoproteins (Part B: Characterization, Cell Biology, and Metabolism)
Edited by JOHN J. ALBERS AND JERE P. SEGREST

VOLUME 130. Enzyme Structure (Part K)
Edited by C. H. W. HIRS AND SERGE N. TIMASHEFF

VOLUME 131. Enzyme Structure (Part L)
Edited by C. H. W. HIRS AND SERGE N. TIMASHEFF

VOLUME 132. Immunochemical Techniques (Part J: Phagocytosis and Cell-Mediated Cytotoxicity)
Edited by GIOVANNI DI SABATO AND JOHANNES EVERSE

VOLUME 133. Bioluminescence and Chemiluminescence (Part B)
Edited by MARLENE DELUCA AND WILLIAM D. MCELROY

VOLUME 134. Structural and Contractile Proteins (Part C: The Contractile Apparatus and the Cytoskeleton)
Edited by RICHARD B. VALLEE

VOLUME 135. Immobilized Enzymes and Cells (Part B)
Edited by KLAUS MOSBACH

VOLUME 136. Immobilized Enzymes and Cells (Part C)
Edited by KLAUS MOSBACH

VOLUME 137. Immobilized Enzymes and Cells (Part D)
Edited by KLAUS MOSBACH

VOLUME 138. Complex Carbohydrates (Part E)
Edited by VICTOR GINSBURG

VOLUME 139. Cellular Regulators (Part A: Calcium- and Calmodulin-Binding Proteins)
Edited by ANTHONY R. MEANS AND P. MICHAEL CONN

VOLUME 140. Cumulative Subject Index Volumes 102–119, 121–134

VOLUME 159. Initiation and Termination of Cyclic Nucleotide Action
Edited by JACKIE D. CORBIN AND ROGER A. JOHNSON

VOLUME 160. Biomass (Part A: Cellulose and Hemicellulose)
Edited by WILLIS A. WOOD AND SCOTT T. KELLOGG

VOLUME 161. Biomass (Part B: Lignin, Pectin, and Chitin)
Edited by WILLIS A. WOOD AND SCOTT T. KELLOGG

VOLUME 162. Immunochemical Techniques (Part L: Chemotaxis and Inflammation)
Edited by GIOVANNI DI SABATO

VOLUME 163. Immunochemical Techniques (Part M: Chemotaxis and Inflammation)
Edited by GIOVANNI DI SABATO

VOLUME 164. Ribosomes
Edited by HARRY F. NOLLER, JR., AND KIVIE MOLDAVE

VOLUME 165. Microbial Toxins: Tools for Enzymology
Edited by SIDNEY HARSHMAN

VOLUME 166. Branched-Chain Amino Acids
Edited by ROBERT HARRIS AND JOHN R. SOKATCH

VOLUME 167. Cyanobacteria
Edited by LESTER PACKER AND ALEXANDER N. GLAZER

VOLUME 168. Hormone Action (Part K: Neuroendocrine Peptides)
Edited by P. MICHAEL CONN

VOLUME 169. Platelets: Receptors, Adhesion, Secretion (Part A)
Edited by JACEK HAWIGER

VOLUME 170. Nucleosomes
Edited by PAUL M. WASSARMAN AND ROGER D. KORNBERG

VOLUME 171. Biomembranes (Part R: Transport Theory: Cells and Model Membranes)
Edited by SIDNEY FLEISCHER AND BECCA FLEISCHER

VOLUME 172. Biomembranes (Part S: Transport: Membrane Isolation and Characterization)
Edited by SIDNEY FLEISCHER AND BECCA FLEISCHER

VOLUME 173. Biomembranes [Part T: Cellular and Subcellular Transport: Eukaryotic (Nonepithelial) Cells]
Edited by SIDNEY FLEISCHER AND BECCA FLEISCHER

VOLUME 174. Biomembranes [Part U: Cellular and Subcellular Transport: Eukaryotic (Nonepithelial) Cells]
Edited by SIDNEY FLEISCHER AND BECCA FLEISCHER

VOLUME 175. Cumulative Subject Index Volumes 135–139, 141–167

VOLUME 176. Nuclear Magnetic Resonance (Part A: Spectral Techniques and Dynamics)
Edited by NORMAN J. OPPENHEIMER AND THOMAS L. JAMES

VOLUME 177. Nuclear Magnetic Resonance (Part B: Structure and Mechanism)
Edited by NORMAN J. OPPENHEIMER AND THOMAS L. JAMES

VOLUME 178. Antibodies, Antigens, and Molecular Mimicry
Edited by JOHN J. LANGONE

VOLUME 179. Complex Carbohydrates (Part F)
Edited by VICTOR GINSBURG

VOLUME 180. RNA Processing (Part A: General Methods)
Edited by JAMES E. DAHLBERG AND JOHN N. ABELSON

VOLUME 181. RNA Processing (Part B: Specific Methods)
Edited by JAMES E. DAHLBERG AND JOHN N. ABELSON

VOLUME 182. Guide to Protein Purification
Edited by MURRAY P. DEUTSCHER

VOLUME 183. Molecular Evolution: Computer Analysis of Protein and Nucleic Acid Sequences
Edited by RUSSELL F. DOOLITTLE

VOLUME 184. Avidin–Biotin Technology
Edited by MEIR WILCHEK AND EDWARD A. BAYER

VOLUME 185. Gene Expression Technology
Edited by DAVID V. GOEDDEL

VOLUME 186. Oxygen Radicals in Biological Systems (Part B: Oxygen Radicals and Antioxidants)
Edited by LESTER PACKER AND ALEXANDER N. GLAZER

VOLUME 187. Arachidonate Related Lipid Mediators
Edited by ROBERT C. MURPHY AND FRANK A. FITZPATRICK

VOLUME 188. Hydrocarbons and Methylotrophy
Edited by MARY E. LIDSTROM

VOLUME 189. Retinoids (Part A: Molecular and Metabolic Aspects)
Edited by LESTER PACKER

VOLUME 190. Retinoids (Part B: Cell Differentiation and Clinical Applications)
Edited by LESTER PACKER

VOLUME 191. Biomembranes (Part V: Cellular and Subcellular Transport: Epithelial Cells)
Edited by SIDNEY FLEISCHER AND BECCA FLEISCHER

VOLUME 192. Biomembranes (Part W: Cellular and Subcellular Transport: Epithelial Cells)
Edited by SIDNEY FLEISCHER AND BECCA FLEISCHER

VOLUME 193. Mass Spectrometry
Edited by JAMES A. MCCLOSKEY

VOLUME 194. Guide to Yeast Genetics and Molecular Biology
Edited by CHRISTINE GUTHRIE AND GERALD R. FINK

VOLUME 195. Adenylyl Cyclase, G Proteins, and Guanylyl Cyclase
Edited by ROGER A. JOHNSON AND JACKIE D. CORBIN

VOLUME 196. Molecular Motors and the Cytoskeleton
Edited by RICHARD B. VALLEE

VOLUME 197. Phospholipases
Edited by EDWARD A. DENNIS

VOLUME 198. Peptide Growth Factors (Part C)
Edited by DAVID BARNES, J. P. MATHER, AND GORDON H. SATO

VOLUME 199. Cumulative Subject Index Volumes 168–174, 176–194

VOLUME 200. Protein Phosphorylation (Part A: Protein Kinases: Assays, Purification, Antibodies, Functional Analysis, Cloning, and Expression)
Edited by TONY HUNTER AND BARTHOLOMEW M. SEFTON

VOLUME 201. Protein Phosphorylation (Part B: Analysis of Protein Phosphorylation, Protein Kinase Inhibitors, and Protein Phosphatases)
Edited by TONY HUNTER AND BARTHOLOMEW M. SEFTON

VOLUME 202. Molecular Design and Modeling: Concepts and Applications (Part A: Proteins, Peptides, and Enzymes)
Edited by JOHN J. LANGONE

VOLUME 203. Molecular Design and Modeling: Concepts and Applications (Part B: Antibodies and Antigens, Nucleic Acids, Polysaccharides, and Drugs)
Edited by JOHN J. LANGONE

VOLUME 204. Bacterial Genetic Systems
Edited by JEFFREY H. MILLER

VOLUME 205. Metallobiochemistry (Part B: Metallothionein and Related Molecules)
Edited by JAMES F. RIORDAN AND BERT L. VALLEE

VOLUME 206. Cytochrome P450
Edited by MICHAEL R. WATERMAN AND ERIC F. JOHNSON

VOLUME 207. Ion Channels
Edited by BERNARDO RUDY AND LINDA E. IVERSON

VOLUME 208. Protein–DNA Interactions
Edited by ROBERT T. SAUER

VOLUME 209. Phospholipid Biosynthesis
Edited by EDWARD A. DENNIS AND DENNIS E. VANCE

VOLUME 210. Numerical Computer Methods
Edited by LUDWIG BRAND AND MICHAEL L. JOHNSON

VOLUME 211. DNA Structures (Part A: Synthesis and Physical Analysis of DNA)
Edited by DAVID M. J. LILLEY AND JAMES E. DAHLBERG

VOLUME 212. DNA Structures (Part B: Chemical and Electrophoretic Analysis of DNA)
Edited by DAVID M. J. LILLEY AND JAMES E. DAHLBERG

VOLUME 213. Carotenoids (Part A: Chemistry, Separation, Quantitation, and Antioxidation)
Edited by LESTER PACKER

VOLUME 214. Carotenoids (Part B: Metabolism, Genetics, and Biosynthesis)
Edited by LESTER PACKER

VOLUME 215. Platelets: Receptors, Adhesion, Secretion (Part B)
Edited by JACEK J. HAWIGER

VOLUME 216. Recombinant DNA (Part G)
Edited by RAY WU

VOLUME 217. Recombinant DNA (Part H)
Edited by RAY WU

VOLUME 218. Recombinant DNA (Part I)
Edited by RAY WU

VOLUME 219. Reconstitution of Intracellular Transport
Edited by JAMES E. ROTHMAN

VOLUME 220. Membrane Fusion Techniques (Part A)
Edited by NEJAT DÜZGÜNEŞ

VOLUME 221. Membrane Fusion Techniques (Part B)
Edited by NEJAT DÜZGÜNEŞ

VOLUME 222. Proteolytic Enzymes in Coagulation, Fibrinolysis, and Complement Activation (Part A: Mammalian Blood Coagulation Factors and Inhibitors)
Edited by LASZLO LORAND AND KENNETH G. MANN

VOLUME 223. Proteolytic Enzymes in Coagulation, Fibrinolysis, and Complement Activation (Part B: Complement Activation, Fibrinolysis, and Nonmammalian Blood Coagulation Factors)
Edited by LASZLO LORAND AND KENNETH G. MANN

VOLUME 224. Molecular Evolution: Producing the Biochemical Data
Edited by ELIZABETH ANNE ZIMMER, THOMAS J. WHITE, REBECCA L. CANN, AND ALLAN C. WILSON

VOLUME 225. Guide to Techniques in Mouse Development
Edited by PAUL M. WASSARMAN AND MELVIN L. DEPAMPHILIS

VOLUME 226. Metallobiochemistry (Part C: Spectroscopic and Physical Methods for Probing Metal Ion Environments in Metalloenzymes and Metalloproteins)
Edited by JAMES F. RIORDAN AND BERT L. VALLEE

VOLUME 227. Metallobiochemistry (Part D: Physical and Spectroscopic Methods for Probing Metal Ion Environments in Metalloproteins)
Edited by JAMES F. RIORDAN AND BERT L. VALLEE

VOLUME 228. Aqueous Two-Phase Systems
Edited by HARRY WALTER AND GÖTE JOHANSSON

VOLUME 229. Cumulative Subject Index Volumes 195–198, 200–227

VOLUME 230. Guide to Techniques in Glycobiology
Edited by WILLIAM J. LENNARZ AND GERALD W. HART

VOLUME 231. Hemoglobins (Part B: Biochemical and Analytical Methods)
Edited by JOHANNES EVERSE, KIM D. VANDEGRIFF, AND ROBERT M. WINSLOW

VOLUME 232. Hemoglobins (Part C: Biophysical Methods)
Edited by JOHANNES EVERSE, KIM D. VANDEGRIFF, AND ROBERT M. WINSLOW

VOLUME 233. Oxygen Radicals in Biological Systems (Part C)
Edited by LESTER PACKER

VOLUME 234. Oxygen Radicals in Biological Systems (Part D)
Edited by LESTER PACKER

VOLUME 235. Bacterial Pathogenesis (Part A: Identification and Regulation of Virulence Factors)
Edited by VIRGINIA L. CLARK AND PATRIK M. BAVOIL

VOLUME 236. Bacterial Pathogenesis (Part B: Integration of Pathogenic Bacteria with Host Cells)
Edited by VIRGINIA L. CLARK AND PATRIK M. BAVOIL

VOLUME 237. Heterotrimeric G Proteins
Edited by RAVI IYENGAR

VOLUME 238. Heterotrimeric G-Protein Effectors
Edited by RAVI IYENGAR

VOLUME 239. Nuclear Magnetic Resonance (Part C)
Edited by THOMAS L. JAMES AND NORMAN J. OPPENHEIMER

VOLUME 240. Numerical Computer Methods (Part B)
Edited by MICHAEL L. JOHNSON AND LUDWIG BRAND

VOLUME 241. Retroviral Proteases
Edited by LAWRENCE C. KUO AND JULES A. SHAFER

VOLUME 242. Neoglycoconjugates (Part A)
Edited by Y. C. LEE AND REIKO T. LEE

VOLUME 243. Inorganic Microbial Sulfur Metabolism
Edited by HARRY D. PECK, JR., AND JEAN LEGALL

VOLUME 244. Proteolytic Enzymes: Serine and Cysteine Peptidases
Edited by ALAN J. BARRETT

VOLUME 245. Extracellular Matrix Components
Edited by E. RUOSLAHTI AND E. ENGVALL

VOLUME 246. Biochemical Spectroscopy
Edited by KENNETH SAUER

VOLUME 247. Neoglycoconjugates (Part B: Biomedical Applications)
Edited by Y. C. LEE AND REIKO T. LEE

VOLUME 248. Proteolytic Enzymes: Aspartic and Metallo Peptidases
Edited by ALAN J. BARRETT

VOLUME 249. Enzyme Kinetics and Mechanism (Part D: Developments in Enzyme Dynamics)
Edited by DANIEL L. PURICH

VOLUME 250. Lipid Modifications of Proteins
Edited by PATRICK J. CASEY AND JANICE E. BUSS

VOLUME 251. Biothiols (Part A: Monothiols and Dithiols, Protein Thiols, and Thiyl Radicals)
Edited by LESTER PACKER

VOLUME 252. Biothiols (Part B: Glutathione and Thioredoxin; Thiols in Signal Transduction and Gene Regulation)
Edited by LESTER PACKER

VOLUME 253. Adhesion of Microbial Pathogens
Edited by RON J. DOYLE AND ITZHAK OFEK

VOLUME 254. Oncogene Techniques
Edited by PETER K. VOGT AND INDER M. VERMA

VOLUME 255. Small GTPases and Their Regulators (Part A: Ras Family)
Edited by W. E. BALCH, CHANNING J. DER, AND ALAN HALL

VOLUME 256. Small GTPases and Their Regulators (Part B: Rho Family)
Edited by W. E. BALCH, CHANNING J. DER, AND ALAN HALL

VOLUME 257. Small GTPases and Their Regulators (Part C: Proteins Involved in Transport)
Edited by W. E. BALCH, CHANNING J. DER, AND ALAN HALL

VOLUME 258. Redox-Active Amino Acids in Biology
Edited by JUDITH P. KLINMAN

VOLUME 259. Energetics of Biological Macromolecules
Edited by MICHAEL L. JOHNSON AND GARY K. ACKERS

VOLUME 260. Mitochondrial Biogenesis and Genetics (Part A)
Edited by GIUSEPPE M. ATTARDI AND ANNE CHOMYN

VOLUME 261. Nuclear Magnetic Resonance and Nucleic Acids
Edited by THOMAS L. JAMES

VOLUME 262. DNA Replication
Edited by JUDITH L. CAMPBELL

VOLUME 282. Vitamins and Coenzymes, Part L (in preparation)
Edited by DONALD B. MCCORMICK, JOHN W. SUTTIE, AND CONRAD WAGNER

VOLUME 283. Cell Cycle Control (in preparation)
Edited by WILLIAM G. DUNPHY

VOLUME 284. Lipases (Part A: Biotechnology) (in preparation)
Edited by BYRON RUBIN AND EDWARD A. DENNIS

VOLUME 285. Cumulative Subject Index Volumes 263, 264, 266–289 (in preparation)

VOLUME 286. Lipases (Part B: Enzyme Characterization and Utilization) (in preparation)
Edited by BYRON RUBIN AND EDWARD A. DENNIS

VOLUME 287. Chemokines (in preparation)
Edited by RICHARD HORUK

VOLUME 288. Chemokine Receptors (in preparation)
Edited by RICHARD HORUK

VOLUME 289. Solid-Phase Peptide Synthesis (in preparation)
Edited by GREGG B. FIELDS

[1] Total Chemical Synthesis of Chirally Pure (6S)-Tetrahydrofolic Acid

By STEVEN W. BAILEY and JUNE E. AYLING

Introduction

The properties of the reduced folates, whether as enzyme cofactors or as substrates for transport or binding proteins, are considerably influenced by the chirality of the carbon at the 6-position. The unnatural isomers can either be less active, inactive, or inhibitory relative to their natural counterparts. Thus, physiologically relevant characterization of folate enzymes or transporters often requires the natural tetrahydrofolate diastereomer [e.g., (6*S*)-FH_4, **1**; see Scheme I]. Useful mechanistic information can also be obtained by comparison to the pure unnatural 6-epimer. The availability of (6*S*)-5-methyl- and 5-formyltetrahydrofolates is of particular interest for their therapeutic and nutritional uses. The natural 6-isomers of tetrahydrofolates have been produced by enzymatic[1–3] or stereoselective chemical[2,4] reduction of folic acid, chromatographic separation,[5] and fractional crystallization (either directly[6] or of derivatives[3,7]). The total stereospecific synthesis of tetrahydropterins including tetrahydrofolic acid has been accomplished.[8] This approach has the advantages of scalability and applicability to both isomers, and is particularly suited for production of labeled compounds or analogs that might not be amenable to other methods. In its most basic form, the use of oxidative cyclization to produce a quinoid dihydropteridine intermediate, for example, **2** (Scheme I), can be

[1] C. Temple, Jr. and J. A. Montgomery, *in* "Folates and Pterins" (R. L. Blakley and S. J. Benkovic, eds.), Vol. 1, p. 97. Wiley-Interscience, New York, 1985.

[2] L. Rees, E. Valente, C. J. Suckling, and H. C. S. Wood, *Tetrahedron* **42,** 117 (1986).

[3] Y. Kuge, K. Inoue, K. Ando, T. Eguchi, T. Oshiro, K. Mochida, T. Uwajima, T. Sugaya, J. Kanazawa, M. Okabe, and S. Tomioka, *J. Chem. Soc. Perkin Trans.* **1,** 1427 (1994).

[4] P. H. Boyle and M. T. Keating, *J. Chem. Soc. Chem. Commun.* **10,** 375 (1974); S. Kwee and H. Lund, *Bioelectrochem. Bioeng.* **7,** 693 (1980).

[5] B. T. Kaufman, K. O. Donaldson, and J. C. Keresztesy, *J. Biol. Chem.* **238,** 1498 (1963); R. J. Mullin and D. S. Duch, *J. Chromatogr.* **555,** 254 (1991).

[6] H. R. Mueller, R. Moser, M. Ulmann, and T. Ammann, European Patent Appl. No. EP 95-106285 (1995); H. R. Mueller, M. Ulmann, J. Conti, and G. Muerdel, European Patent Appl. No. 89-108451 (1989); P. Jequier and F. Marazza, European Patent Appl. No. EP 93-119354 (1993).

[7] J. Owens, L. Rees, C. J. Suckling, and H. C. S. Wood, *J. Chem. Soc. Perkin Trans.* **1,** 871 (1993).

[8] S. W. Bailey, R. Y. Chandrasekaran, and J. E. Ayling, *J. Org. Chem.* **57,** 4470 (1992).

0076-6879/97 $25

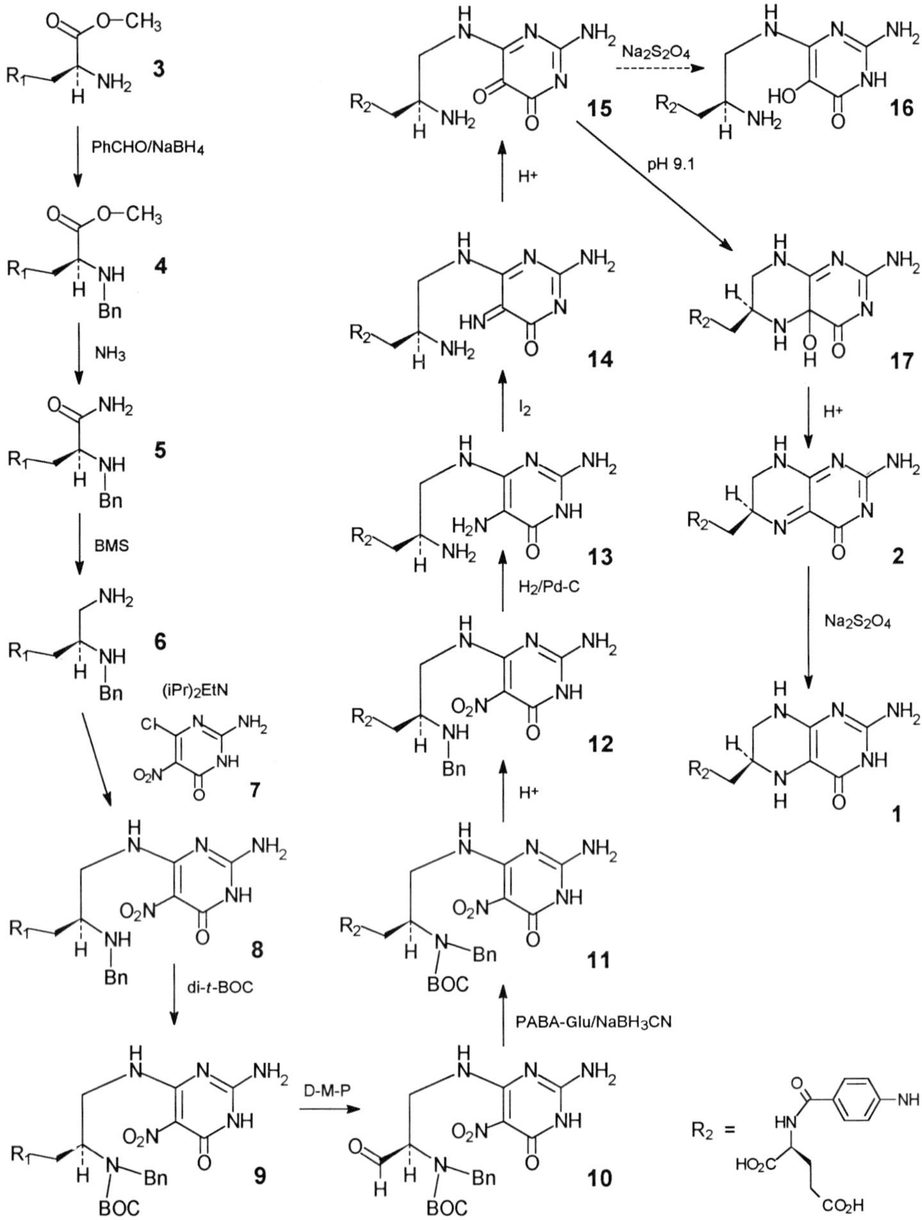
3
PhCHO/NaBH4
4
NH3
5
BMS
6
(iPr)2EtN
7
8
di-t-BOC
9
D-M-P
10
PABA-Glu/NaBH3CN
11
H+
12
H2/Pd-C
13
I2
14
H+
15
Na2S2O4
16
pH 9.1
17
H+
2
Na2S2O4
1
R2 =

SCHEME I

extended to a variety of substitution patterns so long as the pteridine possesses at least one effective electron-donating group at either the 2- or 4-position.[9] In this chapter the synthesis of (6*S*)-tetrahydrofolic acid is presented. The (6*R*)-isomer is produced by this same procedure, except that the synthesis begins with D-serine methyl ester.

Materials and Methods

^{1}H and ^{13}C nuclear magnetic resonance (NMR) spectra are acquired in deuterated dimethyl sulfoxide (DMSO-d_6) at 300 and 75 MHz, respectively, and chemical shifts are referred to $(CH_3)_4Si$, unless otherwise specified. Ultraviolet extinction coefficients (ε) are in units of M^{-1} cm^{-1}. Unless otherwise specified, all pH values of primarily nonaqueous solutions are measured using an Orion (Boston, MA) 81-75 electrode after 10-fold dilution in water. Yields of crude or partially purified products are determined by comparison to purified materials by quantitative high-performance liquid chromatography (HPLC). The format for chromatographic conditions is as follows: [stationary phase, mobile phase, flow rate (ml/min), detection wavelength(s) (nm), electrochemical voltage (vs Ag|AgCl)]. Except as noted, column dimensions are 25 $\times$ 0.46 cm, packing materials are 5 μm, mobile phase buffer molarities refer to the cation concentration, and elution temperature is ambient. Columns used for HPLC analysis or for purification of **1** or **13** (see Scheme I) are pretreated with fresh 1 *M* $Na_2S_2O_4$ to remove adsorbed oxygen. Solvents are removed from the reaction mixtures by using a rotary evaporator with a room temperature bath and vacuum pump, and products are dried under high vacuum over P_2O_5.

Materials

All solvents are spectrophotometric and/or HPLC grade. Tetrahydrofuran (THF) is dried by distillation under argon from $LiAlH_4$; CH_2Cl_2 and dimethylformamide (DMF) are dried over 4-Å sieves; benzaldehyde is distilled under argon. Unless otherwise specified all reagents are ACS grade, or the 99% purity grade available from Aldrich (Milwaukee, WI) or Fluka (Ronkonkoma, NY), and used as received.

2-Amino-6-chloro-5-nitro-4-pyrimidinone is prepared by nitration of 2-amino-6-chloro-4-pyrimidinone (Polyorganix, Newburyport, MA) as previously described,[10] except that the product is precipitated from the sulfuric/nitric acid mixture by pouring slowly onto ice (6 g/ml H_2SO_4). After filtra-

[9] S. Vasudevan, S. W. Bailey, and J. E. Ayling, *Abstr. Papers Am. Chem. Soc.* **212,** 119-ORGN (1996).

[10] S. W. Bailey and J. E. Ayling, *Biochemistry* **22,** 1790 (1983).

tion and washing with a small volume of ice-cold water, the precipitate is thoroughly dried under high vacuum. This material invariably contains a water of crystallization that promotes hydrolysis to 2-amino-4,6-dihydroxy-5-nitropyrimidine at room temperature. This can be largely prevented by storage with desiccant at −80°. Any contaminating dihydroxy compound is not soluble in hot ethanol and should be filtered out before condensation with diamine **6** (see Scheme I).

Optical rotation specifications of commercial preparations of amino acid esters have not proved to be a reliable indicator of freedom from low percentages of the opposite enantiomer. The chiral purity of the L-serine methyl ester is determined by hydrolysis of a sample in 0.5 ml of 1 *M* HCl at 55° for 1 hr. After removal of solvent, the resulting serine is treated with 5-(dimethylamino)-1-naphthalenesulfonyl (dansyl) chloride, and the derivative separated by reversed-phase HPLC [Kromasil ODS, 20 m*M* NH_4OH plus formic acid (pH 3.0)–methanol (1:1, v/v), 1.0, 325]. Chiral chromatography [Cyclobond I (Astec, Whippany, NJ), 50 m*M* sodium acetate (pH 5.5)–CH_3CN (9:1, v/v), 0.6, 250] of the purified material can indicate the presence of less than 0.5 D-dansylserine.

Dess–Martin periodinane [1,1,1-tris(acetoxy)-1,1-dihydro-1,2-benziodoxol-3(1*H*)-one] is easily prepared in two steps from 2-iodobenzoic acid.[11] The original procedure is modified by passage of dry inert gas through the head space of the acetylation reaction to ensure its completion. Alternatively, the use of catalytic *p*-toluenesulphonic acid has been reported to give high yields, rapidly, of the periodinane.[12] In addition to NMR, a reliable indication of fully acetylated material is at least a 0.6 *M* solubility in CH_2Cl_2.

*N-Benzyl-L-serine Methyl Ester Hydrochloride (**4**, R_1 = OH)*

A solution of 20 g (0.129 mol) of L-serine methyl ester hydrochloride (**3**) in 100 ml of methanol is adjusted to pH 8 (measured after a 10-fold dilution of a sample into water) with 12.9 ml of 10 *M* NaOH, thus precipitating NaCl.[13] Benzaldehyde (15.0 g, 0.141 mol) is added and stirred well for 30 min. After cooling to 4°, 2.43 g (0.064 mol) of $NaBH_4$ is added gradually over 1 hr. The reaction is then warmed to ambient temperature and stirred for another hour. Analysis by HPLC [Spherisorb CN, 1-propanol–heptane (1:3, v/v), 1.0, 245] typically shows that aside from the desired product, identified by coelution with material obtained by esterification of *N*-benzyl-L-serine, the only other product observed is benzyl alcohol. The reaction is filtered, concentrated, and dried. The thick residue is dissolved in 120

[11] D. B. Dess and J. C. Martin, *J. Org. Chem.* **48,** 4156 (1983).
[12] R. E. Ireland and L. Liu, *J. Org. Chem.* **58,** 2899 (1993).
[13] Adapted with permission from Ref. 8. Copyright © 1992 American Chemical Society.

ml of acetone, refiltered to remove more salt, adjusted to between pH 1 and 1.5 with HCl gas, 750 ml of $(C_2H_5)_2O$ added, product collected by centrifugation, and dried, giving 28.1 g of colorless hygroscopic solid. This has been shown to contain 91.6 mmol (71%) by chromatographic comparison to purified material.[14] Mass spectrometry (MS) (E.I., direct insertion) *m/z* (percentage relative abundance) 210(MH^+)(5), 178(55), 150(64), 118(6), 106(29), 91(100); HRMS (+DCI, CH_4) *m/z* calculated for $C_{11}H_{16}NO_3$(MH^+) 210.113, found 210.112.

A small contaminant (1–2%) of tetrahydrofolic acid with the 6-epimer is occasionally observed beyond that due to the chiral impurity of the starting L-serine methyl ester. This may possibly arise during the initial Schiff base condensation. Exposure of the *N*-benzylideneserine methyl ester to room temperature should, therefore, be minimized.[15]

*N-Benzyl-L-serinamide (**5**, R_1 = OH)*

N-Benzyl-L-serine methyl ester hydrochloride (28.1 g, 80% pure, 91.5 mmol) is dissolved in 450 ml of methanol saturated with NH_3 at 0°. The cold solution is transferred to a pressure bottle, capped, and allowed to warm to room temperature. The flask is resaturated with NH_3 at 2-day intervals. Analysis by HPLC (as for **4**) typically indicates 86% conversion to product after 6 days, at which time solvent is removed, and product is dried. The resulting 29.0 g of crude material is extracted into 500 ml of ethyl acetate–methanol (3:2, v/v), filtered, and evaporated to a gum. This is reextracted with 200 ml of 1-propanol, 50 ml of ethyl acetate is added, and the suspension is then centrifuged. After removing the solvent from the supernatant, the residue is dried to give 24.3 g of a thick oil of crude **5** (R_1 = OH), hydrochloride salt (74 mmol, 81%). A sample can be purified by crystallization from methanol–acetone–$(C_2H_5)_2O$: ^1H NMR δ 3.70 (m, 1H, *C*H–CH_2), 3.87 (m, 2H, –CH_2–), 4.15 (s, 2H, benzyl), 5.58 (br s, 1H, OH), 7.3–7.6 (m, 5H, Ar), 7.67 (s, 1H, $CONH_a$), 8.03 (s, 1H, $CONH_b$); MS (E.I., direct insertion) *m/z* (percent relative abundance) 195(MH^+)(2), 163(7), 150(51), 106(17), 91(100); HRMS (+DCI, CH_4) *m/z* calculated for $C_{10}H_{15}N_2O_2$(MH^+) 195.113, found 195.113.

*3-Amino-(2R)-benzylamino-1-propanol (**6**, R_1 = OH)*

N-Benzyl-L-serinamide hydrochloride (45.7 mmol, 15.0 g of crude material) is almost entirely dissolved in 800 ml of dry THF under argon. The

[14] E. Hardegger, F. Szabo, P. Liechti, C. Rostetter, and W. Zankowska-Jasinska, *Helv. Chim. Acta* **51,** 78 (1968).

[15] With most precursors to unconjugated 6-alkyltetrahydropteridines formation and even isolation of *N*-α-benzylidene amino acid amides do not lead to racemization.[8]

mixture is heated, and 46 ml of 10.1 *M* borane methyl sulfide (0.465 mol) is added via syringe with stirring over 20 min and simultaneous distillation of $(CH_3)_2S$, and then refluxed for 2 hr. After cooling to room temperature, 11 ml of 6 *N* HCl is added dropwise to give pH 4.5 to 5.0 (measured after a 10-fold dilution of a sample into water), and the reaction stirred for 30 min. Water is added to produce a clear solution that is extracted once with 300 ml of $(C_2H_5)_2O$. The aqueous layer is adjusted to pH 12 with solid NaOH, product is extracted with a total of 3.8 liters of $(C_2H_5)_2O$, and the extract is dried (Na_2SO_4). After removing solvent, drying under vacuum over NaOH pellets typically gives an oil weighing 8.35 g and containing 25.6 mmol of **6**, R_1 = OH (56%) (HPLC, as for **4**). Another 6.3 mmol remains in the aqueous layer. A chromatogram [Spherisorb ODS2, 10 m*M* KPO_4 (pH 6.5)–methanol (3:7, v/v), 1.0, fluorescence 320_{ex}, 460_{em}] of the *o*-phthalaldehyde (OPT) derivative[16] of crude product usually shows greater than 99.5% of the fluorescence response in a single peak.

A sample of the oil can be further purified by distillation at 95° and approximately 50 torr, and precipitation of the dihydrochloride salt from methanol–HCl with $(C_2H_5)_2O$: ^1H NMR δ 3.2–3.7 (m, 3H, –CH_2–NH_2, *CH*–NH), 3.86 (m, 2H, O–CH_2), 4.29 (m, 2H, benzyl), 7.4–7.8 (m, 5H, Ar); ^{13}C NMR δ ~39 (submersed under DMSO), 47.7, 55.7, 56.5, 128.5, 128.8, 130.2, 131.7; MS (E.I. direct insertion) *m*/*z* (percent relative abundance) 181(MH^+)(1), 164(1), 150(87), 106(6), 91(100); HRMS (+DCI, CH_4) *m*/*z* calculated for $C_{10}H_{17}N_2O(MH^+)$ 181.134, found 181.135.

At this point the diamine should be free of other primary amines (as evidenced by lack of other OPT derivatives), otherwise these may react with chloropyrimidine **7** to produce adducts that are difficult to purify from **8**.[17]

*2-Amino-6-([(2′R)-benzylamino-3′-hydroxypropyl]amino)-5-nitro-4(3H)-pyrimidinone (**8**, R_1 = OH)*

Chloropyrimidine **7** (14 g) is partially dissolved in 900 ml of hot absolute ethanol and filtered. To the filtrate, determined to contain 40 mmol of the pyrimidine, *N*,*N*-diisopropylethylamine is added to achieve pH 8 (measured after a 10-fold dilution of a sample into water), and the mixture is taken to reflux. A solution of 38 mmol of **6**, R_1 = OH (12.3 g of crude material, but free of other primary amines), in 100 ml of ethanol is added all at once, and refluxed for 2 hr with stirring while monitored by HPLC [Partisil SCX 10 μm, 0.1 *M* ammonium acetate (pH 4.8)–CH_3CN (9:1, v/v), 0.8, 290 and

[16] P. Lindroth and K. Mopper, *Anal. Chem.* **51,** 1667 (1979).

[17] In the synthesis of unconjugated tetrahydropterins with moderately hydrophobic side analogs, e.g., 6-propyl, (**8**, R = C_2H_5) can be purified by crystallization from organic solvent.

330]. Solvent is then distilled from the mix until 700 ml remains, which is then refrigerated overnight. A first crop of crude **8**, R_1 = OH, collected by filtration typically weighs, after drying, 9.56 g (containing 21.3 mmol), with 12.8 mmol remaining in the filtrate (90% total). Further concentration of the filtrate produces two more crops with a combined weight of 4.44 g (containing 6.06 mmol of **8**, R_1 = OH). All three crops are purified by suspension in ice-cold water. The collected precipitates are dried to give 12.86 g (containing 26.5 mmol) of light yellow powder (70%).

Crude product can be purified by further resuspension in water at 0° and then filtered. The precipitate is partially dissolved in boiling ethanol (300 ml/g), filtered while still warm, and dried to give a light yellow powder: ultraviolet (UV) spectra λ_{max} (ε) (0.1 *N* HCl) 334 nm (15,100), 286 nm [shoulder (sh)], 236 nm (13,400); (0.1 *M* KPO_4, pH 6.5) same as 0.1 *N* HCl; (0.1 *N* NaOH) 347 nm (16,600), 232 nm (sh); ^{1}H NMR δ 2.85 (m, 1H, –NH–), 3.3–3.75 (m, 5H, O–CH_2CHNCH_2N–), 3.85 (m, 2H, benzyl), 7.1–7.5 (m, 7, Ar, NH_2), 9.83 (br, 1H, CONH); ^{13}C NMR δ 41.3, 49.8, 56.9, 60.7, 110.5, 126.8, 128.0, 128.2, 139.5, 154.0, 156.3, 159.0; MS (E.I., direct insertion) *m/z* (percent relative abundance) 335(MH^+)(0.5), 303(2), 167(8), 150(75), 91(100); HRMS (+DCI, CH_4) *m/z* calculated for $C_{14}H_{19}N_6O_4(MH^+)$ 335.147, found 335.146.

The amine base, e.g., $(C_3H_7)_2N\text{-}C_2H_5$, used to promote complete condensation often coprecipitates as its salt with the desired product. This must be largely removed (as determined either by quantitative UV spectrum or NMR) by the above precipitations, otherwise it can promote undesired tautomerization of quinoid dihydropterin **2** to the 7,8-dihydropterin during the final cyclization reaction. For this reason it is also best to ensure that the starting chloropyrimidine **7** be as free as possible from H_2SO_4 remaining from its nitration reaction.

*2-Amino-6-[[(2′R)-(N-tert-Butoxycarbonylbenzylamino)-3′-hydroxypropyl]amino]-5-nitro-4(3H)-pyrimidinone (**9**, R_1 = OH)*

A slurry of 10.3 mmol of **8** (5 g of crude material) in 200 ml of dioxane is cooled on ice, and 1 *M* NaOH is added slowly to give pH 9.5 (26 ml). Di-*tert*-butyl dicarbonate, 97% (2.5 g, 11.1 mmol), is added, and the clear solution is warmed to room temperature. At 3, 5.5, and 23 hr the pH is readjusted with 1 *M* NaOH and an additional 0.3 g of dicarbonate is added. HPLC [Kromasil ODS, 10 m*M* NH_4OH plus H_3PO_4 (to pH 2.8)–CH_3CN (7:3, v/v), 1.0, 330] typically shows 4% **8**, R_1 = OH remaining at 46 hr, and the reaction is then filtered, and carefully evaporated to remove the dioxane. The aqueous solution is adjusted to pH 7 with glacial acetic acid, and kept overnight on ice. The precipitate is collected by centrifugation, resuspended in 10 ml of ice-cold water, recentrifuged, and dried to give typically 5.64 g of light yellow powder (containing 9.6 mmol of **9**, R_1 = OH, 93%), with an additional 0.3 mmol remaining in the aqueous supernatants.

A sample in DMF is purified by preparative HPLC on Bakerbond (J. T. Baker, Phillipsburg, NJ) high hydrophobicity (C_{18}), 40 μm (37 × 2.1 cm) equilibrated with distilled H_2O, and eluted at 10 ml/min with H_2O (40 ml), 1 m*M* formic acid–methanol (52 : 48, v/v) (270 ml), and finally with H_2O–methanol (30 : 70, v/v) to elute product, which on reanalysis typically shows 98% of the 330-nm absorbance in a single peak: UV λ_{max} (ε) [0.1 *M* HCl, 0.1 *M* NH_4PO_4 (pH 2.8), 0.1 *M* KPO_4 (pH 6.5)] 335 nm (13,300), ~288 nm (sh), 260 nm [minimum (min)], ~233 nm (sh); (0.1 *M* NaOH) 346 nm (15,600); ^{1}H NMR (90 MHz) δ 1.33 (br s, 9H, $C(CH_3)_3$), 3.0–3.9 (5H, *CH*–N, O–CH_2, N–CH_2), 4.39 (s, 2H, benzyl), 7.8–8.4 (br s, 7H, Ar, NH_2), 10.4 (br, 1H, CONH); MS (E.I., direct insertion) *m*/*z* (percent relative abundance) 435(MH^+)(0.25), 410(0.4), 404(2), 360(4), 303(1), 250(9), 194(15), 167(12), 150(75), 91(100); HRMS (+DCI, CH_4) *m*/*z* calculated for $C_{19}H_{27}N_6O_6(MH^+)$ 435.199, found 435.197.

This protection with BOC improves solubility in CH_2Cl_2 and allows use of Dess–Martin periodinane in the next step.

*N-[4-[(2S)-(N-tert-Butoxycarbonylbenzylamino)-3-[(2-amino-5-nitro-4(3H)-oxopyrimidin-6-yl)amino]propylamino]benzoyl]-L-glutamic Acid (**11**)*

To a solution of 5.0 g of crude **9** (8.51 mmol) in 125 ml of dry CH_2Cl_2 is added 8.49 g of 1,1,1-tris(acetoxy)-1,1-dihydro-1,2-benziodoxol-3(1*H*)-one (19.2 mmol plus some acetic acid) in 32 ml of dry CH_2Cl_2 while stirring at room temperature under argon. The alcohol **9**, R_1 = OH usually is completely consumed by 70 min as determined by HPLC [Kromasil ODS, 10 m*M* NH_4OH plus H_3PO_4 (to pH 2.8)–CH_3CN (3 : 2, v/v), 1.0, 330], and the reaction is then poured into a mixture of 1 *M* $NaHCO_3$ (76 ml) and 0.125 *M* $Na_2S_2O_3$ (500 ml), and stirred for 15 min. The organic layer is washed with 250 ml of H_2O, dried briefly (Na_2SO_4), the solvent is removed, and the residue dried to give 5.01 g of yellow semisolid. A minimum yield of the aldehyde **10** can be determined by reduction of a small sample with $NaBH_4$ in DMF–H_2O (1 : 1, v/v), which typically gives 93% of the expected alcohol **9**, R_1 = OH. Ultraviolet λ_{max} (pH 2.7 HPLC eluant) 333 nm, 288 nm (sh), 258 nm (min), 233 nm (sh). The product of this reaction, which is free of reagent by-products, is utilized immediately for reductive alkylation.

Crude aldehyde **10** (7.91 mmol) is dissolved in 28.5 ml of dry DMF and cooled to 0°. To this is added rapidly, with stirring, an ice-cold solution of 12 g of *p*-aminobenzoyl-L-glutamic acid (PABA-Glu) (45 mmol) dissolved in 40 ml of DMF. After 5 min, a cold solution of 0.57 g $NaBH_3CN$ (95%, 8.6 mmol) in 8.2 ml DMF was quickly added, and the mixture stirred for 1 hr. Analysis of the reaction by HPLC (as for **10**) typically shows 5.92 mmol of **11** (75%), no aldehyde **10**, and 0.45 mmol of alcohol **9**, R_1 = OH. Solvent is removed, and ice cold water (90 ml) added. The resulting slurry is adjusted to pH 2.5 with concentrated HCl, centrifuged, and the precipitate dried. The supernate contains 0.45 mmol of product and the bulk of the excess PABA-Glu. The resulting 6.27 g of yellow powder is almost entirely dissolved in 350 ml of absolute ethanol at 60°, and filtered. The filtrate is concentrated, rewarmed, and 2.5 vol of water is added. After cooling, the precipitate is collected by centrifugation and dried, giving 4.56 g of light yellow powder containing 4.35 mmol of **11** (51% from **9**, R_1 = OH).

A sample can be purified by semipreparative HPLC on Spherisorb ODS2 (25 × 1 cm), eluted with H_2O–formic acid (pH 2.7), followed by this eluant mixed with CH_3CN (75:25, v/v) at 6 ml/min. Analytical HPLC of the collected product usually shows that greater than 99% of the UV absorbance area at either 295 or 330 nm is contained in a single peak: UV λ_{max} (ε) (0.1 *M* NH_4PO_4, pH 2.8) 301 nm (23,900), 335 (sh); (0.1 *M* KPO_4, pH 6.5) 297 nm (23,300), 335 nm (sh); (0.1 *M* NaOH) 299 nm (23,200), 347 nm (16,000); ^1H NMR (90 MHz) δ 1.29 (br s, 9H, $C(CH_3)_3$), 2.00 (m, 2H, β-CH_2), 2.33 (t, 2H, γ-CH_2), 3.0–4.0 (5H, C*H*–NH, Pry–NH–CH_2, Ph–NH–CH_2), 4.0–4.7 (3H, benzyl, α-CH), 6.53 (d, 2H, 3′,5′-H), 7.22 (br s, 5H, Ar), 7.66 (d, 2H, 2′,6′-H), 8.10 (d, 1H, CON*H*CH), 9.45 (br s, 1H, CONH); HRMS [fast atom bombardment (FAB), Xe, glycerol–K-trifluoroacetate] *m/z* calculated for $C_{31}H_{36}N_8O_{10}K(MK^+)$ 719.219, found 719.233.

HPLC analysis reveals that although the alcohol **9**, R_1 = OH elutes as a typically shaped peak, the chromatographic profile of the aldehyde consists of an early eluting peak, presumably of the hydrate, connected by a continuous bridging plateau of absorbance to a later eluting form suggesting an on-column equilibration.[18]

The reductive amination of aldehyde **10** requires an excess of PABA-Glu in order to prevent significant production of alcohol **9**, R_1 = OH. The unreacted PABA-Glu is easily recovered in the water wash and reused. Alternatively, an efficient reductive alkylation of amines including *p*-aminobenzoic acid with aldehydes and ketones has been reported.[19] This method, which uses stoichiometric quantities of amine and carbonyl compound in

[18] Swern oxidation may be an alternative to Dess–Martin periodinane for the production of aldehyde **10** from **9**, R_1 = OH.

[19] A. F. Abdel-Magid, K. G. Carson, B. D. Harris, C. A. Maryanoff, and R. D. Shah, *J. Org. Chem.* **61,** 3849 (1996).

dichloroethane or THF with sodium triacetoxyborohydride as reductant, may be useful when used in conjunction with diesters of *p*-aminobenzoyl-L-glutamic acid, especially if label is to be introduced at this point. The di-*tert*-butyl ester of PABA-Glu[20] is easily deprotected by acid and could, for example, be removed with slight modification of the conditions at the same time as the BOC group to produce **12**.

*N-[4-[(2S)-Benzylamino-3-[(2-amino-5-nitro-4(3H)-oxopyrimidin-6-yl)amino]propylamino]benzoyl]-L-glutamic Acid (**12**)*

A slurry of 3.87 g of crude **11** (3.69 mmol) in 39 ml of THF plus 156 ml of 1 *M* HCl is heated to 58° with stirring, resulting after 45 min in a clear solution. Analysis by HPLC (as for **9**, but detection at 295 nm) has shown the $t_{1/2}$ of hydrolysis to be 19 min. At 150 min (with 99.5% conversion) the reaction is cooled, the solvent removed with care to minimize foaming, and the residue dried to give 3.51 g (containing 3.65 mmol of **12**). This material is extracted at 60° with water (110 ml), leaving a residue containing 0.57 mmol of product. The extract is taken to pH 3.1 with 10 *M* NaOH and cooled for 18 hr at 4°. The resulting precipitate is collected and dried to give typically 2.17 g of light yellow powder (containing 3.04 mmol of **12**) (3.61 mmol total, 98%). HPLC analysis of both the unextracted residue and the pH 3.1 precipitate usually show more than 98% of the absorbance area at 295 nm within a single peak.

A sample can be further purified by semipreparative HPLC on Spherisorb ODS2 (25 × 1 cm), eluted at 6 ml/min initially with water–formic acid (pH 3.0), followed by this mixed with CH_3CN (65:35, v/v): UV λ_{max} (ε) (0.1 *M* NH_4PO_4, pH 2.8) 292 nm, 332 nm; (0.1 *M* KPO_4, pH 6.5) 290 nm, 332 nm; (0.1 *M* NaOH) 300 nm (22,300), 346 nm (16,500); ^1H NMR δ 2.02 (m, 2H, β-CH_2), 2.34 (t, 2H, γ-CH_2), 2.8–3.8 (5H, C*H*–NH, Pyr–NH–C*H*$_2$, Ph–NH–C*H*$_2$), 3.90 (s, 2H, benzyl), 4.32 (m, 1H, α-CH), 6.60 (d, 2H, 3′,5′-H), 7.15–7.5 (5H, Ar), 7.67 (d, 2H, 2′,6′-H), 8.14 (d, 1H, CON*H*CH), 9.88 (br s, 1H, CONH); HRMS [FAB, Xe, glycerol/polyethylene glycol (PEG) 400] calculated for $C_{26}H_{31}N_8O_8(MH^+)$ 583.226, found 583.228.

*N-[4-[(2S)-Amino]-3-[(2,5-diamino-4(3H)-oxopyrimidin-6-yl)amino]propylamino]benzoyl]-L-glutamic Acid (**13**)*

To a clear solution of 1.05 g of partially purified **12** (1.47 mmol) in 45 ml of DMF and 180 ml of 0.1 *M* HCl is added 1.05 g of 10% (w/w) Pd/C, and the mixture stirred vigorously under 45 psi H_2. The reduction of the

[20] J. B. Hynes, S. J. Harmon, G. G. Floyd, M. Farrington, L. D. Hart, G. R. Gale, W. L. Washtien, S. S. Susten, and J. H. Freisheim, *J. Med. Chem.* **28,** 209 (1985).

5-nitro group is usually complete before significant hydrogenolysis of the benzyl group occurs. Analysis at 25 hr by HPLC [Partisil SCX 10 μm, 0.15 *M* NH_4OH plus formic acid (pH 3.0), 1 m*M* Na_2EDTA–CH_3CN (97:3, v/v), 1.0, 275, +0.3] usually shows complete consumption of the starting material and less than 1% of the benzyl derivative of **13**. With the reaction still under an atmosphere of H_2, the mixture is pulled through an in-line filter [Cole–Parmer (Chicago, IL) L-06621-05 fitted with a Whatman (Clifton, NJ) GFA 427 glass filter pad], using reduced pressure, into an ice-cooled flask. Deaerated fresh 0.01 *M* HCl (10 ml) is added to the hydrogenation vessel and pulled through the filter to wash remaining product from the catalyst. The clear light yellow filtrate is quickly sparged with argon. Solvent is removed, and product redissolved in deaerated methanol–H_2O (2:1, v/v). HPLC analysis of this material typically shows that 99% of the electrochemical response in the chromatogram is located in a single peak, and titration with dichlorophenol indophenol indicates 1.34 mmol of product. This 91% yield is confirmed by UV spectrum in 0.1 *M* HCl.

A sample can be purified semipreparatively on Spherisorb ODS2 (25 × 1.0 cm) eluted initially at 3 ml/min with water, and after 100 ml, with H_2O–CH_3CN (98:2, v/v). The collected fraction is acidified to pH 1 with HCl, evaporated to remove CH_3CN, and lyophilized: UV λ_{max} (ε) (0.1 *M* HCl) 214 nm, 274 nm (23,500), 293 nm (sh); HRMS (FAB, Xe, glycerol–PEG 400–DMF) calculated for $C_{19}H_{27}N_8O_6(MH^+)$ 463.205, found 463.208.

The speed of formation of **13** is highly dependent on the purity of the starting material. HPLC-purified preparations can be fully reduced within as little as 4 hr, whereas typical reactions require 20 to 30 hr. Careful HPLC monitoring of this reaction is advisable, because overreduction of **12** can occur, especially if highly purified.[21] A standard for the nitro-reduced, but not yet debenzylated, intermediate, which is useful for interpretation of the chromatogram, can be quickly generated by addition of $Na_2S_2O_4$ (5 mol/mol) to a solution of **12** in 1:1 (v/v) DMF–H_2O under argon at 80°.

Complete removal of catalyst minimizes premature oxidation by air. For small reaction volumes filtration through a Gelman (Ann Arbor, MI) polytetrafluoroethylene (PTFE) 0.45-μm Acrodisc is more effective than the glass filter pads. If immediate oxidative cyclization of the triaminopyrimidine **13** is not desired, the residue left after removal of solvent can be taken up in minimal argon-sparged methanol, quickly precipitated with $(C_2H_5)_2O$, and dried under vacuum. This white to slightly purple powder can be stored indefinitely under argon at −20°.

[21] With precursors to simple unconjugated 6-alkyltetrahydropteridines, overexposure to reducing conditions is not detrimental.

*(6S)-Tetrahydrofolic Acid (**1**)*

A solution of crude **13** (1.2 mmol) in 60 ml of methanol–H_2O (2:1, v/v) in a thermometer-equipped 250-ml flask is warmed to 27°, and 2.5 ml of 0.5 *M* I_2 in methanol (1.25 mmol) added all at once with vigorous stirring, and maintained at 27°. After 2.0 min the mixture is put into a dry ice–ethanol bath and cooled quickly to less than −60°. The cold solution is titrated to pH 9.1 (measured after a 10-fold dilution of a sample into water) with 1.0 *M* NaOH in methanol. The reaction is then transferred to a 35° bath, rapidly warmed with vigorous agitation over 1 min to −10°, transferred to a water–ice bath to complete the transition to 0°, and sparged with argon. After 6.0 min at 0°, 4.8 ml of 1.0 *M* $Na_2S_2O_4$ (freshly dissolved in deaerated water) is added all at once with stirring, followed 2.0 min later by 0.68 ml of 2-mercaptoethanol (9.7 mmol). Analysis by HPLC [Partisil SCX 10 μm, 30 m*M* NH_4OH plus formic acid (pH 3.0) plus 1 m*M* Na_2EDTA, 1.0, 270, +0.3] typically indicates 0.63 mmol of **1** (52%). Contaminating impurities include small amounts of 7,8-dihydrofolic acid, **16** (λ_{max} 290 nm, pH 3.0), PABA-Glu, and a trace of N^5,N^{10}-methylene-**1**.

The enantiomeric purity of this material is established by collecting the entire peak of **1** from the above HPLC system, and reinjecting it onto a chiral column [Cyclobond I, 1 *M* sodium acetate (pH 5.5)–CH_3CN–3 m*M* HCHO (1:3:16, v/v/v), 1.0, 285, +0.3]. Comparison of the resulting 5,10-methylene derivative with the peaks produced by racemic (6*R*,*S*)-L-tetrahydrofolic acid (R_s = 1.5) shows 97.0 to 97.5% enantiomeric purity.

For purification from salts and most by-products, all of the crude product from a reaction of the scale described above is pumped directly onto a 24 × 10 cm column of Whatman DE52 preequilibrated with deaerated 0.1 *M* 2-mercaptoethanol neutralized to pH 7 with NaOH. The column is then eluted at 10 ml/min with the same solution (500 ml), 0.1 *M* 2-mercaptoethanol in 0.01 *M* HCl (200 ml), and then 0.1 *M* 2-mercaptoethanol in 0.05 *M* HCl. Analysis by HPLC typically shows 90% recovery of tetrahydrofolic acid along with a small amount of 7,8-dihydrofolic acid. Eluant is removed after addition of 5 ml of DMF by evaporation to a volume less than 5 ml, and precipitation of product with $(C_2H_5)_2O$. The chromatographic elution has been shown to be identical to authentic tetrahydrofolic acid. A sample can be purified further by precipitation from aqueous solution by adjustment to pH 3.4 with NaOH, followed by recrystallization from distilled water: UV λ_{max} (ε) (1.0 *M* HCl) 269 nm (25,100), 292 nm (20,600); (0.05 *M* KPO_4, pH 7.0) 218 nm (31,900), 297 nm (30,100). Analysis: Calculated for $C_{19}H_{23}N_7O_6 \cdot 1.3H_2O$: C, 48.67; H, 5.50; N, 20.91. Found: C, 48.71; H, 5.52; N, 20.89.

Addition of some water to the oxidation solvent facilitates hydrolysis of the 5-iminoquinoid pyrimidine **14**. Although a 2:1 mixture of methanol–

water is adequate for this purpose, a lesser amount of water may also prove sufficient.[22] The concentration of the reaction should be such that on cyclization the resulting quinoid dihydrofolic acid is less than 30 m*M*, owing to a propensity toward autocatalysis of tautomerization to 7,8-dihydrofolate.

The amount of iodine required for a particular batch of triaminopyrimidinone **13** can be initially approximated by the color of an oxidized sample. Underoxidized material is purple, whereas overoxidation shows the color of dilute iodine; the optimal I_2 stoichiometry produces a light color between these extremes.[23]

After hydrolysis, the rate of cooling in the dry ice bath is not overly critical, as once below 0° quinones **15** in acid are stable for several hours. Most significantly, the titration to pH 9.1 (the pH of optimal cyclization yield[24]) is best performed after the reaction mixture has equilibrated to the dry ice bath temperature, at which significant cyclization does not occur for many hours. Furthermore, pH 9.1 solutions of quinones **15** can be stored indefinitely in liquid nitrogen prior to cyclization. The temperature/time profile of the cyclization reaction is important for a high yield of clean tetrahydrofolate. A reaction to which reductant is added too early after warming to 0° will produce the divicine derivative **16**. An HPLC standard of **16** can be generated by direct addition of cold pH 9.1 quinone **15** to an aqueous solution of $Na_2S_2O_4$. However, delayed addition of reductant results in accumulation of 7,8-dihydrofolate and cleavage of PABA-Glu. To test the adequacy of a technique for rapid and accurate changes in temperature, a trial should be run with the flask and solvent volume to be used for the reaction. For dilute reactions (~10 m*M*) that are free of extraneous general acid or base catalysts, it is usually safer to err on the side of longer cyclization times. Reagents, such as hydrides, that are capable of reducing 7,8-dihydropterins can lower the final chiral purity of the product. Sodium dithionite when used at 0° and not in great excess does not

[22] For synthesis of unconjugated tetrahydropterins 100% methanol is a good choice for the oxidation solvent. Even freshly opened bottles of reagent-grade methanol have sufficient water to support rapid hydrolysis within 2 min at 27°. Furthermore, the presence of greater than 50% water can lead to increased contamination of the product with 7-tetrahydropterin.[24]

[23] Direct oxidation of triaminopyrimidines such as **13** with halogen generally requires a slight excess of oxidant.[8] A mixture of stoichiometric Br_2 with catalytic ferricyanide has been used for the generation of 4a-hydroxy adducts of unconjugated tetrahydropterin analogs of **17** that are relatively free of halogen adducts.[24] On dehydration and reduction these 4a-hydroxy adducts give high yields, typically greater than 90%, of the corresponding tetrahydropterin. The bromine/ferricyanide system cannot be used for the synthesis of tetrahydrofolates, as by-products are produced despite a rapid rate of iminoquinoid pyrimidine (**14**) formation.

[24] S. W. Bailey, I. Rebrin, S. C. Boerth, and J. E. Ayling, *J. Am. Chem. Soc.* **117,** 10203 (1995).

significantly reduce 7,8-dihydrofolate. Ascorbic acid is an alternative reductant, although this produces slightly lower yields of (6*S*)-tetrahydrofolic acid.

Acknowledgments

Supported by NIH Grant NS-26662. Parts of this chapter were adapted with permission from *J. Org. Chem.* **57,** 4470–4477 (1992). Copyright 1992 American Chemical Society.

[2] Analysis of Folates Using Combined Affinity and Ion-Pair Chromatography

By PAMELA J. BAGLEY and JACOB SELHUB

Principle

The method described allows simultaneous measurement of intact one-carbon and polyglutamyl derivatives of folate from tissues or food using a two-step procedure. The first step is folate purification by affinity chromatography using immobilized bovine milk folate-binding protein (FBP). The second step uses ion-pair high-performance liquid chromatography (HPLC) to separate the folate derivatives with diode array detection to identify and quantify the folate derivatives.

Procedures

Preparation of Affinity Columns

Preparation of Folate–Sepharose Gel. The activation of the Sepharose and the subsequent reaction with diaminohexane[1,2] should be conducted using rubber gloves in a well-ventilated hood. The fumes from cyanogen bromide are very poisonous and precautions should be taken to avoid inhaling them.

Wash Sepharose 4B, 400-ml bed volume, with 2 liters of water and suspend the washed Sepharose in 400 ml of water in a 1-liter beaker containing a magnetic stir bar. Place the beaker in a hood over a magnetic stirrer with a pH meter electrode set up so that it is in the suspension. Also set up a 50- or 100-ml buret containing 5 *M* NaOH.

[1] J. Selhub, O. Ahmad, and I. H. Rosenberg, *Methods Enzymol.* **66,** 686 (1980).
[2] J. Selhub, B. Darcy-Vrillon, and D. Fell, *Anal. Biochem.* **168,** 247 (1988).

0076-6879/97 $25

While stirring the suspension add 30 g of cyanogen bromide crystals to the suspension. For the next 15 to 30 min continuously add NaOH solution from the buret to maintain the pH between pH 10 and 11. While adding the NaOH, periodically add ice to the stirring suspension to maintain the temperature at 20°.

When activation is complete, as indicated when the CNBr crystals are completely dissolved and the pH becomes stable around pH 10, filter the suspension through a 600-ml sintered glass funnel (coarse), using vacuum from a water aspirator. Wash the activated Sepharose with 1.5 to 2 liters of cold (4°) 0.1 *M* $NaHCO_3$. Dissolve 1.8 g of 1,6-diaminohexane in 400 ml of 0.1 *M* $NaHCO_3$ and adjust to pH 9.0 with NaOH. At 4°, mix the diaminohexane solution with the activated Sepharose and stir for 48 hr. Then filter the suspension through the sintered glass funnel and wash with 0.1 *M* $NaHCO_3$ and resuspend in 400 ml of the 0.1 *M* $NaHCO_3$ solution.

Suspend 4 g of folic acid in 50 ml of water and add NaOH dropwise until the folic acid is dissolved. Add the folic acid solution to the Sepharose suspension, stir, and adjust to pH 8.5 with NaOH. While stirring at room temperature, add 4 g of 1-ethyl-3-(3-dimethylaminopropyl)carbodiimide hydrochloride over a period of 45 min. Continue stirring the suspension for 3 hr at room temperature, then stir for an additional 48 hr at 4°. To remove the nonbound folate, sequentially wash the suspension with 4 liters of 2 *M* NaCl containing 0.05 *M* potassium phosphate buffer (pH 7.0), 2 liters of water, 1 liter of 0.2 *M* acetic acid, and 2 liters of water. Suspend and store the folate–Sepharose gel in 0.3% (w/v) sodium azide at 4°.

Purification of Folate-Binding Protein. Bovine milk FBP is isolated from dried whey, a milk product that is available from bakery suppliers.[1] Suspend a 50-lb bag of whey, either sweet or acid, in 40 liters of water. A plastic garbage pail makes a convenient container for the suspension. Neutralize the resulting suspension with 5 *M* sodium hydroxide (about 1 liter) and store the neutralized suspension overnight at 4°. Centrifuge the whey suspension in batches at 15,000 *g* for 10 min at 4° and filter the supernatant fraction over glass wool sandwiched between silkscreening fabric. It is unnecessary to centrifuge the bottom third of the suspension because it contains so much precipitate. Each 50-lb bag of whey yields between 20 and 22 liters of filtered supernatant. The FBP activity is measured in an aliquot of the filtered supernatant fraction and in a 1 : 10 diluted aliquot (method described below).

Folate–Sepharose gel, prepared as described in the previous section, is added to the filtered whey supernatant and the suspension is stirred at 4° overnight, during which time the FBP binds to the folate–Sepharose gel. The folate–Sepharose gel, with bound FBP, is separated from the whey extract by filtration over a water aspirator vacuum using a 2-liter coarse-sintered glass funnel. The FBP activity is measured in a 1:10 diluted aliquot of the filtrate. The folate–Sepharose gel should bind about 70% of the FBP activity in the whey supernatant. The FBP bound to the gel can be stored at 4° in 0.03% (w/v) sodium azide without loss of activity for several months. The folate–Sepharose gel has a high capacity to bind FBP and can be used to extract FBP from four to five 50-lb bags of whey before it is necessary to elute the FBP.

Before extracting the FBP from the folate–Sepharose matrix, wash the matrix in a sintered glass funnel (coarse) sequentially with 2 liters of 0.05 *M* HEPES containing 1 *M* sodium chloride and 0.03% (w/v) sodium azide, 1 liter of 1 *M* potassium phosphate buffer (pH 7), and 2 liters of water. Suspend the Sepharose–folate matrix in 1 liter of 0.2 *M* acetic acid and stir for 2 hr at 4°. Transfer the Sepharose–folate matrix to a sintered glass funnel and elute the FBP by filtration over a water aspirator vacuum. To extract the FBP completely, the Sepharose–folate matrix is suspended in 0.1 *M* acetic acid and stirred for 2 hr at 4° and the FBP is eluted by filtration, three more times. The filtrates are neutralized with potassium carbonate and pooled. The FBP activity is measured in aliquots of the filtrate (undiluted, and diluted 1:10, 1:100, and 1:1000).

Finally, the FBP is concentrated using a 3 × 15 cm Sepharose–folate column. Elute the FBP with 0.2 *M* acetic acid, dcollecting the column eluent in 10- to 20-ml fractions. Fractions containing FBP can be distinguished because of the yellow color of the FBP. The FBP activity can be measured in diluted aliquots of the fractions. Pool the appropriate fractions and neutralize with potassium carbonate. Measure the FBP activity in aliquots diluted 1:10, 1:100, and 1:1000. Lyophilize the pooled fractions to dryness and store at −70°.

Determination of Folate-Binding Protein Activity. Folate-binding protein activity is measured at each step in the isolation of FBP from whey. To a 12 × 100 mm test tube, add 10 or 25 μl of the solution containing the FBP, 1 ml of 0.1 *M* potassium phosphate buffer (pH 7), 50 μl of 0.1% (w/v) bovine serum albumin, and 50 μl of 1 μ*M* [^{3}H]folic acid (26.7 Ci/mmol, containing 0.5 μCi/ml). Blank measurements are made in the same manner except that no sample is added. After incubating the tube for 10 min at room temperature, empty the tube contents into the funnel of a vacuum filter with a 2.5-cm nitrocellulose membrane filter, 0.45-μm pore size, that has been soaked in water. Rinse the test tube twice with 0.1 *M* potassium phosphate, pH 7, adding the rinse solution to the filter funnel. Filter the combined incubation mixture and rinse solution through the nitrocellulose membrane filter. The FBP, with bound folate, will adsorb to the membrane with about 40% efficiency. Wash unbound folate through the membrane by twice filtering 15 ml of 0.1 *M* potassium phosphate, pH 7, through the membrane. Dry the nitrocellulose membrane(s) by pinning it to a piece of cardboard and heating it in a 100° oven for 10–15 min, until the edges of the membrane start to brown. Transfer the dry membrane to a scintillation vial containing 10 ml of toluene–POPOP scintillation fluid and measure the radioactivity with a scintillation counter. Toluene–POPOP scintillation fluid is made by combining 950 ml of toluene with 50 ml of a stock solution made with 60 g of PPO (2,5-diphenyloxazole), 1 g of POPOP [1,4-bis(phenyl-2-oxazolyl)benzene; 2,2′-*p*-phenylenebis(5-phenyloxazole)], and 500 ml of toluene. The stock solution should be mixed overnight when first prepared and mixed thoroughly just before diluting in the toluene. Calculate FBP activity as follows:

$$\text{FBP activity} = \text{dpm} \times \text{SV}(\text{SVA} \times \text{DF} \times \text{SA})^{-1} \qquad (1)$$

where disintegrations per minute (dpm) is the sample membrane radioactivity minus the blank membrane radioactivity, SV is the total sample volume (ml), SVA is the sample volume or diluted sample volume applied to the membrane (ml), DF is the sample dilution factor (if any), and SA is the specific activity of [^{3}H]folic acid (dpm/pmol). One unit of FBP activity is the amount of FBP that binds 1 pmol of folic acid.

Preparation of FBP–Sepharose. The final capacity of the FBP-Sepharose[1] to bind folate should be at least 25–50 kilounits (kU) (30–60 nmol) per milliliter of bed volume. Dried whey contains about 250 U/g, of which about 50% is recovered in the purified fraction. Thus the purified FBP from ten 50-lb bags of whey should contain about 25,000 to 28,000 kU.

Activate an amount of Sepharose 4B, not exceeding a 1-ml bed volume per 70 to 100 kU of FBP, using the same conditions for activating with CNBr with subsequent cold $NaHCO_3$ washes as previously described. Suspend the washed, activated Sepharose in 50 ml of 0.1 *M* $NaHCO_3$. Suspend the purified, lyophilized FBP in 250 ml of 0.1 *M* $NaHCO_3$ and add the resulting cloudy suspension to the activated Sepharose. Stir the FBP-activated Sepharose suspension for 2 hr at room temperature, then stir for an additional 48 hr at 4°. Then transfer the Sepharose–FBP matrix to a coarse-sintered glass funnel and wash it sequentially with 2 liters of 2 *M* NaCl containing 0.05 *M* Tris-HCl (pH 7.4), 1 liter of water, 2 liters of 0.02 *M* trifluoroacetic acid, 2 liters of 0.05 *M* potassium phosphate buffer (pH 7.4), and finally with 1 liter of water. Suspend the Sepharose–FBP in an equal volume of 0.3% (w/v) sodium azide and store at 4°.

Folate Extraction

Extraction of Folate from Tissues. Folate is extracted from tissues[1,3] under conditions that protect against oxidative degradation and γ-Glu-*X*

[3] G. Varela-Moreiras, E. Seyoum, and J. Selhub, *J. Nutr. Biochem.* **2,** 44 (1991).

carboxypeptidase (EC 3.4.19.9) degradation. Although fresh tissues are preferable, tissues that have been frozen in liquid nitrogen immediately after harvesting and stored at −70° can be used. It is important that frozen tissues not be allowed to thaw, as extensive hydrolysis of the folate polyglutamates will occur. Mince fresh or unthawed frozen tissue and transfer into 10 vol of hot extraction buffer [2% (w/v) sodium ascorbate, 10 m*M* 2-mercaptoethanol, and 0.1 *M* Bis–Tris (pH 7.8)] in tubes in a boiling water bath. If using frozen tissue, add 0.05 *M* potassium tetraborate to the extraction buffer instead of 0.1 *M* Bis–Tris buffer, pH 7.8. Boil the capped tubes for 15 min, then cool in ice water and homogenize the contents with a Polytron (Brinkmann, Westbury, NY) homogenizer. Centrifuge the homogenate at 36,000 *g* for 15 min at 4°. Filter the supernatant fraction through glass wool, immediately inject the filtered supernatant into a Vacutainer tube (Becton Dickinson, Mountain View, CA), and store at −70° until the folate is analyzed.

Extraction of Folate from Food. Before extraction, food is broken up by mashing, grinding, or chopping.[4] The broken-up food is suspended in a solution containing 2% (w/v) sodium ascorbate, 10 m*M* 2-mercaptoethanol, and 100 m*M* Bis–Tris buffer (pH 7.8), and capped and heated in an autoclave for 30 min at 120° (15 psi). The sample is then cooled in an ice–water bath, homogenized in a Waring blender for 5 min, and centrifuged at 30,000 *g* for 15 min at 4°. Folate is measured in the supernatant fraction.

Chromatography of Extracted Folates

Preparation of Affinity Column. All work[3,5] to prepare and use affinity columns should be done at 4°. The affinity columns are prepared by transferring a 1-ml bed volume of FBP–Sepharose gel to a column and allowing the matrix to pack by gravity. A glass Pasteur pipette with glass wool packed into the neck can be used as the column. Before the first use, wash the column sequentially with 3 ml of 20 m*M* trifluoroacetic acid, 10 ml of 1 *M* potassium phosphate (pH 7), and two 3-ml volumes of water. The column can be reused until capacity (as monitored by recovery of tracer folate) is diminished. Because the column matrix is not stable at low pH, a 1 *M* potassium phosphate, pH 7, wash should immediately follow any application of trifluoroacetic acid.

Folate Purification by Affinity Chromatography. An aliquot of the folate extract[2] containing 2 to 15 nmol of total folate is mixed with 0.2 μCi of [^{3}H]folic acid tracer (26.7 Ci/mmol) and radioactivity in an aliquot of the

[4] E. Seyoum and J. Selhub, *J. Nutr. Biochem.* **4,** 488 (1993).
[5] J. Selhub, *Anal. Biochem.* **182,** 84 (1989).

mixture is measured. The remaining portion of the mixture is applied to the affinity column. The column is washed with 10 ml of 1 *M* potassium phosphate, pH 7, and washed twice with 3 ml of water. The adsorbed folate is eluted from the column with 3 ml of 20 m*M* trifluoroacetic acid containing 10 m*M* dithioerythritol. The acid eluate is promptly neutralized with 60 μl of 1 *M* piperazine. Once eluted, the folate is unstable and should immediately be analyzed by HPLC. Radioactivity in a 50-μl aliquot of the eluate is measured to calculate recovery over the column. Recovery should be at least 85%. If it is less than 85%, either too much extract was applied to the column or the column needs replacement. (Note that [^{3}H]folate is often impure and impurities are not retained by the affinity column. The affinity column can be used to assess the purity of the [^{3}H]folate or can be used to purify the [^{3}H]folate.)

The affinity column is regenerated by application of 10 ml of 1 *M* potassium phosphate, pH 7, and two 3-ml washes with water. The column should be stored at 4°.

Analysis of Folate Distribution by High-Performance Liquid Chromatography. Folate distribution in the affinity-purified samples is measured using ion-pair HPLC with diode array detection. A Bio-Sil ODS-5S (150 × 4 mm) analytical column with an ODS-5S Micro-Guard cartridge (Bio-Rad, Hercules, CA) is used. Two solutions are used to make the gradient for the mobile phase. Solution A and solution B both contain 5 m*M* tetrabutylammonium phosphate (TBAP), 0.5 m*M* dithioerythritol, and 25 m*M* sodium chloride. Solution B also contains 65% (v/v) acetonitrile and 1 m*M* monobasic potassium phosphate. Before using, filter and degas both solutions.

A flow rate of 1 ml/min is maintained throughout the run. Before injection of sample, the column is equilibrated in 90% A : 10% B for 10 min. After injection, the mobile phase is run isocratically for the first 5 min with 90% A : 10% B. The percentage of solution B is then increased linearly to reach 36% at 15 min, 50% at 35 min, and 60% at 52 min.

Diode array detection is used to monitor folate. The detector is set to monitor absorption at 280, 350, and 258 nm.

Folate Identification and Quantification. Elution of folates from the HPLC column consists of clusters of folates with increasing numbers of glutamate residues. Folates within each cluster contain the same number of glutamate residues.

A combination of retention time and spectral characteristics is used to identify the individual folates within each cluster. The number of glutamate residues attached to the folate molecule is identified by retention time. The lower limit of detection of the diode array detector is about 25 pmol/peak.

Spectral Analysis. In clusters containing mono- and diglutamyl folates, there is separation between the 10-formyl, tetrahydro, and dihydro derivatives. In clusters with longer glutamate chain length, these folate derivatives coelute.

Folate concentration in peaks containing single derivatives can be determined on the basis of the integrated peak area and the coefficients given in Table I. Each folate derivative has a characteristic absorption spectrum. By monitoring absorption at 280, 350, and 258 nm, folate concentration of individual derivatives can be calculated in peaks that contain more than

TABLE I
MOLAR PEAK COEFFICIENTS EXPRESSED AS INTEGRATED PEAK AREA OF FOLIC ACID AND ITS DERIVATIVES[a]

Compound	Molar peak coefficient (unit area/nmol)[b]		
	280 nm	350 nm	258 nm
Pteroylglutamate	875	210	472
Dihydrofolate	787 (0.899)	152 (0.173)	321 (0.367)
THF[c]	682 (0.779)	7 (0.008)	266 (0.304)
10-Formyl-THF	323 (0.369)	0 (0)	513 (0.586)
5-Methyl-THF	787 (0.899)	0 (0)	314 (0.314)
5-Formyl-THF	812 (0.928)	0 (0)	323 (0.370)

[a] The data were obtained at various wavelengths using [^{3}H]folic acid before and after conversion to various reduced forms. Aliquots containing about 1, 3, and 10 nmol were each injected into the column and the column was subsequently eluted under the same conditions described in the HPLC analysis section. Fractions of 0.5 ml were collected to determine folate content on the basis of radioactivity counts. The molar peak coefficient (unit area per nanomole) was calculated from the amount of folate that was eluted and the corresponding peak area determined at the wavelength indicated. Each value represents the average obtained from three aliquots. Standard errors (not shown) were in each case less than 5%. Numbers in parentheses represent molar peak coefficient ratios of the various folates at the indicated wavelengths versus the molar peak coefficient of pteroylglutamate at 280 nm. [Reprinted with permission from J. Selhub, *Anal. Biochem.* **182,** 84 (1989).]

[b] The values were derived from the data obtained under the chromatographic conditions employed in this study. Different conditions, such as different flow rates, different flow cell size, and other chromatographic variables, will result in different values.

[c] THF, Tetrahydrofolate.

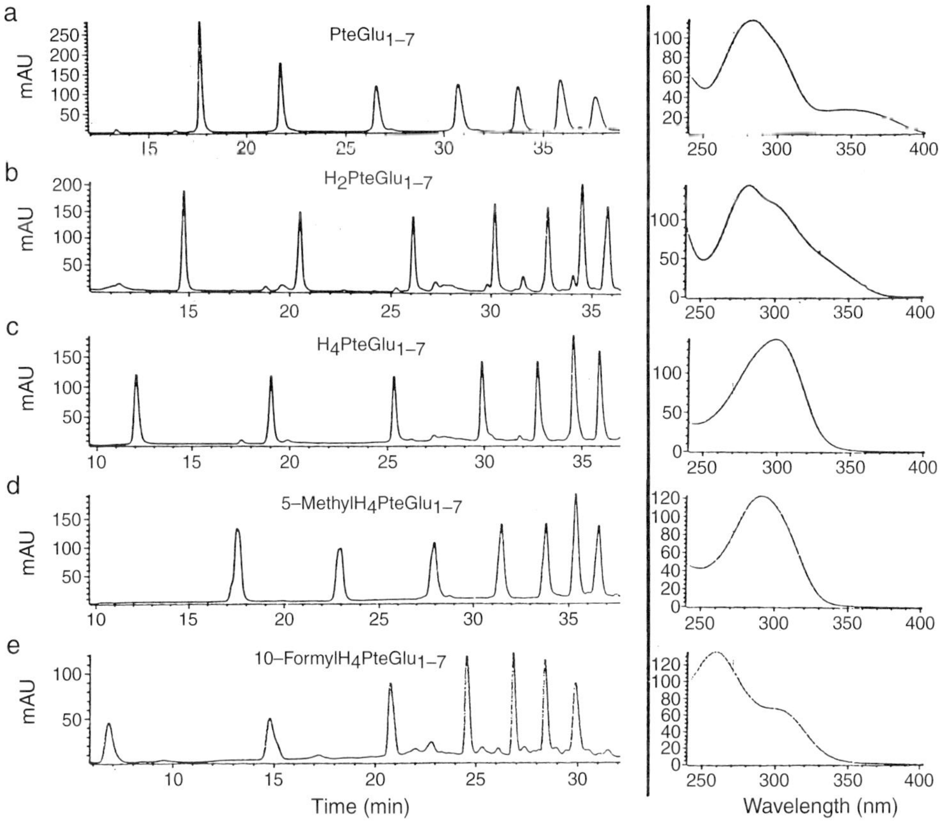

FIG. 1. Ion-pair chromatography of a pteroylglutamate$_{1-7}$ (PteGlu$_{1-7}$) mixture before and after conversion to reduced forms. Mixtures containing 2–5 nmol of each of the corresponding folate derivatives were subjected to chromatography as described in text. *Left:* Chromatograms provided by monitoring the effluents for UV absorption at 280 nm. *Right:* Representative spectra of the activity peaks for each series. [Reprinted with permission from J. Selhub, *Anal. Biochem.* **182,** 84 (1989).]

one derivative. To resolve pteroylglutamate (P) and 5-methyl tetrahydrofolate (M), calculate as follows:

$$[P] = A_{350}/E_{p350} \tag{2}$$

$$[M] = [A_{280} - ([P]E_{p280})]/E_{m280} \tag{3}$$

where A_{350} and A_{280} are the absorption values, in integrated area units, at the wavelengths indicated in subscripts. E_{p350}, E_{p280}, and E_{m280} are the molar peak coefficients, in integrated area units, for the folate derivative and wavelength indicated by the subscript.

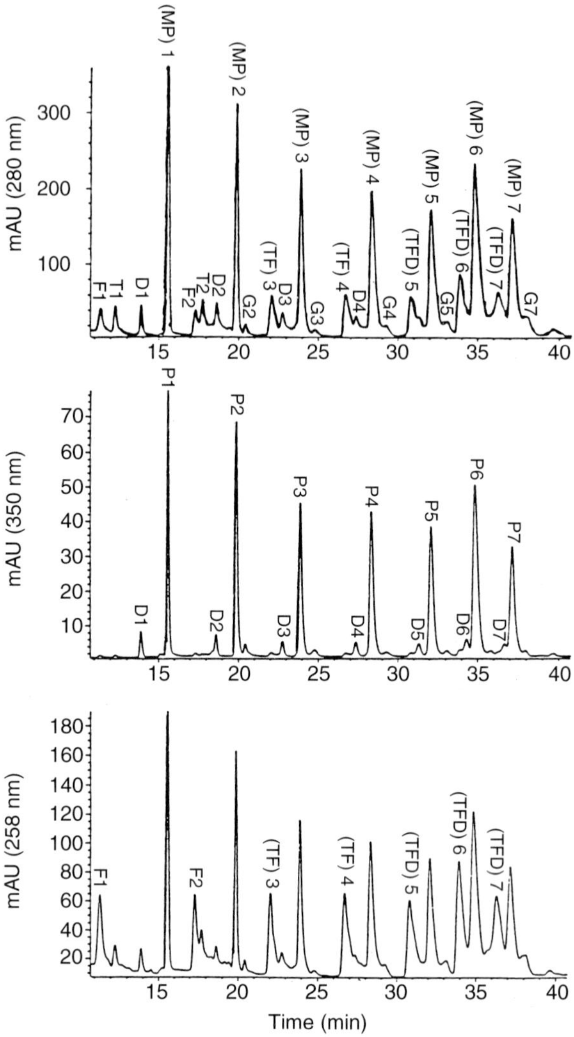

FIG. 2. Chromatography of a mixture that contained pteroylglutamate$_{1-7}$ and its various reduced forms, a total of 35 derivatives. Activities shown represent absorption values at (*top*) 280 nm, (*middle*) 350 nm, and (*bottom*) 258 nm. Symbols and concentrations are as follows: F, 10-formyltetrahydrofolate$_{1-7}$ (1.8–2.0 nmol); T, tetrahydrofolate$_{1-7}$ (0.9–1.2 nmol); D, dihydrofolate$_{1-7}$ (0.4–0.6 nmol); M, 5-methyltetrahydrofolate$_{1-7}$ (1.0–1.2 nmol); and P, pteroylglutamate$_{1-7}$ (2.4–2.7 nmol). Subscripts following the symbols refer to the number of glutamate residues. [Reprinted with permission from J. Selhub, *Anal. Biochem.* **182,** 84 (1989).]

To resolve tetrahydrofolate (T) and 10-formyltetrahydrofolate (F), for clusters 3 and 4, calculate as follows:

$$A_{280} = [\mathrm{T}]E_{\mathrm{t}280} + [\mathrm{F}]E_{\mathrm{f}280} \tag{4}$$
$$A_{258} = [\mathrm{T}]E_{\mathrm{t}258} + [\mathrm{F}]E_{\mathrm{f}258} \tag{5}$$

The symbols are the same as described above.

To resolve dihydrofolate (D), tetrahydrofolate (T), and 10-formyltetrahydrofolate (F) in clusters 5, 6, and 7, calculate as follows:

$$A_{280} = [\mathrm{F}]E_{\mathrm{f}280} + [\mathrm{T}]E_{\mathrm{t}280} + [\mathrm{D}]E_{\mathrm{d}280} \tag{6}$$
$$[\mathrm{D}] = A_{350}/E_{\mathrm{d}350} \tag{7}$$
$$[\mathrm{F}] = A_{258} - ([\mathrm{D}]E_{\mathrm{d}258} + [\mathrm{T}]E_{\mathrm{t}258}) \tag{8}$$

The symbols are the same as described above. [Equations (2)–(8) are reprinted with permission from J. Selhub, *Anal. Biochem.* **182,** 84 (1989).]

Column Calibration. Absolute retention times and extinction coefficients will vary between individual HPLC systems. During development of this method, tetrahydrofolate$_{1-7}$, 5-methyl tetrahydrofolate$_{1-7}$, 5- and 10-formyl tetrahydrofolate$_{1-7}$, and dihydrofolate$_{1-7}$ were synthesized from pteroylglutamate$_{1-7}$, respectively (see Fig. 1). The subscripts indicate the number of glutamate residues. The syntheses of these reduced folates from pteroylglutamate are described by Selhub.[5] Although useful, it is not essential to synthesize these reduced folates to calibrate the column. As shown in Fig. 1, the last peak in each cluster is the folic acid$_n$/5-methyl tetrahydrofolate$_n$ peak, where n is the number of glutamate residues in the cluster (see Fig. 2). The column can be calibrated with a pteroylglutamate$_{1-7}$ standard because the upper bound of the elution time for each cluster will be defined by the pteroylglutamate elution time for its particular cluster. Combining this information about glutamate chain length with the spectral characteristics of the peak will allow identification of all the folate peaks.

Molar extinction coefficients can be determined using the pteroylglutamate as the standard. Known amounts of pteroylglutamate are injected onto the column and the resulting peak areas at 280, 350, and 258 nm are obtained. These peak areas are used to calculate the molar peak coefficients, in units of area per nanomole, for pteroylglutamate at 280, 350, and 258 nm. The molar peak coefficient of the reduced folates at each wavelength can be calculated by multiplying the measured molar peak coefficient of pteroylglutamate at 280 nm by the respective ratio found in parentheses in Table I for the various reduced folates at the various wavelengths.

[3] Molar Quantitation of Folates by Gas Chromatography–Mass Spectrometry

By Cheruppolil R. Santhosh-Kumar and J. Fred Kolhouse

Introduction

Different folate coenzymes serve as donors and acceptors of one-carbon fragments in a number of critical enzymatic reactions including the synthesis of purines, pyrimidines, serine, the methyl group of *S*-adenosylmethionine and the catabolism of purines and histidine.[1] In humans, overt folate deficiency causes megaloblastic anemia and subclinical deficiency is believed to be associated with several diseases including development of neural tube and other birth defects, accelerated atherosclerosis, and initiation or promotion of malignancy.[2,3] Most laboratories "estimate" the relative folate content of a tissue or blood sample using a radioisotope dilution competitive binding assay.[2,4] Methods that employ the principle of competitive binding for assay of folates using limiting amounts of folate-binding protein are prone to inaccuracy because folates exist in tissues in as many as eight different coenzyme forms with different polyglutamate chain lengths (Fig. 1), resulting in differences in affinities for folate-binding protein[4] as well as obvious differences in molecular weight. Other methods of folate measurement include microbiological assays, which depend on auxotrophy of certain microorganisms for folate, high-performance liquid chromatography (HPLC)-based methods, and enzyme-linked immunosorbent assays (ELISAs), which use external standards. Folate levels determined by the above methods are expressed as nanograms per mililiter, which is a misrepresentation because it is based on the weight of a single folate coenzyme form used in the standard curve and ignores the abovementioned variations in molecular weight and relative affinities of various coenzyme forms of folate with various degrees of polyglutamation. For example, folic acid monoglutamate (molecular weight 441) is frequently used in the standard curve while other forms of folate are measured in

[1] T. Brody, B. Shane, and E. L. R. Stokstad, *in* "Handbook of Vitamins" (L. J. Machlin, ed.), p. 459. Marcel Dekker, New York, 1984.

[2] J. Lindenbaum and R. H. Allen, *in* "Folate in Health and Disease" (L. B. Bailey, ed.), p. 43. Marcel Dekker, New York, 1995.

[3] J. B. Mason, *in* "Folate in Health and Disease" (L. B. Bailey, ed.), p. 361. Marcel Dekker, New York, 1995.

[4] C. R. Gilois and D. R. Dunbar, *Med. Lab. Sci.* **44,** 33 (1987).

0076-6879/97 $25

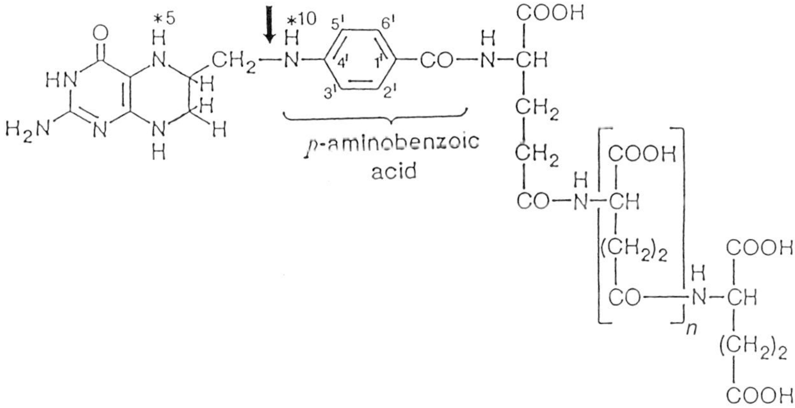

FIG. 1. The structure of tetrahydrofolate polyglutamates ($H_4PteGlu_n$). In metabolically important folates, C_1 substitution occurs at N-5 (*) to give 5-formyl and 5-methyl derivatives and at N-10 (*) to give 10-formyl derivatives. 5,10-Derivatives occur as methylene and methenyl bridge structures. Oxidation at the 5- and 6-positions of the pteridine ring yields dihydrofolate. The *p*-aminobenzoic acid moiety is common to all folate coenzymes. Arrow indicates site of C-9–N-10 cleavage.

samples such as serum, where 5-methyltetrahydrofolate monoglutamate (molecular weight 459) predominates, or erythrocytes, where polyglutamates of 5-methyltetrahydrofolate (molecular weight up to 1500) predominate. Although one attempts to remove the polyglutamates in erythrocyte assays, the completeness of the deglutamation is not routinely measured. Thus, when nanograms are converted to moles using a molecular weight of 441 (folic acid) an error is inherently introduced. The degree of error is obviously more difficult to discern when analyzing other tissues in which multiple coenzyme forms of folate with various different molecular weights are quantitated.

This chapter describes a gas chromatography–mass spectrometry (GC–MS) method that can directly measure the actual mole amounts of folate in a sample.[5]

Principle

The common feature of all folate coenzymes is the *p*-aminobenzoic acid (PABA) portion of the molecule attached to a pterin ring on the NH_2 end and one or more glutamic acids in a γ-amide linkage at the carboxyl end.

[5] C. R. Santhosh-Kumar, J. C. Deutsch, N. M. Kolhouse, K. L. Hassell, and J. F. Kolhouse, *Anal. Biochem.* **225,** 1 (1995).

In the GC–MS method, a known quantity of a mixture of a bacterially synthesized stable isotope-labeled ([$^{13}C_6$]) folate internal standard is added to the blood or tissue sample and the folate coenzymes are specifically purified using a large excess of folate-binding protein to bind all folates and to remove any free PABA, *p*-aminobenzoyl glutamate (PABA-Glu), or *p*-aminobenzoyl polyglutamate (PABA-Glu_n). The C-9–N-10 linkage (arrow, Fig. 1) of PABA to the pterin moiety is chemically cleaved to pterin and PABA-Glu_n. Simultaneously PABA-Glu_n is cleaved to PABA and glutamic acids so that each mole of any folate results in 1 mol of PABA. The resulting PABA derived from folates is quantitated by GC–MS. Because all the ring carbons of the PABA moiety of the bacterial folate internal standard are labeled with ^{13}C, the resulting mass increase of six can be used to differentiate PABA derived from natural sources and from the added [$^{13}C_6$]folate internal standard by MS and a ratio of abundance of natural PABA to [$^{13}C_6$]PABA can be used for molar quantitation. The internal standard is also useful to correct for recovery because it behaves in a manner identical to that of natural folates in terms of binding to various resins and folate-binding protein and with regard to cleavage to PABA and glutamic acids.

Folates are released from tissue-binding sites by heating in detergent and freshly prepared ascorbate. Prior to raising the pH for binding of the folates to a strong anion-exchange resin, the sample is cooled and 2-mercaptoethanol is added because ascorbate is rapidly inactivated as a reducing agent at pH 9.5 (C. R. Santhosh-Kumar and J. F. Kolhouse, unpublished observations, 1996). After application of the crude sample, the strong anion-exchange resin is washed and eluted with 2 *M* sodium chloride and folate-binding protein. Recovery is maximized by warming the resin to 37° for 30 min prior to elution. This step serves to purify the folates in a concentrated solution of 0.5 ml.

Materials

The following materials are utilized in these experiments. *p*-Amino[$^{13}C_6$]benzoic acid ([$^{13}C_6$]PABA, 99.1% ^{13}C) and *p*-amino[2,6-$^2H_2$]benzoic acid ([2,6-$^2H_2$]PABA, 98.7% ^2H) are custom synthesized by MSD Stable Isotopes (Montreal, Quebec, Canada). *Lactobacillus arabinosum* is obtained from the American Type Culture Collection (Rockville, MD) (ATCC 8014, auxotrophic for PABA).[5] Culture medium is mixed as described,[6] except that PABA is replaced with [$^{13}C_6$]PABA. Strong amino-exchange resin, AG MP-1 chloride form (SAX), and strong cation-exchange

[6] T. Shiota, *Arch. Biochem. Biophys.* **80,** 155 (1959).

resin, AG MP-50 hydrogen form (SCX), are obtained from Bio-Rad Laboratories (Hercules, CA). The C_{18} resin is obtained from YMC *Gel (Morris Plains, NJ). Centricon-30 filters with an M_r 30,000 cutoff are obtained from Amicon (Beverly, MA). Filters (0.20-μm pore size, 25-mm diameter) are obtained from Life Science Products, Inc. (Denver, CO). Autosampler vials and plastic disposable columns are obtained from Alltech (Deerfield, IL). Acetonitrile, methanol, and buffers are from Fisher Scientific (Pittsburgh, PA). *N*-Methyl-*n*-*tert*-butyldimethylsilyltrifluoroacetamide is obtained from Regis Chemical (Morton Grove, IL). Propionic anhydride is from Sigma Chemical Company (St. Louis, MO). Rubber ring-sealed plastic tubes for high-temperature acid hydrolysis are obtained from Sarstedt (Newton, NC). Folate-binding protein greater than 99.9% folate free is obtained from Metabolite Laboratories, Inc. (Denver, CO).

Preparation of Standards

Ten milliliters of semidefined medium as described in Materials is inoculated with *L. arabinosum.* The organisms are grown to confluency at 22° with loose lids in 20-ml sterile screw-cap tubes over 24 hr. The culture is split 1:100 five times with a final split of 1:100 into 2 liters of the medium. After this culture grows to confluency (24 hr), the entire 2 liters of culture medium is added to 200 liters of fresh medium. The culture is maintained at pH 6.8 with 5 *N* NaOH. Aeration is at 1 liter/min with agitation at 10–15 rpm. After 24 hr the cells are collected with a sharpel (continuous-flow centrifuge) and the paste weighing 990 g is washed by centrifugation at 4° in 3 vol of phosphate-buffered saline [0.006 *M* sodium phosphate (pH 7.40), 0.004 *M* potassium phosphate (pH 7.50), and 0.14 *M* sodium chloride] at 1500 *g* for 15 min at 4°. The supernatant is decanted and the washing repeated two more times. The cell pellets are then removed from the containers and stored at −20° until further use. The bacterially synthesized mixture of folates is partially purified using 50 g of frozen bacteria with 200 ml of a solution containing 0.1 *M* ammonium bicarbonate buffer (pH 9.6) and 0.3 *M* 2-mercaptoethanol. The suspension of bacteria is heated at 98° for 30 min followed by centrifugation at 30,000 *g* for 30 min at 4°. The yellow supernatant is decanted off and 1-ml aliquots are maintained frozen at −20° until use.

Quantitation of Bacterial Folates Used as Internal Standard

The total $^{13}C_6$-labeled folate content of the purified bacterial folate mixture is determined on triplicate samples at sample volumes ranging from 25 to 100 μl. In another set of triplicate assays, 1 nmol of 5-formyltetrahydrofolate monoglutamate (the quantity determined on the basis of a molar extinction coefficient of 31,500 at 287 nm at pH 7) is added to the same quantities of $^{13}C_6$-labeled folate internal standard. The samples are analyzed on the basis of the gas chromatography–mass spectrometry (GC–MS) method described below and are found to contain 11.55 nmol/ml of [$^{13}C_6$]folate internal standard.

Purification of Folates from Whole Blood

Ten to 200 μl of whole blood is added to a mixture containing in final concentration 0.6% (w/v) ascorbic acid, 1% (v/v) Triton X-100, and 50 pmol of [$^{13}C_6$]folate internal standard in a stoppered glass tube (final volume 2 ml) and heated at 120°. At the end of 60 min the mixture is cooled and centrifuged at 600 *g* for 10 min at 22°. The supernatant is added to glass tubes containing 50 μl of 2-mercaptoethanol followed by 200 μl of 1 *M* ammonium bicarbonate, pH 9.6. The sample is mixed and passed through a 3-ml plastic column (Alltech) containing 50 mg of strong anion-exchange resin (AG MP-1; Bio-Rad). The resin is washed with 10 ml of 0.05 *M* ammonium bicarbonate (pH 9.6), containing 0.1 *M* 2-mercaptoethanol. The columns are sealed with plastic caps prior to submerging them in a 37° water bath for 30 min. The resin is incubated at this temperature in a solution containing 100 μl of bovine folate-binding protein (folic acid-binding ability, 12 μg/ml; >99.9% folate free) in 0.1 *M* ammonium bicarbonate buffer (pH 9.6), 2 *M* NaCl, and H_2O to a final volume of 500 μl. Columns are mixed by vortexing each 10 min of incubation. At the end of the incubation, the caps are removed from the end of the column, the columns are drained through a 0.20-μm (pore size) filter, and the resin is washed with 1 ml of buffer containing 0.1 *M* ammonium bicarbonate (pH 9.6) containing 2 *M* NaCl. The combined eluate and wash is passed through a 0.2-μm (pore size) filter and further filtered through a Centricon-30 concentrator which does not allow passage of the folate-binding protein. The Centricon-30 is centrifuged at 5000 *g* for 30 min at 22°. Centricon-30 filters are further washed with 2 ml of 0.05 *M* ammonium bicarbonate (pH 9.6) in 2 *M* NaCl twice and then twice again with 2 ml of water each time. The purified folates devoid of any contaminating free PABA or PABA-Glu_n are eluted from the Centricon-30 with 1 ml of 0.2 *M* trifluoroacetic acid–0.15 *M* HCl. The eluate is dried in 2-ml plastic tubes in a vacuum centrifuge.

Cleavage of Folates to p-Aminobenzoic Acid, Pterin, and Glutamic Acid and Purification of p-Aminobenzoic Acid

HCl (6 *N*, 0.2 ml) is added to the plastic tubes containing the purified folates and the tubes are capped and heated at 120° for 60 min, cooled, and diluted with 1 ml of water. Plastic tubes used in this manner tolerate the heating without difficulty. The mixture is passed through a column containing 100 mg of C_{18} resin and the column is washed with 1 ml of 1 *N* HCl. The effluent and wash are combined and passed through a 3-ml plastic column containing 50 mg of strong cation-exchange resin (see Materials). The resin is washed with 12 ml of 0.01 *M* acetic acid in methanol followed by 12 ml of water and eluted with 1.0 ml of 5 *M* ammonium hydroxide in methanol into autosampler vials, which are then vacuum dried.

Derivatization

The dried samples in autosampler vials are incubated with 50 μl of propionic anhydride for 10 min to form the propionamide of PABA and again taken to dryness in a vacuum centrifuge. Further derivatization prior to GC–MS analysis is achieved by incubation in a mixture of 12 μl of *N*-methyl-*n*-*tert*-butyldimethylsilyltrifluoroacetamide (MTBSTFA; Regis Technologies) and 18 μl of acetonitrile at 90° for 60 min.

Gas Chromatography–Mass Spectrometry Analysis

Instrumentation

Gas chromatography is performed on a Hewlett-Packard (Palo Alto, CA) 5890A gas chromatograph using helium as carrier through a 10 m × 0.25 mm (internal diameter) SPB-1 fused-silica capillary column (Supelco, Bellefont, PA). The column head pressure is maintained at 50 kPa and the source temperature is set at 300°. The injector port temperature is maintained at 250° and the initial column temperature is 80°. A temperature ramp of 30°/min is applied to a final temperature of 300°. Ionization is by electron impact at 70 eV. A dwell time of 10 msec is used. Mass spectrometry is performed using a Hewlett-Packard 5971A mass detector or any similar instrument. The electron multiplier is set at 1500 V for pure standards and at 1600 to 2200 V for biological samples. Spectra of standards are determined in the scan mode and quantitation is carried out by selected ion monitoring (SIM). Two microliters of derivatized sample is injected onto the capillary column using an automatic falling needle injector (model 7673A autosampler).

Gas Chromatography–Mass Spectrometry Characterization of p-Aminobenzoic Acid

Propionated PABA has two derivatization sites with MTBSTFA, which yields an M^+ of 421. With MTBSTFA derivatization, it is common for the major ion to be *m/z* $[M - 57]^+$. The ratio of $[M - 57]^+$ ions of derivatized PABA (*m/z* 364) to that of derivatized $[^{13}C_6]$PABA (*m/z* 370) is used for quantitation. An example of the structure of the $[M - 57]^+$ ion generated from MTBSTFA-derivatized propionated *p*-aminobenzoic acid is shown in Fig. 2. The total ion chromatograms of PABA, $[^2H_2]$PABA, and $[^{13}C_6]$PABA purified from the internal standard are shown in Fig. 3. The mass spectra of PABA, $[^2H_2]$PABA, and $[^{13}C_6]$PABA are shown in Fig. 4. The differences in mass between PABA, $[^2H_2]$PABA, and $[^{13}C_6]$PABA are shown in each of the ion fragments as differences of 2 and 6 Da, respectively, as expected. The $[^2H_2]$PABA is unnecessary for folate quantitation and is shown only as an example of how PABAs with a variety of stable isotope masses can be measured simultaneously.

Selected Ion Monitoring

Quantitation of PABA is based on the ratio of the $[M - 57]^+$ ion (*m/z* 364 for PABA) to the $[M - 57]^+$ ion *m/z* of known quantities of stable isotope-labeled compounds (*m/z* 366 for $[^2H_2]$PABA, *m/z* 370 for $[^{13}C_6]$PABA) using SIM. An SIM computer program can be written to analyze PABA, $[^2H_2]$PABA, and $[^{13}C_6]$PABA simultaneously following a single injection of 2 μl of derivatized sample.

Calculations

The following formula is used for quantitation of folate from 200 μl of whole blood to which 0.05 nmol of $^{13}C_6$-labeled folate internal standard is added at the outset of sample preparation:

$$\text{Folate (nmol/g Hb)} = 0.05\ (A/B)/C$$

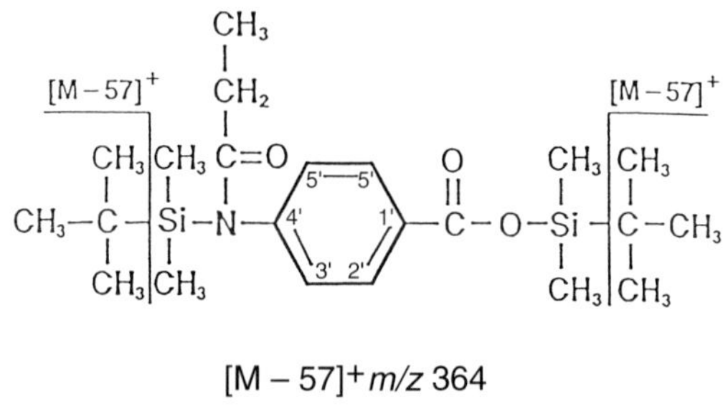

FIG. 2. The chemical structure of *tert*-butyldimethylsilyl derivative of propionated *p*-aminobenzoic acid. Loss of *tert*-butyl group from either amino or carboxyl end results in $[M - 57]^+$ ion (*m/z* 364). The labels indicated as 1′–6′ (inside the benzene ring) are sites of ^{13}C substitution in the $^{13}C_6$-labeled compound.

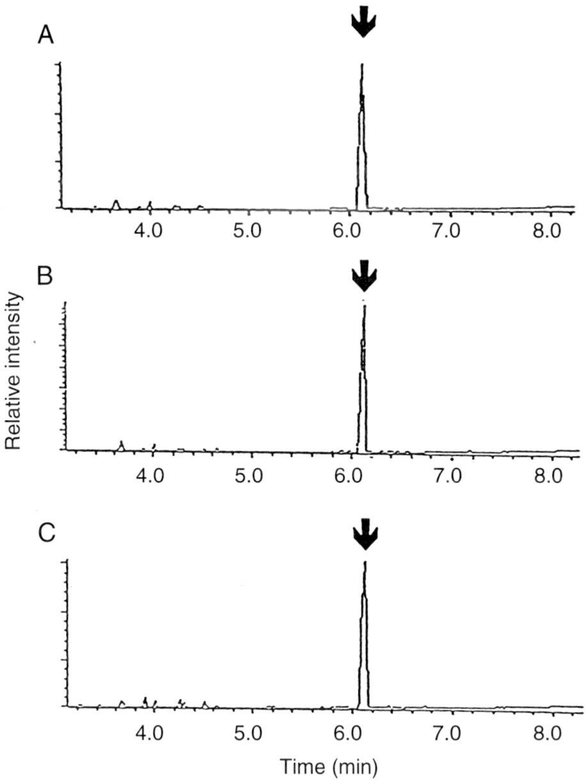

FIG. 3. Total ion chromatograms of MTBSTFA-derivatized compounds generated from (A) PABA, (B) [$^{2}H_2$]PABA, and (C) [$^{13}C_6$]PABA.

where *A* is the isotopic abundance of ion $[M - 57]^+$ m/z 364 (PABA), *B* is the isotopic abundance of ion $[M - 57]^+$ m/z 370 ([$^{13}C_6$]PABA), 0.05 is the nanomole amount of the $^{13}C_6$-labeled folate internal standard added per 200 μl of whole blood, and *C* is the actual amount of hemoglobin (in grams) present in 200 μl of whole blood obtained by venipuncture. Hemoglobin is determined by autoanalyzer.

Calibration curves of increasing amounts of PABA to fixed amounts of [$^{13}C_6$]PABA in aqueous solution generally show linearity with a correlation coefficient (r) of 0.999. Calibration curves are derived by plotting the ratio of the abundances of m/z 364 (PABA) to m/z 370 ([$^{13}C_6$]PABA), representing measured PABA against increasing quantities of added PABA. A similar calibration curve plotted for increasing quantities of whole blood against fixed quantities of $^{13}C_6$-labeled folate internal standard is shown in Fig. 5.

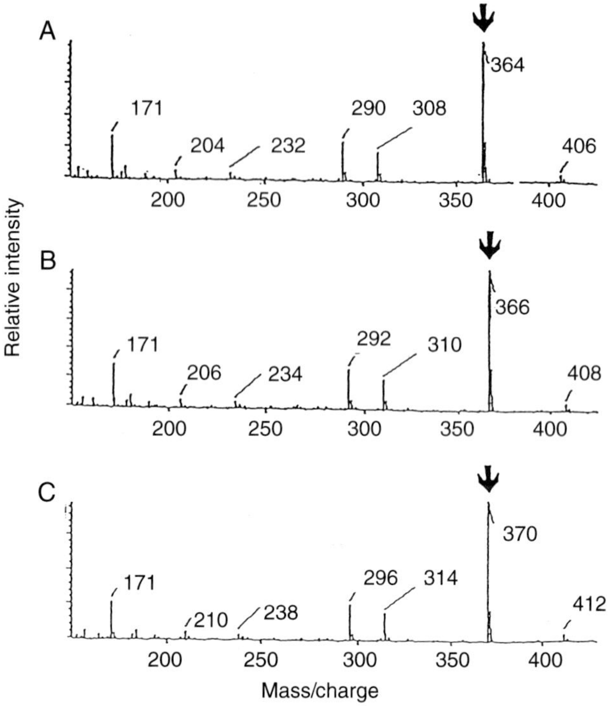

FIG. 4. Mass spectra of MTBSTFA-derivatized compounds generated from (A) PABA, (B) [$^2H_2$]PABA, and (C) [$^{13}C_6$[PABA. Arrows indicate $[M - 57]^+$ ions, abundances of which are used for quantitation.

The ratio of measured folates to increasing amounts of whole blood is linear at the tested range with a correlation coefficient (r) of 0.997. Similar calibration curves plotted for increasing quantities of folic acid or 5-methyltetrahydrofolate, added to fixed amounts of a blood sample, against fixed quantities of $^{13}C_6$-labeled folate internal standard are shown in Fig. 6. The ratio of measured folates to increasing amounts of whole blood is linear at the tested range (50–400 μl of whole blood) with a correlation coefficient (r) of 0.998.

Reference Range

A reference range of whole-blood folates in 121 normal individuals between the ages of 18 and 65 years is shown in Fig. 7 along with folate values in five individuals with clinically confirmed folate deficiency. The mean folate level in the normal individuals is 3.3 nmol/g of Hb with a 0.95

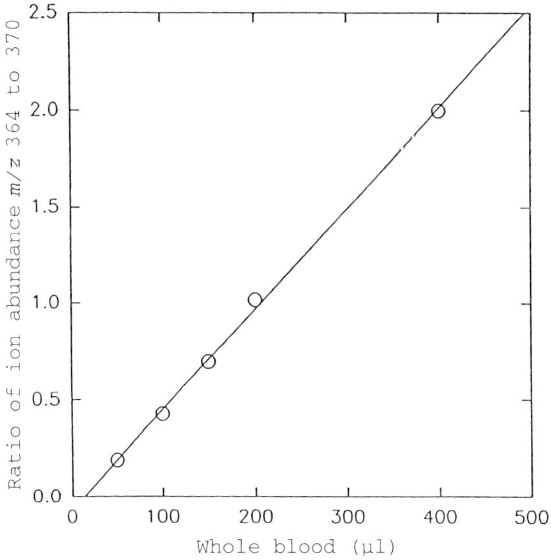

FIG. 5. Calibration curve for increasing amounts of whole blood and fixed amounts of $^{13}C_6$-labeled folate internal standard. Ratios of the ion abundance *m/z* 364 (generated from RBC folates) to *m/z* 370 (generated from $^{13}C_6$-labeled folate internal standard) are plotted against quantities of whole blood. The hematocrit (i.e., packed erythrocyte volume) was 45% and the hemoglobin was 15.1 g/dl; 0.05 nmol of internal standard folate was added.

reference interval ranging from 1.69 to 6.69. The folate levels are less than 0.7 nmol/g Hb in all of the subjects with clinical folate deficiency.

Precision

The coefficient of variation for intrassay precision is 4.7%.[5] The coefficient of variation for interassay precision is 4.5% when folates from a single sample stored at −20° are measured 10 times over a 6-month period.[5]

Comparison with Commercial Radioisotope Dilution Assay

Commercial radioisotope dilution assays (RIDAs) have been performed on all of the 121 normals discussed in the previous section. The correlation coefficient (0.71) is poor. In unpublished studies, we have detected three "normal individuals" with low GC–MS folate assays but normal or near-normal homocysteine and normal RIDA serum and erythrocyte (RBC) folate values. The RBC folates increase in these individuals on supplemental folate (1 mg/day) and homocysteine levels decrease by 40–45%, even when they are within the normal range, within 3 days. Interestingly, two of the

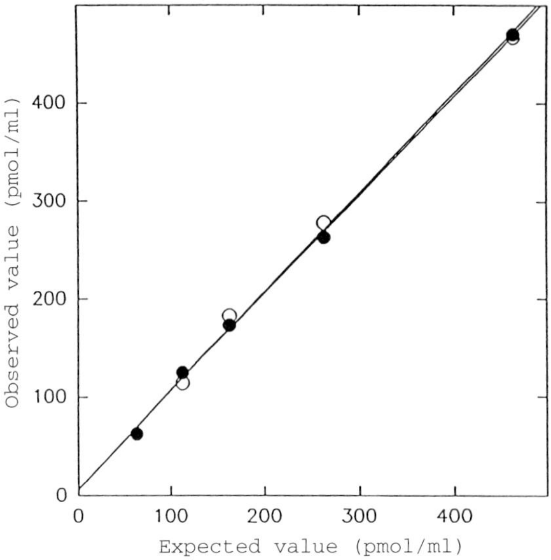

FIG. 6. Calibration curves for increasing amounts of folic acid (●) or 5-methyltetrahydrofolate (○) added to fixed amounts of whole blood and $^{13}C_6$-labeled folate internal standard.

three individuals are young women who are infertile but who have subsequently become pregnant after therapy of their subclinical folate deficiency and have delivered normal babies. Further studies are in progress to study this observation.

Comments

The specificity of the GC–MS analysis for folates is due to the folate-binding protein that is used for purification. Furthermore, the blood folates are copurified with a folate internal standard. Possible contaminants such as PABA, PABA-Glu, or PABA-Glu$_n$ do not copurify with folates in this system.[5] The final analysis is by GC–MS, which adds specificity by combining a separation method with an analytical method that detects specific ion fragments of native and stable isotope-labeled compounds. The method is sensitive to levels 1000-fold less than physiological levels and allows for measurement of even minute amounts of a single coenzyme form of folate. Another advantage of this method is that the use of excess amounts of essentially folate-free folate-binding protein allows quantitative recovery of folates from a sample and avoids competition for binding between folates of different affinities to folate-binding protein. The use of a folate internal

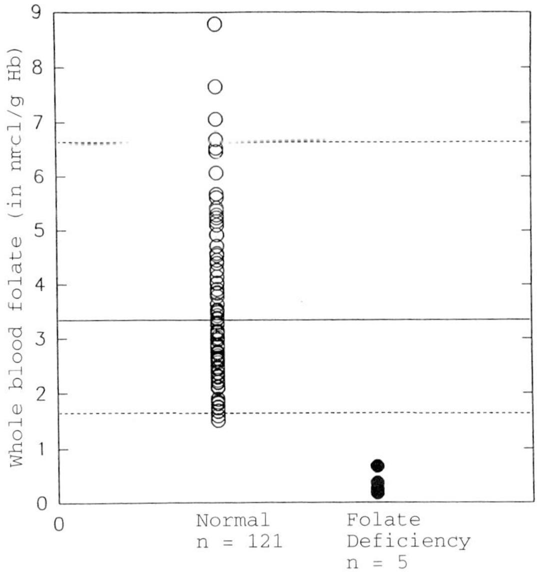

FIG. 7. Molar whole-blood folate levels in 121 normal individuals and 5 subjects with clinically confirmed folate deficiency. Straight line indicates mean and dotted lines indicate limits of 0.95 reference interval.

standard allows for correction for sample losses during processing. The inter- and intraassay coefficient of variation for the assay in our hands is less than 5%.

Even though the methodology appears complex, one can assay 100 samples per day per GC–MS instrument because most of the labor involved is in purification of samples to the point of GC–MS analysis. The GC–MS analysis itself is fully automated and several steps in sample purification can also be automated. The method can be easily adapted for use in other biological fluids or tissues or foodstuffs.

The direct measurement in nanomoles of total folates *in vitro* in tissue samples using an internal standard of physiologically active coenzymes of varying polyglutamate chain length has not been previously possible. Measurement in nanomoles implies the actual number of folate molecules per unit of tissue. Others[7] have reported nanomoles of folates per liter of RBCs after performing a standard competitive binding assay with folic acid monoglutamate as the external standard in the standard curve. Using a molecular weight of 441 for folic acid monoglutamate, a value for nanomoles

[7] L. E. Daly, P. N. Kirke, A. Molloy, D. G. Weir, and J. M. Scott, *JAMA* **274,** 1698 (1995).

of folate per liter of RBCs was obtained even though RBC folates are composed mostly of 5-methyltetrahydrofolate monoglutamate, and of tetra-, penta-, and hexaglutamates (molecular weight of 849, 979, or 1109, respectively).[8,9] Although such estimations of folate content are likely to be reasonable approximations when comparing samples, it is obvious how inaccuracies can occur because of the vast differences in affinity between different folate compounds for folate-binding protein,[4,10] which is used in limiting amounts in competitive binding assays; and because of differences in the molecular weight of measured folate coenzymes compared to the folate used in the standard curve.

The RBC or whole-blood folate assays are conceptually excellent reflections of tissue folate stores because of the relatively long half-life of RBCs. However, these assays have been criticized because of inaccuracies of competitive binding assays. It is hoped that the RBC/tissue folate assay described here provides a degree of accuracy such that clinicians and investigators will once again rely on RBC or whole-blood folates to assess tissue folate sufficiency.

Acknowledgment

The authors thank Academic Press for providing permission to adapt this chapter from Santhosh-Kumar *et al. Anal. Biochem.* **225,** 1–9 (1995).

[8] T. Tamura, Y. S. Shin, M. A. Williams, and E. L. R. Stokstad, *Anal. Biochem.* **49,** 517 (1972).

[9] J. Perry, M. Lumb, M. Laundy, E. H. Reynolds, and I. Chanarin, *Br. J. Hematol.* **32,** 243 (1976).

[10] P. C. Elwood, M. A. Kane, R. M. Portillo, and J. F. Kolhouse, *J. Biol. Chem.* **261,** 15416 (1986).

[4] Microbiological Assay of Folates in 96-Well Microtiter Plates

By DONALD W. HORNE

Our understanding of the dietary requirement for, and the metabolism of, the vitamin folic acid rests, mainly, on our ability to determine the amount present in serum, erythrocytes, and other tissues. Several procedures are currently in use for estimating folate concentrations; these include microbiological, radiochemical, and electrochemical assays. These proce-

0076-6879/97 $25

dures have been reviewed by Tamura,[1] who concluded that the microbiological assay is still the method of choice for the assay of the complex mixture of folates present in biological samples.

Two developments, which we have refined, make the microbiological assay of folates more reproducible, much easier, and much less time consuming than in the past. These developments are the use of glycerol-cryoprotected *Lactobacillus casei* as the inoculum for the assay[2–4] and the use of 96-well microtiter plates and microcomputer analysis of the data.[5,6] The 96-well microtiter plate assay procedure is described in this chapter.

We also point out a problem, apparently inherent to clear-wall plates and turbidimetric assays, which results in overestimation of folate concentration of samples in the two perimeter rows of 96-well plates (rows A and H according to standard designation). We present a solution, which is the use of microtiter plates with opaque, black walls.

Bacteria Cultures and Reagents

- Lyophilized cultures of *L. casei* subspecies *rhamnosis* (ATCC 7469): American Type Culture Collection (Rockville, MD)
- Folinic acid calcium salt pentahydrate [(6*RS*)-5-formyltetrahydrofolate]: Fluka BioChemika (Ronkonkoma, NY)
- L-Ascorbic acid, sodium salt
- 2-Mercaptoethanol
- Glycerol
- Microtiter plates: Microtest III, Falcon No. 3072 (Fisher Scientific, Norcross, GA) or UV96PSB black-wall 96-well plates and US10 polystyrene lids (Polyfiltronics, Rockland, MA)
- Folic acid (*L. casei*) medium: Difco (Detroit, MI)

Instruments

- Microplate absorbance reader: We use a Bio-Rad (Hercules, CA) model 2550 EIA reader
- Microcomputer interfaced with reader: We use Macintosh SE and Bio-Rad MacReader software; not essential for the assay

[1] T. Tamura, *in* "Folic Acid Metabolism in Health and Disease" (M. F. Piccaino, E. L. R. Stokstad, and J. F. Gregory, III, eds.), pp. 121–137. Wiley-Liss, New York, 1990.

[2] N. Grossowicz, S. Waxman, and C. Schreiber, *Clin. Chem.* **27,** 745 (1981).

[3] S. D. Wilson and D. W. Horne, *Clin. Chem.* **28,** 1198 (1982).

[4] S. D. Wilson and D. W. Horne, *Methods Enzymol.* **122,** 269 (1986).

[5] E. M. Newman and J. F. Tsai, *Anal. Biochem.* **154,** 509 (1986).

[6] D. W. Horne and D. Patterson, *Clin. Chem.* **34,** 2357 (1988).

Memowell microplate memory device [Matrix Technologies (Lowell, MA); a light-box device for keeping track of pipetting position on the plates] or similar device; not essential for the assay, but recommended.

Glycerol-Cryoprotected *Lactobacillus casei*

Glycerol-cryoprotected *L. casei* for the inoculum is prepared as described by Wilson and Horne[3] as modified by Horne and Patterson.[6] Dissolve 9.4 g of folic acid *L. casei* medium (Difco) and 50 mg of sodium ascorbate in 200 ml of distilled water. Add 120 ng of (6*RS*)-5-formyltetrahydrofolate and sterilize by filtration (0.22-μm pore size). Suspend the lyophilized *L. casei*, in its shipping vial, in 1 ml of the medium. Transfer 0.25 ml of the suspension to the remaining medium and incubate at 35–37° for about 18 hr. Cool in an ice bath and add an equal volume of cold, 80% (v/v) glycerol (sterilized by autoclaving at 121° for 15 min). Store 4-ml aliquots in sterile tubes at about −70°.

Preparation of Tissue Extracts

Rat pancreas cytosol is used here as an example. Male Sprague-Dawley rats are anesthetized with ketamine and xylazine [80 and 11 mg/kg, respectively, intraperitoneally (i.p.)], and the pancreas of each is removed, trimmed of fat, weighed, and homogenized in 3 vol of ice-cold 0.25 *M* sucrose, 10 m*M* HEPES (pH 7.4), 10 m*M* 2-mercaptoethanol, and 10 m*M* sodium ascorbate. All further steps are carried out at 4°. The homogenate is centrifuged at 18,000 rpm in a Sorvall (Norwalk, CT) RC5B centrifuge for 30 min. The supernatant is removed and 0.2 vol of 5× extract buffer {10% (w/v) sodium ascorbate, 1 *M* 2-mercaptoethanol, 0.25 *M* HEPES, 0.25 *M* CHES [2-(*N*-cyclohexylamino)ethanesulfonic acid], pH 7.85} is added. [We use 0.2 vol of 5× extract buffer so that the final concentration of extract buffer ingredients will be ~1× (see below) to protect labile folates during heating and subsequent conjugase treatment.] The solution is heated in a boiling water bath for 5 min and centrifuged in an Eppendorf microcentrifuge for 5 min to remove precipitated protein. The supernatant is treated with rat plasma conjugase (γ-glutamyl hydrolase) to hydrolyze folylpolyglutamates to the monoglutamates.[4,7]

For routine assay of total folates by the microbiological assay, we extract tissue by mincing in 1× extract buffer [2% (w/v) sodium ascorbate, 0.2 *M* 2-mercaptoethanol, 0.05 *M* HEPES, 0.05 *M* CHES (pH 7.85)], heating in

[7] D. W. Horne, D. Patterson, and R. J. Cook, *Arch. Biochem. Biophys.* **270,** 729 (1989).

a boiling water bath for 10 min, homogenizing using a Polytron (Brinkmann, Westbury, NY), microcentrifuging, and treating with conjugase (as described above). For liquid samples (e.g., serum or plasma) we add 0.2 vol of 5× extract buffer (see above), heat in boiling water bath for 5 min, and microcentrifuge to remove precipitated proteins. Serum and plasma contain only the monoglutamate derivatives; therefore, conjugase treatment is not necessary. These extraction procedures have been shown to preserve the reduced folates present in tissues and to lead to minimal interconversion of individual derivatives.[8,9]

Assay Procedure

Sufficient folic acid *L. casei* medium (Difco) for the anticipated number of assays is prepared according to the package insert. The medium is heated until just boiling, cooled, and filtered through a 0.22-μm (pore size) filter.

The working buffer is composed of 1.6 g of sodium ascorbate, 0.5 ml of 1 *M* potassium phosphate buffer (pH 6.1), and 9.5 ml of distilled water. The solution is filtered through a 0.22-μm (pore size) filter.

To prepare a stock solution of folate standard, (6*RS*)-5-formyltetrahydrofolate, calcium salt is dissolved in distilled water to give a concentration of about 5 m*M*. The actual concentration is determined by measuring the absorbance at 285 nm (of appropriate dilutions in 0.1 *M* potassium phosphate, pH 7), using the molar extinction coefficient of 37.2×10^3 cm liter mol^{-1}.[10] The solution is then diluted with distilled water to 2 m*M*; this is the stock solution of the standard. Because *L. casei* grows only on the (6*S*)-isomer, the concentration of the active isomer is, therefore, 1 m*M*. The stock solution is aliquoted into microcentrifuge tubes and stored at −70° until needed. The working standard is prepared by diluting 15 μl of the stock solution to a final volume of 5 ml with distilled water.

The *L. casei* inoculum is prepared by diluting the thawed, cryoprotected bacterial suspension 1:2.5 (v/v) with sterile sodium chloride solution (9 g/liter). This is the dilution of the bacteria we presently use. This dilution is determined by preparing a series of standard curves using different dilutions of the cryoprotected *L. casei*. We choose the highest dilution of the bacteria that gives the highest absorbance for all points on the standard curve. This dilution is used for routine assays.

[8] S. D. Wilson and D. W. Horne, *Proc. Natl. Acad. Sci. U.S.A.* **80,** 6500 (1983).

[9] S. D. Wilson and D. W. Horne, *Anal. Biochem.* **142,** 529 (1984).

[10] R. L. Blakley, "The Biochemistry of Folic Acid and Related Pteridines." North-Holland, Amsterdam, 1969.

Eight microliters of the working buffer is pipetted into each well of 96-well plates. The standard curve is constructed, in duplicate, from 0 to 180 fmol of 5-formyltetrahydrofolate, with the maximum volume being 60 μl. The unknown sample volume may be up to 122 μl. The volume of standards and samples is adjusted to 122 μl with sterile water, 20 μl of *L. casei* inoculum is added, and 150 μl of folic acid *L. casei* medium is added. The plates are covered with the polystyrene cover and incubated for approximately 18 hr at 37° in a humidified oven.

The absorbance at 600 nm is read in a microtiter plate reader. (It is not necessary to resuspend the bacteria prior to reading; see Comments.) In our case, the reader is interfaced with a computer that reduces the data.

The growth of *L. casei* on standard 5-formyltetrahydrofolate in both clear-wall (Falcon Microtest III) plates and black-wall (Polyfiltronics UV96PSB) plates was determined. While both showed good reproducibility (coefficient of variability was 6.2% or less), black-wall plates gave consistently higher absorbance readings (107% at 15 fmol to 140% at 150 fmol) than clear-wall plates. The reason for this is not known. However, the clear plastic wall may act as a light pipe (similar to transmission in fiber optic cable), thereby transmitting light through the wall in addition to that not scattered by the bacteria. This would lead to lower apparent absorbance. The walls of black-wall plates would not permit this light piping because they are opaque, thereby resulting in greater apparent absorbance than is provided by clear-wall plates.

We had noticed that values of folate obtained from bacteria grown in the wells located on the perimeter of clear-wall microtiter plates were about 20% higher than values for the same samples grown in interior wells. This was confirmed by measuring folate in rat pancreas cytosol extract. Each microtiter plate had a standard curve (columns 1 and 2) and in all other wells a 20-μl aliquot of a 1:200 (v/v) dilution of rat pancreas cytosol extract was assayed. By inspection of the results, it appeared that the values determined in perimeter rows A and H were higher than those of internal rows C–G in clear-wall plates. Therefore, we compared the average folate concentration calculated for each row [rows A–H (omitting values from column 12, since it is on the outside and absorbance values in this column are higher than in columns 2–11)], using one-way analysis of variance (ANOVA). We found that the folate concentrations calculated from rows A and H using clear-wall plates were indeed higher than the values for rows B, C, D, E, F, and G. Using black-wall plates, the results showed no such systematic difference in values from any row. The reason for this is not known for certain. However, the construction of the plate gives some clue. The bottom of the plate over the area of the wells is solid. Looking down on the top of the plate, all of the inside wells have a similar air-to-plastic surface area around the wells. However, the outside wells (those of rows A and H, and columns 1 and 12) have a larger air-to-plastic surface area. One may speculate that more of the light scattered toward the walls would be lost from the outer wells because of the greater air-to-plastic surface. This cannot happen with black-wall plates.

Comments

Previously, we recommended suspending the bacteria by repeatedly aspirating ~150 μl using a pipette. We have found that this step is not necessary, presumably because the light path is up through the well and, because the bacteria have settled to the bottom, essentially equal light scattering occurs whether or not the bacteria have been resuspended.

We recommend that black-wall plates be used for the microbiological assay of folates using *L. casei* and this would, presumably, also apply to assays using *Streptococcus faecium* and *Pediococcus cerevisiae*, and to all turbidimetric assays.

Acknowledgments

The technical assistance of Rosalind S. Holloway is gratefully acknowledged. This work was supported by the Department of Veterans Affairs and by NIH Grant DK32189.

[5] Microbiological Assay for Serum, Plasma, and Red Cell Folate Using Cryopreserved, Microtiter Plate Method

By Anne M. Molloy and John M. Scott

The routine analysis of serum and erythrocyte folate has been the subject of numerous reviews, comments, and controversies. The microbiological methods used originally were difficult to set up, difficult to maintain, and slow in obtaining results. Thus many laboratories changed over to newly developed radioassays when they became available. However, the microbiological method remains for most investigators the "gold standard" and the method of choice. What many investigators do not realize is that it has now been refined and automated to such an extent that it is easy to perform, reliable to maintain, and considerably less costly than the radioassay method, particularly where large numbers of samples are involved. Three important changes in the assay technology have contributed to this. First,

0076-6879/97 $25

the development of antibiotic-resistant strains of organisms allows the use of disposable laboratory ware and removes the need for autoclaving. Second, the ability to cryopreserve the inoculum in multiple individual vials results in standardized growth curves that are reproducible for hundreds of assays. Last, automated microtiter plate technology and its associated computerized analysis packages, which were developed for enzyme-linked immunosorbent assays (ELISAs), are ideally suited to measuring the turbidity of microbiological growth.

The microbiological assay performed in our laboratory is adapted from the method published by O'Broin and Kelleher.[1] We use microtiter plates and a chloramphenicol-resistant strain of *Lactobacillus casei* maintained by cryopreservation. This chapter describes the method in sufficient detail that it can be set up *ab initio* by virtually any laboratory. It thus emulates a similar paper by Scott *et al.*,[2] which described in detail the traditional microbiological assay.

Materials

Laboratory Equipment

The assay requires any cabinet-type 37° incubator and a microtiter plate reader with optional attachment to a PC-controlled data-handling package. We use a Labsystems (Helsinki, Finland) Multiscan Plus plate reader attached to a 486 computer with ELISA$^+$ software (Meddata, Inc., New York, NY). For storage of the cryopreserved inoculum, access to a −80° refrigerator or a liquid nitrogen storage unit is essential. Other general laboratory equipment includes refrigerator and freezer facilities, drying oven, bench centrifuge, autoclave, Bunsen burner or hot plate with magnetic stirrer, and test tube racks suitable for the disposable tubes described below and for whatever blood collection tubes are in use. The following pipettes should be reserved exclusively for the assay; one eight-arm multichannel pipette with 50- to 200-μl dispensing range (such as the Finnpipette digital MCP; Labsystems), one adjustable repetitive sampling pipette with attachments to give an overall dispensing range from 10 to 5000 μl (such as the Socorex stepper; Socorex, Lausanne, Switzerland), a set of four adjustable pipettes to give dispensing ranges of 5–50 μl, 50–200 μl, 200–1000 μl, and 1–5 ml (e.g., Socorex). The manufacturers supply disposable pipette tips that are suitable for the pipettes described. A plate-sealing roller (Flow Laboratories, Bioggio, Switzerland) and two metal spatulas are also re-

[1] S. O'Broin and B. Kelleher, *J. Clin. Pathol.* **45,** 344 (1992).

[2] J. M. Scott, V. Ghanta, and V. Herbert, *Am. J. Med. Technol.* **40,** 125 (1974).

tained for the assay. The spatulas are soaked overnight in general laboratory detergent and rinsed thoroughly before each assay.

Glass and Disposable Ware

Two 500-ml conical flasks, one 500-ml beaker (of wide enough diameter to insert the stepper pipette), one 500-ml measuring cylinder, two 100-ml volumetric flasks, and one 1-liter volumetric flask should be reserved for the assay and routinely soaked overnight in a bath of cleaning mixture to remove any residual folate that may adsorb to glassware and cause contamination in the next assay. The cleaning mixture is prepared by adding 2.5 liters of concentrated sulfuric acid to 2.5 liters of water, then adding approximately 100 g of potassium dichromate. This solution is reusable for up to 6 months and is discarded by diluting into at least 20 vol of water before washing down a foul water drain. The soaked glassware should be rinsed thoroughly and oven dried before use. Some disposable glass 20-ml broth tubes are required for autoclaving glycerol solutions (see Preparation of Cryopreserved Inoculum, below). All other laboratory ware consists of disposable polypropylene or polystyrene and is supplied by Sarstedt, Ltd. (Essen, Germany). This includes 5-ml tubes into which serum or plasma can be dispensed after centrifugation, 4-ml tubes for the initial dilution of assay samples, 1.5-ml minitubes for storing whole-blood lysates and quality control aliquots, 1.5-ml cryotubes for *L. casei* aliquots, polystyrene storage boxes, petri dishes, 30-ml sterile universal tubes (107 × 25 mm) and 96-well (flat) microtiter plates. Adhesive microtiter plate sealers are obtained from Titertek (ICN Biomedicals, Costa Mesa, CA).

Chemicals

Ascorbic acid, sodium ascorbate, potassium dichromate, manganese sulfate, sodium hydroxide, glycerol, and Tween 80 are purchased from Sigma Chemical Co. (Poole, UK). Chloramphenicol is obtained from Parke Davis & Co. (Pontypool, Wales, UK). Concentrated sulfuric acid, folic acid standard, and vitamin folic acid test broth are supplied by Merck (Darmstadt, Germany). We monitor our stocks of broth and verify that new batches give backgrounds and growth curves similar to those of our current medium. We then buy in bulk sufficient broth to last for up to 2 years. All routine chemicals are purchased or subaliquoted and stored in multiple small amounts (25 to 100 g) rather than in large consignments such as 500 g or 1 kg. Thus, if contamination of chemicals, media, etc., with folic acid arises, all in-use chemicals are discarded and new bottles are opened. (We never try to eliminate the source of contamination, we simply start again with unopened reagents.)

Water. The quality of water used in the assay is important because it is a major source of high blanks and poor growth curves. All water must therefore be ultrapure [e.g., Millipore (Bedford, MA) Milli-RO Plus reverse osmosis] or double distilled. If the purification standard is not of this quality, purchase purified water from a reputable supplier.

Microorganism

The chloramphenicol-resistant strain of *L. casei* (NCIB 10463) can be obtained initially as a freeze-dried preparation from the National Collections of Industrial and Marine Bacteria, Ltd. (NCIB, Torry Research Station, Aberdeen, Scotland, UK). After purchases of a stock culture and cryopreserved subcultures have been made, these may be used indefinitely in the assay. (We will supply, subject only to the cost of preparation and shipping, samples of the cryopreserved organism for researchers who have difficulty in obtaining or setting up the method from the freeze-dried organism.)

Methods

Separation of Blood

The use of serum is usually recommended for folate analysis; however, in research studies where one is limited to a single blood sample, the blood may be collected in K_3EDTA Vacutainers (Becton Dickinson, Mountain View, CA), mixed carefully, and sampled for erythrocyte folate analysis as described below. The sample is then centrifuged (at 1500 *g* for 15 min, room temperature) to separate the plasma. Plasma and serum samples may be stored at $-20°$ until assayed. For long-term storage 5 mg of solid ascorbic acid per ml of sample can be added to preserve folates, although this is probably not necessary if samples are not repeatedly thawed and refrozen. We analyze K_3EDTA plasma folate as for serum folate, except for the addition of manganese sulfate to the culture medium as described in detail under Assay Procedure.

Preparation of Whole-Blood Sample for Erythrocyte Folate Analysis

Whole blood is mixed carefully, preferably with a blood cell suspension mixer, particularly if tubes are full. Mixing can be by hand but must be thorough. In a 1.5-ml minitube, 100 μl of whole blood is added to 900 μl of 1% ascorbic acid solution, freshly prepared by adding 1 g of ascorbic acid (not sodium ascorbate) to 100 ml of distilled water. No pH adjustment is made. It is essential that prior to assay the mixture be left at room temperature for 30 min to allow serum conjugase to convert folate polyglutamates released from the lysed erythrocytes to the assayable monoglutamate forms. These ascorbic acid lysates may then be stored at −20° until assayed.

The packed cell volume (PCV) is determined from the same whole-blood sample either on a Coulter (Hialeah, FL) counter or by taking a hematocrit, using a capillary tube and hematocrit reader. If it is not possible to determine the PCV in this way, it can be estimated by assaying the hemoglobin concentration in the ascorbic acid lysates and converting the values to PCV values.[3]

Preparation of Folic Acid Standard

All operations using folic acid should be carried out well away from the microbiological workstation. Folic acid (20 mg) is weighed on a balance accurate to at least four decimal places and dissolved in approximately 4 ml of 0.1 *M* NaOH in a sterile 30-ml disposable universal tube. The solution is transferred and the liquid residue washed into a 1-liter volumetric flask and made up to the mark with water. This stock solution is divided into 20-ml aliquots in 30-ml disposable universal tubes and stored at 4°. New stock solutions are prepared at regular intervals (every 8–12 weeks or if the standard curve shows a trend toward flattening at higher concentrations). A standard curve is always run on new standards before the old standard is discarded. The working standard is prepared fresh for each assay from this stock standard (see under Assay Procedure).

Preparation of Medium

It is important that weigh boats or weighing tissues, spatulas, and balance area are free from any possibility of folate contamination. To make 100 ml of medium, which is enough for four plates, we use 5.7 g of folic acid broth (Merck). Although this is less than the quantity recommended by the manufacturers, O'Broin and Kelleher[1] have found this amount to be optimal for the assay medium using the cryopreserved chloramphenicol-resistant *L. casei.* The 5.7 g of broth powder is weighed and transferred to a 500-ml conical flask. Using a measuring cylinder, 100 ml of water is added. Chloramphenicol (3 mg) is then added and the mixture is heated. Before boiling but when the solution is hot, 30 μl of Tween 80 is added. It is then brought to the boil and boiling is continued until all of the broth powder has fully dissolved. The broth is then cooled to room temperature and a petri dish cover is placed over the flask. (The petri dish base is retained for aliquoting medium with the multichannel pipette in the assay.) Just

[3] G. O'Connor, A. M. Molloy, L. Daly, and J. M. Scott, *J. Clin. Pathol.* **47,** 78 (1994).

before use, when the assay is set up, 75 mg of ascorbic acid (not sodium ascorbate) is dissolved in the medium and, finally, 200 μl of thawed *L. casei* cryopreserved suspension is added. The medium is mixed thoroughly. If EDTA plasma is to be assayed instead of serum it is essential also to add a solution of manganese sulfate (15 mg/ml; 89 mM) to the medium (1 ml/100 ml of medium) to chelate excess EDTA, otherwise residual EDTA interacts with divalent cations in the medium and inhibits growth.[4] We have not found with Merck medium such well-defined inhibition as Tamura *et al.*[4] described for Difco (Detroit, MI) medium; however, leaving out the manganese sulfate results in poor replication between 50- and 100-μl aliquots of analyticals.

Preparation of Cryopreserved Inoculum

Cryopreserved *L. casei* is stored at $-80°$ as a glycerol suspension. The preparation of this suspension is carried out essentially as reported by Wilson and Horne[5] but with modifications to optimize the growth response by freezing cells in the log phase. This modification was developed by Kelleher *et al.*[6] for cryopreservation of *Lactobacillus leichmannii* and is described in detail by Kelleher and O'Broin[7] for use in a microtiter plate assay for vitamin B_{12}. In this instance, medium is prepared using 7.6 g of folic acid broth and 40 mg of chloramphenicol in 200 ml of ultrapure or double-distilled water. (Note that 7.6 g dissolved in 200 ml is half-strength medium according to the recommendation of the manufacturer. This follows the published method for cryopreserving *L. casei.*[5]). As before, the mixture is heated, then 80 μl of Tween 80 is added and the mixture is boiled. The solution is cooled, then 50 mg of ascorbic acid is added. Finally, 50 ng (113 pmol) of folic acid standard is added to the medium. [This is done by adding 50 μl of stock 20-mg/liter (0.045 mM) folic acid standard to 450 μl of 0.5% (w/v) sodium ascorbate and aliquoting 25 μl of the diluted standard to the medium.] The medium is mixed well and 20-ml aliquots are dispensed into sterile 30-ml universal tubes and stored at $-20°$ until required. Glycerol (80%, v/v) is prepared by mixing 32 ml of glycerol with 8 ml of water, dividing into two glass 20-ml broth tubes, autoclaving for 10 min at 121°, and cooling.

To start the culture, one of the containers of stored medium is brought to room temperature and 1 ml of medium is added to a lyophilized vial of *L. casei* obtained from the NCIB. The contents are transferred back into

[4] T. Tamura, L. E. Freeberg, and P. E. Cornwell, *Clin. Chem.* **36,** 1993 (1990).
[5] S. D. Wilson and D. W. Horne, *Clin. Chem.* **28,** 1198 (1982).
[6] B. P. Kelleher, J. M. Scott, and S. D. O'Broin, *Med. Lab. Sci.* **47,** 90 (1990).
[7] B. P. Kelleher and S. D. O'Broin, *J. Clin. Pathol.* **44,** 592 (1991).

the tube of medium. Alternatively, if the assay has already been set up and cryopreserved organism is available, 10 μl of cryopreserved inoculum is dispensed into a 20-ml container brought to room temperature. In either event, the culture is incubated at 37° for 42 hr. It is not necessary to shake the culture during incubation. A 10-μl aliquot is then taken into a second tube of medium thawed to room temperature and cultured for 24 hr. This 24-hr procedure is repeated a further two times, at which stage the cells have achieved a rapid growth rate. At this point, 1 ml of culture is transferred to each of a further two thawed 20-ml tubes of medium and incubated at 37°. After 7 hr the contents of the two tubes (i.e., 40 ml) are transferred to a beaker along with 40 ml of the autoclaved glycerol solution. The suspension is mixed thoroughly, aliquoted (500 μl) into 1.5-ml cryotubes, and frozen to −80° in a polystyrene box. The procedure gives enough inoculum for 160 assays with 8–10 plates per assay. This cryopreserving technique ensures that each assay is carried out with inoculum containing the same number of cells, all at the same stage of growth, thus maintaining day-to-day growth curve reproducibility.

Preparation of Quality Control Samples

Blood is collected from healthy volunteers and pooled to give approximately 200 ml of serum (S), EDTA plasma (P), or whole-blood ascorbic acid lysate (R) depending on whichever of these is to be analyzed. [The whole-blood ascorbic acid lysate fraction is obtained by diluting 20 ml of whole blood to 200 ml with freshly prepared 1% (w/v) ascorbic acid as described in the previous section for preparation of erythrocyte folate.] The pooled samples (designated S1, P1, or R1) are then centrifuged, filtered through Whatman (Clifton, NJ) No. 1 paper, and six aliquots of each are analyzed on three separate days to obtain a baseline mean, standard deviation, and coefficient of variation. These samples usually give folate concentrations well into the normal range. Each of these three pools is then divided in three to give two further pools (e.g., the serum sample is divided into the original S1, plus S2, plus S3; similarly for the plasma and whole-blood lysate samples). S2 and S3, P2 and P3, and R2 and R3 are then appropriately diluted with 0.9% (w/v) sodium chloride to give calculated borderline and deficient levels. Six aliquots from each of these new pools are analyzed on three separate days to obtain means, standard deviations, and coefficients of variation. The three pools for each blood fraction (S1, S2, S3; P1, P2, P3; R1, R2, R3) are then dispensed into 1.5-ml minitubes in 150-μl aliquots, which is sufficient for one duplicate assay, and stored in boxes at −20°. For quality control, depending on the blood fraction to be analyzed, an aliquot from each of the three pools of the relevant blood fraction is run in duplicate with every assay.

Assay Procedure

In the 500-ml beaker a solution of 0.5% (w/v) sodium ascorbate is prepared by dissolving 2.5 g of sodium ascorbate (not ascorbic acid) in 500 ml of water. This solution is used for preparing the working standard and for all assay dilutions. The working standard is prepared in the following way: A universal tube containing stock standard solution is brought to room temperature and a 50-μl aliquot is taken and diluted to 100 ml with 0.5% (w/v) sodium ascorbate in a volumetric flask. This solution is mixed thoroughly, then 5 ml is taken and diluted to 100 ml with 0.5% (w/v) sodium ascorbate in a second volumetric flask to give a final working standard concentration of 500 ng/liter (1.13 n*M*). Duplicate 50-μl aliquots of serum or plasma or duplicate 25-μl aliquots of whole-blood lysate are pipetted into labeled 4-ml polypropylene tubes. Using the adjustable repetitive sampling pipette the aliquots are diluted to a total of 1 ml with 0.5% (w/v) sodium ascorbate. (Our experience with studies of normal populations is that this level of dilution generally gives a growth reading toward the higher, flatter part of the curve. For studies on nondeficient populations we routinely dilute our samples by a further factor of two; i.e., 50 μl of serum or 25 μl of whole-blood lysate is diluted to 2 ml, to ensure that the sample gives readings on the linear part of the curve.) In addition to unknowns, duplicate samples of the three appropriate quality control pools (e.g., S1, S2, S3) are included with each assay depending on whether serum, plasma, or erythrocyte folates are being analyzed.

For each diluted sample, duplicate aliquots of 100 and 50 μl are transferred to four separate wells of a 96-well microtiter plate, again using the repetitive sampling pipette. Fifty microliters of 0.5% (w/v) sodium ascorbate is added to any wells containing 50 μl of analytical sample to bring the total volume in all wells to 100 μl. This results in an overall use of eight wells for every serum specimen analyzed (Fig. 1a). For example, 50-μl

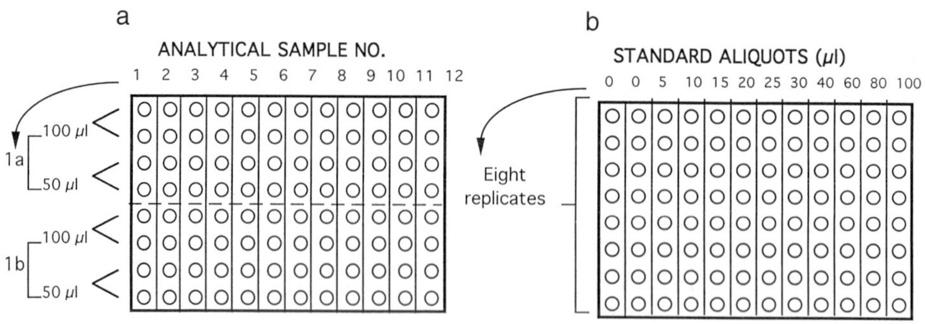

FIG. 1. Diagram of the procedure for aliquoting samples (a) and standards (b) into microtiter plates.

aliquots of serum sample 1 have been pipetted into two polypropylene tubes labeled 1a and 1b and each tube diluted to give a total volume of 1 ml. From each of these tubes, two 100-μl and two 50-μl aliquots are placed in microtiter plate wells and in the latter 50 μl of 0.5% (w/v) sodium ascorbate is also added.

A portion (20 ml) of the working standard is transferred to a 30-ml sterile universal tube and the repetitive sampling pipette is used to aliquot standard amounts into a separate microtiter plate (12 columns with 8 replicates for each standard; see Fig. 1b). Columns 1 and 2 contain 0 μl of standard (i.e., blanks), columns 3 to 12 contain 5 μl (2.5 pg; 5.7 fmol), 10 μl (5.0 pg; 11.3 fmol), 15 μl (7.5 pg; 17.0 fmol), 20 μl (10.0 pg; 22.7 fmol), 25 μl (12.5 pg; 28.3 fmol), 30 μl (15.0 pg; 34.0 fmol), 40 μl (20.0 pg; 45.3 fmol), 60 μl (30.0 pg; 68.0 fmol), 80 μl (40.0 pg; 90.6 fmol), and 100 μl (50.0 pg; 113.3 fmol) of standard, respectively. Appropriate additions of 0.5% (w/v) sodium ascorbate are made to each well, from 100 μl in columns 1 and 2 to 0 μl in column 12, such that the final volume in all wells is 100 μl.

Finally, 200 μl of prepared innoculated folate medium is added to all wells using the eight-arm multichannel pipette. To achieve this it is necessary to pour small amounts of medium into the base of a petri dish and pipette from this. To avoid contamination by folate in the samples and standards the order of addition of medium should be zero standards (columns 1 and 2, standard plate) followed by analyticals (all other plates) followed by remaining standards. This procedure removes the need to change pipette tips for each addition but it is essential that the tips do not touch the liquid in the analytical wells.

Each plate is covered with a plastic plate sealer and a roller is used to ensure that all wells are individually sealed and all air bubbles removed. Plates are incubated at 37° for about 42 hr. After this time each plate is inverted and mixed carefully to produce an even cell suspension and the plate sealer is removed. The plates are read at 590 nm in a microtiter plate reader and a standard curve is drawn from the optical density printout of the plate reader. A typical standard curve is shown in Fig. 2. Alternatively, the plate reader may be linked to a computer using a commercial ELISA program to collect data, draw a standard curve, and calculate the concentration of folate in each well.

Appropriate dilution factors are used to calculate concentrations. Reading the absorbances of analyticals against a growth curve set up as in Fig. 2, the estimated folic acid concentrations per well for the 100- and 50-μl aliquots are multiplied by 200 and 400, respectively, for serum or plasma and by 400 and 800 for whole blood to give results in nanograms per milliliter. (If analyticals have initially been diluted to 2 ml for the assay rather than 1 ml, a further multiplication factor of 2 is applied.) For each

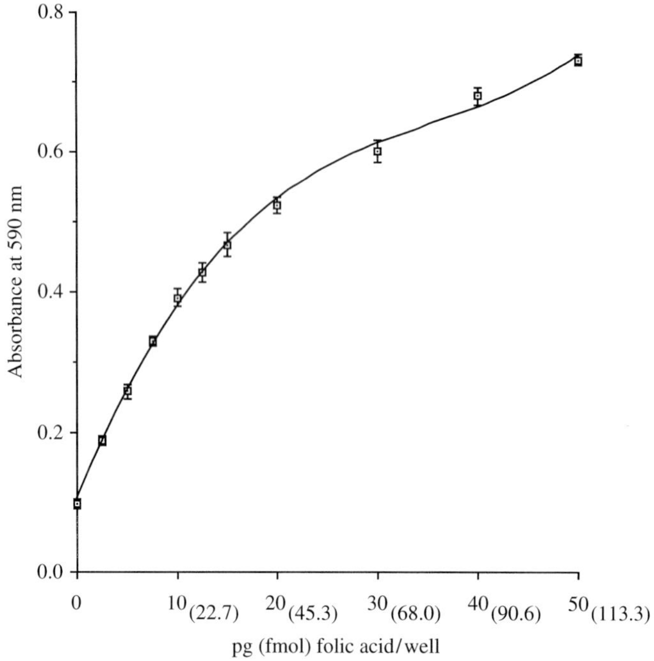

FIG. 2. Standard curve based on the growth response of folic acid to *L. casei.* Values shown are the means ± SD of 8 replicates in the case of folic acid-containing standards and 16 replicates for the zero standard.

sample, the concentration to be reported is calculated by averaging the final results (i.e., ng/ml) from the two 100-μl and the two 50-μl aliquots. The 50-μl average is then checked against the 100-μl average and the result is accepted if there is no more than a 10% difference between the two. The mean of these two averages is then taken as the result for replicate a. The same procedure is carried out on replicate b. Finally, the two results for replicate a and replicate b are checked against each other. If there is agreement within ±5% of the mean the result for that serum or plasma sample is accepted.

The result for each of the three quality control samples is checked against the acceptable limits for each sample on the quality control chart. The limits are ±2 standard deviations from the baseline mean obtained when setting up the quality control pools and the entire assay is repeated if the control samples are outside this range. The quality control charts are monitored for nonrandom trends, such as drifting and a run of results above or below the mean.[8] In such instances, the assay of routine samples is halted until possible causes are checked and corrected. As a typical example of quality controls, our current plasma and whole-blood ascorbic acid lysate controls have the following values (mean ± SD):

P1: 0.95 ± 0.23 ng/ml (2.15 ± 0.52 pmol/ml)
P2: 2.51 ± 0.55 ng/ml (5.69 ± 1.25 pmol/ml)
P3: 6.10 ± 0.78 ng/ml (13.82 ± 1.77 pmol/ml)
R1: 53 ± 4 ng/ml (120 ± 9 pmol/ml)
R2: 89 ± 9 ng/ml (202 ± 20 pmol/ml)
R3: 176 ± 12 ng/ml (399 ± 27 pmol/ml)

To obtain a value for erythrocyte folates (as concentration per milliliter of erythrocytes) the measured whole-blood value is divided by the PCV obtained from an automated (Coulter) counter or hematocrit, or by an estimate of the PCV derived from analysis of hemoglobin in the ascorbic acid lysate samples (O'Connor *et al.*[3]).

[8] E. Mullins, *Analyst* **119,** 369 (1994).

[6] Determination of Tetrahydrobiopterin Biosynthetic Activities by High-Performance Liquid Chromatography with Fluorescence Detection

By ERNST R. WERNER, HELMUT WACHTER, and GABRIELE WERNER-FELMAYER

Introduction

Tetrahydrobiopterin [6*R*-6(L-*erythro*-1,2-dihydroxypropyl)-5,6,7,8-tetrahydropterin] serves as cofactor of phenylalanine 4-monooxygenase, tyrosine 3-monooxygenase, glyceryl-ether monooxygenase, and nitric oxide synthase.[1] Because intracellular tetrahydrobiopterin concentrations affect the amount of metabolites of, e.g., nitric oxide synthase formed by intact cells,[2] regulation of the biosynthesis of tetrahydrobiopterin is of interest. Assays of the three biosynthetic enzymes (Fig. 1) involved in the formation of tetrahydrobiopterin from guanosine 5′-triphospate (GTP) based on high-performance liquid chromatography (HPLC) with fluorescence detection

[1] S. Kaufman, *Annu. Rev. Nutr.* **13,** 261 (1993).
[2] G. Werner-Felmayer, E. R. Werner, D. Fuchs, A. Hausen, G. Reibnegger, and H. Wachter, *J. Exp. Med.* **172,** 1599 (1990).

0076-6879/97 $25

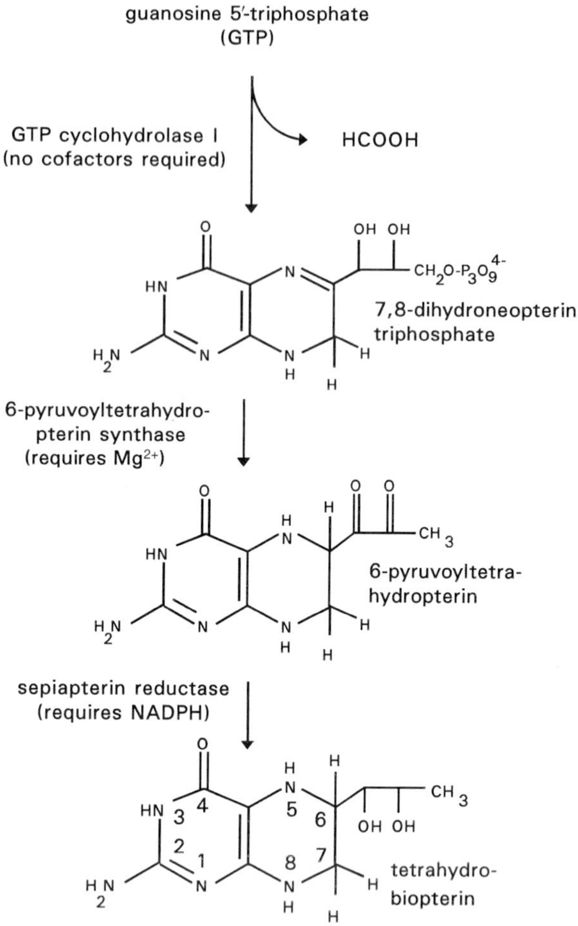

FIG. 1. Biosynthesis of tetrahydrobiopterin from guanosine 5′-triphosphate.

are detailed in this chapter. While the assay of GTP cyclohydrolase I, the first enzyme of the pathway that is regulated by cytokines,[3] as well as the assay for sepiapterin reductase, can be run with materials that are commercially available, 6-pyruvoyltetrahydropterin synthase assays are performed with two purified enzymes, GTP cyclohydrolase I and sepiapterin reductase, to synthesize the labile substrate and to help convert the unstable product to a stable metabolite.

[3] E. R. Werner, G. Werner-Felmayer, D. Fuchs, A. Hausen, G. Reibnegger, G. Wels, J. J. Yim, W. Pfleiderer, and H. Wachter, *J. Biol. Chem.* **265,** 3189 (1990).

Preparation of Extracts from Mammalian Cells and Tissues for Measurement of Cytosolic Activities of Tetrahydrobiopterin Biosynthetic Enzymes

All three assays are performed with extracts from cultured cells prepared in the same manner: 10^7 cells are collected and frozen at $-80°$ in 1 ml of distilled water containing 5 mM dithioerythritol (Serva, Heidelberg, Germany) and phenylmethylsulfonyl fluoride (10 μg/ml; Serva). Cells are broken by thawing (normally one cycle is sufficient) and the amount of broken cells estimated by trypan blue staining. Tissue specimens are best homogenized with a pestle in the same solution; 1 g wet weight per 5 ml of solution is used.

Crude extracts are centrifuged for 15 min at 13,000 *g* and 4°, and the supernatant collected. The protein concentrations of the extracts thus prepared should be in the range of 1 to 5 mg/ml. For further processing, extracts are subjected to gel filtration in the respective assay buffer: 500 μl of cell extract is loaded onto an NAP-5 column (Pharmacia, Uppsala, Sweden) and eluted with 1000 μl of assay buffer. Together with providing the correct buffer for the respective enzyme incubations, this step removes low molecular weight substances, in particular pteridines, which interfere with the assay or may alter the enzyme activities, e.g., via feedback inhibition mechanisms.

Assay for GTP Cyclohydrolase I

Principle

The procedure described here is adapted from work by Viveros *et al.*[4] Extracts containing GTP cyclohydrolase I activity are incubated with GTP (K_m ~50 μM) in the presence of EDTA, the 7,8-dihydroneopterin triphosphate formed is then oxidized to neopterin triphosphate with iodine in acidic solution, the excess iodine is destroyed with ascorbic acid, the mixture is rendered slightly alkaline, and the triphosphate group is cleaved off by alkaline phosphatase. Neopterin is then determined by HPLC with fluorescence detection. The presence of EDTA inhibits further conversion of the product 7,8-dihydroneopterin triphosphate because the enzyme 6-pyruvoyltetrahydropterin synthase requires Mg^{2+} to react.

Enzyme Reaction and Conversion of Product to Neopterin

The assay buffer used is 50 mM Tris-HCl, pH 7.8, containing 300 mM KCl, 2.5 mM EDTA, 10% (v/v) glycerol. NAP-5 eluate (250 μl; see Prepara-

[4] O. H. Viveros, C. L. Lee, M. M. Abou-Donia, J. C. Nixon, and C. A. Nichol, *Science* **213,** 349 (1981).

tion of Extracts) in this buffer is incubated with 2 mM GTP for 30 to 90 min at 37° in the dark in a final volume of 300 μl. The reaction is started by addition of GTP and stopped with 40 μl of a mixture of 20 μl of 1 N HCl and 20 μl of 0.1 M I_2 dissolved in 0.25 mM KI. After a 45- to 60-min incubation in the dark at room temperature, the precipitate is removed by centrifugation (2 min, 13,000 g, 25°); the clear supernatant should have a yellow color due to excess iodine, otherwise the iodine concentration must be increased. On subsequent addition of 20 μl of 0.1 M freshly dissolved ascorbic acid the yellow color should disappear. The pH is then adjusted to pH 7.5–9 by addition of 20 μl of 1 N NaOH and the mixture is incubated with 10 μmol of alkaline phosphatase per minute (ca. 3000 μmol mg^{-1} min^{-1}) at 37° for 60 min in the dark. Incubation mixtures are then stored at −20° until assayed by HPLC for concentration of neopterin.

High-Performance Liquid Chromatography Determination of Neopterin and Biopterin with Fluorescence Detection

Assay mixtures are analyzed by HPLC with fluorescence detection as follows: 10 μl of reaction mixture is injected onto a column (250 mm long, 4-mm i.d.) containing octadecyl-modified silica gel (ODS, 5-μm particle size, e.g., Merck Lichrospher RP-18; Merck, Darmstadt, Germany) protected with a 4-mm guard column filled with the same material. Pteridines are isocratically eluted at a flow rate of 0.8 ml/min with 15 mM potassium phosphate buffer, pH 6.0, and quantified using a fluorescence detector (excitation at 350 nm, emission at 440 nm). Depending on the sensitivity of the fluorescence detector, a detection limit of 10 fmol of neopterin at a signal-to-noise ratio of 5:1 can be reached. Column and injection loop are rinsed daily with 3 ml of distilled water followed by 15 ml of methanol or 2-propanol and 3 ml of water, at a flow rate of 0.3 ml/min. Care should be taken in the preparation of standards of neopterin and biopterin, because the oxidized pterins show only limited solubility in water. We stir 10 mg of neopterin in 10 liters of distilled water neutralized with NaOH containing 0.1 mM dithioerythritol overnight to ensure quantitative solution.

Notes

While the assay is straightforward for use with tissues with high activities such as rat liver (about 1 pmol mg^{-1} min^{-1}), some points should be considered when trying to measure activities in cultured cells, which even when treated with cytokines may be one or two orders of magnitude lower. We observed that the quality of the GTP (some batches contain traces of dihydroneopterin phosphates) and of the alkaline phosphatase (some batches contain traces of GTP cyclohydrolase I activity) is crucial, thus reagent blanks should be run before precious samples are used. Repeated thawing of the incubation mixtures should be avoided, because this leads to loss of analyte at low concentrations (<10 nM neopterin). Ascorbic acid should be prepared freshly and not exposed to light, otherwise fluorescent compounds are formed, which interfere with neopterin determination. As an alternative to phosphatase cleavage, direct determination of neopterin phosphates with ion-pair HPLC has been used by Blau and Niederwieser.[5] This has the advantage of fewer steps required, but poses problems at very low activities owing to the chemical instability of the neopterin phosphates.

Assay for 6-Pyruvoyltetrahydropterin Synthase

Principle

The assay described here is adapted from work by Shintaku *et al.*[6] Extracts containing 6-pyruvoyltetrahydropterin synthase activity are incubated with 7,8-dihydroneopterin triphosphate (K_m 10 μM) in the presence of Mg^{2+}, sepiapterin reductase, and NADPH. 7,8-Dihydroneopterin triphosphate is prepared immediately before the assay, using purified GTP cyclohydrolase I and GTP. Mg^{2+} is required by 6-pyruvoyltetrahydropterin synthase to operate.[7] Sepiapterin reductase reduces the labile 6-pyruvoyltetrahydropterin product with NADPH to 5,6,7,8-tetrahydrobiopterin. This is subsequently oxidized by iodine in acid to the fluorescent biopterin, and the excess iodine is destroyed by addition of ascorbic acid. Biopterin is quantified by HPLC with fluorescence detection. Owing to the presence of a high excess of fluorescent neopterin phosphates originating from the substrate, the HPLC determination requires special methods to separate these from the small amounts of biopterin.

Sources of GTP Cyclohydrolase I and Sepiapterin Reductase

Several protocols for the purification of the two enzymes required, GTP cyclohydrolase I and sepiapterin reductase, have been published, and both have been cloned and expressed. Both enzymes are also active as fusion proteins, thus for their use in these assays the recombinant fusion proteins need not necessarily be cleaved. GTP cyclohydrolase I may be best purified from *Escherichia coli.*[8] Purification is simplified to a one-column procedure

[5] N. Blau and A. Niederwieser, *Anal. Biochem.* **128,** 446 (1983).
[6] H. Shintaku, A. Niederwieser, W. Leimbacher, and H. C. Curtius, *Eur. J. Pediatr.* **147,** 15 (1988).
[7] S. I. Takikawa, H. C. Curtius, U. Redweik, W. Leimbacher, and S. Ghisla, *Eur. J. Biochem.* **161,** 295 (1986).
[8] J. J. Yim and G. M. Brown, *J. Biol. Chem.* **251,** 5087 (1976).

when *E. coli* strains overexpressing GTP cyclohydrolase I are used.[9–11] Sepiapterin reductase is best purified from erythrocytes[12] and is also available in recombinant form.[13,14]

Generation of 7,8-Dihydroneopterin Triphosphate

In a first step, a solution of 7,8-dihydroneopterin triphosphate is prepared from GTP. The actual amount prepared will depend on the number of assays performed subsequently. The following concentrations are used: 0.1 *M* Tris-HCl (pH 8.5), 0.1 *M* KCl, 2.5 m*M* EDTA, 500 μM GTP, and sufficient GTP cyclohydrolase I (>10 nmol min^{-1} ml^{-1}) to metabolize virtually all GTP in 120 min at 37°. If GTP cyclohydrolase I is abundantly available, concentrations might be increased and incubation times reduced. This solution is used directly as substrate, because the presence of GTP cyclohydrolase I does not interfere with the 6-pyruvoyltetrahydropterin synthase reaction.

Enzyme Reaction and Conversion of Product to Biopterin

The assay buffer used for 6-pyruvoyltetrahydropterin synthase is 100 m*M* Tris-HCl (pH 7.4), containing 20 m*M* $MgCl_2$. Seventy microliters of the NAP-5 eluate (see Preparation of Extracts, above) in this assay buffer is incubated with 2 m*M* NADPH, sepiapterin reductase (2 nmol min^{-1}), and 100 μM freshly prepared 7,8-dihydroneopterin triphosphate in a final volume of 100 μl for 30 to 90 min in the dark. The reaction is started by addition of 7,8-dihydroneopterin triphosphate and stopped with 20 μl of a mixture of 10 μl of 1 *N* HCl and 10 μl of 0.1 *M* I_2 dissolved in 0.25 *M* KI. After a 45- to 60-min incubation in the dark at room temperature and subsequent removal of solid by centrifugation (13,000 *g*, 2 min, 25°) 10 μl of freshly prepared ascorbic acid is added and the mixture stored at −20° until analyzed for biopterin content by HPLC.

[9] G. Katzenmeier, C. Schmid, and A. Bacher, *FEMS Microbiol. Lett.* **54,** 231 (1990).

[10] M. Gütlich, E. Jaeger, K. P. Rucknagel, T. Werner, W. Rodl, I. Ziegler, and A. Bacher, *Biochem. J.* **302,** 215 (1994).

[11] H. Ichinose, T. Ohye, E. Takahashi, N. Seki, T. Hori, M. Segawa, Y. Nomura, K. Endo, H. Tanaka, S. Tsuji, K. Fujita, and T. Nagatsu, *Nature Genet.* **8,** 236 (1994).

[12] T. Sueoka and S. Katoh, *Biochim. Biophys. Acta* **717,** 265 (1982).

[13] B. A. Citron, S. Milstien, J. C. Gutierrez, R. A. Levine, B. L. Yanak, and S. Kaufman, *Proc. Natl. Acad. Sci. U.S.A.* **87,** 6436 (1990).

[14] H. Ichinose, S. Katoh, T. Sueoka, K. Titani, K. Fujita, and T. Nagatsu, *Biochem. Biophys. Res. Commun.* **179,** 183 (1991).

High-Performance Liquid Chromatography Analysis of Biopterin in Presence of Abundant Neopterin Phosphates with Fluorescence Detection

The analysis of the small amounts of biopterin formed (nmol/liter) in the presence of 100 μM neopterin phosphates by fluorescence detection requires the use of solid-phase extraction[15,16] or column-switching techniques.[6,17] We use a convenient combination of solid-phase extraction and on-line elution HPLC, which enables enrichment of the analyte and extraction in one step. Rippin[18] has adapted this procedure for conventional solid-phase extraction setups.

For solid-phase extraction with on-line elution HPLC, 100 μl of the acidic reaction mixture is loaded onto a strong cation-exchange solid-phase extraction column (SCX, 2.1 × 17 mm; Varian, Palo Alto, CA) that has been preequilibrated with two 500-μl washes with 0.1 M H_3PO_4. The solid phase does not bind the negatively charged neopterin phosphates, but quantitatively extracts biopterin. This is then eluted by an automated processor using a high-pressure tight chamber (AASP; Varian) directly onto the HPLC column. Elution of biopterin is facilitated by a pulse of 100 μl of 0.6 M potassium phosphate buffer, pH 6.8, which is delivered with the injection valve of the AASP from a 100-μl loop continuously refilled with the buffer using a peristaltic pump.[15] Biopterin is then separated on a reversed-phase system with fluorescence detection, similar to the system described above (see Assay for GTP Cyclohydrolase I, above) for the determination of neopterin; the detection limit for biopterin is 100 fmol.

Notes

Activities in human cells are comparatively low, ranging from ~0.1 to 10 pmol mg^{-1} min^{-1}. For accurate determinations, the concentration of 100 μM 7,8-dihydroneopterin triphosphate, which is 10 times the K_m, should not be underscored even if GTP cyclohydrolase I for the preparation of the substrate is scarce. It is also essential to run reagent controls, because purified sepiapterin reductase may contain residual activities of 6-pyruvoyl-tetrahydropterin synthase. 7,8-Dihydroneopterin triphosphate has also been prepared with recycling of GTP cyclohydrolase I.[19] This compound

[15] E. R. Werner, D. Fuchs, A. Hausen, G. Reibnegger, and H. Wachter, *Clin. Chem.* **33,** 2028 (1987).

[16] E. R. Werner, G. Werner-Felmayer, D. Fuchs, A. Hausen, G. Reibnegger, G. Wels, J. J. Yim, W. Pfleiderer, and H. Wachter, *J. Chromatogr.* **570,** 43 (1991).

[17] A. Niederwieser, W. Staudenmann, and E. Wetzel, *J. Chromatogr.* **290,** 237 (1984).

[18] J. J. Rippin, *Clin. Chem.* **38,** 1722 (1992).

[19] J. Ferre, E. W. Naylor, and K. B. Jacobson, *Anal. Biochem.* **176,** 15 (1989).

is exceptionally labile, so that prolonged storage of these solutions at −80° is feasible only when the 7,8-dihydroneopterin triphosphate content is monitored before each assay. Some protocols use HPLC with electrochemical detection of tetrahydrobiopterin[7]; these show a selectivity for the tetrahydro compound but are less robust than fluorescence detection.

Sepiapterin Reductase Assay

Principle

The assay presented here is adapted from Ferre and Naylor.[20] Extracts containing sepiapterin reductase activities are incubated with the artificial substrate sepiapterin, which is converted by sepiapterin reductase to 7,8-dihydrobiopterin with the aid of NADPH. In the presence of 7,8-dihydrofolate reductase activity in the extract, this is further converted to 5,6,7,8-tetrahydrobiopterin. Both compounds are then oxidized to the fluorescent biopterin by iodine in acid, and excess iodine is reduced with ascorbic acid. Biopterin is determined by HPLC with fluorescence detection.

Enzyme Reaction and Conversion of Product to Biopterin

The assay buffer used is 100 m*M* potassium phosphate, pH 6.4. Seventy microliters of NAP-5 eluate (see Preparation of Extracts, above) prepared in this buffer is incubated with 1 m*M* sepiapterin (K_m ~10 μM) in the presence of 2 m*M* NADPH for 10 min at 37°. The reaction is stopped by adding 20 μl of a mixture of 10 μl of 1 *N* HCl and 10 μl of 0.1 *M* I_2 dissolved in 0.25 *M* KI. After incubation for 45 to 60 min at room temperature in the dark, solid is removed by centrifugation (13,000 *g*, 2 min, 25°), and the mixture is stored at −20° until analyzed for biopterin content by HPLC.

Reversed-Phase High-Performance Liquid Chromatography Analysis of Biopterin with Fluorescence Detection

An HPLC setup similar to the one described above for neopterin is used (see Assay for GTP cyclohydrolase I, above). Depending on the sensitivity of the fluorescence detector used, a detection limit of 20 fmol for biopterin may be reached. Usually sepiapterin incubation mixtures must be diluted 10- to 100-fold to ensure that the signals do not exceed the maximal capacity of the fluorescence detector. Alternatively, the sensitivity range of the fluorescence detector may be changed accordingly.

[20] J. Ferre and E. W. Naylor, *Biochem. Biophys. Res. Commun.* **148,** 1475 (1987).

Notes

Compared to GTP cyclohydrolase I and 6-pyruvoyltetrahydropterin synthase, human cells contain higher sepiapterin reductase activities (~500 pmol mg^{-1} min^{-1}), which can be measured easily without approaching the sensitivity limits of HPLC with fluorescence detection. If material is abundant and activities are high, reduction of sepiapterin may alternatively be monitored by decrease in absorbance at 420 nm,[21] e.g., in a microplate photometer.

Owing to the use of the artificial substrate sepiapterin, the activity may not necessarily be identical to the activity toward the labile natural substrate, 6-pyruvoyltetrahydropterin. For comparative purposes, however, the assay using sepiapterin as substrate is useful.

Acknowledgment

Experimental work described here was supported by the Austrian Research Funds zur Foerderung der Wissenschaftlichen Forschung, Project 11301.

[21] S. Katoh, *Arch. Biochem. Biophys.* **146,** 204 (1971).

[7] Microtiter Plate Assay for Biopterin Using Cryopreserved *Crithidia fasciculata*

By ROBERT J. LEEMING

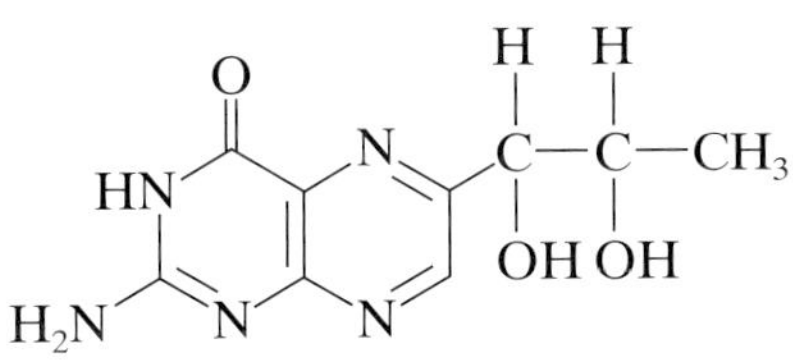

L-*erythro*-Biopterin
(M_r 237)

The protozoan *Crithidia fasciculata* requires biopterin for normal metabolism. Growth in a biopterin-free culture medium is proportional to added biopterin. The defined medium and the culture conditions used in this assay

0076-6879/97 $25

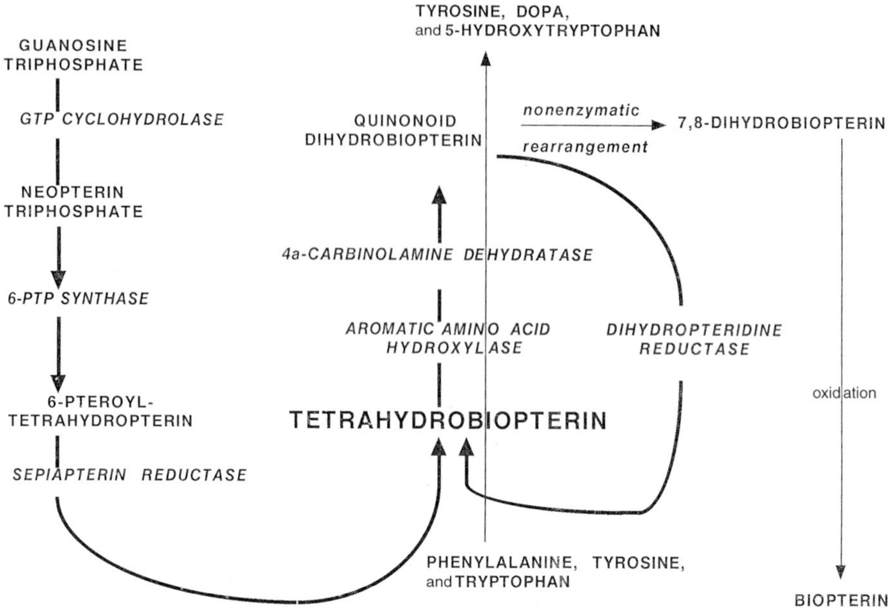

FIG. 1. Synthesis of tetrahydrobiopterin.

have been derived from previously published work[1–3] and from correspondence with S. J. Hutner (Haskins Laboratories at Pace University, New York, NY), who also donated the original culture of *Crithidia* that is still in use. The term *total biopterin* includes biopterin, dihydrobiopterin, and tetrahydrobiopterin.

Tetrahydrobiopterin is synthesized from guanosine triphophosphate and is the essential cofactor in the hydroxylations of phenylalanine, tyrosine, and tryptophan as shown in Fig. 1. It is also required in the synthesis of nitric oxide. Concentrations of tetrahydrobiopterin are maintained by homeostatic mechanisms, including feedback inhibition on guanosine triphosphate cyclohydrolase.

Reagents

Crithidia fasciculata ATCC 12857 (American Type Culture Collection, Rockville, MD)

[1] J. Cowperthwaite, M. Weber, L. Packer, and S. H. Hutner, *Ann. N.Y. Acad. Sci.* **56,** 972 (1953).
[2] R. J. Leeming, J. A. Blair, V. Melikian, and D. J. O'Gorman, *J. Clin. Pathol.* **29,** 444 (1976).
[3] R. J. Leeming, S. K. Hall, H. Friday, P. Hurley, and A. Green, *in* "Advances in Experimental Medicine and Biology" (J. E. Ayling, M. G. Nair, and C. M. Baugh, eds.), Vol. 338, p. 267. Plenum, New York, 1993.

Phosphate buffer (0.2 *M*, pH 5.0)

Potassium dihydrogen orthophosphate (KH_2PO_4)	26.891 g
Disodium hydrogen orthophosphate ($Na_2HPO_4 \cdot 2H_2O$)	0.425 g
Distilled water	to 1000 ml

To prevent contamination phosphate buffer is boiled and then cooled immediately before use

Cycloheximide 5%, w/v): Dissolve 0.25 g of cycloheximide in 2.0 ml of acetone, then add 3.0 ml of distilled water. Store below 6°

Sterile glycerol (80%, v/v)

Glycerol	160 ml
Distilled water	40 ml

Sterilize by autoclaving at 121° for 15 min

Folic acid: Dissolve 40 mg of folic acid in 20% (v/v) ethanol–0.001 *M* NaOH. Store below 6° in the dark. Dilute 1 : 1000 in distilled water immediately before use

Hemin: Dissolve 50 mg of hemin in 10 ml of 50% (v/v) triethanolamine

Vitamin mix

Adenine	1.0 g
Biotin	0.001 g
Calcium pantothenate	0.3 g
Pyridoxamine dihydrochloride	0.1 g
Riboflavin	0.06 g
Thiamin hydrochloride	0.6 g
Nicotinic acid	0.3 g

Grind together in a pestle and mortar. Store dry below 6°

Assay culture medium

L-Arginine hydrochloride	1.0 g
L-Histidine	0.6 g
DL-Leucine	0.2 g
DL-Methionine	0.2 g
DL-Tryptophan	0.16 g
DL-Valine	0.1 g
L-Glutamic acid	2.0 g
DL-Isoleucine	0.2 g
L-Lysine hydrochloride	0.8 g
DL-Phenylalanine	0.12 g
L-Tyrosine	0.12 g

Magnesium sulfate ($MgSO_4 \cdot 7H_2O$)	1.3 g
Manganese sulfate ($MnSO_4 \cdot 4H_2O$)	0.28 g
Zinc sulfate ($ZnSO_4 \cdot 7H_2O$)	0.1 g
Tripotassium orthophosphate ($K_3PO \cdot 4H_2O$)	0.3 g
Ethylenediaminetetraacetic acid	1.2 g
Distilled water	600 ml

Mix on a heated plate until all of the preceding are dissolved. Cool, then add the following on a magnetic stirrer:

Cobaltous sulfate ($CoSO_4 \cdot 7H_2O$)	0.005 g
Ammonium ferrous sulfate [$(NH_4)_2SO_4FeSO_4 \cdot 6H_2O$]	0.002 g
Calcium chloride ($CaCl_2$)	0.001 g
Cupric sulfate ($CuSO_4 \cdot 5H_2O$)	0.005 g
Boric acid (H_3BO_3)	0.001 g
Sucrose	30 g
Triethanolamine (*must* be added before hemin)	5.0 ml
Hemin, 5 mg/ml in 50% (v/v) triethanolamine	10.0 ml
Vitamin mix (see above)	0.1 g
Vitamin-free Casamino Acids (Difco, Detroit, MI)	20.0 g
Folic acid (see above)	5.0 ml

When dissolved make up to 1600 ml with distilled water. Adjust to pH 7.5 with concentrated hydrochloric or sulfuric acid. Then add and dissolve 0.16 g of chloramphenicol. Filter through a presterilized filter (0.3-μm pore size, 50-mm diameter) into a sterile container; complete filter units ready for use are easily available. Store at $-20°$ in 50-ml aliquots in sterile 60-ml containers. *Note:* The addition of Casamino Acids (hydrolyzed casein) appears out of place in an otherwise strictly defined medium but it has been shown by the author and by other workers to enhance growth by up to 70%. Optimum conditions are required in any bioassay when the sample is not a pure solution of analyte and other substances present may generate a nonspecific growth response

Nondefined startup medium

Yeast extract (Oxoid, London, UK)	3 g
Trypticase (BBL, Cockeysville, MD)	3 g
Sucrose	2.5 g
Liver fraction L (Nutritional Chem. Co., Cleveland, OH)	0.1 g
Distilled water	1000 ml

Dissolve and add 5 ml of hemin solution (see above). Adjust to pH 7.2 with concentrated hydrochloric or sulfuric acid and add

0.1 g of chloramphenicol. Autoclave at 121° for 15 min or pass it through a sterilizing filter. Store 20-ml aliquots in 30-ml bottles below 6°. *Note: Crithidia* is not exacting in its requirements and many yeast extracts and tryptic digests (e.g., polypeptone) used in a microbiological laboratory would replace the yeast extract and trypticase specified here. The author has not tried any other liver extract simply because the amount used is so small that the original bottle is still in use

Maintenance medium

Assay culture medium	160 ml (3 of the 50-ml frozen aliquots are used)
Biopterin (200 ng/liter)	1 ml
Phosphate buffer (0.2 *M*)	25 ml
Distilled water	14 ml

Mix and sterilize by autoclaving or filtration as described above

Biopterin standard: Dissolve 2 mg of biopterin in 1000 ml of distilled water containing 5 g of ascorbic acid. If there is any difficulty with solubility a drop of concentrated ammonia should help. Freeze in 1-ml aliquots at −20°. For use make 100 μl to 100 ml in boiled phosphate buffer and further dilute in phosphate buffer to give standards containing 0, 4, 10, 20, 30, 40, 50, 60, 80, 100, 140, and 200 ng/liter.

Preparation of Samples

Most of the diagnostic criteria and normative data available for the *Crithidia* assay has been obtained by automated tube assay without chloramphenicol. This requires heating to remove protein and although heat treatment is not necessary for the microtiter plate assay, the practice has continued for comparative, diagnostic purposes.

Plasma samples are diluted 1:30 (50 μl into 1.45 ml) and whole-blood samples are diluted 1:40 (50 μl into 1.95 ml) in phosphate buffer. Dried blood spots have an 8-mm circle punched out into 1 ml of phosphate buffer (calculated as 1/60 whole blood). Samples are then heated in a steamer for 10 min (any container with means of boiling water and sufficient space above for test tubes would be acceptable; standing the tubes in a boiling water bath would also work), centrifuged, and passed through 0.2-μm (pore size) filters to clarify. When there are reasons to suspect the biopterin level is raised, for example in phenylketonuria, further dilutions should be prepared in phosphate buffer after filtration.

Preparation of Cryopreserved *Crithidia fasciculata*

Inoculate a 20-ml aliquot of startup medium with *Crithidia fasciculata.* Incubate at 29° for 48 hr. Check for contamination microscopically; the characteristic shape, size, and movement of *Crithidia* are distinctive even to an inexperienced microscopist. Then take approximately 0.5 ml of the grown culture and inoculate 200 ml of maintenance medium and incubate for a further 48 hr, shaking occasionally. Centrifuge, decant off most of the supernatant, and resuspend in 80 ml of assay culture medium, then add 20 ml of sterile 80% (v/v) glycerol and 20 ml of boiled (and cooled) phosphate buffer. Mix thoroughly, distribute in 1.0-ml quantities in sealed containers, and freeze in an upright position at −70°.

Assay Protocol

Thaw a vial of cryopreserved *C. fasciculata* and a 50-ml bottle of assay culture medium. Add 200 μl of *Crithidia* (titration may be needed to fine-tune the quantity when a new batch is prepared) to the culture medium and mix well by inverting. Use flat-bottomed microtiter plates with lids. Standard 96-well plates are numbered 1–12 along the long side and A–H down the short side. Pipette 50 μl of standards and samples in quadruplicate into the wells, followed by 200 μl of inoculated assay culture medium, using a multidispensing pipette.

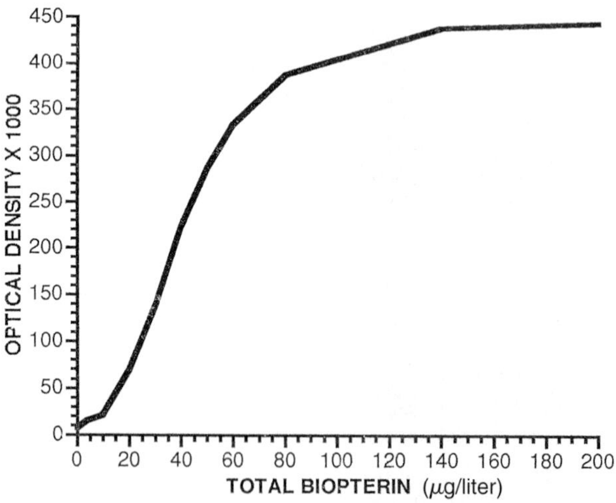

FIG. 2. A typical standard curve of biopterin concentration versus optical density.

Use one well of each quadruplicate as a control by adding 10 μl of 5% (w/v) cycloheximide, which will inhibit growth but not affect optical density. It is best to arrange samples so that the two center horizontal rows D and E are blanks; then, when adding cycloheximide, the covers can be used to shield rows A–C and F–H from aerosol of the cycloheximide, which can affect the growth and alter the results. If added in this way there is no need to mix further. Secure the lids in place with elastic bands at either end, place in a humidity chamber (at a minimum this can be a plastic container lined freshly on each occasion with damp filter paper). Incubate at 29° for 72 hr.

Place the microtiter plates on a shaker for 5–10 min, avoiding spilling. Measure growth by absorbance at 595 nm using a microtiter plate reader. Subtract the blank from the average of the three readings. Draw the standard curve, read off the sample results, and calculate for dilutions. Data reduction can easily be performed by appropriate software. A typical standard curve is shown in Fig. 2.

Limitations of Assay

Pteridines With Greater Than 0.1% Activity

Biopterin	100%
Dihydrobiopterin	100%
Sepiapterin	100%
Tetrahydrobiopterin	80%
L-Neopterin	56%
Diacetylbiopterin	18%
D-Neopterin	0.2%
Pteroic acid	0.2%

Pteridines With Less Than 0.1% Activity

Folic acid
5-Methyltetrahydrofolic acid
5-Formyltetrahydrofolic acid
10-Formyltetrahydrofolic acid
Xanthopterin
Isoxanthopterin
Pterin

Sensitivity and Precision

The lowest standard contains 4 ng/liter and values for total biopterin of 0.5 μg/liter in blood spots and 0.2 μg/liter in plasma can be detected with certainty.

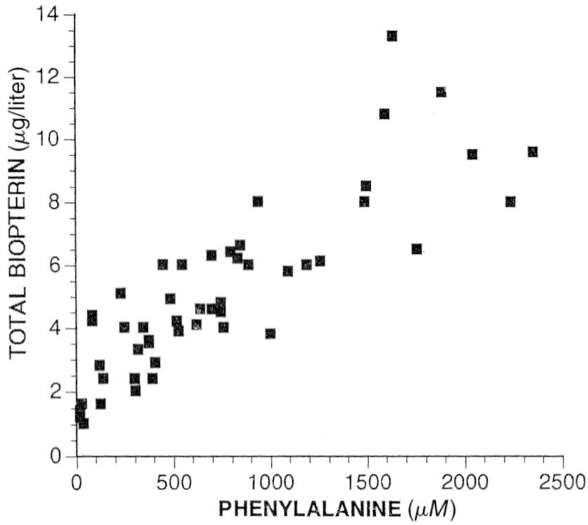

FIG. 3. Total plasma biopterin versus plasma phenylalanine in 50 cases of phenylketonuria.

Precision with whole blood and plasma within the normal range (below) gave a 4.8% between-batch coefficient of variation. However, the quality of blood spots can be variable and in normal subjects the within-batch coefficient of variation for blood spots was 8.8% and between batches it was 13.2%. In hyperphenylalaninemia, for which the blood spot total biopterin was between 7.2 and 10.6 μg/liter the within-batch coefficient of variation was 8.5%, and between batches it was 12.7%. The convenience of blood spots and the ease of discrimination between phenylketonuria and tetrahydrobiopterin-deficient hyperphenylalaninemia make blood spots preferable in initial screening.

Normal Range

In a normal subject the plasma total biopterin lies between 1.0 μg/liter (4.2 nmol) and 3.0 μg/liter (12.7 nmol). The concentration is raised when the phenylalanine increases,[4] except in tetrahydrobiopterin synthesis deficiency. Those reported to date and confirmed by direct enzymatic measurement are 6-pyruvoyltetrahydropterin synthase and GTP cyclohydrolase I (EC 3.5.4.16) deficiencies. The correlation between plasma biopterin and plasma phenylalanine in classic phenylketonuric patients with various degrees of dietary control is shown in Fig. 3. The erythrocyte total biopterin is normally about double the plasma concentration. The erythrocyte-to-

[4] N. Arai, K. Narisawa, H. Hayakawa, and K. Tada, *Paediatrics* **70,** 426 (1982).

plasma ratio is increased severalfold in phenylketonuria, probably because of increased intracellular synthesis during hemopoiesis. Some enzymes in the biosynthetic pathway for tetrahydrobiopterin can also be found in young circulating erythrocytes. The increased concentration of total biopterin in erythrocytes makes possible the use of blood spots on neonatal screening blood test cards of specified absorbency/paper, in the differentiation of phenylketonuria from tetrahydrobiopterin synthesis deficiency. Because biopterin passes out of erythrocytes slowly, whereas the plasma phenylalanine concentration may alter more quickly, the correlation of the erythrocyte biopterin concentration with plasma phenylalanine concentration is poorer (Fig. 4) than that between plasma biopterin and plasma phenylalanine. It does not, however, diminish diagnostic discrimination in the neonate. With a plasma phenylalanine value greater than 300 μmol a blood spot total biopterin value of 5 μg/liter (21.1 nmol) and above is unequivocally phenylketonuria. In tetrahydrobiopterin synthesis deficiency the plasma or blood spot biopterin is usually less than 0.5 μg/liter (2.1 nmol) even when the phenylalanine is elevated and neither responds to subsequent changes in phenylalanine concentration. In dihydropteridine reductase deficiency the total biopterin is increased in both plasma and erythrocyte and may be between 5 μg/liter (21.1 nmol) and 10 μg/liter (42.2 nmol) even when the phenylalanine is within normal limits. This is because *de novo* synthesis of tetrahydrobiopterin increases but the salvage of quinonoid dihydrobiopterin is ineffective and rearrangement to 7,8-dihydrobiopterin occurs.

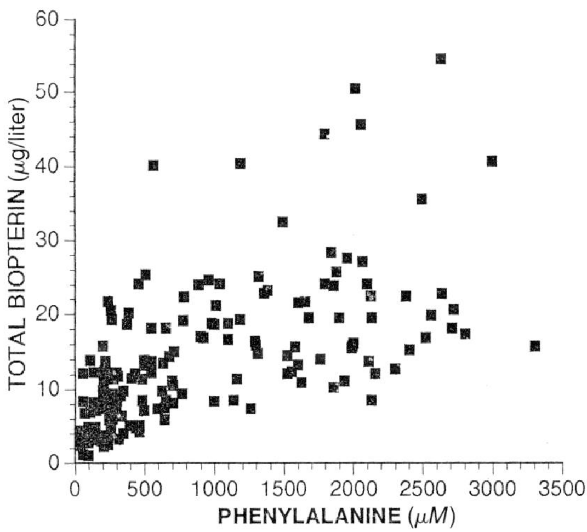

FIG. 4. Blood spot biopterin versus plasma phenylalanine in 171 cases of hyperphenylalaninemia.

When the phenylalanine is elevated the total biopterin in dihydropteridine reductase deficiency can reach 30–50 μg/liter (156.6–211 nmol) and this is principally 7,8-dihydrobiopterin. In the absence of direct measurement of dihydropteridine reductase, these findings are strongly indicative of its inactivity. However, dihydropteridine reductase can also be measured on neonatal screening blood test cards of specified absorbency/paper cards.[4]

Elevated plasma and blood spot biopterin concentrations may arise when patients have received drugs that inhibit dihydropteridine reductase and dihydrofolate reductase. These include triamterin, trimethoprim, methotrexate, and antimalarial drugs.

Urinary concentrations are 1000 times greater than plasma concentrations; therefore, in renal dysfunction the plasma concentration is greatly elevated. Biopterin concentrations may change in other groups of patients with disorders not described above.[5] Most of these changes are not diagnostic although they can have indicative values in expert hands.

In the hands of the author, plasma concentrations are higher in unheated samples, which suggests that some of the total biopterin is lost during the procedure. As total biopterin occurs normally, largely as tetrahydrobiopterin, some oxidation and/or loss from the total biopterin pool will occur during analysis but this does not invalidate the usefulness of the method as described, which is easy to set up, reliable, reproducible, and inexpensive.

Alternative Methods

Alternative methods for analyzing biopterin concentrations include high-performance liquid chromatography[6] and radioimmunoassay.[7]

[5] R. J. Leeming and J. A. Blair, *Clin. Chim. Acta* **108,** 103 (1980).
[6] A. Niederwieser, W. Staudenmann, and E. Wetzel, *J. Chromatogr.* **290,** 237 (1984).
[7] H. Rokos and A. Hey, *in* "Chemistry and Biology of Pteridines" (H.-Ch. Curtius, S. Ghisla, and N. Blau, eds.), p. 151. de Gruyter, Berlin, 1990.

[8] Identification and Assay of Folate Catabolites in Human Urine

By JOE MCPARTLIN and JOHN SCOTT

The obligatory requirement for folate may result from a regular catabolic process involving the cleavage of the C-9–N-10 bond of the molecule.[1]

[1] J. M. Scott, *in* "Folates and Pterins: Chemistry and Biochemistry of Folates" (R. L. Blakley and S. J. Benkovic, eds.), Vol. I, p. 307, John Wiley & Sons, New York, 1984.

0076-6879/97 $25

It is likely that some folate deficiency states such as pregnancy may be caused by an accelerated rate of catabolism accompanying increased metabolic activity.[2] Because the identification and measurement of urinary folate catabolites present the opportunity of determining the rate of utilization of folate in animals and humans, a laboratory-based assessment of daily requirement becomes possible. Urinary analysis of the pteridine product of the cleavage mechanism poses the problem of lability, particularly when 24-hr urine collections are required. In addition, more than one pteridine may be involved depending on the monoglutamate substitution. The most intractable difficulty with choosing pteridines as the indicator of catabolism, however, is that the folate-derived pteridines may not be quantitatively excreted, and those that are excreted need to be distinguished from nonfolate derivatives in urine. Therefore, the relatively stable and quantitatively excreted N-10 products of the cleavage process, *p*-aminobenzoylglutamate (*p*-ABGlu) and its acetamido derivative, *p*-acetamidobenzoylglutamate (A-*p*-ABGlu), were regarded as more appropriate indicators of folate catabolism.

The high-performance liquid chromatography (HPLC) assay requires a number of essential sample preparation steps involving batch-elution ion-exchange chromatography followed by diazotization, extraction on C_{18} solid phase, and reconversion with zinc. The first step removes most of the urinary salt and urea, along with most of the other primary aromatic compounds in urine, including *p*-aminohippurate and *p*-aminobenzoate, which are at substantially higher concentrations than *p*-ABGlu. The remaining steps produce a preparation consisting chiefly of diazotizable material, including the catabolites, for HPLC assay. The number of stages involved in sample preparation requires the addition to the urine sample of internal standards, high specific activity [^{3}H]*p*-ABGlu and [^{3}H]A-*p*-ABGlu.

Reagents and Materials

Unless stated otherwise reagents including ion-exchange resin, buffer constituents, ascorbic acid, and *p*-aminobenzoylglutamate (sodium salt) are purchased from Sigma Chemical Company (Poole, UK). Scintillation fluid (Ecolite) is purchased from ICN (Costa Mesa, CA). HPLC-grade methanol is purchased from Labscan Limited (Dublin, Ireland). HPLC solvents are prepared with reverse osmosis-ultrapure water.

[2] J. McPartlin, A. Halligan, J. Scott, M. Darling, and D. G. Weir, *Lancet* **341,** 148 (1993).

Preparation and Standardization of *p*-Acetamidobenzoylglutamic Acid

p-Acetamidobenzoylglutamic acid is prepared according to the method of Baker *et al.*[3] To a solution of 1.00 g of *p*-ABGlu in 10 ml of 50% aqueous acetic acid is added 1.5 ml (16 mmol) of acetic anhydride. After allowing the mixture to stand for 24 hr, it is filtered and the product is washed with water. Material is dissolved in 10 ml of dilute acetic acid (0.05 *M*). To remove unreacted *p*-ABGlu the mixture is diazotized as described below (see Synthesis of *p*-Amino[3′,5′-^{3}H]benzoylglutamic Acid), using reagents at 10 times the concentration described. Diazotized *p*-ABGlu is retained on the Sep-Pak column (Waters, Milford, MA) used, while A-*p*-ABGlu elutes through and is retained and checked for homogeneity by HPLC as described later (see Quantitation by High-Performance Liquid Chromatography).

Standardization of Catabolites

A stock solution of *p*-ABGlu (1 mg%) is standardized using published extinction coefficient data.[4] A stock solution of freshly synthesized A-*p*-ABGlu is standardized as follows. To 500 μl of a solution of the material (approximately 1 mg%) is added a known amount of [^{3}H]A-*p*-ABGlu (1.0×10^4 dpm) and HCl to a final concentration of 0.2 *M*. The material is deacetylated by boiling for 1 hr. An aliquot of the reaction mixture is then chromatographed by HPLC as previously described and the *p*-ABGlu produced by the reaction is quantitated by reference to *p*-ABGlu standard material. The [^{3}H]*p*-ABGlu coeluting with the spectral peak is counted and related back to the total disintegrations per minute in the reaction mixture to determine the stock concentration of A-*p*-ABGlu.

Preparation of Radioisotopes

Synthesis and Purification of p-Amino[3′,5′-^{3}H]benzoylglutamic Acid

Although not listed in their current catalog Moravek Biochemicals (Brea, CA) is prepared to supply high specific activity [3′,5′-^{3}H]*p*-ABGlu. Alternatively, both catabolites may be prepared from [3′,5′,7,9-^{3}H]pteroylglutamic acid ([3′,5′,7,9-^{3}H]PteGlu) purchased from either Amersham (Amersham, UK) (TRK 212, specific activity 10–50 Ci/mmol) or Moravek (MT783, specific activity 10–30 Ci/mmol). [^{3}H]*p*-ABGlu is prepared from [^{3}H]PteGlu as follows. To 100 μl of the supplied preparation of [^{3}H]PteGlu

[3] B. L. Baker, D. V. Santi, P. I. Almaula, and W. C. Werkheiser, *J. Med. Chem.* **7,** 24 (1964).
[4] R. L. Blakley, "The Biochemistry of Folic Acid and Related Pteridines," p. 94. North-Holland, Amsterdam, 1969.

in a 1.5-ml polypropylene microcentrifuge tube is added 50 μl of HCl (5 *M*) and 50 μl of zinc dust (British Drug House, Poole, UK) suspension [1 g of zinc per 2 ml of gelatin (0.5%, w/v)]. The mixture is capped, vortexed intermittently for 15 min at room temperature, then centrifuged to remove excess zinc. The supernatant is carefully removed and purified by HPLC.

Whether [^{3}H]*p*-ABGlu is purchased directly or is synthesized from [^{3}H]PteGlu it is essential for quantitation purposes that homogeneous material is used as the internal standard. Stored tritiated material is liable to undergo tritium exchange in aqueous solution. A special HPLC column should be set aside for preparative purposes. Although retention times for the column must be determined with authentic catabolite standards, avoid applying quantities greater than 500 ng for this purpose to prevent carryover contamination of radioisotope material. Following marking of the column with standards, it should be flushed by injecting about 1 ml of methanol. Aliquots of radiolabeled material up to 200 μl are purified on a 100 × 8 mm Radial Pak C_{18} cartridge column (Waters) contained within either a Z-Module radial compression system or an RCM 8 × 10 Radial-Pak cartridge holder (both Waters). A precolumn module containing a disposable C_{18} insert (RCSS Guard Pak; Waters) is attached between the injection port and column. The column is eluted isocratically at 2 ml/min with citrate–phosphate buffer (0.1 *M*, pH 4.0)–acetic acid (1%, v/v)–methanol (80 : 14 : 6, by volume). An early radiolabeled peak corresponding to pteridine material is collected and disposed of safely. [^{3}H]*p*-ABGlu is eluted at approximately 8.5 min and is collected by fraction collector. A portion of this purified label is reserved for the assay of urinary *p*-ABGlu. The remainder is used as substrate for the synthesis of [^{3}H]A-*p*-ABGlu.

Synthesis of p-Acetamido[3',5',-^{3}H]benzoylglutamic Acid

The preparation of the acetamido derivative requires that the [^{3}H]*p*-ABGlu starting material prepared above be concentrated by solid-phase adsorption (C_{18} Sep-Pak; Waters). This is done by converting the [^{3}H]*p*-ABGlu preparation to its diazotized derivative by coupling with naphthylethylenediamine (the Bratton–Marshall reaction) as follows. To 5–10 ml of [^{3}H]*p*-ABGlu (25 μCi) is added 100 μl of HCl (5 *M*) and 100 μl of sodium nitrite (1%, w/v). At 5-min intervals add in sequence 100 μl of ammonium sulfamate (5%, w/v) and 100 μl of *N*-1-naphthylethylenediamine (1%, w/v). After a 30-min incubation at ambient temperature the diazotized radiolabeled derivative is applied by syringe to a preactivated (3 ml of methanol followed by 5 ml of water) C_{18} Sep-Pak. The diazotized material, indicated by a faint purple band retained at the head of the Sep-Pak column, is washed with 10 ml of HCl (0.05 *M*). The column is then dried by evacuating it with an empty syringe, after which the visible diazotized material is eluted slowly with 3 ml of acidified methanol [50 μl of HCl (5 *M*) to 25 ml of methanol]. [^{3}H]*p*-ABGlu is regenerated by the addition of 50 μl of HCl (5 *M*) and 50 μl of zinc dust to the methanolic eluate. The extract is vortexed intermittently for 5 min, then centrifuged to clarify the [^{3}H]*p*-ABGlu-containing supernatant. To prepare the acetamido derivative the supernatant is transferred to a 10-ml conical glass tube and evaporated to dryness under N_2 at 37°. To the dried extract is added 34 μl of acetic acid (50%, v/v) and 5 μl of acetic anhydride. The reactants are mixed by vortexing gently, and the tube is plugged with cotton wool and incubated at 22° for 24 hr. The reaction mixture is diluted with 100 μl of distilled H_2O, the entire volume (139 μl) is applied to the preparative HPLC column, and [^{3}H]A-*p*-ABGlu is fractionated as described above for [^{3}H]*p*-ABGlu, except that the acetamido radiolabel is eluted at approximately 25 min.

Urine Collection

Urine samples (24 hr) are collected in 5-liter plain plastic containers containing ascorbic acid (2 g). Total volume is recorded and a 300-ml aliquot is retained and stored at −20° before assay.

Assay of Urinary *p*-Aminobenzoylglutamic Acid and *p*-Acetamidobenzoylglutamic Acid

A simple bench-top support for an array of glass columns is shown in Fig. 1. It consists simply of a supported plank of wood with holes at regular intervals through which the glass columns are suspended above collection bottles. Alternatively, polypropylene columns (Econo-Pac 1.5 × 12 cm), column racks, and funnels are available from Bio-Rad Laboratories (Hercules, CA).

Preparation of Ion-Exchange Columns

A modification of the method of Tompsett[5] is used to fractionate the urine sample. One column per 25-ml urine sample is prepared as follows. Glass columns (200 × 15 mm, with a 200-ml reservoir) are plugged with glass wool and filled with a slurry of Dowex 50W (50X8-400; Sigma) cation-exchange resin to a height of approximately 4.5 cm (7 g/10 ml of H_2O).

[5] S. L. Tompsett, *Clin. Chim. Acta* **5,** 415 (1960).

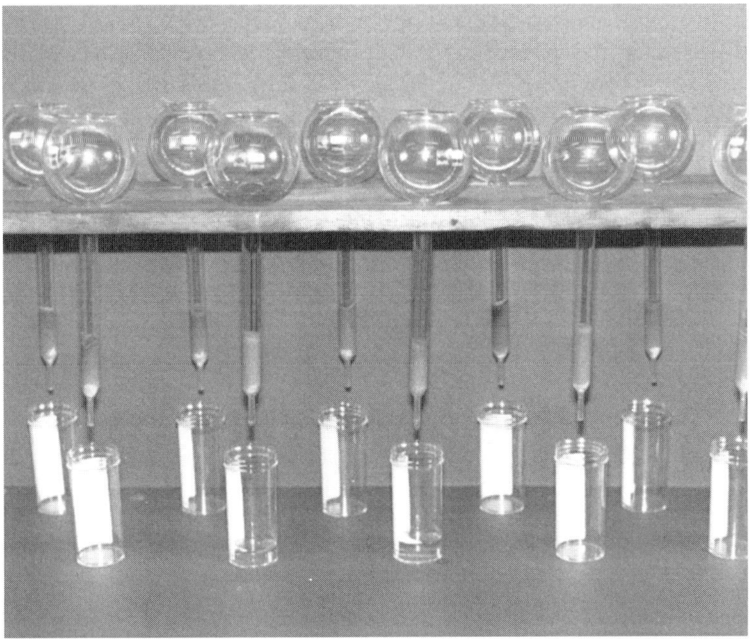

FIG. 1. A simple arrangement for ion-exchange chromatography of urine samples.

The column is activated with 50 ml of HCl (6 *M*), washed with 50 ml of H_2O, and equilibrated with 10 ml of HCl (0.1 *M*).

Fractionation of Urine by Ion-Exchange Chromatography

To duplicate 25-ml aliquots of urine is added HCl (6.0 *M*) to a final concentration of 0.1 *M*, as well as [^{3}H]A-*p*-ABGlu (8.0×10^4 dpm) and [^{3}H]*p*-ABGlu (6.0×10^4 dpm). The urine aliquots are then applied to ion-exchange columns. A radiolabeled fraction containing A-*p*-ABGlu, not retained by the column, is collected along with a wash of 50 ml of 0.1 *M* HCl. A subsequent fraction eluted with 0.3 *M* HCl (50 ml) is discarded. A second radiolabeled fraction containing *p*-ABGlu is eluted in 100 ml of 0.6 *M* HCl and retained for subsequent solid-phase extraction and assay. The column is then reactivated, washed, and equilibrated as described above for reuse following deacetylation of A-*p*-ABGlu material.

Deacetylation and Rechromatography of Urinary p-Acetamidobenzoylglutamic Acid Fraction

The combined A-*p*-ABGlu fraction prepared above (75 ml) is deacetylated by boiling for 1 hr in 0.2 *M* HCl. The solution is cooled to room temperature before further chromatography. The deacetylated fraction is reapplied to the same ion-exchange column. The sample eluate, as well as subsequent eluates of 50 ml of 0.2 *M* HCl and 50 ml of 0.3 *M* HCl, are discarded and the *p*-ABGlu-containing material is eluted with 100 ml of 0.6 *M* HCl.

Diazotization and C_{18} Sep-Pak Extraction

Both [^{3}H]*p*-ABGlu-containing isolates (100 ml each) are independently processed by an identical procedure as follows. Each isolate is diazotized as described above except that 1-ml volumes of diazotization reagents are used. The purple color is allowed to develop overnight. The diazotized solutions are cleaned up further and concentrated by application to a C_{18} Sep-Pak column, using a Waters Sep-Pak rack attached to a vacuum pump. The retained material is washed with 10 ml of H_2O and 3 ml of dilute acidified methanol [25% (v/v) in H_2O]. Radiolabeled diazotized material is eluted into 25-ml polystyrene universal containers with 100% methanol and the solution is evaporated to dryness at 56° in a water bath.

Regeneration of p-Aminobenzoylglutamic Acid

The residue is reconstituted in 250 μl of H_2O. *p*-ABGlu is regenerated by the addition of 25 μl of HCl (5 *M*) and 25 μl of zinc dust suspension. The reaction mixture is transferred to a microcentrifuge tube and centrifuged for 5 min on a Beckman (Palo Alto, CA) microcentrifuge. The supernatant is removed from the zinc pellet, retained, and stored at −20° prior to HPLC analysis.

Quantitation by High-Performance Liquid Chromatography

A 100-μl aliquot of this extract is analyzed by HPLC as described above and the radiolabeled fraction corresponding to *p*-ABGlu is collected. Some samples such as those from postpartum pregnancy and collections from some geriatric subjects do not readily chromatograph satisfactorily on a Radial-Pak column. Instead, we find that a steel column such as a LiChrospher 100 RP-18 (5-μ particle size, 25 × 4 mm; Merck, Darmstadt, Germany) eluted with buffer at a flow rate of 1 ml/min is adequate for the purpose. In this case, the retention time of *p*-ABGlu is 16 min. *p*-ABGlu in the injected sample is quantitated by peak height and related to a standard plot of 10–100 ng. [A known amount of [^{3}H]*p*-ABGlu (about 4000 dpm) is added to each standard prior to injection and collected postcolumn to determine recovery of standards from the HPLC.] The calculated value for each urine extract is adjusted by the final percentage recovery of the radiolabels, either [^{3}H]*p*-ABGlu or [^{3}H]A-*p*-ABGlu, added to the initial 25-ml aliquot of urine. This value in turn is adjusted for total urinary volume to give a daily output value of catabolite.

Note on Sample Collection

Urinary folate catabolites consist of an aggregrate of catabolites generated endogenously as well as those derived directly from dietary sources. To avoid including dietary-derived material it was found necessary to maintain subjects on a catabolite-free regime by providing enteral liquid diets such as Fortisip (Cow & Gate Nutricia, Trowbridge, Wiltshire, UK) or Ensure (Abbott, Chicago, IL) not only for the duration of the urine collection but also for at least 8 hr beforehand.

[9] Enzymatic Determination of Folylpolyglutamate Pools

By VERNE SCHIRCH

Introduction

Polyglutamate derivatives are the *in vivo* form of tetrahydrofolate (THF), with three to eight glutamate residues linked as amides through the γ-carboxyl group. Different organisms have different relative concentrations of glutamate chain length. To describe accurately the number of glutamate residues the nomenclature system is related to tetrahydropteroate (H_4Pte), which has no glutamate residues. Thus THF is called H_4PteGlu and the pentaglutamate form is $H_4PteGlu_5$.

In addition to a varying number of glutamate residues there are six different one-carbon derivatives of $H_4PteGlu_n$. Two of these derivatives, 5-forminino- and 5,10-methenyl-$H_4PteGlu_n$, exist in low concentrations and probably do not contribute significantly to the total intracellular pool of $H_4PteGlu_n$. This leaves 10-CHO-, 5-CHO-, 5-CH_3-, and 5,10-CH_2- one-carbon derivatives and $H_4PteGlu_n$ as the major forms of reduced folates in the cell. Under certain conditions dihydrofolate ($H_2PteGlu_n$) may also be a significant part of the folate pool. Determining the concentration and relative distribution of $H_4PteGlu_n$ compounds in cells has proved useful in analyzing the role of one-carbon metabolism. A variety of methods have been developed to address these questions. Each method has its strengths and weaknesses. Described here is a rapid and sensitive method of determining concentrations of 5-CH_3-$H_4PteGlu_n$, 5-CHO-$H_4PteGlu_n$, and the com-

0076-6879/97 $25

bined pools of $H_4PteGlu_n$, 5,10-CH_2-$H_4PteGlu_n$, and 10-CHO-$H_4PteGlu_n$. The key advantages of this method are its sensitivity and rapidity of determination. The major disadvantages are the need for several enzymes and the fact that separate pool sizes for three of the folate compounds cannot be determined.

Materials

Glycine, L-serine, $NADP^+$, 2-mercaptoethanol, L-methionine, $NADP^+$, MgATP, and dihydrofolate reductase (DHFR) can all be purchased from several biochemical companies. Serine hydroxymethyltransferase (SHMT), C_1-tetrahydrofolate synthase (C_1-THF synthase), 5,10-methenyltetrahydrofolate synthetase (CH^+-THF synthase,5-formyltetrahydrofolate cyclo-ligase), and 10-formyltetrahydrofolate dehydrogenase (FTD) can be easily purified in large amounts from rabbit liver by methods published in [19] in this volume.[1] Methionine synthase can be purified from an overexpressing clone according to Matthews.[2] $H_4PteGlu_n$ and its one-carbon derivatives can be synthesized from $PteGlu_n$ by methods also published in this volume (see [10]).[3]

Extraction of $H_4PteGlu_n$

Reduced folates are unstable, especially with respect to oxidation by oxygen and other mild oxidizing agents. Both 2-mercaptoethanol and ascorbate help stabilize these reduced folates. In addition, several one-carbon derivatives can be altered during extraction. 10-CHO-$H_4PteGlu_n$ can be converted to both 5,10-CH^+-$H_4PteGlu_n$ and 5-CHO-$H_4PteGlu_n$ during extraction at pH values below pH 7. These reactions are accelerated in the presence of phosphate. Also, 5-10-CH_2-$H_4PteGlu_n$ is converted to $H_4PteGlu_n$ during extraction by loss of formaldehyde.

The tissue to be extracted (about 10 to 200 mg) is minced or homogenized in 500 μl of 50 m*M* HEPES, pH 7.6, containing 0.2 m*M* EDTA, 50 m*M* 2-mercaptoethanol, and 2% (w/v) sodium ascorbate. This homogenate is placed in a boiling water bath for about 30 sec. After rapid cooling the slurry is centrifuged to remove precipitated macromolecules. The supernatant is diluted with an equal volume of 20 m*M* KP_i, pH 6.8, containing 1 m*M* $MgCl_2$ and 10 m*M* $(NH_4)SO_4$. The amount of material to extract is

[1] V. Schirch, *Methods Enzymol.* **281,** [19], 1997 (this volume).

[2] J. C. González, R. V. Banerjee, S. Huang, J. S. Summer, and R. G. Matthews, *Biochemistry* **31,** 6045 (1992).

[3] V. Schirch, *Methods Enzymol.* **281,** [10], 1997 (this volume).

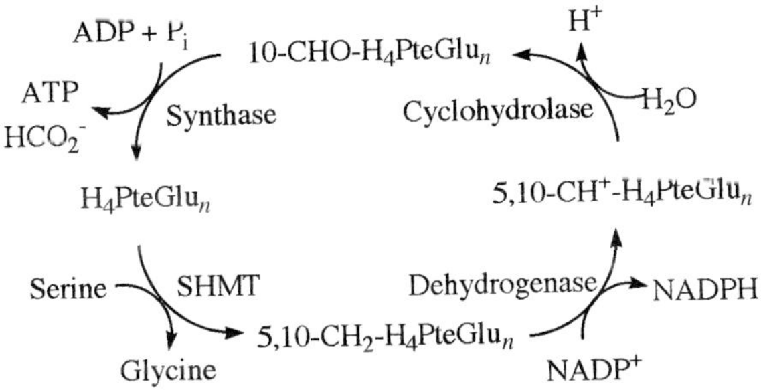

SCHEME I. Reaction for the analysis of the combined pools of $H_4PteGlu_n$, 10-CHO-$H_4PteGlu_n$, and 5,10-CH_2-$H_4PteGlu_n$. SHMT and the three activities of C_1-THF synthase catalyze the four reactions in this cycle. The three activities of C_1-THF synthase are 10-CHO-THF synthase (synthase); 5,10-CH^+-THF cyclohydrolase (cyclohydrolase); and 5,10-CH_2-THF dehydrogenase (dehydrogenase).

determined by the level of folates in the cells. This can be determined either by having some previous knowledge of reduced folate levels or by trial and error.

Determination of Combined $H_4PteGlu_n$, 10-CHO-$H_4PteGlu_n$, and CH_2-$H_4PteGlu_n$ Pools

The coupled enzymatic reactions of SHMT and the three activities of C_1-THF synthase shown in Scheme I catalyze a metabolic cycle in which $H_4PteGlu_n$ is regenerated during each cycle. In the presence of excess serine, $NADP^+$, MgADP, and phosphate a serine molecule is converted to glycine and formate and $NADP^+$ is converted to NADPH. The formation of NADPH during each cycle serves as a sensitive method for following the rate of the cycle, using a spectrophotometer at 340nm.[4] Using excess concentrations of the enzymes results in the rate of the cycle being determined by the catalytic levels of $H_4PteGlu_n$ or its one-carbon derivatives involved in the cycle. Using nmol levels of the two enzymes the rate of the cycle, as determined by the linear increase in A_{340}, is directly proportional to the levels of $H_4PteGlu_n$ in the 10- to 200 pmol range. Any of the folate molecules involved in the cycle can be used with the same sensitivity. This cycle operates only if the number of glutamate residues on $H_4PteGlu_n$ is three or greater. The monoglutamate forms do not give linear rates. This is due at least in part to the much lower affinity of the monoglutamate forms for the active sites of the two enzymes.

[4] H. L. Kruschwitz, D. McDonald, E. A. Cossins, and V. Schirch, *J. Biol. Chem.* **269,** 28757 (1994).

A typical assay includes 800 μl of cell extract, 1 mM MgADP, 0.2 mM NADP$^+$, 5 mM L-serine, and 150 μg of SHMT at 30°. The absorbance is monitored at 340 nm for about 30 sec to establish a baseline. The reaction is started by the addition of 100 μg of C_1-THF synthase and the increase in A_{340} is followed for about 1 min. The background rate is subtracted from the reaction rate to determine the $H_4PteGlu_n$-dependent rate. As noted above, this rate is linear with the concentration of $H_4PteGlu_n$ between 10 and 200 pmol.

Determination of 5-CH_3-$H_4PteGlu_n$ and 5-CHO-$H_4PteGlu_n$ Levels

The determination of the two $H_4PteGlu_n$ derivatives 5-CH_3-$H_4PteGlu_n$ and 5-CHO-$H_4PteGlu_n$ requires a preincubation of the cell extract with an additional enzyme. For 5-CH_3-$H_4PteGlu_n$ two equal volumes of cell extract are incubated for 15 min at 37° with 0.2 mM DL-homocysteine. To one of the extracts is added 10 μg of methionine synthase, which converts the 5-CH_3-$H_4PteGlu_n$ to $H_4PteGlu_n$.[2] After preincubation, the other reagents are added as described above. The rate of the cycle with the extract lacking methionine synthase represents the total concentration of $H_4PteGlu_n$, 10-CHO-$H_4PteGlu_n$, and CH_2-$H_4PteGlu_n$ as discussed above. It was observed that homocysteine inhibits by about 10% the rate of the cycle in the absence of methionine synthase.[4] The rate exhibited by the extract containing the methionine synthase will be greater by the amount of 5-CH_3-$H_4PteGlu_n$ that was converted to $H_4PteGlu_n$. The difference between these two rates represents the amount of 5-CH_3-$H_4PteGlu_n$ that was converted to $H_4PteGlu_n$.

The level of 5-CHO-$H_4PteGlu_n$ is determined by preincubating the cell extract for 2 min at 37° with 1 mM MgATP and 10 μg of CH^+-THF synthase, which converts this substrate to 5,10-CH^+-$H_4PteGlu_n$. This compound hydrolyzes nonenzymatically to 10-CHO-$H_4PteGlu_n$ in a few minutes. After the preincubation, the other reagents are added as described above for determining $H_4PteGlu_n$ levels. The increased rate of the cycle after these preincubations compared to the direct determination for $H_4PteGlu_n$ represents the amount of 5-CHO-$H_4PteGlu_n$ in the extract.

Because the calculations for 5-CH_3-$H_4PteGlu_n$ and 5-CHO-$H_4PteGlu_n$ require taking the difference between two numbers, the level of either 5-CH_3-$H_4PteGlu_n$ or 5-CHO-$H_4PteGlu_n$ cannot be determined with reasonable accuracy if they are less than 10% of the total size of the folate pool.

It should also be possible to use this method to determine the level of $H_2PteGlu_n$ pools by preincubating the cell extract with DHFR and NADPH. Levels of the polyglutamate form of this folate occur only as the result of inhibition of DHFR. This method has not been developed.

Conclusions

The enzymes used in this procedure can be purified rapidly from rabbit liver, as discussed in [19] in this volume.[1] In 1 week enough of these enzymes can be purified from a few livers to perform several hundred assays. Also, each of the enzymes is stable for months and need not be highly purified for this procedure. Assays can be performed in triplicate with ease, resulting in the ability to determine the total folate pools in a cell extract with accuracy at the pmole level in a few hours. This is the key advantage of this procedure. Another advantage is that the amount of manipulation of the cell extract in the determination is minimal, which decreases the chance that folates will be interconverted during isolation and analysis.

[10] Synthesis and Interconversion of Reduced Folylpolyglutamates

By VERNE SCHIRCH

Introduction

Tetrahydrofolate (THF) is the coenzyme for several enzymes involved in one-carbon metabolism. The physiological form of THF contains a chain of glutamate residues linked through the γ-carboxyl group.[1] To accommodate the naming of these derivatives they are named as polyglutamate derivatives of tetrahydropteroate (H_4Pte), which has no glutamate residue. Tetrahydrofolate then becomes H_4PteGlu and the pentaglutamate form $H_4PteGlu_5$. This nomenclature system will be used throughout this chapter.

The physiologically active one-carbon derivatives of $H_4PteGlu_n$ are not commercially available and are unstable, being easily oxidized by a variety of oxidizing agents. These two facts have hindered the study of the role of these physiological forms of $H_4PteGlu_n$ in one-carbon metabolism.

This chapter describes a method for synthesizing 5-formyl-$H_4PteGlu_n$, where n is 1 to 6, and the interconversion of this derivative to the other one-carbon derivatives. 5-Formyl-$H_4PteGlu_n$ is the only stable folate derivative and can be stored under a variety of conditions at either neutral or alkaline pH values without degradation.

[1] V. Schrich and W. B. Strong, *Arch. Biochem. Biophys.* **269,** 371 (1989).

0076-6879/97 $25

Materials

Serine, glycine, α-ketoglutarate, sodium hydrosulfite, sodium ascorbate, glucose 6-phosphate, glutamate dehydrogenase, glucose-6-phosphate dehydrogenase, NADPH, dihydrofolate reductase (DHFR), BioGel P-2 (Bio-Rad, Richmond, CA), [^{13}C]formate, and [^{14}C]formate, about 50 mCi/mol, can be purchased from many biochemical supply companies. Potassium *N,N*-bis(2-hydroxyethyl)-2-aminoethane sulfonate (KBES) and potassium 2-(*N*-morpholino)ethane sulfonate (KMES) can also be purchased from most biochemical supply companies. Pteroylpolyglutamates ($PteGlu_n$) can be purchased from the laboratory of B. Schircks (Jona, Switzerland). Serine hydroxymethyltransferase (SHMT), C_1-tetrahydrofolate synthase (C_1-THF synthase), and methenyltetrahydrofolate synthetase (CH^+-THF synthetase, 5-formyltetrahydrofolate cyclo-ligase) can be easily purified from rabbit liver as described in [19] in this volume.[2]

Quantitative Measurements

Pteroylglutamate concentrations are determined using an ε_{282} of 27 mM^{-1} cm^{-1}, at pH 7.0.[3] The concentration of dihydropteroylpolyglutamates ($H_2PteGlu_n$) is determined by measuring the decrease in total absorbance at 340 nm following the addition of 20 μg of DHFR to a solution containing 100 μ*M* NADPH (ε_{340} of 6240 M^{-1} cm^{-1}) and 5 to 20 μmol of $H_2PteGlu_n$ in 1 ml of 50 m*M* KBES, pH 7.0, containing 5 m*M* 2-mercaptoethanol. The concentration of $H_4PteGlu_n$ is determined by measuring the increase in absorbance at 340 nm following the addition of 20 μg of C_1-THF synthase to a solution of $H_4PteGlu_n$, 0.1 m*M* $NADP^+$, 20 m*M* L-serine, 0.2 mg of SHMT in 1 ml of KBES, pH 7.3, containing 5 m*M* 2-mercaptoethanol. In this determination the $H_4PteGlu_n$ is first converted to 5,10-CH_2-$H_4PteGlu_n$ by SHMT and serine and then to 5,10-CH^+-$H_4PteGlu_n$ by the CH_2-THF dehydrogenase activity of C_1-THF synthase. The $H_4PteGlu_n$ concentration is obtained by dividing the total absorbance change at 340 nm by the extinction coefficient of 7200 M^{-1} cm^{-1}, which is the result of the combined absorbance at 340 nm of NADPH and 5,10-methenyltetrahydrofolate at pH 7.3. This extinction coefficient of 7200 M^{-1} cm^{-1} changes with pH, so it is important that the pH be pH 7.3.

The 5,10-CH^+-$H_4PteGlu_n$ concentration is determined by spectral analysis using an ε_{360} of 25.1 mM^{-1} cm^{-1} at pH 2.0. The concentrations of 10-

[2] V. Schirch, *Methods Enzymol.* **281,** [19], 1997 (this volume).

[3] C. Temple and J. A. Montgomery, *in* "Folates and Pterins" (R. L. Blakely and S. J. Benkovic, eds.), Vol. 1, p. 80. John Wiley & Sons, New York, 1984.

CHO-H_4PteGlu$_n$ and 5-CHO-H_4PteGlu$_n$ are determined following their conversion to 5,10-methenyl-H_4PteGlu$_n$ by incubation at pH 3.0 for several hours at room temperature or overnight at 4°. The concentrations of both 5-CHO-H_4PteGlu$_n$ and 5-CH_3-H_4PteGlu$_n$ can also be determined from the formation of a ternary complex of SHMT (1 mg/ml) and glycine (20 mM) in KBES, pH 7.3. Both of these compounds form a complex that absorbs intensely at 502 nm with an ε_{502} of 40,000 M^{-1} cm^{-1}.[4] In addition, if each H_4PteGlu$_n$ derivative is pure, the concentration can be determined from its spectrum using standard molar extinction coefficients.

Synthesis of H_2PteGlu$_n$ from PteGlu$_n$

Pteroylglutamate is reduced to H_2PteGlu$_n$ by sodium hydrosulfite. Twenty milligrams of PteGlu$_n$ is dissolved in 350 μl of degassed, argon-purged H_2O containing 100 mM sodium ascorbate in a 16 × 100 mm test tube. The pH is adjusted to pH 11 by the addition of 150 μl of 1 N NaOH. Argon is bubbled through the solution for the entire procedure. Sodium hydrosulfite (500 μmol) is added to the solution and pH 11 is maintained by the addition of 1 N NaOH. This solution is incubated at about 30° for 20 min. The pH of the solution is then slowly dropped to pH 5.5 with microliter aliquots of 3 N HCl and incubated on ice for 20 min to destroy the excess hydrosulfite. Ten microliters of 2-mercaptoethanol is added and the solution applied to a 1.5 × 60 cm Bio-Gel P-2 column equilibrated with 5 mM KBES, pH 7.6, and 50 mM 2-mercaptoethanol. The flow rate is maintained at about 1 ml/min with a peristaltic pump. Two milliliter fractions are collected and the spectrum of each fraction recorded. H_2PteGlu$_n$ is the first compound to elute from the column and exhibits an absorption maximum at 282 nm. If it is clear which fractions contain the H_2PteGlu$_n$ these can be pooled. If the spectra do not clearly establish the fractions containing H_2PteGlu$_n$ each fraction can be assayed with DHFR and NADPH. With the pooled fractions the total amount of H_2PteGlu$_n$ is determined by enzymatic assay. If the fractions are contaminated with PteGlu$_n$ there will be some absorbance at 350 nm. The yield of H_2PteGlu$_n$ is about 95%. This compound is usually used immediately for the next step in the purification. It can be lyophilized and stored as a dry powder under vacuum or argon at −70° for several weeks.

[4] P. Stover and V. Schirch, *Anal. Biochem.* **202,** 82 (1992).

Synthesis of (6S)-5-CHO-H_4PteGlu$_n$

Dihydropteroylglutamate (H_2PteGlu$_n$) is converted to (6*S*)-5-CHO-H_4PteGlu$_n$ in a single reaction vial without purification according to the reactions shown in Scheme I.[4] The H_2PteGlu$_n$ is dissolved in 1.5 ml of degassed water in a 1.5 × 10 cm Pyrex centrifuge tube containing 20 m*M* 2-mercaptoethanol, 2 m*M* L-serine, and 0.2 m*M* α-ketoglutarate and adjusted to pH 7.5 with KOH if necessary. The following reagents are added: DHFR, 1 mg; C_1-tetrahydrofolate synthase (C_1-THF synthase), 1 mg; SHMT, 5 mg; NADPH, 10 nmol (1 μl of a 10 m*M* solution). This solution results in the conversion of H_2PteGlu$_n$ to an equilibrium mixture of (6*R*)-5,10-CH^+-H_4PteGlu$_n$ and (6*R*)-10-CHO-H_4PteGlu$_n$. Note that substitution on N-10 results in a change of priority numbers so that (6*S*)-H_4PteGlu$_n$ becomes the (6*R*) N^{10}-substituted derivative. This reaction solution is incubated for 1 hr at 35°. Argon is streamed gently over the solution during the reaction. After 1 hr, an additional 10 nmol of NADPH is added and the reaction allowed to incubate for an additional 20 min. The remaining NADPH is depleted by the addition of 1 mg of L-glutamate dehydrogenase in ammonium sulfate and the reaction mixture incubated for another 15 min. This last step blocks the conversion of any 10-CHO-H_4PteGlu$_n$ to 5,10-CH_2-H_4PteGlu$_n$. The solution is adjusted to 1.8 ml by addition of water

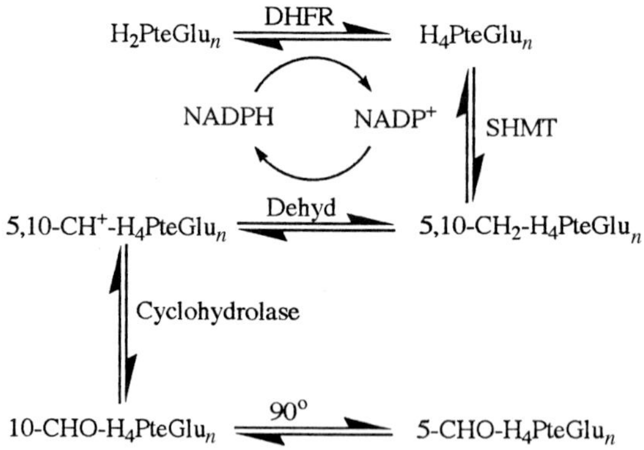

SCHEME I. Reactions involved in the conversion of H_2PteGlu$_n$ to (6*S*)-5-CHO-H_4PteGlu$_n$. DHFR, SHMT, and C_1-THF synthase activities convert H_2PteGlu$_n$ to 10-CHO-H_4PteGlu$_n$. The two activities of C_1-THF synthase utilized in this scheme are 5,10-CH_2-THF dehydrogenase (Dehyd) and 5,10-CH^+-THF cyclohydrolase (Cyclohydrolase). The (6*R*)-10-CHO-H_4PteGlu$_n$ formed by these enzymes is converted to (6*S*)-5-CHO-H_4PteGlu$_n$ nonenzymatically by increasing the phosphate concentration and heating to 90° for 3 to 5 hr.

and made 100 mM in KP_i, pH 6.9, by the addition of 0.2 ml of a 1 M solution of this buffer. The pH should be checked with a strip of pH paper to ensure that the pH is between pH 6.5 and 7.0. The solution is then incubated at 90° for 3–5 hr to precipitate the added proteins and to convert the 10-CHO-H_4PteGlu$_n$ to 5-CHO-H_4PteGlu$_n$. The precipitated proteins are removed by centrifugation and the supernatant diluted 10-fold with 10 mM K_2P_i and added to a 4 × 6 cm column of DE-52 cellulose equilibrated with 10 mM KP_i, pH 8.0. The column is washed with 50 ml of equilibration buffer and eluted with a linear gradient containing 150 ml each of equilibration buffer and 500 mM NaCl in 20 mM KP_i, pH 6.7. Fractions of 5 ml are collected, with the 5-CHO-H_4PteGlu$_n$ eluting near the end of the gradient. The spectrum of each fraction is taken and those containing material with an absorption maximum at 287 nm are pooled and lyophilized to dryness. The dried material is dissolved in 1–2 ml of H_2O and applied to a Bio-Gel P-2 column (1.5 × 60 cm) equilibrated with H_2O. The 5-CHO-H_4PteGlu$_n$ elutes with H_2O and emerges as the first absorbing material. Fractions containing 5-CHO-H_4PteGlu$_n$ are pooled and concentrated by evaporation under vacuum. Spectral analysis shows a compound with an absorbance maximum at 287 nm and a minimum absorbance at 242 nm, with a 287/242 nm ratio of 4.9. The purity of 5-CHO-H_4PteGlu$_n$ is greater than 95%, with a yield of about 90% for each polyglutamate form of the coenzyme. The 5-CHO-H_4PteGlu$_n$ can be stored as a solution at −70° for many months without loss of material. This compound is then used as the starting material for all of the other H_4PteGlu$_n$ derivatives as described in the following sections.

Synthesis of 5,10-CH$^+$-H_4PteGlu$_n$ and 10-CHO-H_4PteGlu$_n$

Both 5,10-CH$^+$-H_4PteGlu$_n$ and 10-CHO-H_4PteGlu$_n$ can be easily prepared from 5-CHO-H_4PteGlu$_n$ in 100% yield without purification of the products. Purified 5-CHO-H_4PteGlu$_n$ (10–500 nmol) is adjusted to pH 1.5 with 6 N HCl. This solution is incubated at 4° for 12–24 hr, resulting in the quantitative formation of 5,10-CH$^+$-H_4PteGlu$_n$. This compound is stable for weeks at −20° if kept under argon. It is important that the pH of the solution be maintained below pH 2.0 to prevent the formation of hydrated forms of the compound. The monoglutamate derivative forms yellow crystals under these conditions.

The 5,10-CH$^+$-H_4PteGlu$_n$ is converted to 10-CHO-H_4PteGlu$_n$ by making the solution 50 mM in 2-mercaptoethanol and adjusting the pH to pH 8.5 with 2 N KOH. This solution is purged with argon and incubated at room temperature for 30 min. The yield of 10-CHO-H_4PteGlu$_n$ is 95 to 100%.

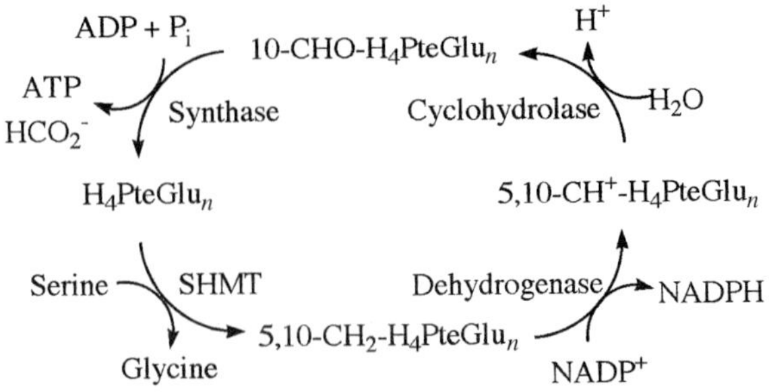

SCHEME II. Reactions for the conversion of 5,10-CH^+-H_4PteGlu$_n$ to H_4PteGlu$_n$. SHMT and the three activities of C_1-THF synthase catalyze the four reactions in this cycle. The three activities of C_1-THF synthase are as follows: 10-CHO-THF synthase (Synthase); 5,10-CH^+-THF cyclohydrolase (Cyclohydrolase); and 5,10-CH_2-THF dehydrogenase (Dehydrogenase).

The 10-CHO-H_4PteGlu$_n$ can be stored for several weeks at −70° if it is layered with mineral oil and stored in sealed vials flushed with argon.

Synthesis of 5-CH_3-H_4PteGlu$_n$ from 5,10-CH^+-H_4PteGlu$_n$

This procedure has been previously described for the monoglutamate form of the coenzyme.[5] A solution of 2.5–5 μmol of 5,10-CH^+-H_4PteGlu$_n$ at pH 1.5 and 4° is made 50 m*M* in 2-mercaptoethanol and 50 m*M* KP_i, pH 7.3, and the pH adjusted to pH 7.0 with KOH. Twenty milligrams of $NaBH_4$ is added and the solution incubated on ice for 90 min. Excess $NaBH_4$ is removed after the incubation by the addition of 20 μl of glacial acetic acid and the incubation continued for an additional 10 min. This solution is applied to a 1.5 × 60 cm Bio-Gel P-2 column equilibrated with argon-purged 5 m*M* KBES containing 10 m*M* 2-mercaptoethanol. The 5-CH_3-H_4PteGlu$_n$ is the first absorbing material to elute from the column and can be identified by its ability to form a 502 nm-absorbing ternary complex with SHMT and glycine. Both the yield and purity are usually greater than 90% for each polyglutamate form of this derivative.

Synthesis of H_4PteGlu$_n$ from 5,10-CH^+-H_4PteGlu$_n$

This reaction is performed using two enzymes, as shown in Scheme II. 5,10-CH^+-H_4PteGlu$_n$ (1–2 μmol) is added to a 500-μl solution of 50 m*M* KBES, pH 7.6, containing 20 m*M* 2-mercaptoethanol, 50 m*M* glycine, and

[5] I. Chanarin and J. Perry, *Biochem. J.* **105,** 633 (1967).

10 mM glucose 6-phosphate, and the pH is adjusted to pH 7.6 with KOH if necessary. One milligram each of C_1-THF synthase and glucose-6-phosphate dehydrogenase, and 10 nmol of NADPH, are added to the reaction mixture and incubated for 15 min at 30° to reduce all of the 5,10-CH^+-$H_4PteGlu_n$ to 5,10-CH_2-$H_4PteGlu_n$. The conversion of 5,10-CH_2-$H_4PteGlu_n$ is initiated by the addition of 1 mg of SHMT. The solution is incubated for an additional 15 min. The pH is then adjusted to pH 5.0 with glacial acetic acid and the solution placed in a 45° bath for 10 min to precipitate proteins, which are then removed by a brief centrifugation (in a microfuge, 5 min at room temperature).

The $H_4PteGlu_n$ is purified by chromatography on a DE-52 column (1 × 6 cm), using the same buffers and procedure as described in the purification of 5-CHO-$H_4PteGlu_n$, except that all buffers are made 50 mM in 2-mcercaptoethanol. This column separates $H_4PteGlu_n$ from residual $NADP^+$ and NADPH. Fractions containing $H_4PteGlu_n$ are pooled and the concentration determined from its A_{298}, using an ε of 28.5 mM^{-1} cm^{-1}. The spectrum suggests that the $H_4PteGlu_n$ is greater than 90% pure and the yield from 5,10-CH^+-$H_4PteGlu_n$ is about 75%. This solution is divided into several small serum bottles, layered with mineral oil, purged with argon, and stored at −70°. If a higher concentration of $H_4PteGlu_n$ is required the solution can be lyophilized in the dark to reduce the volume. We have found it best not to lyophilize to complete dryness.

Synthesis of Labeled One-Carbon Derivatives of $H_4PteGlu_n$

The C-11 position of 5,10-CH^+-$H_4PteGlu_n$ can be labeled with either ^{13}C or ^{14}C by cycling the 5,10-CH^+-$H_4PteGlu_n$ with the reactions shown in Scheme II. The following reagents are added at a final concentration to a solution containing 10 nmol of 5,10-$CH^+H_4PteGlu_n$ in 1–2 ml of KBES (pH 7.5)–20 mM 2-mercaptoethanol: 0.2 mM α-ketoglutarate; 50 mM glycine; 0.3 mM MgATP; 1 mg of SHMT; 1 mg of C_1-THF synthase; and 15 nmol of NADPH. This solution is incubated at 30° for 20 min, which results in the conversion of the 5,10-CH^+-$H_4PteGlu_n$ to $H_4PteGlu_n$. One milligram of glutamate dehydrogenase is then added to remove the excess NADPH. Labeled formate is added in excess (15 nmol), which results in conversion of the $H_4PteGlu_n$ to 10-CHO-$H_4PteGlu_n$. After 20 min at 30° the solution is made 100 mM in KP_i with a stock solution of 1 M KP_i, pH 6.8. The solution is flushed with argon, capped, and incubated at 90° for 3–5 hr to convert the 10-CHO-$H_4PteGlu_n$ to 5-CHO-$H_4PteGlu_n$. The 5-CHO-$H_4PteGlu_n$ is then purified as described above. The labeled one-carbon group has the same specific activity as the labeled formate. This labeled compound can be converted to labeled 5,10-CH^+-$H_4PteGlu_n$, 10-CHO-$H_4PteGlu_n$, and 5-CH_3-$H_4PteGlu_n$ as described above.

[11] Chemical Synthesis of (6S)-5-Formyl-5,6,7,8-tetrahydropteroylpoly-γ-L-glutamates

By ANTHONY L. FITZHUGH and RHONE K. AKEE

Introduction

The active folate content of most cells is stored in the form of 5- and/or 10-substituted 5,6,7,8-tetrahydropteroyldi- to octa-γ-L-glutamate derivatives, which participate as coenzymes in the biosynthesis of certain precursors of RNA and DNA.[1] As a consequence of this important role, the binding affinity of these derivatives to folate-dependent enzymes has been extensively investigated. The pattern emerging from these studies is that up to a point, binding affinity is closely tied to the degree of polyglutamylation. For example, 5- and/or 10-substituted 5,6,7,8-tetrahydropteroyltetra- to hexa-γ-L-glutamyl derivatives routinely exhibit binding affinities that are several orders of magnitude lower than those of their mono-L-glutamyl homologs. However, beyond the hexa-γ-L-glutamyl stage of polyglutamylation only marginal gains in binding affinity are observed. Nevertheless, because of several problems associated with the synthesis and storage of poly-γ-L-glutamyl derivatives,[2,3] the far less active but more readily accessible mono-L-glutamyl forms of these coenzymes continue to be widely used in biological assays of folate-dependent enzymes.

In contrast to these methods, we have developed a wholly chemical approach to a series of (6*S*)-5-formyl-5,6,7,8-tetrahydropteroylpoly-γ-L-glutamate derivatives.[4] It consists of a simple, four-step sequence of reactions. The new method is presented in detail below.

Principle

The overall approach to the synthesis of the (6*S*)-5-formyl-5,6,7,8-tetrahydropteroylpoly-γ-L-glutamates is shown in Fig. 1. In step i (Fig. 1), the

[1] D. J. Cichowicz, S. K. Foo, and B. Shane, *Mol. Cell. Biochem.* **39,** 209 (1981); R. L. Kisliuk, *Mol. Cell. Biochem.* **39,** 331 (1981); J. E. Ayling, M. Gopal Nair, and C. M. Baugh (eds.), "Advances in Experimental Medicine and Biology: Chemistry and Biology of Pteridines and Folates," Vol. 338, pp. 629–675. Plenum, New York, 1993.

[2] R. L. Blakely, *in* "Frontiers of Biology: The Biochemistry of Folic Acid and Related Pteridines," Vol. 13, p. 58. American Elsevier, New York, 1969.

[3] L. D'Ari and J. C. Rabinowitz, *Methods Enzymol.* **113,** 169 (1985).

[4] A. L. Fitzhugh, G. N. Chmurny, and J. R. Klose, *Bioorg. Med. Chem. Lett.* **1,** 155 (1991); A. L. Fitzhugh, R. K. Akee, F. C. Ruei, J. Wu, J. R. Klose, and B. A. Chabner, *J. Chem. Soc. Perkin Trans.* **1,** 897 (1994).

0076-6879/97 $25

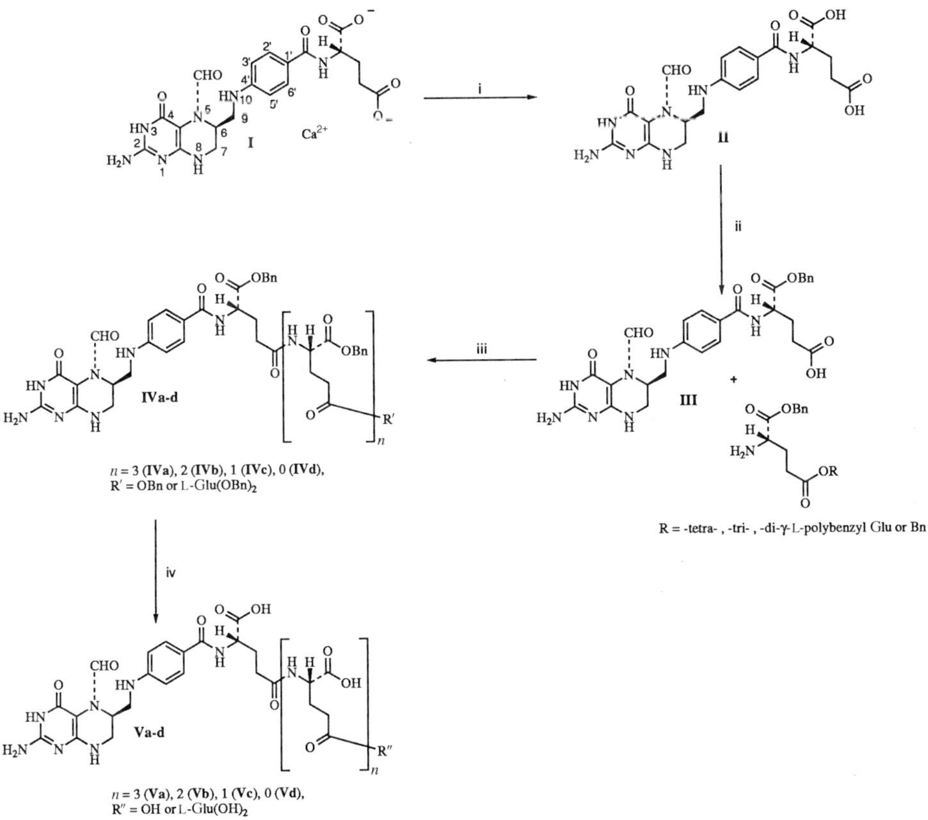

FIG. 1. *Conditions:* Step i, pH adjusted to pH 4.5 with 88% (v/v) formic acid; step ii, benzyl bromide (BnBr), $NaCO_3$ in dimethyl sulfoxide (DMSO); step iii, triethylamine (Et_3N), diphenylphosphoroazide $(PhO_2)P(=O)N_3$, and polybenzyl esters of mono-, di-, tri-, and tetra-γ-L-glutamic acid in dimethylformamide (DMF); step iv, H_2, palladium black in DMF–water (8:2, v/v).

calcium salt of (6*S*)-5-formyl-5,6,7,8-tetrahydropteroylmono-L-glutamate (**I**) is converted to its free acid form (**II**) with 88% formic acid. In step ii (Fig. 1) the (6*S*)-5-formyl-5,6,7,8-tetrahydropteroylmono-L-glutamic acid (**II**) is preferentially converted to the α-benzyl monoester derivative (**III**) in the presence of 1.2 mol equivalents of sodium carbonate and benzyl bromide. In step iii (Fig. 1) the α-benzyl monoester derivative of (6*S*)-5-formyl-5,6,7,8-tetrahydropteroylmono-L-glutamic acid (**III**) is coupled in four separate reactions to preformed mono-, di-, tri-, and tetra-γ-L-glutamate polybenzyl esters in the presence of diphenylphosphorylazide to give the (6*S*)-5-formyl-5,6,7,8-tetrahydropteroylpenta-, -tetra-, -tri-, and -di-γ-

L-glutamate polybenzyl ester derivatives (**IVa–d**), respectively. In step iv (Fig. 1), the benzyl ester protecting groups of **IVa–d** are removed in four separate reactions via catalytic hydrogenation with palladium black to give the desired (6*S*)-5-formyl-5,6,7,8-tetrahydropteroylpenta-, -tetra-, -tri-, and -di-γ-L-glutamic acid derivatives (**Va–d**).

Materials

The following materials are obtained from the commercial sources listed.

High-performance liquid chromatography (HPLC)-grade water, methanol, ethanol, 2-propanol, chloroform, and *N*,*N*-dimethylformamide (DMF) are procured from EM Science (Gibbstown, NJ). Cellulose F and silica gel 60 $WF_{254}S$ TLC plates (0.1 mm, 20 × 20 cm) are supplied by EM Separations (Gibbstown, NJ)

Formic acid (88%, v/v) and glacial acetic acid are purchased from Fisher Scientific (Pittsburgh, PA)

Absolute methyl sulfoxide and *N*,*N*-dimethylformamide over molecular sieves, sodium carbonate, benzyl bromide, triethylamine, diphenylphosphorylazide, palladium black, and silica gel 60 are all obtained from Fluka (Ronkonkoma, NY)

L-Glutamic acid dibenzyl ester *p*-tosylate is purchased from Schweizerhall, Inc. (South Plainfield, NJ)

The C_{18} column (4.6 mm i.d. × 250 mm, Partisil 10 ODS) used in HPLC analyses is purchased from Whatman (Clifton, NJ)

Other materials are prepared by published procedures or are provided as gifts:

The di-, tri-, and tetra-γ-L-glutamate polybenzyl esters are prepared as hydrogen chloride or *p*-tosylate salts according to the procedure of Piper *et al.*[5]

The calcium salt of (6*S*)-5-formyl-5,6,7,8-tetrahydropteroylmono-L-glutamic acid (**I**, Fig. 1) is the kind gift of F. Marraza (SAPEC S.A. Fine Chemicals, Lugano, Switzerland)

Methods

Melting points are measured on a Fisher–Johns melting point apparatus and are uncorrected. The HPLC analyses are performed on a high-performance liquid chromatographic system (Hewlett-Packard, Avondale, PA). The components of the HPLC system include a model 1050 pumping system, a 1050 ultraviolet–visible (UV–Vis) variable wavelength detector, and a

[5] J. R. Piper, G. S. McCaleb, and J. A. Montgomery, *J. Med. Chem.* **26,** 291 (1983).

3396A integrator. ^{1}H nuclear magnetic resonance (NMR) spectra are obtained on a Varian (Palo Alto, CA) VXR-500S spectrometer with a SUN 4/110 workstation. Values for J are given in hertz (Hz). The NMR assignments of the partial structures for the protons at positions 6, 7, 8, 9, and 10 are based on chemical-shift comparisons of this area to that in Ref. 6. These proton assignments are obtained from the following two-dimensional (2D) heteronuclear multiple bond coherence (HMBC) correlations: 7-H^b to C-6, C-8, and C-9; 6-H to 5-CHO and C-9; and 9-H^b to C-4′, C-6, and C-7. Mass spectra are obtained for all of the (6S)-5-formyl-5,6,7,8-tetrahydropteroylmono- and -poly-γ-L-glutamate polyester derivatives (except the penta-γ-L-glutamate hexabenzyl ester derivative). By contrast, none of the (6S)-5-formyl 5,6,7,8-tetrahydropteroylpoly-γ-L-glutamic acid derivatives give such spectra. Presumably, the analysis of these derivatives is prevented by their low volatility.

Synthesis and Purification

*(6S)-Formyl-5,6,7,8-tetrahydropteroylmono-L-glutamic Acid (**II**).* A total of 5.0 g (8.3 mmol) of the calcium salt of (6S)-5-formyltetrahydropteroylmono-L-glutamic acid (**I**) is added to 50 ml of water and stirred until the compound is completely dissolved. The solution is chilled to 0–5° in an ice–water bath and 88% (v/v) formic acid added dropwise until the pH measures pH ~4.5 whereupon a dense, off-white precipitate rapidly forms. The precipitate is centrifuged (20 min at 8000 rpm, 4°) and the supernatant decanted. The pellet is subjected to two more cycles of washing with ice-cold water (10 ml, pH adjusted to pH ~4.5 with formic acid)/centrifugation/decantation. The moist precipitate is lyophilized overnight (16 hr at 25°, 0.1 mmHg) to give 2.6 g (55%) of **II**, melting point (m.p.) > 300°.

*α-Benzyl Ester of (6S)-5-Formyl-5,6,7,8-tetrahydropteroylmono-L-glutamic Acid (**III**).* A total of 500 mg (0.76 mmol) of **II** is placed in a 25-ml round-bottom flask fitted with a drying adapter. The compound is then dried *in vacuo* (0.1 mmHg) in two stages; first for 1 hr at 105° in an oil bath followed by 30 min at room temperature. Next, the drying adapter is closed and the flask transferred to an N_2 glove box where 56 mg of Na_2CO_3 (0.53 mmol) and 15 ml of anhydrous dimethyl sulfoxide (DMSO) are added. The mixture is stirred until the solution turns clear and 63 μl of benzyl bromide (0.53 mmol) is introduced into the flask. The drying adapter is reattached, closed, and the flask removed from the N_2 glove box. Next, an N_2 bubbler is attached to the drying adapter and the reaction stirred under N_2 at room temperature for 24 hr.

[6] M. Poe, *Methods Enzymol.* **66,** 483 (1980).

Work-up and purification proceed as follows: Excess DMSO is removed on a rotary evaporator (5 hr at 60°, 0.1 mmHg) and the resulting crude product purified by flash chromatographic techniques on a silica gel column {30 g of silica gel 60, eluted isocratically with (7:1, v/v) $CHCl_3$–CH_3OH [containing 3% (v/v) acetic acid]}. Fractions (50 ml) are collected and analyzed by thin-layer chromatography (TLC) {5 × 10 cm silica gel $WF_{254}S$ plates, (2:1, v/v) $CHCl_3$–CH_3OH [containing 3% (v/v) acetic acid]}. Product-containing fractions (R_f of 0.49) are pooled and excess solvent removed on a rotary evaporator (40 min at 40°, 5 mmHg); followed by drying *in vacuo* (12 hr at 40°, 0.1 mmHg) to give 392 mg (66%) of **III**, m.p. > 300° (found: C, 54.5; H, 5.15; N, 15.9. $C_{27}H_{29}N_7O_7$ + 0.85 CH_3CO_2H + 0.85 H_2O requires C, 54.72; H, 5.45, N, 15.56%); ν_{max}(KBr)/cm^{-1} 3340 (NH), 1727 (ester CO), 1620 (amide CO), 1325 (CH), 1188 (amide CO) and 765 (aryl CH); δ_H-([$^2H_6$]DMSO; 500 MHz) 1.96 (1 H, cm, β-H), 2.01 (1 H, cm, β-H), 2.19 (2 H, t, *J* 6.5, γ-H), 2.94 (1 H, ddd, *J* 5.0, 7.3, and 13.5, 9-H^b), 3.07 (1 H, ddd, *J* 6.9, 7.9, and 13.5, 9-H^a), 3.13 (1 H, dd, *J* 4.4 and 12.3, 7-H^b), 3.38 (1 H, dd, *J* 4.9 and 12.2, 7-H^a), 4.35 (1 H, cm, α-H), 4.78 (1 H, ddd, *J* 4.9, 7.3, and 7.9, 6-H), 5.11 (1 H, AB, *J* 12.6, OCH_2), 5.13 (1 H, AB, *J* 12.7, OCH_2), 6.29 (1 H, dd, *J* 5.0 and 6.9, 10-NH), 6.53 (2H, d, *J* 8.7, 3′- and 5′-H), 6.59 (2H, br s, 2-NH_2), 6.93 (1 H, d, *J* 4.4, 8-H), 7.28–7.38 (5 H, cm, ArH), 7.28–7.38 (1 H, obs. d, 8-NH), 7.62 (2 H, *J* 8.5, 2′- and 6′-H), 8.81 (1 H, s, 5-CHO), 9.31 (1 H, br s, 3-NH), and 11.12 (1 H, br s, CO_2H); *m/z* 564 (MH^+, 9%) and 185 (100%).

*General Procedure for Preparation of Polyesters **IVa–d***

*Hexabenzyl Ester of (6S)-5-Formyl-5,6,7,8-tetrahydropteroylpenta-γ-L-glutamic Acid (**IVa**).* A total of 200 mg (0.35 mmol) of **III** is placed in a 50-ml round-bottom flask and fitted with a drying adapter. The compound is then dried *in vacuo* (1 hr at 50°, 0.1 mmHg) and transferred to an N_2 glove box. Twenty milliliters of anhydrous DMF, 680 mg (0.85 mmol) of the tetrabenzyl ester of tetra-γ-L-glutamate hydrochloride salt, and 148 μl (1.06 mmol) of triethylamine are added next. A 10-ml graduated addition funnel containing 192 μl (0.89 mmol) of diphenylphosphorylazide in 2 ml of anhydrous DMF and a closed drying adapter are fitted to the flask. Following removal of this apparatus from the N_2 glove box, the drying adapter is opened and attached to an N_2 bubbler. The flask is chilled to 0–5° in an ice–water bath and the diphenylphosphorylazide solution added dropwise over a 15-min period. After stirring for 1 hr at this temperature, the flask is removed from the ice–water bath and allowed to warm gradually to room temperature. The mixture is stirred under N_2 for an additional 16 hr.

Workup and product purification proceed as follows: Excess DMF is removed on a rotary evaporator (3 hr at 40°, 5 mmHg). The remaining oily residue is treated with 15 ml of ice-cold water whereupon a precipitate forms. The precipitate is centrifuged (20 min at 8000 rpm, 4°) and the supernatant decanted. After two more cycles of washing with ice-cold water (10 ml)/centrifugation/decantation, the pellet is freeze-dried overnight (16 hr at 25°, 0.1 mmHg). The resulting crude material is then purified by flash chromatographic techniques on a silica gel column [20 g of silica gel 60, eluted isocratically with (9:1, v/v) $CHCl_3$–CH_3OH]. Fractions (20 ml) are collected and analyzed by TLC [on 5 × 10 cm silica gel 60 $WF_{254}S$ plates, (6:1, v/v) $CHCl_3$–CH_3OH; R_f values for the di, tri, tetra, and penta derivatives are 0.50, 0.52, 0.59, and 0.64, respectively]. Product-containing fractions are pooled and excess solvent removed on a rotary evaporator (40 min at 40°, 5 mmHg). This is followed by drying *in vacuo* (12 hr at 40°, 0.1 mmHg) to give 464 mg (99%) of **IVa**, m.p. 170–172° (found: C, 62.7; H, 5.8; N, 9.15. $C_{82}H_{87}N_{11}O_{19}$ + 3.5 CH_3OH + 0.25 H_2O requires C, 62.47; H, 6.01; N, 9.37%); ν_{max}(KBr)/cm^{-1} 3300 (NH), 1728 (ester CO), 1645 (amide CO), 1335 (CH), 1182 (amide CO) and 764 (aryl CH); δ_H-([$^2H_6$]DMSO; 500 MHz) 1.70–2.60 (20 H, cm, CH_2), 2.84–2.90 (1 H, cm, 9-H^b), 3.00–3.20 (2 H, cm, 9-H^a and 7-H^b), 3.40 (1 H, cm, 7-H^a), 4.29 (4 H, cm, α-H), 4.46 (1 H, cm, α-H), 4.80 (1 H, cm, 6-H), 5.09 (12 H, cm, OCH_2), 6.21 (2 H, br s, 2-NH_2), 6.37 (1 H, X of ABX, 10-NH), 6.61 (2 H, d, *J* 8.3, 3′- and 5′-H), 7.00 (1 H, X of ABX, 8-NH), 7.32 (30 H, cm, ArH), 7.69 (2 H, d, *J* 8.5, 2′- and 6′-H), 8.25 (1 H, d, *J* 7.3, NHCO), 8.25 (1 H, d, *J* 7.3, NHCO), 8.32 (1 H, d, *J* 7.4, NHCO), 8.35 (1 H, d, *J* 7.8, NHCO), 8.40 (1 H, d, *J* 7.1, NHCO), 8.85 (1 H, s, 5-CHO), and 10.24 (1 H, br s, 3-NH); *m/z* not obtainable.

*Pentabenzyl Ester of (6S)-5-Formyl-5,6,7,8-tetrahydropteroyltetra-γ-L-glutamic Acid (**IVb**).* Yield, 89%; m.p. 151.5–152.5° (found: C, 62.5; H, 6.0; N, 10.2. $C_{70}H_{74}N_{10}O_{16}$ + 1.0 CH_3OH + 1.0 H_2O requires C, 62.64; H, 5.92; N, 10.29%); ν_{max}(KBr)/cm^{-1} 3300 (NH), 1730 (ester CO), 1640 (amide CO), 1326 (CH), 1190 (amide CO) and 751 (aryl CH); δ_H-([$^2H_6$]DMSO; 500 MHz) 1.70–2.50 (16 H, cm, CH_2), 2.87 (1 H, td, *J* 5.5 and 13.5, 9-H^b), 3.06 (1 H, ddd, *J* 6.7, 8.6, and 13.5, 9-H^a), 3.13 (1 H, dd, *J* 4.2 and 12.6, 7-H^b), 3.40 (1 H, dd, *J* 5.0 and 12.2, 7-H^a), 4.23–4.34 (3 H, 2 cm, α-H), 4.46 (1 H, ddd, *J* 4.6, 7.6, and 10.1, α-H), 4.80 (1 H, cm, 6-H), 5.06 (2 H, AB, *J* 12.8, OCH_2), 5.08 (3 H, AB, *J* 12.7, OCH_2), 5.10 (3H, AB, *J* 12.7, OCH_2), 5.12 (2 H, AB, *J* 12.7, OCH_2), 6.12 (2 H, brs, 2-NH_2), 6.36 (1 H, X of ABX, 10-NH), 6.61 (2 H, d, *J* 8.9, 3′- and 5′-H), 6.98 (1 H, X of ABX, 8-NH), 7.27–7.37 (25 H cm, ArH), 7.70 (2 H, d, *J* 8.8, 2′- and 6′-H), 8.25 (1 H, d, *J* 7.4, NHCO), 8.28 (1 H, d, *J* 7.5, NHCO), 8.28 (1 H, d, *J* 7.5, NHCO), 8.34 (1 H, d, *J* 7.5, NHCO), 8.39 (1 H, d, *J* 7.3, NHCO), 8.85 (1 H, s, 5-CHO), and 10.26 (1 H, s, 3-NH); *m/z* 1312 (MH^+, 14%) and 327 (100%).

*Tetrabenzyl Ester of (6S)-5-Formyl-5,6,7,8-tetrahydropteroyltri-γ-L-glutamic Acid (**IVc**).* Ethanol is substituted for methanol during the purification; yield, 87%; m.p. 105–108° (found: C, 62.7; H, 5.9; N, 10.5. $C_{58}H_{61}N_9O_{13}$ + 2.0 C_2H_5OH requires C, 62.73; H, 6.21; N, 10.64%); ν_{max}(KBr)/cm^{-1} 3325 (NH), 1734 (ester CO), 1635 (amide CO), 1323 (CH), 1185 (amide CO), and 760 (aryl CH); δ_H-([$^2H_6$]DMSO; 500 MHz) 1.70–2.50 (12 H, cm, CH_2), 2.86 (1 H, td, *J* 5.4 and 13.3, 9-H^b), 3.06 (1 H, ddd, *J* 6.5, 8.2, and 13.3, 9-H^a), 3.12 (1 H, dd, *J* 4.1 and 12.6, 7-H^b), 3.40 (1 H, dd, *J* 4.8 and 12.6, 7-H^a), 4.27 (1 H, ddd, *J* 5.2, 7.5, and 9.4, α-H), 4.31 (1 H, ddd, *J* 5.4, 7.6, and 9.0, α-H), 4.45 (1 H, ddd, *J* 4.6, 7.4, and 10.1, α-H), 4.80 (1 H, cm, 6-H), 5.06 (2 H, AB, *J* 12–13, OCH_2), 5.09 (2 H, AB, *J* 12–13, OCH_2), 5.09 (2 H, AB, *J* 12–13, OCH_2), 5.12 (2 H, AB, *J* 12–13, OCH_2), 6.24 (2 H, br s, 2-NH_2), 6.36 (1 H, X of ABX, 10-NH), 6.61 (2 H, d, *J* 8.7, 3′- and 5′-H), 6.97 (1 H, X or ABX, 8-NH), 7.33 (20 H, cm, ArH), 7.69 (2 H, d, *J* 8.8, 2′- and 6′-H), 8.29 (1 H, d, *J* 7.4, NHCO), 8.30 (1 H, d, *J* 7.4, NHCO), 8.39 (1 H, d, *J* 7.6, NHCO), 8.85 (1 H, s, 5-CHO), and 10.29 (1 H, br s, 3-NH); *m/z* 1093 (MH^+, 20%) and 233 (100%).

*Tribenzyl Ester of (6S)-5-Formyl-5,6,7,8-tetrahydropteroyldi-γ-L-glutamic Acid (**IVd**).* Ethanol is substituted for methanol during the purification; yield, 88%; m.p. 105–108° (found: C, 62.5; H, 6.15; N, 11.6. $C_{46}H_{48}N_8O_{10}$ + 2.0 C_2H_5OH requires C, 62.23; H, 6.27; N, 11.61%); ν_{max}(KBr)/cm^{-1} 3325 (NH), 1730 (ester CO), 1630 (amide CO), 1323 (CH), 1185 (amide CO), and 765 aryl (CH); δ_H-([$^2H_6$]DMSO; 500 MHz) 1.84 (cm, 1 H, CH_2), 1.96 (1 H, cm, CH_2), 2.00 (1 H, cm, CH_2), 2.10 (1 H, cm, CH_2), 2.28 (2 H, cm, CH_2), 2.42 (2 H, cm, CH_2), 2.86 (1 H, td, *J* 5.4 and 13.3, 9-H^b), 3.06 (1 H, ddd, *J* 6.7, 8.8, and 13.3, 9-H^a), 3.13 (1 H, dd, *J* 4.2 and 12.6, 7-H^b), 3.40 (1 H, dd, *J* 5.2 and 12.3, 7-H^a), 4.32 (1 H, ddd, *J* 5.3, 7.4, and 9.1, α-H), 4.43 (1 H, ddd, *J* 4.8, 7.4, and 10.0, α-H), 4.80 (1 H, cm, 6-H), 5.07 (1 H, AB, *J* 12.9, OCH_2), 5.07 (1 H, AB, *J* 12.9, OCH_2), 5.09 (1 H, AB, *J* 12.6, OCH_2), 5.10 (1 H, AB, *J* 12.7, OCH_2), 5.13 (1 H, AB, *J* 12.8, OCH_2), 5.13 (1 H, AB, *J* 12.8, OCH_2), 6.23 (2 H, brs, 2-NH_2), 6.35 (1 H, X of ABX, 10-NH), 6.60 (2 H, d, *J* 8.8, 3′- and 5′-H), 6.97 (1 H, X of ABX, 8-NH), 7.33 (15 H, cm, ArH), 7.68 (2 H, d, *J* 8.8, 2′- and 6′-H), 8.31 (1 H, d, *J* 7.4, NHCO), 8.36 (1 H, d, *J* 7.5, NHCO), 8.85 (1 H, s, 5-CHO), and 10.28 (1 H, br s, 3-NH); *m/z* 873 (MH^+, 14%) and 119 (100%).

*General Procedure for Preparation of Polyacids **Va–d***

*(6S)-5-Formyl-5,6,7,8-tetrahydropteroylpenta-γ-L-glutamic Acid (**Va**).* A total of 186 mg (0.14 mmol) of **IVa**, 200 mg of palladium black, and 10 ml of a DMF–water (8:2, v/v) mixture are added together in a 25-ml round-bottom flask. The suspension is carefully degassed and a latex balloon

containing H_2 is attached to the reaction flask. The mixture is then stirred for 20 hr at room temperature.

Workup and purification proceed as follows: Excess solvent is removed by rotary evaporation (6 hr at 30°, 0.5 mmHg) and the remaining residue is purified chromatographically on a cellulose column [40 g of cellulose, eluted isocratically with (1 : 1, v/v) 0.5 *M* aqueous NH_4HCO_3–2-propanol]. Fractions (20 ml) are collected and analyzed by HPLC on a C_{18} column [(15 : 85, v/v) 2-propanol–0.005 *M* tetrabutylammonium phosphate, detection at 280 nm, and a flow rate of 1 ml/min, t_R 31.27 min]. Product-containing fractions were pooled and excess solvent removed on a rotary evaporator (40 min at 40°, 5 mmHg) until the final volume is ~20 ml. The suspension is then lyophilized overnight (16 hr at 25°, 0.1 mmHg) and then dried *in vacuo* (3 hr at 50°, 0.1 mmHg) to give 122 mg (75%) of **Va**. m.p. 275–285° (decomposition, progressive darkening > 155°) (Found: C, 41.9; H, 6.2; N, 17.3. $C_{40}H_{48}N_{11}O_{19}$ + $6H_2O$ + $3NH_4^+$ requires C, 41.81; H, 6.32; N, 17.07%); ν_{max}(KBr)/cm^{-1} 3366 (NH, OH), 1636 (amide CO), 1331 (CH), 1188 (amide CO), and 770 (aryl CH); δ_H-([$^2H_6$]DMSO; 500 MHz) 1.66–1.82 (5 H, cm, CH_2), 1.85–2.02 (4 H, cm, CH_2), 2.02–2.28 (11 H, cm, CH_2), 2.91 (1 H, t_{AB}, *J* 5.8 and 13.2 9-H^b), 3.07 (1 H, dd, *J* 6.6, 7.6, and 13.2, 9-H^a), 3.13 (1 H, d_{AB}, *J* 3.9 and 12.4, 7-H^b), 3.39 (1 H, d_{AB}, *J* 4.6 and 12.1, 7-H^a), 4.04 (4 H, cm, α-H), 4.19 (1 H, dt, *J* 5.1 and 7.3, α-H), 4.79 (1 H, dt, *J* 3.7 and 7.1, 6-H), 6.23 (1 H, X of ABX, 10-NH), 6.56 (2 H, d, *J* 8.7, 3′- and 5′-H), 6.62 (2 H, br s, 2-NH_2), 6.99 (1 H, X of ABX, 8-NH), 7.62 (2 H, d, *J* 8.6, 2′- and 6′-H), 7.81 (1 H, d, *J* 7.4, NHCO), 7.83 (1 H, d, *J* 7.4, NHCO), 7.84 (1 H, d, *J* 7.4, NHCO), 7.85 (1 H, d, *J* 7.4, NHCO), 7.92 (1 H, d, *J* 6.7, NHCO), 8.20 (1 H, vbr s, 3-NH), and 8.79 (1 H, s, 5-CHO); HPLC, single peak at t_R 31.27 min.

*(6S)-5-Formyl-5,6,7,8-tetrahydropteroyltetra-γ-L-glutamic Acid (**Vb**).* Yield, 70%, m.p. 275–285° (decomposition, progressive darkening > 160°) (Found: C, 43.7; H, 5.6; N, 16.2. $C_{35}H_{43}N_{10}O_{16}$ + $4.5H_2O$ + $1.0NH_4^+$ requires C, 43.98; H, 5.59; N, 16.12%); ν_{max}(KBr)/cm^{-1} 3280 (NH, OH), 1630 (amide CO), 1330 (CH), 1190 (amide CO), and 770 (aryl CH); δ_H-([$^2H_6$]DMSO; 500 MHz) 1.72 (2 H, cm, CH_2), 1.78 (2 H, cm, CH_2), 1.90 (2 H, cm, CH_2), 1.96 (2 H, cm, CH_2), 2.12 (4 H, cm, CH_2), 2.20 (4 H, cm, CH_2), 2.92 (1 H, t_{AB}, *J* 6.1 and 13.5, 9-H^b), 3.06 (1 H, t, *J* 6.8 and 14.1, 9-H^a), 3.13 (1 H, d_{AB}, *J* 4.1 and 12.4, 7-H^b), 3.39 (1 H, d_{AB}, *J* 4.9 and 12.3, 7-H^a), 4.04 (2 H, dt, *J* 4.7 and 8.1, α-H), 4.09 (1 H, q, *J* 7.3, α-H), 4.21 (1 H, dt, *J* 5.0 and 7.8, α-H), 4.79 (1 H, dt, *J* 3.9 and 7.2, 6-H), 6.21 (1 H, X of ABX, 10-NH), 6.56 (2 H, d, *J* 8.8, 3′- and 5′-H), 6.56 (2 H, s, 2-NH_2), 6.95 (1 H, X of ABX, 8-NH), 7.63 (2 H, d, *J* 8.7, 2′- and 6′-H), 7.82 (2 H, d, *J* 7.6, NHCO), 7.85 (1 H, d, *J* 7.7, NHCO), 7.97 (1 H, d, *J* 7.1, NHCO), and 8.80 (1 H, s, 5-CHO); HPLC, single peak at t_R 21.11 min.

(6S)-5-Formyl-5,6,7,8-tetrahydropteroyltri-γ-L-glutamic Acid (**Vc**). Yield, 65%, m.p. 250–260° (decomposition, progressive darkening > 175°) (Found: C, 42.35; H, 5.9; N, 16.0. Calculated for $C_{30}H_{37}N_9O_{13}$ + $6H_2O$ + $0.5NH_4^+$ requires C, 42.45; H, 6.06; N, 15.68%); ν_{max}(KBr)/cm^{-1} 3260 (NH, OH), 1620 (amide CO), 1325 (CH), 1190 (amide CO), and 770 (aryl CH); δ_H-([$^2H_6$]DMSO; 500 MHz) 1.60–2.30 (12 H, cm, CH_2), 2.92 (1 H, td, *J* 6.0 and 13.4, 9-H^b), 3.07 (1 H, td, *J* 6.9 and 13.4, 9-H^a), 3.13 (1 H, dd, *J* 4.2 and 12.6, 7-H^b), 3.39 (1 H, dd, *J* 5.0 and 12.6, 7-H^a), 4.02 (1 H, dt, *J* 4.9 and 8.1, α-H), 4.07 (1 H, q, *J* 7.3, α-H), 4.15 (1 H, dt, *J* 5.6 and 6.6, α-H), 4.79 (1 H, dt, *J* 4.2 and 6.9, 6-H), 6.19 (1 H, X of ABX, 10-NH), 6.51 (2 H, brs, 2-NH_2), 6.56 (2 H, d, *J* 8.6, 3′- and 5′-H), 6.94 (1 H, X of ABX, 8-NH), 7.60 (2 H, d, *J* 8.5, 2′- and 6′-H), 7.72 (1 H, brd, *J* 6.3, NHCO), 7.78 (1 H, cm, NHCO), 7.89 (1 H, cm, NHCO), and 8.80 (1 H, s, 5-CHO); HPLC, single peak at t_R 14.95.

(6S)-5-Formyl-5,6,7,8-tetrahydropteroyldi-γ-L-glutamic Acid (**Vd**). Yield, 77%, m.p. 250–260° (decomposition, progressive darkening > 180°) (Found: C, 44.6; H, 5.4; N, 17.3. $C_{25}H_{30}N_8O_{10}$ + $3.50H_2O$ + $0.5NH_4^+$ requires C, 44.8; H, 5.71; N, 17.24%); ν_{max}(KBr)/cm^{-1} 3230 (NH, OH), 1620 (amide CO), 1320 (CH), 1185 (amide CO), and 800 (aryl CH); δ_H-([$^2H_6$]DMSO; 500 MHz) 1.70 (1 H, cm, CH_2), 1.80 (1 H, cm, CH_2), 1.90 (1 H, cm, CH_2), 2.00 (1 H, cm, CH_2), 2.20 (3 H, cm, CH_2), 2.27 (1 H, cm, CH_2), 2.88 (1 H, td, *J* 6.5 and 13.6, 9-H^b), 3.07 (1 H, ddd, *J* 6.9, 8.1, and 13.6, 9-H^a), 3.13 (1 H, dd, *J* 4.2 and 12.7, 7-H^b), 3.40 (1 H, dd, *J* 4.9 and 12.5, 7-H^a), 4.14 (1 H, q, *J* 7.6, α-H), 4.19 (1 H, dt, *J* 4.8 and 7.6, α-H), 4.79 (1 H, cm, 6-H), 6.28 (2 H, br s, 2-NH_2), 6.30 (1 H, X of ABX, 10-NH), 6.59 (2 H, d, *J* 8.8, 3′- and 5′-H), 6.96 (1 H, X of ABX, 8-NH), 7.63 (2 H, d, *J* 8.7, 2′- and 6′-H), 7.74 (1 H, d, *J* 7.2, NHCO), 8.02 (1 H, d, *J* 6.3, NHCO), and 8.83 (1 H, s, 5-CHO); HPLC, single peak at t_R 10.08 min.

Acknowledgments

This project has been funded in part with federal funds from the Department of Health and Human Services. The content of this publication does not necessarily reflect the views of the Department of Health and Human Services, nor does mention of trade names, commercial products, or organizations imply endorsement by the U.S. government. Portions of this chapter are reproduced from A. L. Fitzhugh *et al., J. Chem. Soc., Perkin Trans.* **1,** 897 (1994), by permission of The Royal Society of Chemistry, Cambridge, UK.

[12] Biotin Derivatives of Folate Compounds: Synthesis and Utilization for Visualization and Affinity Purification of Folate Transport Proteins

By JIANGUO FAN and FRANK M. HUENNEKENS

Eukaryotic cells (and some prokaryotes), unable to synthesize the folate structure *de novo,* depend on transport systems for uptake of folate compounds from the environment. L1210 mouse leukemia cells, which have provided a well-studied paradigm for folate transport, can express two separate systems to mediate this process.[1] The reduced folate/methotrexate (MTX) carrier (also referred to as the micromolar folate transporter) is an integral membrane protein whose primary substrates are 5-methyl- and 5-formyltetrahydrofolate (K_t values, ca. 1 and 5 μM); the antifolate MTX is also a good substrate for this transporter. The folate receptor (also referred to as folate-binding protein or nanomolar folate transporter) is an extracellular protein bound to the membrane (via a glycosylphosphatidylinositol moiety) whose primary substrate is folate (K_t, ca. 1 nM).

Attempts to purify folate transporters have been hampered by the relatively small amounts of these proteins in membranes. Parental L1210 cells, for example, express the micromolar folate transporter at only 0.1 pmol (4×10^{-3} μg)/10^6 cells and have no detectable nanomolar transporter. Although levels of these transporters can be enhanced by growth of cells on low concentrations of folates,[2] by cell selection procedures,[3] or by introduction of transporter cDNAs into cells,[4] purification by multistep, conventional procedures of protein fractionation results in low yields. To overcome this problem, a single-step, affinity purification procedure, utilizing biotin derivatives of MTX and folate, has been devised to isolate both folate transporters in homogeneous form and in high yield from L1210 cells. These biotin–folate compounds are also useful probes for visualizing the folate transporters in individual cells.

[1] J. Fan, K. S. Vitols, and F. M. Huennekens, *Adv. Enzyme Regul.* **32,** 3 (1992).
[2] M. A. Kane, P. C. Elwood, R. M. Portillo, A. C. Antony, V. Najfeld, A. Finley, S. Waxman, and J. F. Kolhouse, *J. Clin. Invest.* **81,** 1398 (1988).
[3] F. M. Sirotnak, D. M. Moccio, and C.-H. Yang, *J. Biol. Chem.* **259,** 13139 (1984).
[4] K. H. Dixon, B. C. Lanpher, J. Chiu, K. Kelley, and K. H. Cowan, *J. Biol. Chem.* **269,** 17 (1993).

0076-6879/97 $25

Principle

Fluorescein derivatives of MTX and folate have been synthesized by linking the γ-carboxyl group of the latter compounds to the fluorophore via a diaminopentane (DAP) spacer. These probes, after activation of the open α-carboxyl by *N*-hydroxysuccinimide (NHS), were used to label covalently the folate transport proteins.[5] In the present procedures, the folate compounds are derivatized with biotin, and the two functional units are joined either with a stable linking group (DAP) or a group that can be cleaved by reduction [DAP plus -CO-$(CH_2)_2$-S-S-$(CH_2)_2$-NH-]. After activation with *N*-hydroxysulfosuccinimide (NHSS), these probes are attached covalently to the transporters *in situ*. Membrane proteins are extracted with detergent, and the labeled transporters are purified from the mixture by adsorption and elution from streptavidin beads. Other procedures allow the labeled transporters to be visualized on individual cells by light, fluorescence, or electron microscopy.

Reagents

Methotrexate (MTX; Lederle, Pearl River, NY)
Folic acid (Sigma, St. Louis, MO)
Biotin (Sigma)
1-Ethyl-3-[3-(dimethylamino)propyl]carbodiimide (EDC; Sigma)
N-Hydroxysuccinimide (NHS; Sigma)
N-Hydroxysulfosuccinimide (NHSS; Polysciences, Warrington, PA)
Diaminopentane (DAP; Aldrich, Milwaukee, WI)
N-Hydroxysulfosuccinimidyl-2-(biotinamido)ethyl-1,3′-dithiopropionate (NHSS-SS-biotin; Pierce, Rockford, IL)
Fluorescein–streptavidin (Life Technologies, Gaithersburg, MD)
Streptavidin–agarose beads (Life Technologies)
DEAE-Trisacryl (IBF Biotechnics, Villeneuve-la-Garenne, France)
PEI-cellulose thin-layer chromatography (TLC) plates (J. T. Baker, Phillipsburg, NJ)
C_{18} reversed-phase silica gel cartridges (Waters, Milford, MA)
Pepstatin A, leupeptin, aprotinin (U.S. Biochemicals, Cleveland, OH)
Phenylmethylsulfonyl fluoride (PMSF; Sigma)
3-[(3-Cholamidopropyl)-dimethyl-ammonio]-1-propanesulfonate (CHAPS), Nonidet P-40 (NP-40) (Pierce)
L1210 mouse leukemia cells (parental line) [American Type Culture Collection (ATCC), Rockville, MD]
L1210 cells upregulated for nanomolar folate transporter (JF subline)[6]

[5] J. Fan, L. E. Pope, K. S. Vitols, and F. M. Huennekens, *Biochemistry* **30,** 4573 (1991).
[6] J. Fan, K. S. Vitols, and F. M. Huennekens, *J. Biol. Chem.* **266,** 14862 (1991).

Synthesis of Biotin Derivatives of Methotrexate and Folate

Biotin–Methotrexate. Biotin–MTX, the prototype (Fig. 1) of the biotin derivatives of folate compounds, is prepared by the following procedure. Biotin (45.3 mg, 0.18 mmol), dried over P_2O_5 *in vacuo,* is activated by treatment (5 hr, room temperature, stirring) with EDCC (42.7 mg, 0.22 mmol) and NHS (25.7 mg, 0.22 mmol) in 1.5 ml of anhydrous dimethyl sulfoxide (DMSO). The resulting solution containing the NHS ester of biotin is added dropwise to MTX–DAP (100 mg, 0.18 mmol), prepared as described previously,[6] in 1.5 ml of DMSO. After 3 hr at room temperature, 40 ml of acetone is added, and the precipitate is collected by centrifugation, washed with acetone, and air dried. The material is dissolved in H_2O (3 ml) and chromatographed on a DEAE-Trisacryl column (1.5 × 30 cm) that has been equilibrated with 0.05 *M* NH_4HCO_3, followed by H_2O. Elution is performed with 0.05 *M* NH_4HCO_3 (1 ml/min; 8-ml fractions). Fractions 51–87 containing biotin–MTX are pooled and lyophilized. Yield: 73.1 mg (53%). Fast atom bombardment mass spectrometry (FABMS), *m/z* 765 $(M + H)^+$. High-performance liquid chromatography (HPLC) (reversed-phase silica gel). Mobile phase: A, 5 m*M* tetrabutylammonium phosphate–10 m*M* $NH_4H_2PO_4$, pH 6.2; B, same buffer in 50% (v/v) CH_3CN, pH 7.2. Flow rate, 1 ml/min. Gradient: 10% B, 3 min; 40% B, 2 min; linear gradient 40–88% B, 20 min and then 88% B, 5 min. Retention time of 16.6 min;

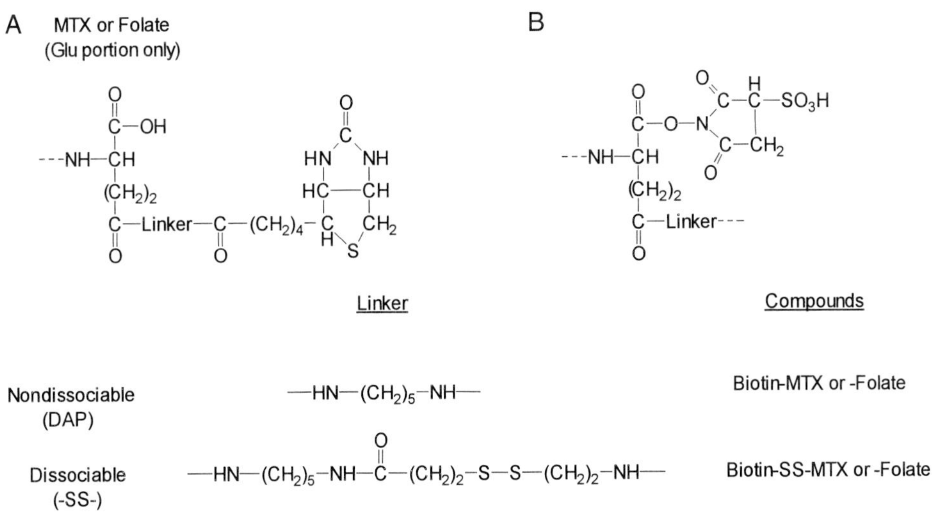

FIG. 1. Structures of (A) biotin–folate compounds with nondissociable or dissociable linkers and (B) NHSS-activated (ester) forms of these compounds.

ε_{mM} of 22.2 at 306 nm (pH 13). Biotin–MTX is stable when stored dry at −20°. It is soluble in aqueous buffers (pH 7) or DMSO.

Biotin–Folate. Biotin–folate is prepared essentially by the above procedure, except that MTX–DAP is replaced by folate–DAP (prepared as described previously[6]). ε_{mM} of 23.8 at 282 nm (pH 13). Biotin–folate is less soluble in DMSO than the MTX counterpart.

Biotin-SS-MTX and Biotin-SS-Folate. Biotin-SS-MTX and biotin-SS-folate in which the linker includes DAP and a disulfide group (Fig. 1), are prepared by procedures[6] similar to those described above, except that the NHS ester of biotin is replaced by the commercially available NHSS-SS-biotin, a compound in which the carboxyl of biotin is derivatized with a linker that contains a disulfide group and terminates in an activated carboxyl group, namely, the *N*-hydroxysulfosuccinimide (NHSS) ester. Biotin-SS-folate is eluted from DEAE-Trisacryl with 0.1 *M* NH_4HCO_3 containing 20% (v/v) CH_3CN. Yields: Biotin-SS-MTX (49%); biotin-SS-folate (36%). HPLC: biotin-SS-MTX, 18.8 and 17.0 min (before and after treatment with 2-mercaptoethanol); biotin-SS-folate, 17.5 min (untreated). Extinction coefficients, see above for counterparts without the disulfide linker.

Activation of Biotin–Folate Compounds with *N*-Hydroxysulfosuccinimide

The following general procedure is used to activate biotin–folate compounds. The unsubstituted α-carboxylate group of the glutamate moiety is converted to the carboxyl form by adsorption on a C_{18} reversed-phase silica gel cartridge, extensive washing with H_2O and 2% (v/v) acetic acid, and elution with 50% (v/v) CH_3CN (MTX compounds) or 50% (v/v) tetrahydrofuran (folate compounds).[6] The carboxyl form, dried over P_2O_5, is dissolved in 0.4 ml of anhydrous DMSO and treated sequentially with a 10-fold molar excess of EDC (room temperature, 0.5 hr) and a 3-fold molar excess of NHSS (room temperature, 3.5 hr). The resulting NHSS esters, whose concentrations are determined spectrophotometrically (see above for extinction coefficients), are used immediately for covalent labeling of the transporters.

Affinity Purification of Folate Transporters

Micromolar Folate Transporter. Parental L1210 cells, which contain only the micromolar folate transporter (0.1 pmol/10^6 cells), provide the starting material. A flow diagram of the purification procedure is shown in Fig. 2.

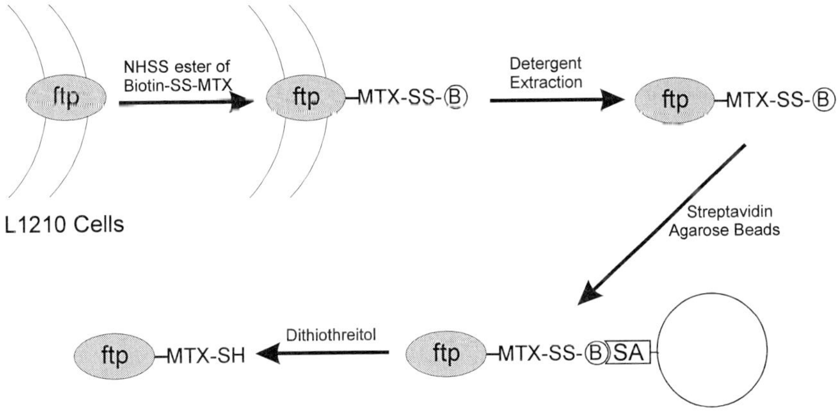

FIG. 2. Flow diagram for affinity purification of L1210 folate transport proteins. ftp, Folate transport protein; B, biotin; SA, streptavidin.

For optimal yields of undamaged transporter, it is essential to adhere closely to the following protocol (especially concerning the inclusion of protease inhibitors where indicated) and to conduct all operations at 4°. L1210 cells, recovered by centrifugation from ca. 6 liters of culture (1.5×10^6 cells/ml), are washed with buffer A (20 m*M* HEPES, pH 7.4, containing 140 m*M* NaCl, 10 m*M* KCl, 2 m*M* $MgCl_2$, and 2 m*M* $CaCl_2$) and then twice with buffer B (160 m*M* HEPES, pH 7.4, containing 5 m*M* $MgCl_2$). Cells (ca. 10^{10}) are suspended in 30 ml of buffer B, and treated (5 min, gentle shaking) with the NHSS ester of biotin-SS-MTX (400 nmol in 10 ml of buffer B). Cells are recovered by centrifugation, washed once with buffer B and three times with buffer A, and suspended in 20 ml of buffer C [1.0 m*M* $NaHCO_3$ (pH 8.2), 20 m*M* $CaCl_2$, 5.0 m*M* $MgCl_2$, and protease inhibitors (20 μg of pepstatin A per milliliter, 50 μg of aprotinin per milliliter, 10 μg of leupeptin per milliliter, and 1 m*M* phenylmethylsulfonyl fluoride)]. After 15 min, cells are collected by centrifugation, suspended in 45 ml of buffer C, and disrupted (20 times, glass tube with zero-clearance Teflon pestle). The plasma membrane fraction is isolated by sucrose density centrifugation,[7] washed three times in buffer D (50 m*M* HEPES, pH 7.5, containing 5.0 m*M* $MgCl_2$), and then suspended in 8 ml of this buffer containing the protease inhibitors (see above) and 1.2% (w/v) CHAPS. After 6 hr, the supernatant is recovered by high-speed centrifugation (25,000 *g*, 50 min). Streptavidin–agarose beads (bed volume, 125 μl), prewashed three times with buffer D containing 0.1% (v/v) Nonidet P-40 and

[7] G. B. Henderson and E. M. Zeveley, *J. Biol. Chem.* **259,** 4558 (1984).

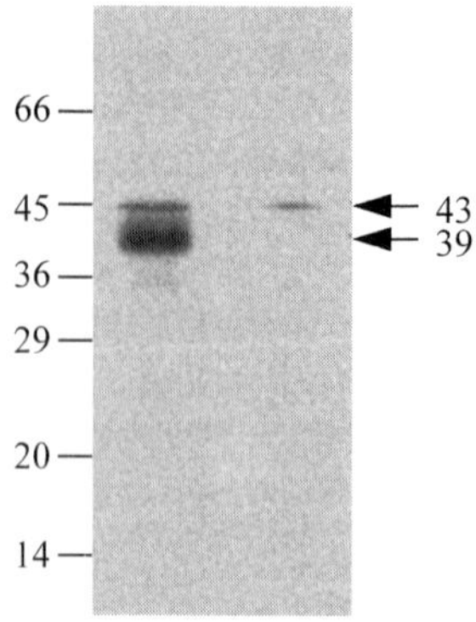

FIG. 3. SDS–PAGE analysis of (*right*) affinity-purified L1210 micromolar folate transport protein and (*left*) the micromolar and nanomolar folate transport proteins. Stain, Ponceau S, after transfer of proteins to polyvinylidine difluoride membranes.

once with the same buffer containing 1.2% (w/v) CHAPS, are added, and the suspension is shaken gently for 6 hr. The beads are recovered by centrifugation, and washed three times with buffer D–0.1% (v/v) Nonidet P-40, twice with buffer E [62 m*M* Tris, pH 7.0, containing 0.2% (w/v) sodium dodecyl sulfate (SDS)], and once with H_2O. The transporter is released by treating the beads three times with 80 μl of buffer E containing 50 m*M* dithiothreitol, and the supernatants, obtained by centrifugation, are combined. Yield, 10–20 μg.

Homogeneity of the isolated micromolar folate transporter is illustrated by sodium dodecyl sulfate-polyacrylamide gel electrophoresis (SDS–PAGE) (Fig. 3, right-hand side). The apparent relative molecular weight of the protein, obtained by this method, is ca. 43,000. This is in agreement with the value of 45,000–48,000 reported by Yang *et al.*,[8] based on gel filtration and SDS–PAGE. A putative cDNA for this transporter, however, corresponds to a 54-kDa protein.[4]

Nanomolar Folate Transporter. The same procedure is employed for purification of the nanomolar folate transporter in L1210 cells, except that the starting material is the JF subline (containing both the micromolar and nanomolar folate transporters—0.1 and 6 pmol/10^6 cells, respectively), and the NHSS ester of biotin-SS-folate is used. Under these conditions, both folate transporters are isolated (Fig. 3, left-hand side). The relative molecular weight of the nanomolar transporter, based on SDS–PAGE, is 39,000. The relative intensities of the two bands reflect their amounts in the JF

[8] C.-H. Yang, F. M. Sirotnak, and L. S. Mines, *J. Biol. Chem.* **263,** 9703 (1988).

subline of L1210 cells, and the diffuse nature of the heavier band is due to the presence of carbohydrate in the nanomolar folate transporter.

The affinity purification procedures described above that utilize biotin-SS-MTX and biotin-SS-folate in conjunction with streptavidin beads should be equally effective for the isolation of folate transport proteins from other cells.

Visualization of Folate Transporters with Biotin–Folate Compounds

The following procedures are representative of the usefulness of biotin derivatives of folate compounds for visualizing the folate transporters in individual cells; in each instance, only one of the transporters is used as an example. Probes without the disulfide spacer should be employed to prevent their release through adventitious reduction.

Fluorescence Microscopy. The L1210 micromolar and nanomolar folate transporters have been visualized previously (via fluorescence microscopy) on cell membranes by covalent and noncovalent labeling with NHS-

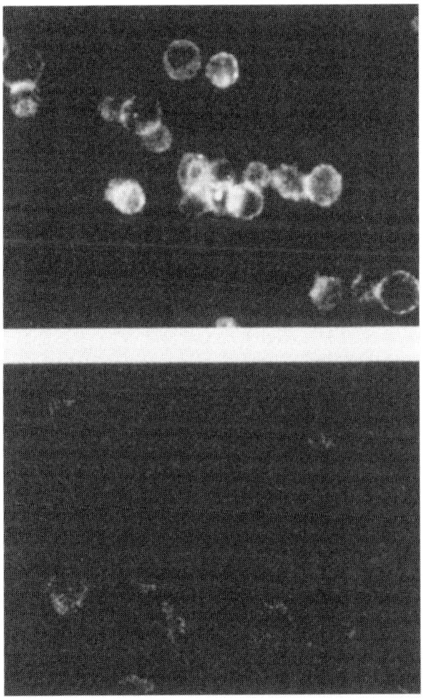

FIG. 4. Fluorescence microscopy of parental L1210 cells labeled with NHSS ester of biotin–MTX. *Top:* Cells labeled as described in text. *Bottom:* Cells labeled in the presence of a 200-fold excess of MTX.

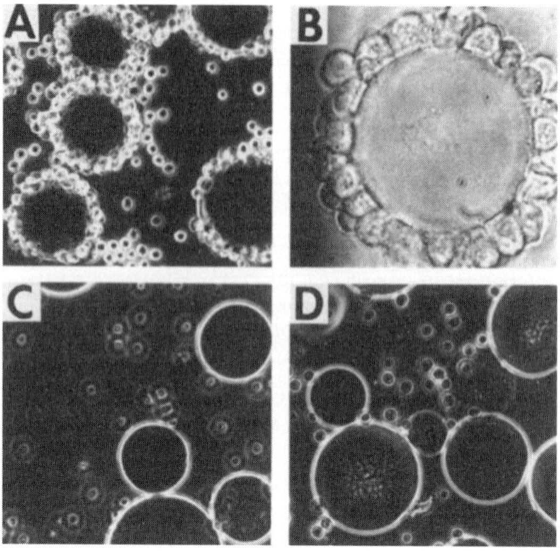

FIG. 5. Light microscopy of parental cells labeled with biotin–MTX and exposed to streptavidin beads. (A and B) Binding of cells to beads. (C) Labeled cells treated with streptavidin prior to exposure to beads. (D) Beads treated with excess biotin prior to exposure to labeled cells. Original magnification: (A, C, and D) ×100; (B) ×250.

activated fluorescein–MTX[5] or fluorescein–folate.[6] Alternatively, these transporters can be labeled with biotin–MTX or biotin–folate in conjunction with fluorescein–streptavidin.

The following operations are conducted at 4°. Parental L1210 cells (5×10^6), propagated as described previously,[5] are washed with buffer B and suspended in 0.2 ml of the same buffer. NHSS-activated biotin–MTX in DMSO (0.5 μl, 2.0 mM) is added to the cell suspension and, after 5 min, the cells are washed three times with buffer and suspended in 0.1 ml of buffer A supplemented with fluorescein–streptavidin (final concentration, 3 μg/ml). After 30 min, the cells are recovered by centrifugation, washed three times with buffer B, and examined by fluorescence microscopy. Labeling on the membrane (Fig. 4, *top*) is specific for the micromolar folate transporter, as shown by the diminished fluorescence when labeling was conducted in the presence of a 200-fold excess of MTX (Fig. 4, *bottom*).

Labeling with biotin–folate compounds provides greater sensitivity than that afforded by fluorescein derivatives because of the presence of multiple fluorescein groups (three to five) per streptavidin. The NHSS ester (rather than the NHS ester) is used in this procedure because the negative charge on the sulfonate group renders the probe more soluble and prevents its passage through the membrane. The nanomolar transporter, because of its high affinity for folate (K_D, ca. 1 nM), can also be labeled noncovalently by biotin or fluorescein derivatives of folate. This accommodating property of the nanomolar folate transporter allows cell populations to be screened rapidly for transporter-positive cells.

Electron Microscopy. Biotin–folate compounds, in conjunction with gold-conjugated streptavidin, have been used to visualize the transporters via electron microscopy. This is illustrated by a previous study[9] in which the nanomolar folate transporter in the JF subline of L1210 cells is labeled noncovalently by the gold-conjugated streptavidin complex of biotin-SS-folate and migration of the transporter from the membrane to the cell interior is followed during the endocytotic process.

Light Microscopy. It has been shown previously[10] that agarose beads containing bound vitamin B_{12} and treated with transcobalamin II (TCII) are able to adsorb L1210 cells with B_{12}–TCII receptors. A similar procedure can be used to detect (and collect) cells with membrane-associated folate transporters. Parental L1210 cells in midlog phase (8×10^6) are washed twice with buffer B, suspended in 0.2 ml of the same buffer, and treated for 5 min with NHSS-activated biotin–MTX (final concentration, 5 μM). Streptavidin–agarose beads (diameter, 100–200 μm; bed volume, 30 μl) that have been washed twice with buffer are added. After 5 min, beads are recovered by low-speed centrifugation, resuspended in buffer, and an aliquot is examined by light microscopy (Fig. 5). Cells are clustered on the beads (Fig. 5A and B), and specificity of the process is indicated by the lack of binding when labeled cells are treated with excess streptavidin, or beads are treated with excess biotin, prior to interaction (Fig. 5C and D).

This rapid visual method may have clinical application in screening transport-deficient cell lines from patients resistant to MTX, and the ease of separating unbound cells by centrifugation would permit the deficient cells to be isolated, characterized, and propagated.

[9] J. Fan, N. Kureshy, and K. S. Vitols, *Oncol. Res.* **7,** 511 (1995).

[10] D. W. Jacobsen, Y. D. Montejano, K. S. Vitols, and F. M. Huennekens, *Blood* **55,** 160 (1980).

[13] Preparation of Stable Isotopically Labeled Folates for *in Vivo* Investigation of Folate Absorption and Metabolism

By CHRISTINE M. PFEIFFER and JESSE F. GREGORY III

Isotopic methods are powerful tools for the study of the absorption, metabolism, and *in vivo* kinetics of vitamins and a wide range of other nutrients in humans. Through the effective use of isotopically labeled compounds, quantitative and qualitative information not readily available by conventional techniques may be obtained. Aside from the inherently greater specificity often provided by the experimental use of isotopically labeled compounds, additional factors favor their use. Benefits include (1) the ability to evaluate the metabolic and physiological fate of an administered compound, (2) the potential to use multiple labels to evaluate several forms of the vitamin simultaneously, and (3) the ability to determine the kinetics of absorption and turnover of an administered compound. Other advantages of stable isotopic tracers are fewer secondary isotopic effects than with analogous radiolabeled compounds, availability of isotopes for labeling that cannot be otherwise achieved (e.g., for nitrogenous compounds), and greater opportunities for repeated use of one or more simultaneous tracers used in the same subject.[1] In spite of these many theoretical and practical advantages of isotopic methods, there are several limitations. A major problem is the lack of commercially labeled compounds, which makes synthesis methods a significant factor. In contrast to the comparative ease of measuring radiolabeled folates and other B vitamins, stable isotope-labeled B vitamins are, at levels encountered in most biological materials, quantifiable only by rather complex mass spectrometric methods.

This chapter examines the current status of stable isotopic methods for *in vivo* research on folates. Methods of synthesizing folates labeled with stable isotopes and of quantifying data are discussed. Important experimental factors related to use of stable isotope-labeled folate for human studies *in vivo* are also explored. Methods of synthesis focus primarily on the introduction of isotopic labeling of the *p*-aminobenzoylglutamate moiety for greatest ease of subsequent analysis using gas chromatography–mass spectrometry (GC–MS) of the folates as a derivative of the *p*-aminobenzoylglutamate following intentional cleavage of the C-9–N-10 bond, as described below in this chapter. In addition, methods are reported for the

[1] D. M. Bier, *Bailliere's Clin. Endocrinol. Metab.* **1,** 817 (1987).

0076-6879/97 $25

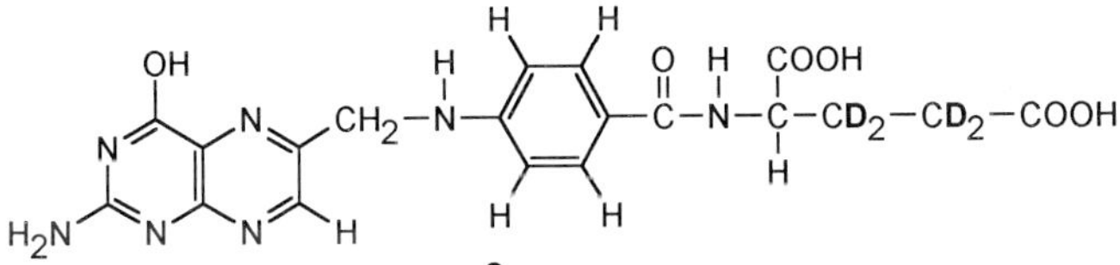

[Glu-$^2H_4$]Folic Acid

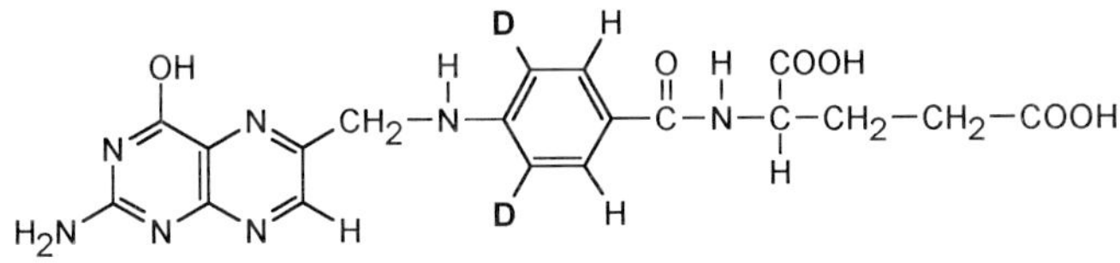

[3′,5′-$^2H_2$]Folic Acid

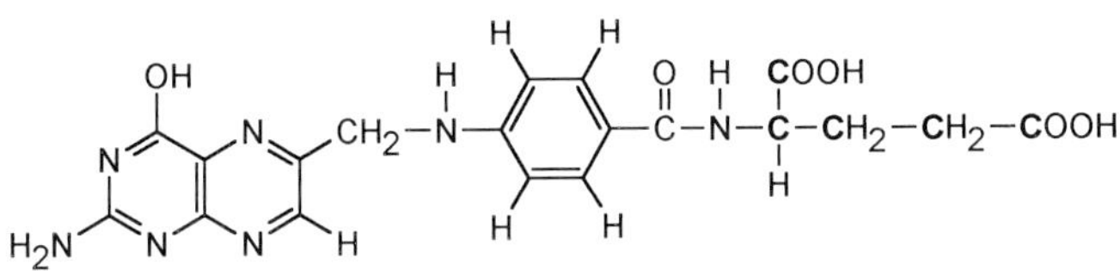

[Glu-$^{13}C_5$]Folic Acid

FIG. 1. Structural formulas of labeled folates.

introduction of several extents of labeling (e.g., M + 2, M + 4, and M + 5) to facilitate the design of *in vivo* protocol with double-label design. Structural formulas of the labeled folates discussed here are shown in Fig. 1.

Preparation of [3′,5′-$^2H_2$] Folic Acid

Several stable isotope-labeled folates have been synthesized for use in kinetic and spectroscopic studies of enzymatic mechanisms, i.e., folate labeled with deuterium at the C-7 position,[2] and (6*S*)-[7,8-$^2H_2$]dihydrofolate.[3] Alternative methods of labeling with deuterium have been used by this research group[4–6] for compatibility of the product with GC–MS analysis and to provide greater metabolic retention of the isotopic label, and for

[2] E. J. Pastore, *Methods Enzymol.* **66,** 538 (1980).
[3] E. J. Pastore, *Methods Enzymol.* **66,** 541 (1980).
[4] J. F. Gregory and J. P. Toth, *J. Labelled Compound Radiopharm.* **25,** 1349 (1988).
[5] J. F. Gregory, *J. Agric. Food Chem.* **38,** 1073 (1990).
[6] J. F. Gregory and J. P. Toth, *Anal. Biochem.* **170,** 94 (1988).

ease of synthesis, relative to other potential methods. Convenient methods for synthesis of [3′,5′-$^2H_2$]folic acid (also termed folic acid-d_2) include catalytic dehalogenation[4,5] and acid-catalyzed deuterium exchange.[7]

Materials and Procedure. Nonlabeled folic acid (Sigma Chemical Co., St. Louis, MO) is converted to 3′,5′-dibromofolic acid by the method of Cosulich *et al.*[8] For this bromination reaction, folic acid is dissolved in concentrated HCl, diluted with water, then reacted by bubbling with bromine vapor in a carrier stream of nitrogen gas, as described. Complete bromination is verified by reversed-phase high-performance liquid chromatography (HPLC)[9] and ultraviolet (UV) spectrophotometry.[8] Sodium deuteroxide (NaOD) and deuterium chloride (DCl) may be obtained from Sigma Chemical Co., and deuterium oxide (D_2O) from Cambridge Isotope Laboratories (Andover, MA) at enrichments of 99+, 99+, and 99.9 atom% excess. A 2-g portion of dry 3′,5′-dibromofolic acid is suspended in 60 ml of D_2O and dissolved by mixing with 2.0 ml of 40% (w/w) NaOD. After approximately 5 min, the 3′,5′-dibromofolic acid is precipitated by addition of 1.6 ml of 35% (w/w) DCl, which yields an approximately pH 2 solution, followed by centrifugation for 15 min at 5000 *g* (4°). The pellet is suspended in 200 ml of D_2O and dissolved by addition of 2.5 ml of 40% (w/w) NaOD. The catalyst (0.25 g of 10% palladium on carbon; Kodak, Rochester, NY) is added without prior hydration, followed by bubbling with nitrogen gas to exclude dissolved oxygen. Catalytic dehalogenation is performed at room temperature using a Parr (Moline, IL) hydrogenation apparatus (model 3911) with a 500-ml reaction bottle agitated at 175 cycles per minute (cpm). Deuterium gas (99.5 atom% deuterium, grade 2.5; Airco Industrial Gases, Riverton, NJ) is typically employed at an initial pressure of ~42 psi, and the pressure monitored throughout the reaction period of approximately 1 hr. Approximately one-third less deuterium labeling occurs in the final product when this step is conducted using NaOH in H_2O, rather than NaOD in D_2O, as the solvent. The mixture is then filtered under vacuum [Whatman (Clifton, NJ) No. 1 paper] to remove the catalyst. Crude [$^2H_2$]folic acid is recovered from the filtrate by acidification to pH 2 with HCl, followed by centrifugation. This material can either be dried in a vacuum desiccator over P_2O_5 and stored dry until needed or dissolved in 0.05 *M* sodium phosphate buffer, pH 7.0 (using addition of dropwise 1 *M* NaOH as needed), and taken for purification. [$^2H_2$]Folic acid is purified by ion-exchange chromatography on a column containing DEAE-Sephadex

[7] D. L. Hachey, L. Palladino, J. A. Blair, I. H. Rosenberg, and P. D. Klein, *J. Labelled Compound Radiopharm.* **14,** 479 (1978).

[8] D. B. Cosulich, D. R. Seeger, M. J. Fahrenbach, *et al., J. Am. Chem. Soc.* **73,** 2554 (1951).

[9] J. F. Gregory, D. B. Sartain, and B. P. F. Day, *J. Nutr.* **114,** 341 (1984).

A-25 [35 × 2.5 cm (i.d.); Pharmacia Fine Chemicals, Piscataway, NJ] with a 2-liter gradient of 0.1–0.8 *M* NaCl in 0.05 *M* sodium phosphate buffer (pH 7.0). Fractions containing pure [$^2H_2$]folic acid, as determined by HPLC,[9] are pooled. The orange product is precipitated by acidification with HCl, recovered by centrifugation, and dried under vacuum over P_2O_5.

The identity and purity of this product (Fig. 1) are confirmed by HPLC,[9] proton nuclear magnetic resonance (NMR),[10] and gas chromatography–mass spectrometry (GC–MS).[11] Proton NMR spectra are determined at 25° with 0.1 *M* NaOD in D_2O as a solvent on a 300-MHz instrument (model NR-300; Nicolet Instrument Corp., Madison, WI) at a concentration of 3 mg of folate/ml. The proton NMR spectrum of [$^2H_2$]folic acid, in which no signal was detected for protons at the 3′- and 5′-positions (3′,5′-protons) (a doublet at ~6.8 ppm in nonlabeled folate), indicates complete deuterium labeling of these positions. Further evidence of labeling of [$^2H_2$]folic acid at the 3′- and 5′-positions is found in the 2′,6′-proton signal, which exists as a doublet for nonlabeled folic acid (~7.7 ppm) but collapses to a singlet for [$^2H_2$]folic acid due to the absence of adjacent protons at the 3′ and 5′ positions. Electron-capture negative-ion (ECN) GC–MS has confirmed the bideutero nature of the [$^2H_2$] product, with [$^2H_2$] species predominating, and [$^2H_1$] and [^1H] present at apparent concentrations of 4.9 and 2.3% of the concentration, respectively, for a typical preparation.

Preparation of [*Glutamate-β,β,γ,γ-*$^2H_4$]Folic Acid

To permit convenient synthesis of a stable isotope-labeled folic acid that contains four atoms of deuterium per molecule, the method of Plante *et al.*,[12] originally used to prepare [*carbonyl-*^{13}C]folic acid, had been modified to synthesize [*glutamate-*$^2H_4$]folic acid from pteroic acid and commercially available L-[$^2H_4$]glutamic acid.[6] The synthesis of [2′,3′,5′,6′-$^2H_4$]folic acid from perdeuterated toluene also has been reported, although this method is somewhat more complex.[13]

Materials and Procedure. Because of the inhibitory effect of water on these reactions, all reagents should be freshly distilled and stored over dry molecular sieve beads (methanol, triethylamine, dimethylformamide) or over dry calcium carbonate (isobutyl chloroformate). *Caution:* isobutyl chloroformate is a strong irritant and potent lachrimator and must be used

[10] M. Poe, *Methods Enzymol.* **66,** 483 (1980).
[11] J. P. Toth and J. F. Gregory, *Biomed. Environ. Mass Spectrom.* **17,** 73 (1988).
[12] L. T. Plante, K. L. Williamson, and E. J. Pastore, *Methods Enzymol.* **66,** 533 (1980).
[13] S. R. Dueker, A. D. Jones, G. M. Smith, and A. J. Clifford, *J. Labelled Compound Radiopharm.* **36,** 981 (1995).

only in a proper fume hood. Glassware should be oven dried, then stored in a desiccator prior to use. Pteroic acid, prepared from folic acid by bacterial cleavage,[14] can be converted to N^{10}-trifluoroacetylpteroic acid using trifluoroacetic anhydride,[15] or this derivative can be obtained commercially (Sigma Chemical Co.). L-[$\beta,\beta,\gamma,\gamma$-$^2H_4$]Glutamic acid (97.6 atom%; MSD Isotopes, Montreal, Quebec, Canada), 101 mg (0.675 mmol), is incubated for 2 hr at 65° in a dry-sealed glass vessel with 0.194 ml (2.67 mmol) of thionyl chloride and 0.54 ml (13.35 mmol) of freshly distilled methanol. A viscous oil of dimethyl[$^2H_4$]glutamic acid hydrochloride is obtained after evaporation of the remaining methanol and thionyl chloride under a stream of nitrogen. This may be stored desiccated at ambient temperature for extended periods of time.

For the coupling reaction, 1.09 g of N^{10}-trifluoroacetylpteroic acid (2.67 mmol), 144 mg of [$^2H_4$]dimethylglutamate (0.668 mmol), 0.53 ml of isobutyl chloroformate (4.0 mmol), 0.56 ml of triethylamine (4.0 mmol), and 20 ml of dimethylformamide are incubated in a sealed flask at 30° for 28 hr. The mixture is concentrated to a viscous liquid in a 60° sand bath under a stream of nitrogen. Saponification is performed in 45 ml of 0.4 *M* sodium deuteroxide in deuterium oxide at 100° for 60 min under a nitrogen headspace.

Purification is performed by chromatography on a 4.4 cm (i.d.) × 85 cm column packed with Whatman CF-11 cellulose isocratically eluted with 0.1 *M* glycine buffer containing 2% (v/v) isoamyl alcohol and 0.15% (w/v) ascorbic acid, adjusted to pH 10.0 with NaOH.[14] The saponified mixture is diluted with an equal volume of the glycine–NaOH buffer, adjusted to pH 10.3 with HCl, and flushed with nitrogen before application to the column. The separation of fluorescent impurities and the orange bands of [$^2H_4$]folic acid and pteroic acid can be visually monitored, and the purity of collected fractions verified using reversed-phase HPLC with a Whatman Partisil 10 ODS-3 column, 0.1 *M* potassium acetate mobile phase containing 15% (v/v) acetonitrile (pH 3.6), and detection monitoring absorbance at 280 nm. Fractions containing pure [$^2H_4$]folic acid are pooled, adjusted to pH 2 with HCl to precipitate the product, then filtered through a fine-mesh sintered glass funnel and washed with water. The product is dried at ambient temperature in a vacuum desiccator. By using a fourfold molar excess of N^{10}-trifluoroacetylpteroic acid in the coupling reaction, the yield of [$^2H_4$]folate is 90% relative to the initial quantity of labeled glutamate.[16]

[14] J. M. Scott, *Methods Enzymol.* **66,** 657 (1980).

[15] C. L. Krumdieck and C. M. Baugh, *Methods Enzymol.* **66,** 523 (1980).

[16] We originally devised this synthesis for the preparation of [$^2H_4$]folic acid using commercial [$^2H_4$]glutamate. Commercial sources of L-[$^2H_4$]glutamic acid are no longer available but have been replaced by [$^2H_5$]glutamate, which yields a functionally equivalent [$^2H_5$]folate derivative.

The identity and purity of the [$^2H_4$]folic acid (Fig. 1) are confirmed by HPLC,[9] NMR,[10] and GC–MS.[11] ^1H NMR spectra, in comparison with that of nonlabeled folic acid, confirm the site and extent of labeling. If fully labeled, no signals corresponding to glutamyl β (~2.0–2.2 ppm) and γ (~2.3 ppm) protons are observed. The lack of splitting of the α-proton signal of [$^2H_4$]folic acid is further evidence of the absence of adjacent protons. Negative-ion mass spectra have confirmed the tetradeutero nature of the product, with 88% of the product as [$^2H_4$]folate species in a typical preparation. This suggests small loss of deuterium, possibly by isotopic exchange during chromatographic purification under alkaline conditions or during derivatization for GC–MS. Any incomplete labeling poses no problem for *in vivo* use in metabolic studies. No evidence of deuterium exchange was observed under milder, nonalkaline analytical or physiological conditions.

Preparation of [*Glutamate*-$^{13}C_5$]Folic Acid

The method used for the synthesis of [$^2H_4$]folic acid is employed with minor modifications to synthesize [*glutamate*-$^{13}C_5$]folic acid using commercial [$^{13}C_5$]glutamic acid.[17] The [$^{13}C_5$]glutamic acid must be methyl-esterified prior to use, as described for the synthesis of [$^2H_4$]folic acid. [$^{13}C_5$]Folic acid may be used in some applications as an alternative to [$^2H_4$]folic acid.

Materials and Procedure. Equal molar amounts of N^{10}-trifluoroacetylpteroic acid (prepared from pteroic acid[15]; from bacterial synthesis or obtained commercially) and dimethyl-L-[$^{13}C_5$]glutamic acid (98 atom%; Cambridge Isotope Laboratories) are coupled as described above to form N^{10}-trifluoroacetyldimethyl[$^{13}C_5$]folic acid. When saponified in 0.1 *M* NaOH, the yield of [$^{13}C_5$]folic acid is typically 75% relative to the initial quantity of labeled glutamate. The overall yield of the synthesis of [$^{13}C_5$]folic acid (Fig. 1) after purification is generally ~40%. GC–MS analysis[11] of the [$^{13}C_5$]folic acid preparation, as the derivatized *p*-aminobenzoylglutamate (*p*-ABG) derivative, has shown the $^{13}C_5$ isotopomer as the predominant species, with an apparent $^{13}C_4$ concentration of 4.8%. No other isotopomers were detected at significant concentrations.

Preparation of [3′,5′-$^2H_2$]Pteroylpolyglutamates

The lack of commercially available stable isotope-labeled *N-tert*-butyloxycarbonyl-L-glutamic acid precludes labeling of the first glutamyl residue in the synthesis of polyglutamates. The approach used in the synthesis of deuterium-labeled pteroylpolyglutamates (PteGlu$_n$) is the coupling of the

[17] C. M. Pfeiffer and J. F. Gregory, unpublished data (1995).

N^{10}-trifluoroacetyl derivative of [3′,5′-$^2H_2$]pteroic acid (pteroic acid-d_2) to a nonlabeled polyglutamyl peptide (prepared by solid-phase synthesis) via a standard mixed-anhydride reaction used in solid-phase synthesis[6] or in a solution phase reaction.[18]

Materials and Procedure. The method used to prepare [$^2H_2$]pteroic acid is similar to that used to synthesize [3′,5′-$^2H_2$]folic acid.[4,5,8] A slurry of pteroic acid (from bacterial synthesis or commercially obtained) in 6 *M* HCl is converted to 3′,5′-dibromopteroic acid. Catalytic dehalogenation in 0.1 *M* NaOD yields [3′,5′-$^2H_2$]pteroic acid with approximately 90% labeling. We have conducted several preparations of resin-bound hexaglutamyl peptide by solid-phase synthesis,[15] to which the N^{10}-trifluoroacetyl derivative of [$^2H_2$]pteroic acid has been coupled via a mixed-anhydride reaction using isobutyl chloroformate.[15]

A convenient alternative to the solid-phase synthesis is now available because of the commercial availability of protected (*N*-benzyloxycarbonyl, *O-tert*-butyl ester) oligoglutamyl peptides, which eliminates the need for the initial synthesis of the polyglutamyl peptide. We have used the method of D'Ari and Rabinowitz[18] to synthesize [$^2H_2$]PteGlu$_5$ using commercially obtained *N*-benzyloxycarbonylpenta-γ-glutamylhexa-*tert*-butyl ester (Schirck Laboratories, Jona, Switzerland). This peptide is first subjected to hydrogenation to remove the benzyloxycarbonyl protecting group, then N^{10}-trifluoroacetyl[$^2H_2$]pteroic acid is immediately coupled by a carbodiimide-mediated process. The product (N^{10}-trifluoroacetyl-PteGlu$_5$-hexa-*tert*-butyl ester) is purified on a silica gel column, followed by deprotection and further purification on DEAE-Sephadex A-25. In many ways, this procedure is a preferable alternative to the more widely used solid-phase method.

Preparation of Other Deuterium-Labeled Folates

The availability of [$^2H_2$]PteGlu, [$^2H_4$]PteGlu, and [$^2H_2$]PteGlu$_n$ provides the potential to synthesize any folate in deuterium-labeled form. For *in vivo* applications of tetrahydrofolates, the naturally occurring C-6 isomer is generally the desired product. Thus, reduction methods using borohydride or catalytic hydrogenation, which produce a racemic mixture of 6*R* and 6*S* species of tetrahydrofolates, would be unsuitable for preparing deuterium-labeled folates for many *in vivo* studies. Enzymatic reduction of deuterium-labeled folates with dihydrofolate reductase permits the preparation of tetrahydrofolates in the biologically active 6*S* form.[19–21]

[18] L. D'Ari and J. C. Rabinowitz, *Methods Enzymol.* **113,** 169 (1985).
[19] F. M. Huennekens, C. K. Mathews, and K. G. Scrimgeour, *Methods Enzymol.* **VI,** 802 (1963).
[20] P. F. Nixon and J. R. Bertino, *Methods Enzymol.* **66,** 547 (1980).
[21] D. W. Horne, W. T. Briggs, and C. Wagner, *Methods Enzymol.* **66,** 545 (1980).

Quantification of Stable Isotopic Labeling

Our primary approach to quantification in stable isotopic studies has been to use HPLC for the determination of urinary folate concentration and GC–MS for the determination of isotopic enrichment (i.e., molar excess of each isotopomer above natural abundance). Although several other modes of mass spectrometry have been examined and may be potentially suitable, ECNI GC–MS in selected-ion monitoring mode is currently the only quantitative method with sufficient sensitivity for use with most applications. Because of their lack of volatility and their thermal instability, folates in their native forms are inherently poor candidates for GC–MS. GC–MS analysis of the *p*-ABG fragment following cleavage of the C-9–N-10 bond[11] has been a highly successful approach to routine mass spectrometry of folates. The precision of mass spectral analysis is facilitated if the mass of the labeled folate is two or more mass units greater than that of the unlabeled compound. Elements critical to MS analysis include purification of folates in urine and other samples by affinity chromatography, procedures for C-9–N-10 cleavage of all folates, and sensitive, specific analysis by GC–MS.

Materials and Procedure. We have performed affinity chromatography by a modification of the procedure of Selhub *et al.*[22] Folate-binding protein isolated from bovine whey and covalently coupled to commercially activated agarose beads (Affi-Gel 10; Bio-Rad Laboratories, Richmond, CA) permits retention and concentration of folates from urine or extracted, deproteinated plasma or other biological materials. Urine or sample extracts are applied at pH 7 and washed to remove most unretained nonfolate components, and then folates are eluted with 0.1 *M* HCl.[6] Affinity-purified folates are subjected to C-9–N-10 bond cleavage procedures[6] based on previously published methods.[23,24] The *p*-ABG product is isolated from the reaction mixture by reversed-phase HPLC with a preparative column (Partisil M9 10/25 ODS-3, 9 × 250 mm; Whatman) and a volatile mobile phase [0.1 *M* formic acid–5% (v/v) acetonitrile].

Prior to GC–MS analysis, derivatives are prepared from *p*-ABG samples by simultaneous acylation and esterification with trifluoroacetic anhydride and trifluoroethanol, yielding a lactam of (*N*-trifluoroacetyl-α-trifluoroethyl)-*p*-aminobenzoylglutamate as the major product (Fig. 2).[11] This derivative is well suited for GC–MS analysis because of its volatility, stability, and behavior during gas chromatography. It also shows excellent stability during anhydrous storage at $-20°$ in dry ethyl acetate prior to analysis.

[22] J. Selhub, O. Ahmad, and I. H. Rosenberg, *Methods Enzymol.* **66,** 686 (1980).
[23] I. Eto and C. L. Krumdieck, *Anal. Biochem.* **115,** 138 (1981).
[24] B. Shane, *Methods Enzymol.* **122,** 323 (1986).

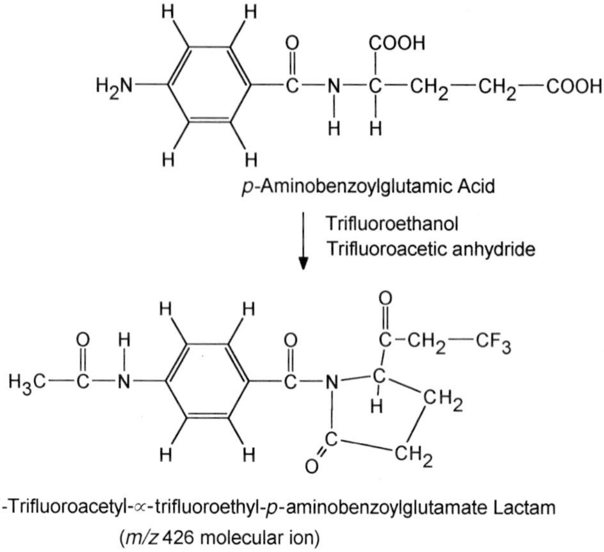

FIG. 2. Preparation of derivatives from *p*-ABG.

The electron-capture negative ionization (ECNI) mode provides the greatest sensitivity and an abundant molecular anion that facilitates quantitative GC–MS analysis.[11] Absolute limits of detection by ECNI GC–MS are highly sensitive to minor variations in instrumental conditions. However, with a reasonably clean, well-tuned instrument, a 200-fmol injection of the *p*-ABG lactam was easily detected, with a signal-to-noise ratio of >10, using ECNI and single-ion monitoring.[11] Such detection limits are more than adequate for *in vivo* studies in humans, using protocols that typically involve excretion of 10–100 nmol of labeled urinary folate over 24 hr. Using selected-ion monitoring, the mass spectrometer serves as a highly specific, sensitive detector for the gas chromatographic (GC) system by monitoring only the intensity of selected ions in the GC effluent. For analysis of the *p*-ABG lactam, selected-ion monitoring is performed at *m/z* 426, 428, and 430 (or 431), respectively, which correspond to the molecular anions of the unlabeled [^{1}H], [$^2H_2$], and [$^2H_4$] (or [$^2H_5$] or [$^{13}C_5$]), respectively, and reflect the molar ratios of those isotopes in the isolated urinary folates. We now generally perform selected-ion monitoring of the entire cluster of masses (i.e., *m/z* 426, 427, 428, 429, 430, and 431), which facilitates assessment of the full isotopomer distribution.

In Vivo Applications of Stable Isotopes in Folate Research

Deuterium-labeled folates have been used in human subjects to develop and evaluate protocols and evaluate short-term bioavailability/metabolism of various chemical forms of the vitamin using dual-label procedures. When administered simultaneously, [$^2H_2$]- and [$^2H_4$]folates exhibit similar rates and/or extents of absorption, metabolism, and urinary excretion.[25,26] The use of prior "saturation" of subjects by chronic administration of a 2-mg/day loading dose of unlabeled folic acid facilitates such studies by enhancing urinary excretion. A larger saturation dose would be required to obtain more complete recovery of the label as intact urinary folate, but this has been avoided to prevent isotopic dilution of the deuterium-labeled urinary folates, which could have reduced the precision of the GC–MS measurement of isotopic enrichment. The saturation dose should not be administered on the same day as the labeled folates, because this would invalidate many bioavailability studies by causing intestinal absorption to occur mainly by passive diffusion, rather than by the physiologically more relevant carrier-mediated saturable process.

Using this type of protocol, studies were performed to assess the relative short-term utilization of deuterium-labeled monoglutamyl- and hexaglutamylfolates in human subjects[27] as well as of deuterium-labeled monoglutamyltetrahydrofolates and folic acid.[28] In additional studies of this type, the folate metabolite *p*-aminobenzoylglutamate (following acid hydrolysis for deacetylation) was found to account for $<10\%$ of total isotopic urinary excretion, relative to labeled folates in humans.[26]

We have conducted several evaluations of the use of [$^{13}C_5$]folic acid for simultaneous *in vivo* use with [$^2H_2$]folic acid in human subjects. Without using a saturation protocol or a flushing dose to enhance folate excretion, isotopic enrichment of plasma folate and of excreted urinary folate could be accurately measured by ECNI GC–MS.[29] The subject-to-subject variation was higher than in protocols using prior folate saturation. As reflected by short-term (24-hr) urinary excretion of labeled folates, *in vivo* handling of the [$^2H_2$]- and [$^{13}C_5$]folates through absorption, transport, metabolism,

[25] J. F. Gregory, L. B. Bailey, J. P. Toth, and J. J. Cerda, *Am. J. Clin. Nutr.* **51,** 212 (1990).

[26] P. A. Kownacki-Brown, C. Wang, L. B. Bailey, J. P. Toth, and J. F. Gregory, *J. Nutr.* **123,** 1101 (1993).

[27] J. F. Gregory, S. D. Bhandari, L. B. Bailey, J. P. Toth, T. G. Baumgartner, and J. J. Cerda, *Am. J. Clin. Nutr.* **53,** 736 (1991).

[28] J. F. Gregory, S. D. Bhandari, L. B. Bailey, J. P. Toth, T. G. Baumgartner, and J. J. Cerda, *Am. J. Clin. Nutr.* **55,** 1147 (1992).

[29] C. M. Pfeiffer and J. F. Gregory, Abstract 4628, *Experimental Biology '96,* Washington, DC, April 14–17, 1996.

and excretion processes was equivalent. However, a slightly greater carryover of [$^{13}C_5$]folates in plasma during the 3-week intervals between studies provided evidence of an isotope effect in which [$^{13}C_5$]folate may be retained longer in the body than is the D_2 isotopomer. Although short-term use of [$^{13}C_5$]folates is fully appropriate, such possible kinetic differences in long-term metabolism require further evaluation.

Chronic supplementation of human subjects with [$^2H_2$]folic acid (1.6 mg/day) was performed in a preliminary study of *in vivo* folate kinetics.[30] A similar protocol was used in which chronic doses of [$^2H_2$]folic acid were evaluated and found to exhibit no effect on zinc nutritional status.[31] Further studies of long-term *in vivo* folate kinetics using chronic consumption of more nutritionally relevant quantities of [$^2H_2$]folic acid are in progress and are based on measuring enrichment of urinary folate, urinary *p*-aminobenzoylglutamate, and urinary *N*-acetyl-*p*-aminobenzoylglutamate.

[30] A. E. von der Porten, J. F. Gregory, J. P. Toth, J. J. Cerda, S. H. Curry, and L. B. Bailey, *J. Nutr.* **122,** 1293 (1992).

[31] G. P. A. Kauwell, L. B. Bailey, J. F. Gregory, D. W. Bowling, and R. J. Cousins, *J. Nutr.* **125,** 66 (1995).

[14] Interconversion of 6- and 7-Substituted Tetrahydropterins via Enzyme-Generated 4a-Hydroxytetrahydropterin Intermediates

By SHELDON MILSTIEN

Almost all known naturally occurring pterins are substituted in the 6-position of the pterin ring. 6*R*-Tetrahydrobiopterin (BH_4) is the natural cofactor for phenylalanine, tyrosine, and tryptophan hydroxylase as well as nitric oxide synthase. During the BH_4-dependent hydroxylation of phenylalanine to tyrosine, the cofactor is stoichiometrically oxidized to a 4a-hydroxycarbinolamine intermediate. The enzyme, 4a-hydroxytetrahydrobiopterin dehydratase (carbinolamine dehydratase, EC 1.14.16.1), catalyzes dehydration of the carbinolamine intermediate to form quinonoid dihydrobiopterin, which is then recycled back to BH_4 by the enzyme dihydropteridine reductase. Children with a mild form of the genetic disease hyperphenylalaninemia were found to excrete a large amount of an unknown

0076-6879/97 $25

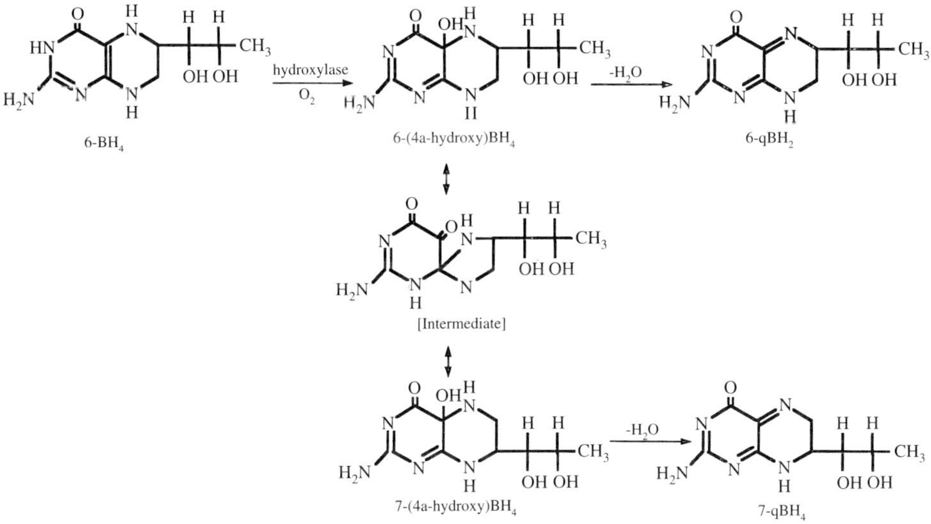

FIG. 1. Scheme for interconversion of 6- and 7-substituted tetrahydrobiopterin. Aromatic amino-acid hydroxylases catalyze tetrahydropterin- and oxygen-dependent hydroxylation of either phenylalanine, tyrosine, or tryptophan. In the reaction, the tetrahydropterin (6-BH_4) is converted to the 4a-carbinolamine intermediate [6-(4a-hydroxy-BH_4], which then loses a water molecule, either spontaneously or catalyzed by 4a-carbinolamine dehydratase, to form quinonoid dihydropterin (6-qBH_2), which can then be reduced back to the tetrahydropterin in a reaction catalyzed by dihydropteridine reductase. In the absence of carbinolamine dehydratase activity, the carbinolamine is converted to an unstable intermediate that can then rearrange to the 7-4a-hydroxy-BH_4 isomer.

pterin in their urine, which was subsequently identified as the 7-isomer of BH_4.[1] Because the excretion of 7-BH_4 was increased following administration of 6-BH_4, it appeared that 6-BH_4 was the precursor. Subsequently, these children were shown to indeed have a deficiency of carbinolamine dehydratase activity[2] and it was demonstrated that 6-BH_4-dependent hydroxylation of phenylalanine in the absence of carbinolamine dehydratase would allow an unstable intermediate to form from the carbinolamine, which could then spontaneously rearrange to the 7-isomer[3,4] (Fig. 1). It should be noted that other 6-substituted tetrahydropterins are also con-

[1] H. C. Curtius, A. Matasovic, G. Schoedon, T. Kuster, P. Guibaud, T. Giudici, and N. Blau, *J. Biol. Chem.* **265,** 3923 (1990).

[2] B. A. Citron, S. Kaufman, S. Milstien, E. W. Naylor, C. L. Greene, and M. D. Davis, *Am. J. Hum. Genet.* **53,** 768 (1993).

[3] H. C. Curtius, C. Adler, I. Rebrin, C. W. Heizmann, and S. Ghisla, *Biochem. Biophys. Res. Commun.* **172,** 1060 (1990).

[4] M. D. Davis, S. Kaufman, and S. Milstien, *Proc. Natl. Acad. Sci. U.S.A.* **88,** 385 (1991).

verted to their 7-isomers in an analogous fashion and 7-BH_4 can be converted back to 6-BH_4. Furthermore, although 7-BH_4 is a poor substrate for hydroxylation reactions, it has been found to be a potent inhibitor of the aromatic amino acid hydroxylases,[5,6] suggesting that a role of 4a-carbinolamine dehydratase, in addition to facilitating aromatic amino acid hydroxylations, may be the prevention of the formation of the toxic 7-BH_4 isomer. Interestingly, 4a-carbinolamine dehydratase is also a nuclear transcription factor dimerization cofactor.[7]

The usual method of identification and quantitation of BH_4 involves the use of reversed-phase high-performance liquid chromatography (HPLC) after oxidation to convert reduced pterins to their fully oxidized, fluorescent forms.[8] Separation of 6- and 7-biopterin can be achieved by reversed-phase HPLC.[1] However, identification of chromatographic peaks on the basis of elution times of standards is not sufficient, especially when working with biological samples. Additional methods must be employed to confirm both identification and quantitation. The "gold standard" method involves the use of mass spectrometry.[1] While gas chromatography–mass spectrometry instrumentation is not readily available in most laboratories, other convenient analytical methods can be used to identify and quantitate mixtures of 6- and 7-biopterin. As described in detail below, these include two-dimensional HPLC; oxidation of pterin side chains with permanganate followed by separation of the pterin carboxylic acids[4]; and periodate–borohydride treatment to cleave the polyhydroxy side chains to corresponding hydroxymethylpterins, which can then be separated by HPLC.[4]

Preparation of Samples

Reagents

H_3PO_4 (1 *M*)
I_2 [1% (w/v) in 2% (w/v) KI]
MnO_2
Potassium phosphate (pH 8.2), 1 *M*
Phenylalanine hydroxylase[9]
L-Phenylalanine (10 m*M*)

[5] M. D. Davis and S. Kaufman, *FEBS Lett.* **285,** 17 (1991).

[6] C. Adler, S. Ghisla, I. Rebrin, J. Haavik, C. W. Heizmann, N. Blau, T. Kuster, and H. C. Curtius, *Eur. J. Biochem.* **208,** 139 (1992).

[7] B. A. Citron, M. D. Davis, S. Milstien, J. Gutierrez, D. B. Mendel, G. R. Crabtree, and S. Kaufman, *Proc. Natl. Acad. Sci. U.S.A.* **89,** 11891 (1992).

[8] T. Fukushima and J. C. Nixon, *Methods Enzymol.* **66,** 429 (1980).

[9] Rat liver phenylalanine hydroxylase was purified to homogeneity as described in Ref. 4.

6*R*-Tetrahydrobiopterin, 6-biopterin, 7-biopterin (0.2 m*M*; Schirck Laboratories, Jona, Switzerland)
Catalase (1 mg/ml; Boehringer Mannheim, Indianapolis, IN)
Dihydropteridine reductase (Sigma, St. Louis, MO)
NADH (1 m*M*)
Glucose 6-phosphate, sodium salt (100 m*M*)
Glucose-6-phosphate dehydrogenase, *Leuconostoc mesenteroides* (Boehringer)
NaOH (1 *M*)
Saturated $KMnO_4$
Potassium periodate, 15 μM in 0.1 *M* potassium bisulfate (pH 2.3)
Glycerol (10%, v/v)
Sodium borohydride, 10 m*M* in 0.1 *M* KOH

Oxidation of Reduced Pterins. Reduced pterins are oxidized to their fluorescent forms with iodine[10] or MnO_2[11] in acidic solution. For iodine oxidation, samples (50 μl) are added to a solution containing 50 μl of 1 *M* H_3PO_4 and 100 μl of 1% (w/v) I_2 in 2% (w/v) KI. After 60 min in the dark, excess I_2 is reduced by the addition of 50 μl of 2% (w/v) ascorbic acid. For MnO_2 oxidation, samples (50 μl) are mixed with 100 μl of 0.2 *M* H_3PO_4 followed by the addition of 10 mg of MnO_2. After 10 min at room temperature, MnO_2 is removed by centrifugation.

Phenylalanine Hydroxylase-Catalyzed Conversion of 6- to 7-Tetrahydrobiopterin. In the phenylalanine hydroxylase reaction carried out in the absence of 4a-carbinolamine dehydratase activity, the 4a-hydroxy-BH_4 intermediate rearranges to the 7-BH_4 isomer. Reactions are carried out at pH 8.2, at which the carbinolamine is more stable, to allow time for the rearrangement to take place. Reactions (1 ml, 25°) contain 30 m*M* potassium phosphate (pH 8.2), 0.2 μM phenylalanine hydroxylase,[9] 1 m*M* L-phenylalanine, 20 μM 6*R*-BH_4, catalase (0.1 mg/ml), excess dihydropteridine reductase, 0.085 m*M* NADH, and an NADH-regenerating system (10 m*M* glucose 6-phosphate plus 0.2 U of *Leuconostoc mesenteroides* glucose-6-phosphate dehydrogenase). After 2 hr, about 15% of the 6-BH_4 is converted to 7-BH_4. The reduced pterins are then oxidized as described above.

Two-Dimensional High-Performance Liquid Chromatographic Analysis of 6- and 7-Biopterin

High-Performance Liquid Chromatography System. Any good-quality HPLC pump, injector, and fluorescence detector can be used.[8] Most of

[10] T. Fukushima and J. C. Nixon, *Anal. Biochem.* **102,** 176 (1980).
[11] N. Sakai, S. Kaufman, and S. Milstien, *Mol. Pharmacol.* **43,** 6 (1993).

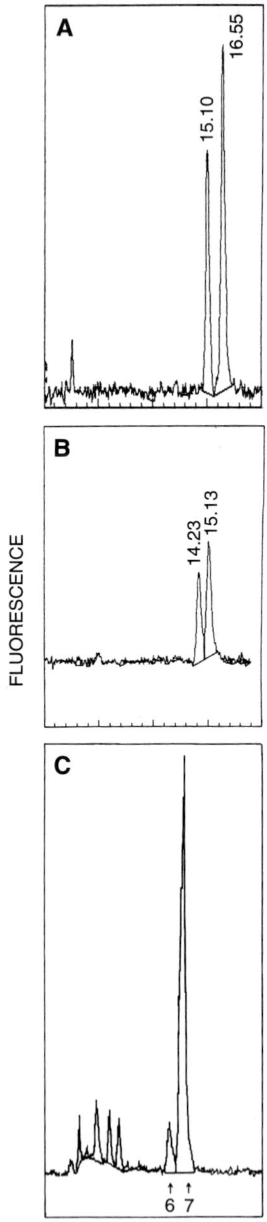

FIG. 2. Separation of 6- and 7-biopterin by HPLC. (A) Reversed-phase separation on a Spherisorb ODS-1 column eluted with 5% methanol–95% water at 1 ml/min. 6-Biopterin and 7-biopterin are normally eluted at 15 and 16.5 min, respectively. (B) Strong cation-exchange

the common fully oxidized pterins have similar fluorescence spectra with excitation and emission maxima at about 350 and 450 nm, respectively. Pterin standards are available from Schirck Laboratories.

Separation of 6- and 7-Biopterin. Although 6- and 7-biopterin can be separated by reversed-phase HPLC, they are not well resolved by chromatography on most C_{18} columns. The best separation has been found using a C_{18} column that is not completely end capped, such as Whatman (Clifton, NJ) ODS-1 or Spherisorb ODS-1 (5 μm, 4.6 × 250 mm), eluted with an isocratic solvent of 5% methanol–95% water at a flow rate of 1 ml/min (Fig. 2A). After several injections, these columns lose their resolving power for 6- and 7-biopterin. However, they can be regenerated by two injections of 100 μl of 1 *M* H_3PO_4 or by washing with 25 ml of 0.5 *M* H_3PO_4. In normal biological samples, such as urine or plasma, the ratio of 6-biopterin to 7-biopterin is approximately 100:1. Thus, it is not possible to quantitate accurately the amount of 7-biopterin present in such samples by direct reversed-phase HPLC analysis. Furthermore, there are other, not yet identified fluorescent materials in biological samples that also elute with retention times similar to that of 7-biopterin. To overcome these problems, a two-dimensional HPLC method has been developed. In this method, samples are separated by reversed-phase HPLC as described above and the fraction eluting at the time corresponding to the elution of standard 7-biopterin is collected and evaporated to dryness by lyophilization or on a Speed-Vac (Savant, Hicksville, NY). After redissolving in 100 μl of 20 m*M* H_3PO_4, the sample is analyzed by HPLC on a strong cation-exchange column (Whatman SCX, 2.6 × 250 mm) eluted with 10 m*M* H_3PO_4–10 m*M* KH_2PO_4–8% methanol–2% acetonitrile (v/v) at a flow rate of 0.7 ml/min (Fig. 2B). The fluorescence is monitored as described above. It should be noted that the fluorescence yield of biopterin is approximately 50% lower in this solvent than in the reversed-phase solvent. The recovery of 7-BH_4 added to urine was 82.6 ± 3.6% (n = 10).

chromatography of 6-biopterin and 7-biopterin standards on a Whatman SCX column eluted with 10 m*M* H_3PO_4–10 m*M* KH_2PO_4–8% methanol–2% acetonitrile at a flow rate of 0.7 ml/min. 6-Biopterin and 7-biopterin are normally eluted at 14 and 15 min, respectively. (C) Two-dimensional HPLC analysis of 7-biopterin concentration in a urine sample from a carbinolamine dehydratase-deficient patient. After oxidation, the sample was chromatographed on a C_{18} reversed-phase column as in (A); the fraction eluting at the same time as 7-biopterin standard was collected, concentrated, and analyzed by cation-exchange HPLC on a Whatman SCX column as in (B).

TABLE I
LEVELS OF 6- AND 7-BIOPTERIN IN NORMAL CHILDREN, CLASSIC PHENYLKETONURICS, AND CHILDREN WITH CARBINOLAMINE DEHYDRATASE DEFICIENCY[a]

Sample	6-Biopterin (nmol/mmol C)	7-Biopterin (%) [100 × 7B/(6B + 7B)]
Controls[b] (n = 10)	1138 ± 161	0.92 ± 0.19 (0.25 − 1.78)
PKU[b] (n = 10)	2199 ± 161	0.91 ± 0.16 (0.29 − 1.75)
CDH deficient[c]	1041, 410	45, 49
CDH deficient[b]	448	52.0

[a] The data are expressed as the mean ± SD. The numbers in parentheses are the ranges. C, Creatinine; PKU, hyperphenylalaninemia due to phenylalanine hydroxylase deficiency; CDH, hyperphenylalaninemia due to 4a-carbinolamine dehydratase deficiency; 6B, 6-biopterin; 7B, 7-biopterin.
[b] Determined by two-dimensional HPLC.
[c] Determined by GC–MS.[6]

Preparation of Derivatives of 6- and 7-Tetrahydrobiopterin Isolated by High-Performance Liquid Chromatography

Oxidation to Corresponding Pterin Carboxylic Acids. The side-chain groups of 6- and 7-substituted pterins are oxidized to the corresponding carboxylic acids by alkaline permanganate oxidation.[4] In brief, HPLC-purified samples of 6- or 7-biopterin are mixed with 0.3 vol of 1 *M* NaOH and 1 vol of saturated $KMnO_4$ and then heated in a boiling water bath for 30 min. After cooling, 1 vol of methanol is added to remove excess $KMnO_4$ and, after centrifugation, the basic supernatant is neutralized by diluting 1:1 with 1 *M* H_3PO_4. The pterin carboxylic acids are analyzed by reversed-phase HPLC on a Spherisorb ODS-1 column eluted with 0.1 *M* KPO_4 (pH 3) in 5% (v/v) methanol at 0.8 ml/min. Pterins are detected by fluorescence (excitation, 358 nm; emission, 444 nm) and pterin-6-carboxylic acid and pterin-7-carboxylic acid are eluted at 11 and 12 min, respectively.

Conversion to Corresponding Hydroxymethyl Pterins. Periodate oxidation followed by borohydride reduction converts 6- and 7-biopterin to their corresponding hydroxymethyl derivatives.[4] In brief, samples are incubated for 1 hr at room temperature with 1 vol of a solution of 15 μM potassium periodate in 0.1 *M* potassium bisulfate (pH 2.3). Excess periodate is removed by addition of 10 μl of 10% (v/v) glycerol, followed by addition of 1 vol of 10 m*M* sodium borohydride in 0.1 *M* KOH. The hydroxymethyl pterins are separated by reversed-phase HPLC on a Spherisorb ODS-1 column eluted with 5% methanol–95% water at a flow rate of 0.8 ml/min and detected by fluorescence; 6-hydroxymethylpterin and 7-hydroxymethylpterin are eluted at 20 and 21 min, respectively.

Levels of 7-Biopterin in Normal, Phenylalanine Hydroxylase-Deficient, and Carbinolamine Dehydratase-Deficient Children

Levels of 7-biopterin were measured in the urine of newborn children by the two-dimensional HPLC method described above (Table I). 7-Biopterin is present at about 1% of the 6-biopterin level even in the absence of phenylalanine hydroxylase activity, suggesting that the 7-biopterin normally present in physiological fluids likely arises during hydroxylation reactions catalyzed by other hydroxylases. The amount of 7-biopterin in urine of carbinolamine dehydratase-deficient patients is increased more than 50-fold (Fig. 2C). Furthermore, it should be noted that normally only an average of 25% of the total fluorescent material eluting with the same retention time as authentic 7-biopterin on reversed-phase HPLC was identified as 7-biopterin by HPLC in the second-dimension, cation-exchange HPLC.

Acknowledgment

This contribution would not have been possible without the keen insight of Professor Sandro Ghisla and especially Professor Hans-Christof Curtius.

[15] Enzymatic Synthesis of 6*R*-[U-^{14}C]Tetrahydrobiopterin from [U-^{14}C]GTP

By MASAAKI HOSHIGA and KAZUYUKI HATAKEYAMA

Introduction

6*R*-L-*erythro*-5,6,7,8-Tetrahydrobiopterin (6*R*-BH_4) is the natural cofactor of phenylalanine hydroxylase (EC 1.14.16.1, phenylalanine 4-monooxygenase), tyrosine hydroxylase (EC 1.14.16.2, tyrosine 3-monooxygenase), tryptophan hydroxylase (EC 1.14.16.4, tryptophan 5-monooxygenase), the glyceryl-ether cleaving enzyme (EC 1.14.16.5, glyceryl-ether monooxygenase), and the three isozymes of nitric oxide synthase (EC 1.14.13.39).[1,2] 6*R*-BH_4 is enzymatically synthesized from GTP *in vivo* and its levels in the cells are actively regulated.[1,3] Radioactive 6*R*-BH_4 is useful for studying the metabolism of this cofactor, especially in degradation and transport.

[1] C. A. Nichol, G. K. Smith, and D. S. Duch, *Annu. Rev. Biochem.* **54,** 729 (1985).
[2] M. A. Marletta, *J. Biol. Chem.* **268,** 12231 (1993).
[3] T. Harada, H. Kagamiyama, and K. Hatakeyama, *Science* **260,** 1507 (1993).

0076-6879/97 $25

Two chemical methods have been reported for producing reduced forms of biopterin.[4–6] However, these also produce an isobiopterin (7-BH_4)[4,5] or diastereoisomer (6*S*-BH_4)[6] as a side product. Enzymatic synthesis yields only the 6*R* form of BH_4.[1] The method described in the next section was developed for the *in vitro* enzymatic preparation of the pteridine in our laboratory.[7] A similar method was subsequently reported by another group.[8] [U-^{14}C]6*R*-BH_4 has been used for whole-body autoradiography to determine the distribution of plasma 6*R*-BH_4 into tissues of rats of different developmental stages.[9] [3′-^{3}H]6*R*-BH_4 has been used for an *in vitro* binding assay to determine the affinity of nitric oxide synthase for the cofactor.[10]

Strategy

6*R*-BH_4 is produced from GTP by the sequential actions of GTP cyclohydrolase I (EC 3.5.4.16), 6-pyruvoyltetrahydropterin synthase (EC 4.6.1.10), and sepiapterin reductase (EC 1.1.1.153).[1] GTP cyclohydrolase I is present in prokaryotes and eukaryotes, differing from the other two enzymes that have been reported only in eukaryotes. *Escherichia coli* GTP cyclohydrolase I is selected for the enzymatic synthesis of 6*R*-BH_4 because of its higher affinity for the substrate GTP (0.02 μM),[11] compared to the corresponding rat enzyme.[12] For every enzyme, purified preparations are used for the *in vitro* synthesis of 6*R*-BH_4 to minimize the production of by-products by possible contaminating enzymes. The enzymatic synthesis of 6*R*-BH_4 presented here is performed in a single tube without isolation of the products of each enzymatic reaction because the reaction products of 6-pyruvoyltetrahydropterin synthase and sepiapterin reductase are too unstable for isolation. Thus the following conditions for the enzymatic synthesis of 6*R*-BH_4 are extensively optimized to maximize efficiency of each enzymatic reaction and also to stabilize the reaction products.[7]

[4] E. L. Patterson, R. Milstrey, and E. L. R. Stokstad, *J. Am. Chem. Soc.* **78,** 5868 (1956).
[5] H. Rembold and H. Metzger, *Z. Physiol. Chem.* **348,** 194 (1967).
[6] E. C. Tayor and P. A. Jacobi, *J. Am. Chem. Soc.* **98,** 2301 (1976).
[7] K. Hatakeyama, M. Hoshiga, S. Suzuki, and H. Kagamiyama, *Anal. Biochem.* **209,** 1 (1993).
[8] E. R. Werner, M. Schmid, G. Werner-Felmayer, B. Mayer, and H. Wachter, *Biochem. J.* **304,** 189 (1994).
[9] M. Hoshiga, K. Hatakeyama, M. Watanabe, M. Shimada, and H. Kagamiyama, *J. Pharmacol. Exp. Ther.* **267,** 971 (1993).
[10] P. Klatt, M. Schmid, E. Leopold, K. Schmidt, E. R. Werner, and B. Mayer, *J. Biol. Chem.* **269,** 13861 (1994).
[11] J. J. Yim and G. M. Brown, *J. Biol. Chem.* **251,** 5087 (1976).
[12] K. Hatakeyama, T. Harada, S. Suzuki, Y. Watanabe, and H. Kagamiyama, *J. Biol. Chem.* **264,** 21660 (1989).

Chemicals and Enzymes

Chemicals

[U-^{14}C]GTP (500 Ci/mol) and [8,5′-^{3}H]GTP (33.2 Ci/mmol) are purchased from Amersham (Arlington Heights, IL) and New England Nuclear (Boston, MA), respectively. All pteridine compounds used are available from Schirck Laboratories (Jona, Switzerland), except for three pterins that are intermediates in the 6*R*-BH_4 biosynthetic pathway. These pterins are enzymatically synthesized and immediately used as reference compounds for chromatography: 6-pyruvoyltetrahydropterin, a product of the second enzyme 6-pyruvoyltetrahydropterin synthase, is synthesized as described previously[13]; 6-lactoyltetrahydropterin and 6*R*-(1′-hydroxy-2′-oxopropyl)tetrahydropterin, products of the third enzyme sepiapterin reductase, are synthesized according to the method of Smith and Nichol.[14]

Enzymes

Escherichia coli GTP cyclohydrolase I is purified from *E. coli* strain B or JM109 according to the method of Yim and Brown.[11] Rat liver 6-pyruvoyltetrahydropterin synthase is purified as described.[13] Rat erythrocyte sepiapterin reductase is purified according to the method of Sueoka and Katoh,[15] with some modifications.[13] The recombinant proteins of these enzymes purified from bacteria work similarly.[16] Rat liver phenylalanine hydroxylase is purified according to the method of Shiman *et al.*[17] The purified preparations of these enzymes are stored at −70° as aliquots to avoid freezing and thawing.

Procedures

General Precautions

Because 6*R*-BH_4 is sensitive to light, the sample should be protected from light throughout the preparation. All operations are carried out under subdued light.

[13] Y. Inoue, Y. Kawasaki, T. Harada, K. Hatakeyama, and H. Kagamiyama, *J. Biol. Chem.* **266,** 20791 (1991).
[14] G. K. Smith and C. A. Nichol, *J. Biol. Chem.* **261,** 2725 (1986).
[15] T. Sueoka and S. Katoh, *Biochim. Biophys. Acta* **717,** 265 (1982).
[16] K. Hatakeyama and T. Yoneyama, unpublished observations (1995).
[17] R. Shiman, D. W. Gray, and A. Peter, *J. Biol. Chem.* **254,** 11300 (1979).

Synthesis of 6R-L-erythro-5,6,7,8-Tetrahydrobiopterin

The reaction mixture (12.5 μl) contains 50 m*M* potassium phosphate (pH 6.8 at 0.5 *M*), 0.2 m*M* GTP, 100 m*M* KCl, 10 m*M* dithioerythritol, 4 m*M* NADPH, 8 m*M* $MgCl_2$, and 0.1 m*M* EDTA. The specific activity of GTP and the specific isotope used are chosen by the investigator. Dithiothreitol can be used instead of dithioerythritol. The enzymes are finally added to the reaction mixture in the following order: 0.2 μg of sepiapterin reductase (specific activity, 6.8 μmol/min/mg protein) 1 μg of 6-pyruvoyl-tetrahydropterin synthase (0.29 μmol/min/mg protein), and 7.7 μg of GTP cyclohydrolase I (0.06 μmol/min/mg protein). These operations are performed on ice. The reaction is then started by raising the temperature to 37°, continued for 30 min, and stopped by the additon of 3 μl of 1.0 *N* trichloroacetic acid containing 25 m*M* dithiothreitol. The solution is kept on ice for 10 min. After centrifugation, the supernatant solution is extracted twice with diethyl ether to remove trichloroacetic acid.

Isolation of 6R-L-erythro-5,6,7,8-Tetrahydrobiopterin

Separation is performed on a high-performance liquid chromatography system. The reaction products are applied to a Whatman (Clifton, NJ) Partisil 10 SCX column (4.6 × 250 mm) equilibrated with 0.1 *M* ammonium acetate buffer (pH 3.8) containing 5% (v/v) methanol and 1 m*M* dithiothreitol.[14,18] The Milli-Q water (Millipore, Bedford, MA) is extensively degassed, acetate is added to the water, and the pH is then adjusted with ammonium hydroxide. After brief degassing and equilibration with H_2 gas, the solution is mixed with methanol and dithiothreitol. Just before the sample is injected, the column is pretreated with a 50-μl, freshly prepared solution of 1 *M* sodium hydrosulfite. The reaction products are eluted isocratically with the equilibration buffer at 20° and a flow rate of 1.5 ml/min. The eluate is preliminarily monitored by electrochemical or radiochemical detection to determine the elution time of 6*R*-BH_4. A typical chromatogram of the reaction products is shown in Fig. 1. The chromatographic profile (Fig. 1A) detected by an electrochemical detector (at +300 mV) shows a single peak of 6*R*-BH_4 that eluted at 14 min, demonstrating the absence of tetrahydropterin compounds in the reaction mixture that are intermediates of the biosynthetic pathway of 6*R*-BH_4. In the chromatographic profile of radioactivity (Fig. 1B), an isolated major peak is also observed at the same retention time as that of authentic 6*R*-BH_4. In this fraction about 75% of the radioactivity applied to the column is recovered. Two additional minor peaks are observed in the radiochromatogram. The first peak, which is

[18] S. W. Bailey and J. E. Ayling, *J. Biol. Chem.* **253,** 1598 (1978).

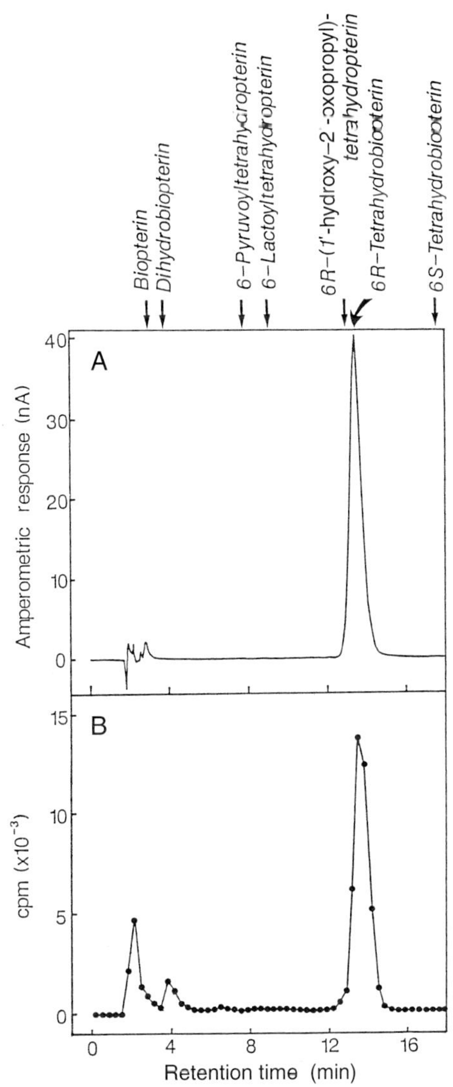

FIG. 1. Chromatograms of the reaction products derived from [U-^{14}C]GTP (25 Ci/mol). The enzyme reaction and chromatography were performed as described in text. The retention time for authentic pteridines is indicated above. (A) The eluate was monitored by electrochemical detection at +300 mV using an LC-4B amperometric detector (Bioanalytical Systems, West Lafayette, IN). (B) The eluate was monitored for radioactivity. The radioactivity of each fraction (500 μl) was quantified with a Packard Tri-Carb spectrometer (model 2200CA, Meriden, CT).

eluted without adsorbing to the column and comprises 10% of the radioactivity applied to the column, probably represents formic acid. Formic acid is a hydrolytic product of GTP in the reaction catalyzed by GTP cyclohydrolase I and is not adsorbed to the cation-exchange column under the conditions used. The second peak at the retention time of 4 min probably represents dihydrobiopterin, an oxidized product of 6*R*-BH_4, because the peak coelutes with authentic dihydrobiopterin and also because biopterin is recovered after acid or alkaline oxidation of a fraction derived from the peak. When the column is not treated with sodium hydrosulfite, this second peak becomes higher; also, the baseline between this peak and the 6*R*-BH_4 peak increases, probably due to continuous oxidation of 6*R*-BH_4 while passing through the column. During the actual preparation, the eluate is not passed through any detector to avoid oxidation of 6*R*-BH_4, and the fraction containing 6*R*-BH_4 is pooled and lyophilized immediately after elution. The lyophilized sample is stored at $-70°$ under N_2 gas, remaining stable for at least 1 year. In our hands the recovery of 6*R*-BH_4 from GTP has been more than 60% in at least 30 preparations.

Characterization of 6R-L-erythro-5,6,7,8-Tetrahydrobiopterin

The specific activity of 6*R*-BH_4 thus obtained is determined on the basis of its radioactivity and the concentration estimated from the absorbance at 266 nm in 0.1 *N* HCl, using the molar extinction coefficient of 16×10^3.[19] A typical preparation starting from [U-^{14}C]GTP with a specific activity of 1×10^6 cpm/nmol gives 1.8 nmol of 6*R*-BH_4 with a specific activity of 9×10^5 cpm/nmol. This specific activity of 6*R*-BH_4 corresponds to 90% of that of the GTP used, a result expected because in the first reaction catalyzed by GTP cyclohydrolase I, one carbon is removed as formic acid from GTP (which contains 10 carbons), and in the subsequent reactions to generate 6*R*-BH_4 no carbon is removed.

The biological activity of 6*R*-BH_4 isolated has been examined for its cofactor activity for phenylalanine hydroxylase. When the K_m and V_{max} values of rat phenylalanine hydroxylase are determined for both authentic 6*R*-BH_4 and [U-^{14}C]6*R*-BH_4, using a purified enzyme preparation, identical values of K_m (13 μM) and V_{max} (3.1 μmol/min/mg protein) are obtained for both cofactors.

Acknowledgments

The original work presented here was carried out at Osaka Medical College. We gratefully acknowledge helpful discussions with Dr. Setsuko Katoh (Meikai University) and Dr. Toshie

[19] T. Fukushima, K. Kobayashi, I. Eto, and T. Shiota, *Anal. Biochem.* **89,** 71 (1978).

Yoneyama (University of Pittsburgh), the generous supply of 6*R*,*S*-BH_4 from Dr. Sadao Matsuura (Fujita Health University) and sepiapterin from Dr. Masahiro Masada (Chiba University), and critical reading of the manuscript by Dr. Sidney M. Morris, Jr. (University of Pittsburgh).

[16] Use of 10-Formyl-5,8-dideazafolate as Substrate for Rat 10-Formyltetrahydrofolate Dehydrogenase

By ROBERT J. COOK

10-Formyltetrahydrofolate (10-HCO-H_4PteGlu; Fig. 1)[1] is susceptible to oxidative degradation[2] and must be protected by reducing agents during synthesis and use in enzyme assays. (6*R*,*S*)-10-HCO-H_4PteGlu, while highly unstable, is easily generated from stable, commercially available (6*R*,*S*)-5-HCO-H_4PteGlu in the presence of 2-mercaptoethanol (2-ME), using the method of Rabinowitz.[3] (6*R*)-10-HCO-H_4PteGlu, the naturally occurring isomer, is much more difficult to prepare and involves the conversion of dihydrofolate to (6*S*)-tetrahydrofolate by dihydrofolate reductase (EC 1.5.1.3) followed by reaction with formic acid to give (6*R*)-5,10-methenyltetrahydrofolate and subsequent conversion to (6*R*)-10-HCO-H_4PteGlu as summarized by Nagy *et al.*[4]

10-Formyl-5,8-dideazafolate (Fig. 1) was originally synthesized as a quinazoline analog of folic acid. It was found to be a modest inhibitor of rat liver dihydrofolate reductase and had activity against L1210 leukemia in mice.[5] Work on glycinamide ribotide transformylase (EC 2.1.2.2; GARTF, phosphoribosylglycinamide formyltransferase) and 5-aminoimidazole-4-carboxamide ribotide transformylase (EC 2.1.2.3; AICARTF, phosphoribosylaminoimidazolecarboxamide formyltransferase) with N^{10}-formyldeazafolate analogs as possible inhibitors showed, surprisingly, that some of these compounds could function as substrates.[6] This led to the finding that (6*R*)-

[1] This work was supported by NIH Grants AM 15289 and DK 49563 and the Medical Research Service of the Veterans Administration.

[2] R. L. Blakley, "The Biochemistry of Folic Acid and Related Pteridines," p. 80. North-Holland, Amsterdam, 1969.

[3] J. C. Rabinowitz, *Methods Enzymol.* **6,** 814 (1963).

[4] P. L. Nagy, A. Marolewski, S. J. Benkovic, and H. Zalkin, *J. Bacteriol.* **177,** 1292 (1995).

[5] J. B. Hynes, D, E. Eason, C. M. Garrett, P. L. Colvin, K. E. Shores, and J. H. Freisheim, *J. Med. Chem.* **20,** 588 (1977).

[6] G. K. Smith, W. T. Mueller, P. A. Benkovic, and S. J. Benkovic, *Biochemistry* **20,** 1241 (1981).

0076-6879/97 $25

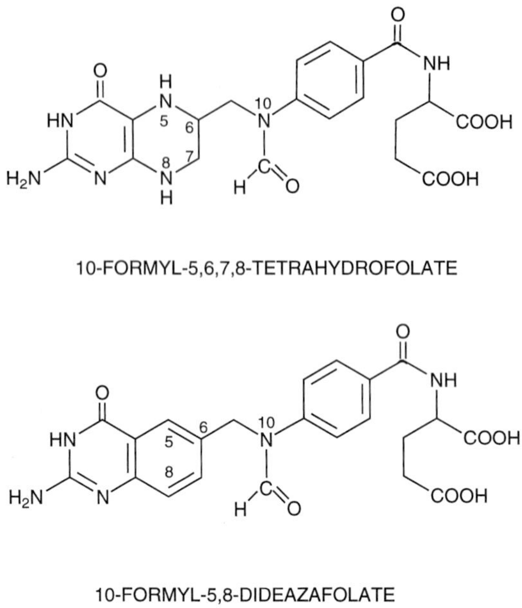

FIG. 1. Structure of 10-formyltetrahydrofolate and 10-formyl-5,8-dideazafolate.

10-HCO-H_4PteGlu was the specific cofactor for GARTF, while the (6*S*)-isomer was a potent inhibitor.[7] In particular, 10-formyl-5,8-dideazafolate was an excellent substrate for GARTF, with a relative rate of 77% when compared to (6*R*)-10-HCO-H_4PteGlu.[6] Since these observations were made 10-formyl-5,8-dideazafolate has become the standard substrate for routine GARTF assays. The current assay procedure for GARTF is detailed by Inglese *et al.*[8]

10-Formyl-5,8-dideazafolate has two major advantages over 10-HCO-H_4PteGlu: it is much more stable to oxidative degradation, because the tetrahydropyrazine ring is replaced by a benzene ring, and it has no asymmetry at the C-6 position. 10-Formyl-5,8-dideazafolate may be synthesized by the method of Acharya and Hynes[9] or may be purchased from J. Hynes (Department of Pharmaceutical Chemistry, Medical University of South Carolina, Charleston, SC).

[7] G. K. Smith, P. A. Benkovic, and S. J. Benkovic, *Biochemistry* **20,** 4034 (1981).

[8] J. Inglese, D. L. Johnson, A. Shiau, J. M. Smith, and S. J. Benkovic, *Biochemistry* **29,** 1436 (1990).

[9] S. A. Acharya and J. B. Hynes, *J. Heterocyclic Chem.* **12,** 1283 (1975).

10-Formyltetrahydrofolate dehydrogenase (EC 1.5.1.6; FTD) is a tetramer of identical 99-kDa subunits and is an abundant protein in rat,[10,11] rabbit,[12] and pig liver cytosol.[11,13] The cloned rat liver FTD has 902 amino acids[14] and has been expressed by inset cell culture techniques.[15] It was observed by Kutzbach and Stokstad[16] that impure pig liver FTD was able to catalyze two reactions:

$$(6R)\text{-10-HCO-H}_4\text{PteGlu} + \text{NADP}^+ \xrightarrow{\text{FTD}} (6S)\text{-H}_4\text{PteGlu} + \text{NADPH} + \text{H}^+ + \text{CO}_2 \quad (1)$$

$$(6R)\text{-10-HCO-H}_4\text{PteGlu} + \text{H}_2\text{O} \xrightarrow{\text{FTD}} (6S)\text{-H}_4\text{PteGlu} + \text{formate} \quad (2)$$

Both reactions use only (6*R*)-10-HCO-H_4PteGlu. (6*S*)-10-HCO-H_4PteGlu neither reacts with nor inhibits FTD.[16] In the absence of $NADP^+$ a hydrolase activity [Eq. (2)], which varied between 15 and 30% of the dehydrogenase activity, was also seen.[16] Subsequently, pure preparations of rat,[14,17] rabbit,[12] and pig[13] liver FTD have shown the same activities. FTD hydrolase activity [Eq. (2)] of native rat liver and recombinant, insect cell-expressed enzymes were both strictly dependent on the presence of, and were proportional to, the concentration 2-mercaptoethanol (2-ME) in the reaction mixture.[18,19] This is also true for rabbit liver FTD.[12,20] On the basis of these observations the hydrolase activity of FTD may be an artifact of the assay mixture and is nonphysiological; however, the reaction is characteristic of FTD and can be used to confirm the presence of the enzyme.

FTD has two functional domains. The N-terminal domain shows identity with GARTF and L-methionyl-tRNA formyltransferase (EC 2.1.2.9; FMT), which also uses (6*R*)-10-HCO-H_4PteGlu to catalyze the addition of a formyl group to L-methionyl-tRNA, a required but not essential step for the initiation of bacterial and mitochondrial protein synthesis.[21] It is the N-terminal domain that catalyzes the 2-ME-dependent hydrolase reaction[12,22] and con-

[10] R. J. Cook and C. Wagner, *Biochemistry* **21,** 4427 (1982).
[11] H. Min, B. Shane, and E. L. R. Stokstad, *Biochim. Biophys. Acta* **967,** 348 (1988).
[12] D. Schirch, E. Villar, D. Barra, and V. Schirch, *J. Biol. Chem.* **269,** 24728 (1994).
[13] E. M. Rios-Orlandi, C. G. Zarkadas, and R. E. MacKenzie, *Biochim. Biophys. Acta* **871,** 24 (1986).
[14] R. J. Cook, R. S. Lloyd, and C. Wagner, *J. Biol. Chem.* **266,** 4965 (1991).
[15] S. A. Krupenko, C. Wagner, and R. J. Cook, *Prot. Expression Purif.* **6,** 457 (1995).
[16] C. Kutzbach and E. L. R. Stokstad, *Methods Enzymol.* **18B,** 793 (1971).
[17] M. C. Scrutton and I. Beis, *Biochem. J.* **177,** 833 (1979).
[18] S. A. Krupenko, C. Wagner, and R. J. Cook, *Biochem. J.* **306,** 651 (1995).
[19] R. J. Cook and C. Wagner, *Arch. Biochem. Biophys.* **321,** 336 (1995).
[20] V. Schirch, *Methods Enzymol.* **281,** [19], 1997 (this volume).
[21] D. Kahn, M. Fromant, G. Fayat, P. Dessen, and S. Blanquet, *Eur. J. Biochem.* **105,** 489 (1980).
[22] S. A. Krupenko, C. Wagner, and R. J. Cook, *J. Biol. Chem.* **272,** 10273 (1997).

tains the folate-binding site.[22] The C-terminal domain shows a 50% identity to NAD-dependent aldehyde dehydrogenase (EC 1.2.1.3),[19] catalyzes the $NADP^+$-dependent oxidative reaction,[12,23] and has a highly conserved active site cysteine (Cys-707).[24] The active site sulfhydryl group of FTD is highly susceptible to oxidation and must be kept in the presence of reducing agents.[19] Thus FTD binds (6*R*)-10-HCO-H_4PteGlu in the N-terminal domain and catalyzes the $NADP^+$-dependent oxidation reaction with the C-terminal domain.

The identity of the N-terminal domain of FTD with GARTF and FMT suggested that 10-formyl-5,8-dideazafolate may be a substrate for FTD. Dehydrogenase assays with recombinant FTD showed the K_m values for 10-formyl-5,8-dideazafolate and (6*R*)-10-HCO-H_4PteGlu, in a (6*R*,*S*)-10-HCO-H_4PteGlu mix, to be 3.2 and 5.5 μM, and the K_m values for $NADP^+$ were 0.86 and 0.88 μM, respectively.[18] The hydrolase reaction gave K_m values of 5.8 and 11.0 μM for 10-formyl-5,8-dideazafolate and (6*R*)-10-HCO-H_4PteGlu, in a (6*R*,*S*)-10-HCO-H_4PteGlu mix.[18] The V_{max} for the dehydrogenase reaction with 10-formyl-5,8-dideazafolate was 55% of that with (6*R*)-10-HCO-H_4PteGlu, in a (6*R*,*S*)-10-HCO-H_4PteGlu mix. The V_{max} values for the hydrolase reactions were essentially identical.[18]

Assay Method

Principle

The oxidation of 10-formyl-5,8-dideazafolate is followed by the production of 5,8-dideazafolate at 295 nm ($\Delta\varepsilon$ of 18.9 mM^{-1} cm^{-1})[6] or NADPH at 340 nm ($\Delta\varepsilon$ of 8.5 mM^{-1} cm^{-1}). The absorption of 5,8-dideazafolate at 340 nm is reflected in the adjusted extinction coefficient for NADPH. The absorption of NADPH at 295 nm is approximately 2% of the extinction coefficient for 5,8-dideazafolate and is ignored. Hydrolase activity is followed by the production of 5,8-dideazafolate at 295 nm ($\Delta\varepsilon$ of 18.9 mM^{-1} cm^{-1}) in the presence of 100 m*M* 2-ME. The assay may be run at room temperature. Assays of crude extracts require the removal of low molecular weight compounds by spin column desalting.

Reagents

Tris-HCl buffer, 0.6 *M*, pH 7.8
2-Mercaptoethanol, 1.0 *M*

[23] S. A. Krupenko, C. Wagner, and R. J. Cook, *J. Biol. Chem.* **272,** 10266 (1997).
[24] S. A. Krupenko, C. Wagner, and R. J. Cook, *J. Biol. Chem.* **270,** 519 (1995).

10-Formyl-5,8-dideazafolate, 6 mM
$NADP^+$, 10 mM

Procedure

Hydrolase Reaction. Pipette into a 1-cm light path cuvette (1.0- to 1.5-ml capacity) 100 μl each of Tris-HCl buffer and 2-ME, 10 μl of 10-formyl-5,8-dideazafolate, plus extract or enzyme to a final volume of 1 ml. Production of 10-formyl-5,8-dideazafolate is followed at 295 nm ($\Delta\varepsilon$ of 18.9 mM^{-1} cm^{-1}). Enzyme activity is expressed as nanomoles per minute.

Dehydrogenase Reaction. The dehydrogenase reaction may be determined after estimation of the hydrolase activity by adding 10 μl of $NADP^+$ to the reaction mixture. The combined hydrolase/dehydrogenase reactions can be followed at 295 nm or the dehydrogenase may be specifically followed at 340 nm ($\Delta\varepsilon$ of 8.5 mM^{-1} cm^{-1}). This continuous assay saves reagents and time and gives a good estimate of FTD activity. To assay the dehydrogenase reaction specifically, add only 10 μl of 2-ME to the hydrolase reaction mix to maintain the active site cysteine. Under these conditions there should only be a very slow rate of hydrolyase activity.[19] The absorption at 295 nm is recorded for 2–3 min to establish a base hydrolytic rate. The dehydrogenase reaction is started by the addition of 10 μl of $NADP^+$ to the reaction mixture and the activity is followed at 295 nm ($\Delta\varepsilon$ of 18.9 mM^{-1} cm^{-1}) or 340 nm ($\Delta\varepsilon$ of 8.5 mM^{-1} cm^{-1}).

Comments

The assay used for FTD was originally designed by Kutzbach and Stokstad,[16] who routinely used 100 mM 2-ME presumably to protect 10-HCO-H_4PteGlu from oxidative degradation. High concentrations of 2-ME facilitate the hydrolase reaction.[19] A 10 mM concentration of 2-ME is adequate to protect 10-HCO-H_4PteGlu and the active site cysteine in the reaction mix during the relatively short time course of the assay (3–5 min). The assays described above can be run with 100–200 μM (6R,S)-10-HCO-H_4PteGlu generated from (6R,S)-5-HCO-H_4PteGlu by the method of Rabinowitz.[3]

A characteristic of FTD is that H_4PteGlu, one of the reaction products, is a potent, competitive inhibitor of the dehydrogenase reaction and a noncompetitive inhibitor of the hydrolase reaction.[16,17] Thus even over short time periods (2–5 min) there is a slowing of the rate due to H_4PteGlu. 5,8-Dideazafolate also appears to act in a similar manner (R. J. Cook, unpublished observations).

There are five enzymes that utilize the formyl group of 10-HCO-H_4PteGlu. GARTF and AICARTF catalyze the addition of one-carbon groups in *de novo* purine biosynthesis. FMT facilitates the addition of a formyl group to L-methionyl-tRNA.[21] FTD oxidizes excess one-carbon groups in the form of 10-HCO-H_4PteGlu to CO_2 and recycles H_4PteGlu. 10-Formyltetrahydrofolate hydrolase (FH) is a bacterial enzyme that monitors the folate pools and supplies formate for purine biosynthesis.[4] FH is described in detail in [25] in this volume.[25] Each of these enzymes uses (6*R*)-10-HCO-H_4PteGlu but is unaffected by (6*S*)-10-HCO-H_4PteGlu, with the exception of GARTF, which is potently inhibited by the (6*S*)-isomer.[7] The effect of (6*S*)-10-HCO-H_4PteGlu on FH has not been tested, as it has been assayed only with (6*R*)-10-HCO-H_4PteGlu and 10-formyl-5,8-dideazafolate.[4] GARTF, FDH, and FH each are able to use 10-formyl-5,8-dideazafolate. It is speculated that FMT should also be able to use 10-formyl-5,8-dideazafolate.

[25] H. J. Zalkin, *Methods. Enzymol.* **281,**]25], 1997 (this volume).

[17] Human Cytosolic Folylpoly-γ-glutamate Synthase

By IAN ATKINSON, TIMOTHY GARROW, ALBERT BRENNER, and BARRY SHANE

Folylpolyglutamates, the major intracellular forms of folate, are the preferred coenzymes of one-carbon metabolism.[1,2] The polyglutamate forms of the vitamin are also required for intracellular retention of folate. Intracellular metabolism of pteroylmonoglutamates to their polyglutamate derivatives is catalyzed by the enzyme folylpolyglutamate synthase [FPGS; tetrahydrofolate: L-glutamate γ-ligase (ADP forming), EC 6.3.2.17]. The human enzyme catalyzes the following reaction:

$$H_4PteGlu_n + MgATP + \text{L-glutamate} \rightarrow H_4PteGlu_{n+1} + MgADP + P_i$$

where $H_4PteGlu_n$ is tetrahydropteroylpoly-γ-glutamate, with n (equals 1 to 11) indicating the number of glutamate moieties. The major folylpolyglutamate species in human cells are hepta- and octaglutamates.

Mammalian FPGS is a low-abundance protein, which has hindered its isolation in sufficient quantities for detailed physical analysis. The pig liver enzyme has been purified to homogeneity but the high enrichment required (more than 200,000-fold) has limited its characterization to kinetic

[1] J. J. McGuire and J. K. Coward, *in* "Folates and Pterins" (R. L. Blakley and S. J. Benkovik, eds.), Vol. 1, p. 135. Wiley Interscience, New York, 1984.
[2] B. Shane, *Vitam. Horm.* **45,** 263 (1989).

0076-6879/97 $25

analyses.[3] The cloning of a cDNA-encoding human FPGS has now allowed overexpression of the protein in bacteria and insect cells and has made possible the production of sufficient purified protein to allow mechanistic studies.[4,5]

Mammalian tissues possess both mitochondrial and cytosolic FPGS isozymes,[6] which are encoded by a single gene that has been localized to the human chromosome 9q34.1 region.[5,7] Exon 1 of this gene contains two ATG codons separated by 126 base pairs (bp). Alternate transcription start sites in this region generate two classes of mRNA containing or lacking the first ATG codon.[5,7] Translation from the first ATG generates the mitochondrial isozyme with a 42-amino acid leader sequence. Translation from the second ATG generates the cytosolic isozyme. Additional splice variants of the FPGS gene, containing alternate exon 1 forms, have been described, one of which may encode an additional mitochondrial isozyme. However, these appear to be less abundant and whether they are translated into functional FPGS proteins remains to be determined.[5]

In this chapter we describe the expression of human FPGS in *Escherichia coli* and the purification and properties of the enzyme.

Folylpolyglutamate Synthase Assay

Folylpolyglutamate synthase activity is normally measured by the incorporation of [^{14}C]glutamate into folylpolyglutamate products using (6*ambo*)-H_4PteGlu as the substrate.[3,8] Assay mixtures normally contain 100 m*M* Tris–50 m*M* glycine buffer (pH 9.75 at 22°), (6*ambo*)-H_4PteGlu (40 μM), L-[^{14}C]glutamate (250 μM, 0.625 μCi), ATP (5 m*M*), $MgCl_2$ (10 m*M*), KCl (20 m*M*), 2-mercaptoethanol (100 m*M*), bovine serum albumin (50 μg), and enzyme in a total volume of 0.5 ml. Assays are normally conducted for 1 hr at 37°. In kinetic studies with pure enzyme, the assay mixtures contain various concentrations of the substrate under investigation and fixed concentrations of (6*ambo*)-H_4PteGlu (40 μM), ATP (1 m*M*), and L-glutamate (2 m*M*), as appropriate. The amount of FPGS is adjusted to ensure that less than 10% of the limiting substrate is converted to product at the lowest substrate concentration used.

[3] D. J. Cichowicz and B. Shane, *Biochemistry* **26,** 504 (1987).

[4] T. A. Garrow, A. Admon, and B. Shane, *Proc. Natl. Acad. Sci. U.S.A.* **89,** 9151 (1992).

[5] L. Chen, H. Qi, J. Korenberg, T. A. Garrow, Y.-J. Choi, and B. Shane, *J. Biol. Chem.* **271,** 13077 (1996).

[6] B.-F. Lin, R.-F. S. Huang, and B. Shane, *J. Biol. Chem.* **268,** 21674 (1993).

[7] S. J. Freemantle, S. M. Taylor, G. Krystal, and R. G. Moran, *J. Biol. Chem.* **270,** 9579 (1995).

[8] J. J. McGuire, P. Hsieh, J. K. Coward, and J. R. Bertino, *J. Biol. Chem.* **255,** 5776 (1980).

The reaction is stopped by the addition of ice-cold-2-mercaptoethanol (30 m*M*, 1.5 ml) containing 10 m*M* glutamate and the mixture is applied to a DEAE-cellulose (DE52; Whatman, Clifton, NJ) column (2 × 0.7 cm), protected by a 3-mm layer of nonionic cellulose, that has been equilibrated with 10 m*M* Tris buffer (pH 7.5) containing 80 m*M* NaCl and 30 m*M*-2-mercaptoethanol. Unreacted glutamate is eluted with the equilibration buffer (three times, 5 ml each) and the labeled folate product is eluted with 0.1 *N* HCl (3 ml).

Expression of Human Folylpolyglutamate Synthase in *Escherichia coli*

An *Nde*I restriction site was introduced at the cytosolic (second) ATG in the human FPGS cDNA (GenBank accession number M98045) by mutagenesis.[4,5] A 2117-bp *Nde*I–*Bam*HI fragment containing the entire open reading frame of human FPGS was ligated into similarly treated pET3A (Novagen, Madison, WI). Plasmid pET3A has the bacteriophage T7 gene 10 promoter and transcriptional termination sequences. This new construct was transformed into *E. coli* JM109(λDE3) for the production of unfused human cytosolic FPGS. JM109(λDE3) is a lambda (λ) lysogen that has the bacteriophage T7 RNA polymerase linked to the *lacUV5* promoter.

JM109(λDE3) freshly transformed with the human FPGS cDNA construct is grown to midlog phase in Luria medium containing ampicillin (50 μg/ml) and 0.5% (v/v) glycerol (15 liter). The cells are collected by centrifugation at 650 *g* for 10 min at 4° and the cells extracted as described below or stored at −20° until use.

Expression of FPGS from this construct is variable but usually represents about 0.5% of the total soluble protein in *E. coli* crude extracts. Use of a modified cDNA in which the first 27 codons are optimized for preferred codon usage in *E. coli* increases expression up to a maximum of about 2% of soluble protein. Isopropyl-β-D-thiogalactopyranoside (IPGT) induction of T7 polymerase reduces or eliminates expression of active soluble human FPGS but does result in high levels of inactive protein in inclusion bodies (greater than 50 mg of protein per liter of culture). Attempts at resolubilizing the nearly pure protein in inclusion bodies have met with only limited success.

Enzyme Purification

A typical purification of the human FPGS protein from *E. coli* extracts is shown in Table I and is outlined below. The purification obtained is from 50- to 600-fold, depending on the starting extract, and the purity of the

TABLE I
PURIFICATION OF HUMAN FOLYLPOLYGLUTAMATE SYNTHASE

Fraction	Volume (ml)	Activity (units/ml)[a]	Protein (mg/ml)	Specific activity (units/mg)	Purification (-fold)	Yield (%)
1. Crude extract	325	2.56	30.6	0.083	1	100
2. Hydroxylapatite	450	1.54	8.90	0.173	2	83
3. Affi-Gel Blue	224	2.66	1.39	1.91	23	71
4. Phenyl-agarose	213	1.18	0.156	7.58	92	30
5. Heparin–agarose	20	7.92	0.671	11.8	143	19
6. DEAE-cellulose	11	12.9	0.701	18.4	223	17

[a] Micromoles of glutamate incorporated into folate product per hour. Assay mixtures contained 5 m*M* ATP, 40 μM H_4PteGlu, and 250 μM glutamate.

final preparation is greater than 99%, as judged by the absence of contaminating bands after sodium dodecyl sulfate-gel electrophoresis.

Step 1. Crude Extract. All buffer solutions are adjusted to the indicated pH at room temperature. Extracts are maintained at 0° and all other procedures are performed at 0–4°. The cell paste (114 g) is resuspended in 100 m*M* Tris-HCl buffer, pH 7.5 (300 ml), containing 2.5 m*M* EDTA, 50 m*M* mercaptoethanol, and 1 m*M* phenylmethylsulfonyl fluoride (PMSF), and the cells disrupted using a Branson (Danbury, CT) sonicator (power setting 8, using a 50% duty cycle for three 8-min cycles). The extract is then centrifuged at 12,000 *g* for 45 min and the supernatant decanted through a double layer of cheesecloth to give the crude extract (fraction 1).

Step 2. Hydroxylapatite Chromatography. The crude extract (325 ml) is applied to a hydroxylapatite (HTP; Bio-Rad, Richmond, CA) column (7.5 × 7.5 cm) that has been equilibrated with 100 m*M* Tris-HCl buffer, pH 7.5, containing 2.5 m*M* EDTA and 50 m*M* mercaptoethanol (buffer A). The column is washed with the equilibration buffer (1000 ml) and eluted with a linear gradient (1500 ml) of potassium phosphate buffer, pH 7.5 (0–150 m*M*) in the same buffer. Fractions containing FPGS activity are pooled (fraction 2).

Step 3. Affi-Gel Blue Chromatography. Fraction 2 enzyme (450 ml) is applied to an Affi-Gel Blue (Bio-Rad) column (20 × 1.5 cm) that has been equilibrated with buffer A. The column is washed with buffer A containing 150 m*M* KCl (300 ml) and a linear gradient (500 ml) of KCl (0.15–1 *M*) in the same buffer is used to elute the enzyme. Fractions containing FPGS activity are pooled (fraction 3).

Step 4. Phenyl-Agarose Chromatography. Fraction 3 enzyme (224 ml) is applied to a phenyl-agarose [Bethesda Research Laboratories (BRL),

Gaithersburg, MD] column (20 × 1 cm) that has been equilibrated with 100 m*M* Tris-HCl buffer, pH 7.5, containing 1 m*M* EDTA, 600 m*M* KCl, and 50 m*M* mercaptoethanol. The column is washed with the equilibration buffer (180 ml) and enzyme activity is eluted with 100 m*M* Tris-HCl buffer, pH 8.2, containing 1 m*M* EDTA, 50 m*M* mercaptoethanol, and 20% (v/v) ethylene glycol. Active fractions are pooled and dialyzed twice against 12 liters of 100 m*M* Tris-HCl buffer, pH 7.5, containing 0.5 m*M* EDTA and 50 m*M* mercaptoethanol.

Step 5. Heparin–Agarose Chromatography. The dialyzed enzyme (fraction 4, 213 ml) is applied to a heparin–agarose (BRL) column (20 × 1 cm) equilibrated with 100 m*M* Tris-HCl buffer, pH 7.5, containing 1 m*M* EDTA and 50 m*M* mercaptoethanol. The column is washed with 150 ml of equilibration buffer and eluted with a linear gradient (300 ml) of KCl (0–500 m*M*) in the same buffer. Fractions containing FPGS activity are pooled and dialyzed twice against 4 liters of 50 m*M* Tris-HCl buffer, pH 8.4, containing 0.5 m*M* EDTA, 50 m*M* mercaptoethanol, and 10% (v/v) ethylene glycol (fraction 5).

Step 6. DEAE-Cellulose Chromatography. Fraction 5 enzyme (20 ml) is applied to a DE52 (Whatman) column (20 × 1 cm) equilibrated with 50 m*M* Tris-HCl buffer, pH 8.4, containing 0.5 m*M* EDTA, 10% (v/v) ethylene glycol, and 50 m*M* mercaptoethanol. The column is washed with equilibration buffer (150 ml) and eluted with a linear gradient (300 ml) of KCl (0–500 m*M*) in the same buffer. Fractions containing FPGS activity are pooled, giving fraction 6 (11 ml).

Fraction 6 enzyme can be stored at 0–4° for several days or at −20° for longer periods. The stability of the purified protein has not been rigorously investigated. However, many of the problems of extreme lability previously encountered with the highly dilute purified preparations of other mammalian FPGS proteins[3] appear to have been alleviated by the higher concentrations of human protein obtained after overexpression.

General Properties

Human FPGS is a monomeric protein. A single band with a relative molecular weight of about 61,000 is obtained after sodium dodecyl sulfate-gel electrophoresis, which is consistent with the amino acid sequence deduced from the cDNA sequence[4] (545 amino acids, M_r 60,128), and the apparent size obtained after native gel chromatography. The general properties of human FPGS are similar to those previously reported for homogeneous pig liver FPGS.[3] Activity requires a monovalent cation. K^+ (20 m*M*) is most effective, followed by NH_4^+ and Rb^+, while Na^+, Li^+, and Cs^+ are ineffective. There is an absolute requirement for a reducing agent (tested using aminopterin as a substrate) and the enzyme displays a high pH optimum (pH 9.6). The K_m for L-glutamate is about 200 μM (pH 9.6, H_4PteGlu substrate) and increases at lower pH values and with other folate substrates. The K_m for MgATP is also about 200 μM (pH 9.6, H_4PteGlu substrate) and is decreased fourfold by the addition of bicarbonate (10 mM), which also increases the k_{cat} by about 20%.

Specificity for Folates and Folate Analogs

Kinetic constants for a variety of folates and folate analogs are shown in Table II. Marked substrate inhibition is observed with many of the more effective substrates for the enzyme. Unsubstituted reduced folates are the preferred substrates for human FPGS. H_2PteGlu is the most effective pteroylmonoglutamate substrate, followed by H_4PteGlu. Both 6*S*- and

TABLE II
KINETIC CONSTANTS OF FOLATES AND FOLATE ANALOGS FOR HUMAN FOLYLPOLYGLUTAMATE SYNTHASE[a]

Substrate	K_m (μM)	k_{cat} (sec^{-1})	k_{cat}/K_m (M^{-1} sec^{-1} $\times 10^{-3}$)
PteGlu	59	0.65	11
$PteGlu_2$	16	0.82	52
$PteGlu_3$	20	0.55	27
$PteGlu_4$	12	0.34	29
$PteGlu_5$	64	0.01	0.16
H_2PteGlu	0.81	0.84	1035
$H_2PteGlu_2$	47	0.95	19
(6*ambo*)-H_4PteGlu	1.6	0.82	518
(6*S*)-H_4PteGlu	4.4	0.99[b]	225
(6*S*)-$H_4PteGlu_2$	3.3	0.81	245
(6*S*)-$H_4PteGlu_3$	1.4	0.19	135
(6*S*)-$H_4PteGlu_4$	1.6	0.10	65
(6*S*)-$H_4PteGlu_5$	1.4	0.006	4.3
(6*R*)-10-Formyl-H_4PteGlu	3.7	0.23	61
(6*R*)-10-Formyl-$H_4PteGlu_2$	2.7	0.30	110
(6*S*)-5-Formyl-H_4PteGlu	105	0.84	8.1
(6*S*)-5-Formyl-$H_4PteGlu_2$	13	0.67	52
(6*S*)-5-Methyl-H_4PteGlu	48	0.82	17
Aminopterin	4.4	1.17	266
Methotrexate (Glu-1)	71	0.93	13
Methotrexate (Glu-2)	50	0.16	3.2
Methotrexate (Glu-3)	148	0.032	0.19

[a] Fixed substrates were ATP (1 mM) and L-glutamate (2 mM).
[b] A k_{cat} value of 0.99 sec^{-1} is equivalent to 59 μmol/hr/mg protein.

$6R$-H_4PteGlu are effective substrates. Oxidized pteroylglutamic acid (PteGlu) is a fairly poor substrate, with a higher K_m value and a reduced on rate (k_{cat}/K_m). Substitution at the 10-position decreases k_{cat} while substitution at the 5-position of folates causes a large increase in K_m values. The maximal catalytic rate of human FPGS (1 sec^{-1}) is about 40% that of the pig liver enzyme.[9]

H_4PteGlu$_n$ derivatives are the most effective polyglutamate substrates. Extension of the glutamate chain decreases K_m values but also causes a decrease in V_{max} with chain lengths beyond the diglutamate. With most folates, the diglutamate derivative is a more effective substrate than the monoglutamate, owing primarily to a decrease in K_m value, a decrease that is most pronounced for 5-formyl-H_4PteGlu$_2$. However, H_2PteGlu$_2$ is a poor substrate owing to a large increase in K_m value.

Aminopterin is a good substrate for the enzyme, with the 4-amino substitution decreasing K_m values by about 15-fold compared to PteGlu, while methotrexate is less effective. As has been found for other mammalian FPGS enzymes, extension of the glutamate chain of 4-aminofolates dramatically decreases substrate effectiveness, primarily due to a drop in k_{cat} values.[10,11]

This specificity for folylpolyglutamate substrates is qualitatively similar to that reported for the pig liver enzyme[9] except that diglutamate derivatives, compared to their respective pteroylmonoglutamates, tend to be somewhat better substrates for the human enzyme. Similarly, the large decrease in substrate effectiveness of 4-aminofolates with extension of the glutamate chain is pronounced at the diglutamate level with the pig liver and other mammalian FPGS enzymes, while activity falls off less sharply with the human enzyme and a large fall-off in activity occurs at the triglutamate level. This relatively improved substrate activity of the diglutamate makes human cells particularly sensitive to methotrexate, as methotrexate accumulation by mammalian cells requires its metabolism to at least the triglutamate derivative.

Acknowledgment

This work was supported in part by National Cancer Institute Grant CA41991.

[9] D. J. Cichowicz and B. Shane, *Biochemistry* **26,** 513 (1987).
[10] S. George, D. J. Cichowicz, and B. Shane, *Biochemistry* **27,** 522 (1987).
[11] R. J. Coll, D. Cesar, J. B. Hynes, and B. Shane, *Biochem. Pharmacol.* **42,** 833 (1991).

[18] Folylpolyglutamate Synthase from Higher Plants

By HELENA C. IMESON and EDWIN A. COSSINS

Although folylpolyglutamate synthase activity (FPGS, EC 6.3.2.17) has been reported from a number of different plant species,[1-4] only that from pea seeds has been purified and characterized.[5,6]

$$H_4PteGlu_n + ATP + \text{L-glutamate} \rightarrow H_4PteGlu_{n+1} + ADP + P_i$$

Assay Method

Principle. The assay is based on the incorporation of tritiated L-glutamate into folylpolyglutamates with tetrahydrofolate monoglutamate [(6*R,S*)-H_4PteGlu], serving as the folate substrate. The procedure is essentially that described by Cossins and Chan[7] for assay of the *Neurospora* enzyme. The isolation of polyglutamate products involves an ion-exchange procedure that is a modification of that used by Pristupa *et al.*[8]

Reagents

Tris-HCl, 0.5 *M*, pH 8.5
ATP, 0.05 *M*
$MgCl_2 \cdot 6H_2O$, 0.1 *M*
(6*R,S*)-H_4PteGlu, 1 m*M*
L-Glutamic acid, 0.01 *M*
2-Mercaptoethanol, 2 *M*

Procedure. (6*R,S*)-Tetrahydrofolate monoglutamate, prepared from folic acid by catalytic hydrogenation,[9] is adjusted to a final concentration of 1 m*M* by the addition of 2 *M* 2-mercaptoethanol. The reaction mixture contains 100 μmol of Tris-HCl (pH 8.5), 5 μmol of ATP, 5 μmol of $MgCl_2$,

[1] P. Y. Chan, J. W. Coffin, and E. A. Cossins, *Plant Cell Physiol.* **27,** 431 (1986).
[2] H. C. Imeson, L. Zheng, and E. A. Cossins, *Plant Cell Physiol.* **31,** 223 (1990).
[3] K. Wu, E. A. Cossins, and J. King, *Plant Physiol.* **104,** 373 (1994).
[4] M. Neuburger, F. Rebeille, A. Jourdain, S. Nakamura, and R. Douce, *J. Biol. Chem.* **271,** 9466 (1996).
[5] H. C. Imeson and E. A. Cossins, *J. Plant Physiol.* **138,** 476 (1991).
[6] H. C. Imeson and E. A. Cossins, *J. Plant Physiol.* **138,** 483 (1991).
[7] E. A. Cossins and P. Y. Chan, *Phytochemistry* **23,** 965 (1984).
[8] Z. B. Pristupa, P. J. Vickers, G. B. Sephton, and K. G. Scrimgeour, *Can. J. Biochem. Cell Biol.* **62,** 495 (1984).
[9] V. E. Reid and M. Freidkin, *Mol. Pharmacol.* **9,** 74 (1973).

0076-6879/97 $25

0.1 μmol of (6*R*,*S*)-H_4PteGlu, 100 μmol of 2-mercaptoethanol, 1.5 μmol of L-glutamate containing 2.5 μCi of L-[^{3}H]glutamate, and enzyme protein in a final volume of 1 ml. Controls lack the folate substrate. Reaction tubes are purged with hydrogen, capped, and incubated at 37° for 2 hr. The reaction is terminated by the addition of 3 ml of 93 m*M* sodium acetate (pH 5.2)[8] and the tubes are incubated at 100° for 5 min to denature protein that is subsequently removed by centrifugation. Tritiated polyglutamates are separated from excess L-[^{3}H]glutamate using Cellex D cellulose (Bio-Rad Laboratories, Richmond, CA). Bio-Rad Econo-Columns are filled, to a 3-cm bed height, with Cellex D and the surface is covered with a 0.5-cm layer of acid-washed sand. The columns are equilibrated with 0.07 *M* sodium acetate (pH 5.2) prior to applying the reaction supernatants. Glutamate is recovered from the columns by washing with 40 ml of equilibration buffer. Labeled folylpolyglutamates are recovered by washing the columns with 5 ml of 0.2 *N* HCl and radioactivity is measured by scintillation counting. The folylpolyglutamate chain lengths of the reaction products are readily determined by high-performance liquid chromatography (HPLC) after cleavage of the folates to *p*-aminobenzoylpolyglutamates.[10]

Definition of Unit and Specific Activity. A unit is defined as the amount of enzyme that incorporates 1 nmol of L-glutamate into folylpolyglutamates per hour under the standard assay conditions. Specific activity is based on units per milligram of protein, with the latter being determined colorimetrically.[11]

Source of Enzyme. Seeds of pea (*Pisum sativum* L. cv. Homesteader) are initially surface sterilized by washing for 5 min in 1% (v/v) sodium hypochlorite solution followed by thorough rinsing in distilled water. The seeds are then soaked in distilled water at room temperature for 18 hr.

Purification Procedure

Step 1. Crude Extract. After removal of the seed coats and embryos, the cotyledons are surface sterilized by rapid washing in 70% (v/v) ethanol followed by a 1-min wash in 1% (v/v) H_2O_2 solution. All subsequent steps are carried out at 2–4°. Approximately 500 g of cotyledons are placed in a blender and homogenized with 500 ml of ice-cold 20 m*M* potassium phosphate buffer (pH 7.4) containing 50 m*M* 2-mercaptoethanol, 1 m*M* phenylmethylsulfonyl fluoride (PMSF), and 2 m*M* benzamidine (extraction buffer). The homogenate is filtered through four layers of cheesecloth and centrifuged at 30,000 *g* for 15 min at 4°.

[10] B. Shane, *Methods Enzymol.* **122,** 323 (1986).
[11] M. Bradford, *Anal. Biochem.* **72,** 248 (1976).

Step 2. Streptomycin Sulfate Fractionation. Streptomycin sulfate [10% (w/v), dissolved in extraction buffer, pH 7.4] is added to the crude extract to give a final concentration of 1% (w/v). After stirring for 1 hr, the precipitate is removed by centrifuging as in step 1.

Step 3. Ammonium Sulfate Fractionation. Solid ammonium sulfate (20.9 g/100 ml) is slowly added with stirring to the step 2 extract to give 35% of saturation. After additional stirring for 1 hr, the suspension is centrifuged as noted above and the pellet is discarded. The resulting supernatant is treated with further ammonium sulfate (6.3 g/100 ml) to give 45% of saturation, stirred for 1 hr, and recentrifuged as noted above. The supernatant is discarded and the pellet is dissolved in 15 ml of extraction buffer.

Step 4. Sephacryl S-200 Chromatography. The step 3 extract is dialyzed against extraction buffer for 1 hr at 2° prior to gel filtration. A 12-ml aliquot of the dialyzed extract is applied to a 2.6 × 80 cm column of Sephacryl S-200 that has been preequilibrated with extraction buffer. The column is washed with this buffer at a flow rate of 0.5 ml/min. Fractions of 6 ml are collected and those with FPGS activity are pooled.

Step 5. Chromatography on DE-52 Cellulose. The pooled fractions from step 4 (approximately 100 ml) are applied to a 2.6 × 52 cm column of Whatman (Clifton, NJ) DE-52 cellulose that has been preequilibrated with 30 m*M* potassium phosphate buffer (pH 7.0) containing 10 m*M* KCl, 50 m*M* 2-mercaptoethanol, 1 m*M* PMSF, and 2 m*M* benzamidine (buffer A). The column is washed with 100 ml of this buffer and FPGS protein is then eluted by applying a linear KCl gradient (300 ml of buffer A in the mixing vessel to 300 ml of buffer A containing 600 m*M* KCl in the reservoir). Fractions of 6 ml are collected and FPGS activity is eluted after the bulk of the applied protein. The FPGS-containing fractions are pooled and protein is precipitated by slow addition of a saturated ammonium sulfate solution in extraction buffer. Protein recovered at 60% of saturation (39 g/100 ml) is then redissolved in 4.5 ml of extraction buffer containing 10% (w/v) ammonium sulfate.

Step 6. Chromatography on Phenyl-Agarose. Protein from step 5 is applied to a 1 × 15 cm column of phenyl-agarose that has been preequilibrated with extraction buffer containing 10% (w/v) ammonium sulfate. The column is washed with 30 ml of this buffer followed by a 250-ml linear gradient of decreasing ammonium sulfate concentration (10–0%, w/v) in extraction buffer. The column is finally washed with extraction buffer to elute FPGS activity. Fractions of 6 ml are collected and those containing FPGS activity are pooled.

The purification procedure is summarized in Table I.

TABLE I
PURIFICATION OF FOLYLPOLYGLUTAMATE SYNTHASE FROM PEA COTYLEDONS[a]

Step	Volume (ml)	Total activity (units)	Specific activity (units/mg)	Yield (%)
1. Crude extract	540	4480	0.9	100
2. Streptomycin sulfate	559	9390	1.7	210
3. $(NH_4)_2SO_4$, 35–45% saturated	38	2044	3.5	46[b]
4. Sephacryl S-200	98	5664	12.3	126
5. DE-52 cellulose	16	938	67.0	21[b]
6. Phenyl-agarose	102	1907	2384.0	43

[a] Reprinted with permission from Ref. 5.
[b] As pea FPGS is inhibited by ammonium sulfate and by chloride ions, recovery of activity at these steps is relatively low.

Enzyme Properties

Stability. The enzyme from step 4 can be stored for 1 month at 4° with only a 10% loss of activity. Additions of albumin (5 mg/ml enzyme) result in preparations that are stable for at least 6 weeks at 4°. In contrast, step 6 enzyme is unstable and loses all activity within 1 to 2 days of storage at 4°. However, addition of glycerol to a final concentration of 50% (v/v) effectively stabilize activity for several days at 4°. Enzyme stability is not improved when the protein is concentrated by ultrafiltration or when additions of dimethyl sulfoxide (DMSO) (20%, v/v) are made prior to storage at 4°.

TABLE II
FOLATE SPECIFICITY OF PEA FOLYLPOLYGLUTAMATE SYNTHASE[a]

Folate substrate provided	Substrate concentration (μM)	Relative activity[b] (%)
(*R,S*)-H_4PteGlu	100	100.0
PteGlu	100	3.4
(*R,S*)-5,10-Methylene-H_4PteGlu	100	99.0
Methotrexate	100	3.4
Aminopterin	50	3.0
(*R,S*)-H_4PteGlu + methotrexate	100 + 100	95.5
(*R,S*)-H_4PteGlu + PteGlu	100 + 100	92.9

[a] Reprinted with permission from Ref. 6.
[b] Standard reaction systems containing (*R,S*)-H_4PteGlu incorporated 5 nmol of L-glutamate into folylpolyglutamates per hour at 37°.

Physical and Catalytic Properties. Sephacryl S-200 chromatography[5] indicates an apparent molecular weight of 68,000. In cotyledon extracts, activity is mainly cytosolic,[1] but in pea leaves activity is associated with mitochondria.[4] At saturating L-glutamate concentrations (1.5 mM), the optimal pH is pH 8.5. At physiological pH (pH 7.4), the reaction rate is about 50% of the optimal rate. Polyglutamate formation is dependent on a folate substrate, MgATP, and L-glutamate. At pH 8.5, the K_m value for H_4PteGlu is 3.5 μM; for ATP the value is 2.3 mM; and for L-glutamate the value is 0.61 mM. The saturating concentrations of these substrates at pH 8.5 are H_4PteGlu (0.1 mM), ATP (5 mM), and L-glutamate (1.5 mM). Under the standard assay conditions, product formation is a linear function of time for at least 120 min and is proportional to protein concentration (up to 20 μg/ml).

Folate Specificity. As summarized in Table II, (6*R*,*S*)-H_4PteGlu and 5,10-methylene-H_4PteGlu are good substrates and, under the standard assay conditions, both support polyglutamate synthesis at similar rates. Folic acid, methotrexate, and aminopterin are poorly utilized by the enzyme. Reaction systems containing folic acid or methotrexate in conjunction with H_4PteGlu support polyglutamate synthesis at about the same rate as H_4PteGlu controls. This higher plant FPGS appears to have a narrower folate specificity than the corresponding proteins of mammalian[8,12] and bacterial origins.[13]

Folylpolyglutamate Products. In standard reaction systems, incubated for 2 hr, the major polyglutamate product is $H_4PteGlu_2$.[6] Longer reaction periods result in the formation of di- and triglutamate derivatives. After 24 hr of reaction, these products account for about 75 and 25% of the recovered polyglutamates, respectively. When (6*R*,*S*)-$H_4PteGlu_2$ (100 μM) is provided for 2 hr, the products are triglutamate (95%) and tetraglutamate (5%). After 24 hr of reaction with $H_4PteGlu_2$ as folate substrate, the tri- and tetraglutamates account for about 90 and 10% of the product label, respectively.[2] The distribution of polyglutamate products is also affected by the concentration of the provided folate substrate. Thus, a nonsaturating concentration (10 μM) of (6*R*,*S*)-H_4PteGlu results in labeling of di- (20%), tri- (45%), and tetraglutamates (35%) following a 24-hr incubation period. The endogenous folate pool of pea cotyledons consists of pentaglutamates (83%) with smaller amounts of the tetra- (11%) and triglutamate (6%) derivatives.[2]

[12] D. J. Cichowicz and B. Shane, *Biochemistry* **26,** 513 (1987).
[13] A. L. Bognar and B. Shane, *J. Biol. Chem.* **258,** 12574 (1983).

[19] Purification of Folate-Dependent Enzymes from Rabbit Liver

By VERNE SCHIRCH

Introduction

Rabbit liver is an unusually rich source of enzymes that utilize tetrahydrofolate (THF) as a coenzyme. Several of these enzymes have proved useful in analyzing THF pools, as well as in understanding the role and mechanism of THF in one-carbon metabolism. In addition to being a rich source of enzymes, rabbit liver can be obtained in relatively large quantities at modest cost from Pel-Freeze Biologicals (Rogers, AR). We have found no differene in yield of enzymes between livers obtained from 6- to 8-month-old rabbits that weigh about 70 g and those from mature rabbit livers that weigh about 90 g. The livers are shipped frozen on dry ice and stored at −70°. All of the enzymes described below are stable for months in the frozen liver, except for serine hydroxymethyltransferase (SHMT). This enzyme activity decreases about 10 to 15% per week and should be purified within a few days after receiving the livers.

The physiological form of THF contains a chain of glutamate residues linked through the γ-carboxyl group. To accommodate the naming of these derivatives they are named as polyglutamate derivatives of tetrahydropteroate (H_4Pte), which has no glutamate residue. Tetrahydrofolate thus becomes H_4PteGlu and the pentaglutamate form becomes $H_4PteGlu_5$. This nomenclature system is used throughout this chapter.

Serine Hydroxymethyltransferase

$$\text{L-Serine} + H_4PteGlu_n \rightleftharpoons \text{glycine} + 5,10\text{-}CH_2\text{-}H_4PteGlu_n \qquad (1)$$

Materials

Serine, H_4PteGlu, yeast alcohol dehydrogenase, and NADH can be obtained from many biological supply companies. DL-*allo*-Threonine can be purchased from Sigma (St. Louis, MO) and [2-^{3}H]glycine (30 mCi/mmol) can be purchased from most companies supplying radiochemicals. A strong cation-exchange resin (AG 50W-X4) and hydroxylapatite (Bio-Gel HTP) are obtained from Bio-Rad (Richmond, CA). C_1-Tetrahydrofolate synthase (C_1-THF synthase) is purified from rabbit liver as described below.

METHODS IN ENZYMOLOGY, VOL. 281
0076-6879/97 $25

Assays

There are several sensitive assays for SHMT (EC 2.1.2.1). For routine purification the assay usually used depends on the ability of the enzyme to cleave L-*allo*-threonine to glycine and acetaldehyde.[1] The rate of formation of acetaldehyde is determined by coupling the reaction to alcohol dehydrogenase and observing the conversion of NADH to NAD^+ at 340 nm. The advantage of this assay, over the one for the physiological reaction, is that it does not require H_4PteGlu, which is unstable in solutions and usually must be made fresh every few days. A typical assay is performed in the following buffer solution; 900 μl of 20 m*M* potassium phosphate (KP_i) containing 5 m*M* 2-mercaptoethanol and 0.1 m*M* EDTA, 0.3 m*M* NADH, 0.07 mg of yeast alcohol dehydrogenase, and 5 to 50 μl of a protein solution containing SHMT (the absorbance is adjusted to zero prior to the addition of NADH). After mixing, the solution is brought to 30° and the absorbance followed for about 30 sec at 340 nm to establish a baseline. There is usually a slow decrease in absorbance when solutions of SHMT from the early steps in the purification procedure are used. The reaction is initiated by the addition of 50 μl of a 0.5 *M* solution of DL-*allo*-threonine and the linear decrease in absorbance at 340 nm with time determined. A unit of enzyme is defined as the formation of 1 μmol of NADH per minute determined by dividing the initial slope at 340 nm by the molar extinction coefficient for NADH of 6.24 mM^{-1} cm^{-1}.

The physiological assay couples the oxidation of the product 5,10-CH_2-H_4PteGlu to 5,10-CH^+-H_4PteGlu by 5,10-methylenetetrahydrofolate dehydrogenase (0.1 mg) and the reduction of $NADP^+$ (0.2 m*M*) to NADPH. This assay is performed in the same assay buffer used for *allo*-threonine [900 μl of 20 m*M* potassium phosphate (KP_i) containing 5 m*M* 2-mercaptoethanol and 0.1 m*M* EDTA] but with 0.3 m*M* $NADP^+$, 20 m*M* L-serine, 0.2 m*M* H_4PteGlu, and a solution containing SHMT. The reaction is initiated with L-serine after a baseline absorbance at 340 nm is established (the absorbance of this solution is near 0.05 when a blank of just the buffer is used).

There are situations when investigators want to know the amount of SHMT in a tissue without consideration of purification. These studies include use of cell culture lines believed to be defective in SHMT activity. The two assays described above are not sensitive enough to determine SHMT levels in these situations. A more sensitive assay for SHMT is to take advantage of its H_4PteGlu-dependent exchange of the α-proton of glycine with solvent.[2] To 50 μl of 20 m*M* KP_i, pH 7.3, containing 5 m*M*

[1] V. Schirch, K. Shostak, M. Zamora, and M. Gautam-Basak, *J. Biol. Chem.* **266,** 759 (1991).

[2] M. S. Chen and L. Schirch, *J. Biol. Chem.* **248,** 3631 (1993).

2-mercaptoethanol and 0.1 mM EDTA, is added 5 μl of a 0.4 mM solution of [2-^{3}H]glycine in H_2O containing about 1.5 $\times$ 10^6 cpm and 20 μl of a 5 mM H_4PteGlu solution. The reaction is started with the addition of microliter amounts of the cell extracts containing SHMT. The resulting solution is incubated at 30° for 5 to 30 min, depending on the level of SHMT in the cell extract. At the end of the incubation time, 10 μl of 10% (w/v) trichloroacetic acid is added to stop the reaction. After centrifugation, 70 μl of the supernatant is added to a 1 $\times$ 2 cm Dowex-50 column equilibrated with 10 mM HCl. After the 70 μl has been absorbed the column is washed with three successive 200-μl aliquots of H_2O, which are collected in one vial containing 7 ml of scintillation fluid. The collected eluants are counted to determine how much ^{3}H has been lost from the glycine. A control lacking H_4PteGlu is used as background. This assay is linear with time and measures 5 to 200 ng of SHMT. The [2-^{3}H]glycine should be purified on a Dowex-50 column as previously described.[2]

Purification of Serine Hydroxymethyltransferase

The enzyme can be purified from 2 to 100 rabbit livers by simply changing the size of the columns described in the following procedure.[3] The method for starting with 20 livers is described because this usually gives an adequate amount of enzyme for most studies. Twenty rabbit livers at −70° are placed in a plastic freezer bag and fractured into small pieces with a hammer. After the pieces partially thaw (usually about 10 min) they are homogenized in a Waring blender for 1 min at room temperature in 2 liters of 30 mM dipotassium phosphate containing 50 mM DL-serine, 0.1 mM pyridoxal-P, 10 mM 2-mercaptoethanol, and 0.2 mM EDTA. This homogenate is placed in either a 4-liter glass or stainless steel beaker and placed in a water bath at about 80°. The homogenate is stirred rapidly and the bath temperature maintained at 70 to 80° until the homogenate reaches 63°. After reaching this temperature the beaker is removed from the hot water bath and placed in an ice–water bath to cool rapidly to 20°. The homogenate is then centrifuged for 25 min at 13,000 g at 4° and the supernatant decanted from the firm pellet. A 20-μl aliquot is taken for assay. All of the following steps can be done at room temperature except for overnight dialysis. However, we found no difference in yield when each step was performed at 0 to 4°. We usually find that the total units of enzyme activity increase in subsequent purification steps, suggesting that there is some inhibitor present in the homogenate (Table I).

[3] P. Stover and V. Schirch, *Anal. Biochem.* **202,** 82 (1992).

TABLE I
PURIFICATION OF RABBIT LIVER CYTOSOLIC SERINE HYDROXYMETHYLTRANSFERASE[a]

Purification step	Volume (ml)	Total protein (mg)	Specific activity (units/mg)	Total units[b]	Purification (-fold)
Homogenate–heat step	1,740	27,840	0.012	329	1.0
$(NH_4)_2SO_4$	180	15,200	0.025	380	2.1
CM-Sephadex	232	441	0.80	353	67
Phenyl-Sepharose	110	265	1.5	405	125

[a] From 20 livers using *allo*-threonine as substrate.
[b] One enzyme unit is defined as the formation of 1 μmol of product per minute under standard assay conditions.

Ammonium sulfate is added to the supernatant to 50% of saturation (313 g/liter) with rapid stirring. After centrifugation, as described for the homogenate, the pellet is dissolved in KP_i, pH 7.3, containing 0.1 m*M* pyridoxal-P, 10 m*M* 2-mercaptoethanol, and 0.2 m*M* EDTA. All subsequent buffers contain the same concentrations of pyridoxal-P, EDTA, and 2-mercaptoethanol and are referred to as buffer A. The dissolved pellet is dialyzed against 4 liters of buffer A until the ammonium sulfate concentration is less than 2% of saturation. This usually can be achieved in 4 hr of dialysis with two changes of buffer. After removal of the enzyme solution from the dialysis bag, ammonium sulfate is added to 33% of saturation (196 g/liter) and centrifuged. The pellet is discarded and ammonium sulfate added to the homogenate to 50% of saturation (107 g/liter). The pellet is dissolved in a minimal amount of buffer A and dialyzed overnight against two 4-liter volumes of 10 m*M* potassium phosphate, pH 6.8, containing pyridoxal-P, EDTA, and 2-mercaptoethanol.

The dialyzed enzyme is loaded onto a CM-Sephadex column (6 × 40 cm; Pharmacia, Piscataway, NJ) equilibrated with 10 m*M* KP_i, pH 6.8, containing EDTA and 2-mercaptoethanol but not pyridoxal-P. The enzyme absorbs to the top of the column and can be seen as a yellow band owing to the SHMT-bound pyridoxal-P. The column is washed with equilibration buffer until the A_{280} of the eluate is less than 0.1. The SHMT is eluted with a linear gradient of 500 ml of equilibration buffer and 500 ml of 0.3 *M* potassium phosphate, pH 7.3, containing EDTA and 2-mercaptoethanol. The enzyme elutes as a yellow band near the end of the gradient and is collected in 10-ml fractions that are assayed for SHMT activity. Another yellow protein elutes just ahead of SHMT. Fractions containing activity are pooled and the protein precipitated by adding ammonium sulfate to 60% of saturation (390 g/liter). The precipitated protein is dissolved in a small amount of buffer A and loaded onto a phenyl-Sepharose column

(4 × 20 cm; Pharmacia) that has been equilibrated with 30% (w/v) ammonium sulfate in buffer A. The enzyme binds to the column and is eluted with a linear decreasing salt gradient of 300 ml each of equilibration buffer and buffer A. The cytosolic SHMT isoenzyme elutes near the end of the gradient as a yellow band of protein. This enzyme is about 90% pure. About 150 to 200 mg of SHMT is obtained in 2 days by this procedure (Table I). The enzyme is precipitated by adding ammonium sulfate to 60% of saturation and collected by centrifugation. The pellet is dissolved in buffer A. If the enzyme is to be used at this stage of purity, it is dialyzed against 20 m*M* potassium phosphate, pH 7.3, containing only EDTA and 2-mercaptoethanol, and 15% (v/v) glycerol. The protein concentration of the dialyzed enzyme is adjusted to 10 to 15 mg/ml and stored in small vials at −70°. The enzyme is stable under these conditions for at least 6 months.

The SHMT eluted from the phenyl-Sepharose column is the cytosolic form. The mitochondrial isoenzyme remains bound to the phenyl-Sepharose column and can be seen as a yellow band near the middle of the column. It can be eluted with 10% (v/v) propylene glycol in buffer A. The fractions containing activity are pooled and concentrated with 60% (w/v) ammonium sulfate. The pellet is dissolved in a minimal amount of buffer and dialyzed at 4° against 20 m*M* KP_i, pH 7.3, containing 5 m*M* 2-mercaptoethanol. After two or three changes of buffer yellow crystals of the mitochondrial enzyme form in the dialysis bag. These are collected by centrifugation and the crystal pellet dissolved at 30° in 300 m*M* KP_i, pH 7.3, with 2-mercaptoethanol. The concentration (in milligrams per milliliter) is determined by dividing the A_{278} by 0.72.[4] The enzyme is frozen at −70° where it remains active for months. About 20 to 30 mg of mitochondrial SHMT is obtained by this procedure.

Further purification of cytosolic SHMT can be achieved by chromatography on a hydroxylapatite column (2 × 10 cm) equilibrated with 20 m*M* potassium phosphate buffer, pH 7.3. The SHMT dialyzed against the same buffer is absorbed onto the column and eluted with a linear gradient of 250 ml of equilibration buffer and 250 ml of 300 m*M* potassium phosphate, pH 7.3. The enzyme elutes near the end of the gradient. The spectra of fractions containing SHMT are recorded and those fractions having an A_{278}/A_{428} ratio of less than 7 are pooled and the protein precipitated with 60% of saturation of ammonium sulfate (390 g/liter). The resulting pellet is dissolved and treated as described above for final storage. This step gives about a 70% yield and increases the purity to greater than 98%. It is used only when pure enzyme is required. The concentration of cytosolic SHMT (in milligrams per milliliter) is determined by dividing the A_{278} by 0.72.

[4] L. Schirch and D. Peterson, *J. Biol. Chem.* **255,** 7801 (1980).

Properties of Purified Serine Hydroxymethyltransferase

All purified SHMTs are yellow, exhibiting an absorption maximum between 420 to 430 nm. For both rabbit liver cytosolic and mitochondrial SHMT the absorption maximum is at 428 nm and the pure enzyme exhibits an A_{280}/A_{428} ratio of 6.5. This ratio can be smaller if not all pyridoxal-P is removed during dialysis or greater if dialysis is extended for long periods of time in the absence of pyridoxal-P. Enzyme that is stored for long periods sometimes accumulates an inactive species that absorbs at 335 nm. This can be removed by adding 20 m*M* L-serine and heating the solution to 60° for a few minutes, followed by removal of the precipitated protein by centrifugation.

The k_{cat} values for the purified enzyme at 30° is 500 min^{-1} for L-serine and 130 min^{-1} for L-*allo*-threonine. The K_m values for H_4PteGlu, serine, and glycine are 15 μM, 0.8 m*M*, and 5 m*M*, respectively. The K_m value for $H_4PteGlu_5$ is 0.2 μM.[5]

C_1-Tetrahydrofolate Synthase

C_1-Tetrahydrofolate synthase is a homodimeric trifunctional enzyme that catalyzes the following three reactions.[6,7] Each subunit is 108 kDa and the enzyme has a p*I* near 7.0. Reaction (2) is catalyzed by 5,10-methylenetetrahydrofolate dehydrogenase (CH_2-THF dehydrogenase; EC 1.5.1.5); reaction (3) is catalyzed by 5,10-methenyltetrahydrofolate cyclohydrolase (CH^+-THF cyclohydrolase; EC 3.5.4.9); reaction (4) is catalyzed by 10-formyltetrahydrofolate synthetase (CHO-THF synthetase; EC 6.3.4.3, formate–tetrahydrofolate ligase).

$$5,10\text{-}CH_2\text{-}H_4PteGlu + NADP^+ \rightleftharpoons 5,10\text{-}CH^+\text{-}H_4PteGlu + NADPH \quad (2)$$
$$5,10\text{-}CH^+\text{-}H_4PteGlu + 2H_2O \rightleftharpoons 10\text{-}CHO\text{-}H_4PteGlu + H_3O^+ \quad (3)$$
$$10\text{-}CHO\text{-}H_4PteGlu + ADP + P_i \rightleftharpoons H_4PteGlu + formate + ATP \quad (4)$$

Materials

Polyethylene glycol 3350 (PEG 3350) and H_4PteGlu are obtained from Sigma. Heparin–agarose and hydroxylapatite are from Bio Rad, and Orange A–Sepharose is from Amicon (Danvers, MA). Potassium *N,N*-bis(2-hydroxyethyl)-2-aminoethane sulfonate (KBES) and potassium 2-(*N*-morpholino)ethane sulfonate (KMES) can be obtained from most

[5] W. B. Strong and V. Schirch, *Biochemistry* **28,** 9430 (1989).
[6] D. W. Hum and R. E. MacKenzie, *Protein Eng.* **4,** 493 (1991).
[7] W. Strong, G. Joshi, R. Lura, N. Muthukumaraswamy, and V. Schirch, *J. Biol. Chem.* **262,** 12519 (1987).

suppliers of biochemicals. Ammonium formate, MgATP, phosphoenol pyruvate, pyruvate kinase, and lactate dehydrogenase can also be purchased from most biochemical suppliers.

Twenty-five milligrams of H_4PteGlu is dissolved in 2.5 ml of 0.2 *M* K_2PO_4, containing 0.2 m*M* EDTA and 50 m*M* 2-mercaptoethanol at 0°. The buffer is purged with argon for 5 min before adding the H_4PteGlu. The slightly yellow solution is divided into smaller vials and layered with mineral oil. This gives an ~12 m*M* solution of (6*R*,*S*)-H_4PteGlu. However, only the 6*S*-isomer is active with these enzymes. (Because of a change in priority number the active H_4PteGlu$_n$ derivatives substituted at N-10 are 6*R*). These H_4PteGlu solutions are stored at −70°. For stock solutions of 5,10-CH_2-H_4PteGlu, formaldehyde is added to a vial of H_4PteGlu in the ratio of 2:1 (mole/mole) formaldehyde:H_4PteGlu, and incubated for 5 min at 23°. 5,10-CH^+-H_4PteGlu is prepared by dissolving the calcium salt of 5-CHO-H_4PteGlu in 20 m*M* KP_i, pH 7.3. The solution remains cloudy as a result of the formation of calcium phosphate. After stirring for 15 min at 23° the solution is centrifuged and the concentration of the dissolved 5-CHO-H_4PteGlu is determined from its spectrum using an ε_{287} of 27 mM^{-1} cm^{-1}. After adjusting the solution to pH 2 by the addition of 6 *M* HCl, the solution is incubated at 4° for at least 12 hr. With the monoglutamate form of the coenzyme a yellow crystalline 5,10-CH^+-H_4PteGlu is formed. With the polyglutamate forms of 5,10-CH^+-H_4PteGlu$_n$ the solution does not form crystals. These solutions can be stored at −70° for several weeks in the absence of 2-mercaptoethanol.

10-CHO-H_4PteGlu is formed from the 5,10-CH^+-H_4PteGlu$_n$ by the addition of 6 *M* NaOH to raise the pH to pH 8.0. Before adjusting the pH fresh 2-mercaptoethanol is added to 20 m*M*. At the increased pH the 5,10-CH^+-H_4PteGlu$_n$ hydrolyzes to 10-CHO-H_4PteGlu$_n$ in 15 min at 23°.

Assays

The enzyme is easily assayed using either reaction (2) or (3). The rate of reaction (2) is followed by incubating at 30° a 1-ml solution of 20 m*M* KP_i, pH 7.4, containing 10 m*M* 2-mercaptoethanol, 0.2 m*M* $NADP^+$, and 0.2 m*M* 5,10-CH_2-H_4PteGlu. An A_{340}-versus-time baseline is determined for about 30 sec (the instrument zero absorbance was for buffer only) and the reaction initiated by the addition of microliter aliquots of enzyme. The increasing slope at A_{340} is used to determine the activity of the preparation.

The substrate 5,10-CH^+-H_4PteGlu is unique for H_4PteGlu derivatives having a maximum absorbance at 360 nm (ε_{360} of 25.1 mM^{-1} cm^{-1}). This derivative is stable in acid but hydrolyzes rapidly above pH 7 to

10-CHO-H_4PteGlu. For this assay the A_{360} of a 0.2 mM solution of 5,10-CH^+-H_4PteGlu in 20 mM KMES, pH 6.0, at 30° and containing 10 mM 2-mercaptoethanol is followed for 30 sec to establish a baseline. A small aliquot of enzyme is added and the rate of decrease in A_{360} is determined.

Reaction (4) can be assayed in either direction. For the forward reaction the product H_4PteGlu is coupled to SHMT and CH_2-THF dehydrogenase, resulting in the conversion of $NADP^+$ to NADPH. The reaction rate is determined as an increase in A_{340}. A small aliquot of a solution containing enzyme is added to 1 ml of KP_i, pH 7.3, containing 10 mM 2-mercaptoethanol, 20 mM L-serine, 0.2 mM $NADP^+$, 10 mM MgADP, and 0.1 mg of SHMT. The reverse of reaction (4) is followed by coupling the formation of ADP to pyruvate kinase and phosphoenol pyruvate. The resulting pyruvate is reduced to lactate with lactate dehydrogenase and NADH. The rate of the reaction is followed by the decrease in A_{340}. A small aliquot of a solution containing enzyme is added to 1 ml of 50 mM KBES, pH 7.3, containing 10 mM 2-mercaptoethanol, 0.2 mM NADH, 20 mM ammonium formate, 10 mM MgATP, 1 mM phosphoenol pyruvate, 0.1 mg of pyruvate kinase, and 0.2 mg of lactate dehydrogenase.

Purification

C_1-Tetrahydrofolate synthase is cold labile and it is best to do all purification steps at room temperature. Those steps performed in the cold must be done with 20% (w/v) sucrose in the buffer, which protects the enzyme from cold denaturation.[7] Six frozen rabbit livers at −70° are placed in a plastic freezer bag and broken into small pieces with a hammer and allowed to thaw for 5 to 10 min. These livers are homogenized in a Waring blender for 1 min in 1.2 liters of 30 mM K_2HPO_4 containing 20 mM 2-mercaptoethanol, 0.1 mM EDTA, and 12% (w/w) PEG 3350. After centrifugation at 12,000 g for 30 min at 4° the supernatant is collected and the PEG 3350 concentration increased to 40% (w/w) by addition of an amount equivalent to (0.388 g × milliliters of supernatant) and stirred for 10 min. This solution is centrifuged as before and the resulting pellet dissolved in 0.5 liter of 20 mM KP_i, pH 7.3, buffer containing 10 mM 2-mercaptoethanol. (The supernatant can be used to isolate CH^+-THF synthetase as described below.) The dissolved pellet is added to a 4 × 10 cm Orange A affinity column (Amicon, Danvers, MA), previously equilibrated with the same buffer. The column should be fitted with a coarse glass frit to ensure a fast flow. The Orange A column is washed with 2 liters of equilibration buffer or until the A_{280} is below 0.1. The C_1-THF synthase is eluted by addition of 500 ml of equilibration buffer containing 2 M NaCl. Fifteen ml fractions are collected and assayed for CH_2-THF dehydrogenase activity. Fractions containing activity are pooled and concentrated by a 60% (w/v) ammonium sulfate precipitation (390 g/liter). The pellet is dissolved in a minimal amount of 20 m*M* KP_i, pH 6.5, containing EDTA and 10 m*M* 2-mercaptoethanol. This solution is dialyzed against 2 liters of the dissolving buffer for 1 hr at room temperature with rapid mixing. Further desalting is achieved by chromatography on a 4 × 20 cm Sephadex G-25 column equilibrated with the same buffer.

The desalted protein solution is added to a heparin–agarose column (5 × 10 cm) equilibrated with the 20 m*M* KP_i, pH 6.5, buffer described above. The column is washed with this buffer until the A_{280} is below 0.1. The enzyme is eluted with a linear gradient of 100 ml of equilibration buffer and 100 ml of 200 m*M* KP_i, pH 7.3. Fractions of 5 ml are collected and both the A_{280} and dehydrogenase activity determined. The activity elutes near the end of the gradient. Fractions exhibiting activity are pooled and precipitated with 60% (w/v) ammonium sulfate. After dissolving the pellet in a small amount of buffer it is dialyzed overnight against 20 m*M* KP_i, pH 7.3, containing 0.1 m*M* EDTA, 10 m*M* 2-mercaptoethanol, and 20% (w/v) sucrose at 4°.

The dialyzed enzyme is centrifuged briefly to remove precipitated protein. The supernatant is added to a hydroxylapatite column (3 × 5 cm) at room temperature equilibrated with the dialysis buffer without the sucrose. The column is washed with equilibration buffer until the A_{280} is below 0.1. The enzyme is eluted with a linear gradient of 100 ml of equilibration buffer and 100 ml of 100 m*M* KP_i, pH 7.3, containing 10 m*M* 2-mercaptoethanol. The enzyme elutes as the first major absorbing material. The fractions, 3 ml, are assayed for dehydrogenase activity and A_{280}. Those fractions containing the highest activity are pooled and the protein precipitated by addition of 60% (w/v) ammonium sulfate followed by centrifugation. The protein pellet is dissolved in minimal amount of equilibration buffer and dialyzed at 4° overnight in this buffer containing 20% (w/v) sucrose. The dialyzed enzyme is stored in small fractions (usually 200 to 500 μl) and frozen at −70°. The concentration of enzyme (in milligrams per milliliter) is determined by dividing the A_{278} by 0.68. This preparation gives about 70 to 90 mg of enzyme that is 85 to 90% pure as determined by sodium dodecyl sulfate-polyacrylamide gel electrophoresis (SDS–PAGE) (Table II).

Properties of C_1-Tetrahydrofolate Synthase

The rabbit enzyme is a dimer of 108-kDa subunits.[6,7] However, each subunit is composed of two independently folded domains that appear to

TABLE II
PURIFICATION OF RABBIT LIVER C_1-TETRAHYDROFOLATE SYNTHASE[a]

Purification step	Volume (ml)	Total protein (mg)	Specific activity (units/mg)	Total units[b]	Yield (%)
Homogenate	1,040	17,800	0.060	1,115	100
Orange A	430	744	1.4	1,115	100
Heparin–agarose	200	204	3.6	725	65
Hydroxylapatite	77	96	5.7	545	49

[a] From six livers, using methylene tetrahydrofolate dehydrogenase activity.
[b] One enzyme unit is defined as the formation of 1 μmol of product per minute under standard assay conditions.

be connected by a solvent-exposed polypeptide chain.[7] Many proteases can cleave this exposed polypeptide chain, resulting in two fragments. The large fragment of 144 kDa contains the activity for 10-CHO-THF synthase and is a dimer of 72-kDa monomers. The smaller, monomeric 36-kDa fragment contains the activity for both the CH_2-THF dehydrogenase and CH^+-THF cyclohydrolase activities. It is currently believed that these two activities share parts of a common active site. The fragments are fully active for their respective activities but are less stable. The addition of $NADP^+$ stabilizes the small fragment and polyglutamate forms of $H_4PteGlu_n$ stabilize the large fragment. These fragments can be separated by high-performance liquid chromatography (HPLC) size-exclusion chromatography. Human C_1-THF synthase has been cloned and sequenced and the large domain has been cloned and expressed.[6]

10-Formyltetrahydrofolate Dehydrogenase

10-Formyltetrahydrofolate dehydrogenase, a tetrameric enzyme (EC 1.5.1.6; FTD), catalyzes the $NADP^+$-dependent oxidation of 10-CHO-$H_4PteGlu$ to CO_2 [Eq. (5)] and the hydrolyses of 10-CHO-$H_4PteGlu$ to formate and $H_4PteGlu$ [Eq. (6)].[8–10] The enzyme also catalyzes the oxidation of small aldehydes such as propanal [Eq. (7)]. A unique property of this enzyme is that it was identified as one of several $H_4PteGlu$-binding proteins in the cytosol.[10] This is an abundant enzyme in mammalian liver, representing as much as 1% of the soluble protein.

[8] E. M. Rios-Orlandi, C. G. Zarkadas, and R. E. MacKenzie, *Biochim. Biophys. Acta* **871,** 24 (1986).
[9] H. Min, B. Shane, and E. L. R. Stokstad, *Biochim. Biophys. Acta* **967,** 348 (1988).
[10] R. J. Cook and C. Wagner, *Biochemistry* **21,** 4427 (1982).

$$10\text{-CHO-H}_4\text{PteGlu}_n + \text{NADP}^+ \rightarrow \text{CO}_2 + \text{NADPH} + \text{H}_4\text{PteGlu}_n \quad (5)$$
$$10\text{-CHO-H}_4\text{PteGlu}_n + \text{H}_2\text{O} \rightarrow \text{formate} + \text{H}_4\text{PteGlu}_n \quad (6)$$
$$\text{Propanal} + \text{NADP}^+ \rightarrow \text{propanoate} + \text{NADPH} \quad (7)$$

Materials

$NADP^+$, L-serine, dithiothreitol (DTT), and propanal can be purchased from most biochemical supply companies. The SHMT is purified as described in this chapter. 5,8-Dideazafolate and 10-formyl-5,8-dideazafolate are purchased from J. Hynes, University of South Carolina (Columbia, SC). The 5,8-dideazafolate affinity column is prepared essentially by the method of Grimshaw *et al.*[11]

Assays

Activity for reaction (5) is measured at 30° in 1 ml of 50 m*M* Tris-HCl, pH 7.7, containing 10 m*M* 2-mercaptoethanol, 0.2 m*M* EDTA, 100 μM (6*R,S*)-10-CHO-H_4PteGlu, 0.2 m*M* $NADP^+$, and 10-CHO-THF dehydrogenase. This reaction shows product inhibition resulting in a slowing of the reaction during the first 30 sec of the assay. A linear assay can be achieved by coupling the reaction to the SHMT conversion of H_4PteGlu to 5,10-CH_2-H_4PteGlu. This is accomplished by including in the assay buffer 2.5 μM rabbit liver SHMT and 20 m*M* L-serine. The rate of reaction is followed at 340 nm and 1 unit of enzyme activity is defined as the formation of 1 μmol of NADPH per minute.

The hydrolase activity [reaction (6)] is determined under similar conditions, except that the buffer contains no $NADP^+$, but has 100 m*M* 2-mercaptoethanol. The rate of formation of 5,10-CH_2-H_4PteGlu, formed from serine and SHMT [Eq. (1)], is followed at 295 nm using an ε_{295} of 25.1 mM^{-1} cm^{-1}. The rate of this reaction is directly proportional to the concentration of 2-mercaptoethanol. Without this thiol the reaction rate is low. For both the dehydrogenase activity [Eq. (5)] and the hydrolase activity [Eq. (6)] 10-formyl 5,8-dideazafolate can be used as the substrate. This substrate analog is stable in solution, unlike the physiological substrate.

The propanal dehydrogenase activity [reaction (7)] is determined at 30° in 60 m*M* sodium pyrophosphate (pH 8.5), 0.2 m*M* EDTA, with 10 m*M* 2-mercaptoethanol, 30 m*M* propanal, 0.2 m*M* $NADP^+$, and enzyme. The reaction rate is determined from the linear formation of NADPH as determined by A_{340}.

[11] C. E. Grimshaw, G. B. Henderson, G. G. Soppe, G. Hansen, E. J. Mathur, and F. M. Huennekens, *J. Biol. Chem.* **259,** 2728 (1984).

Purification

All purification steps can be performed at room temperature except for overnight dialysis. Two frozen rabbit livers are placed in a plastic freezer bag and broken into small pieces with a hammer. After partial thawing they are homogenized in a Waring blender for 2 min in 400 ml of 50 m*M* KP_i, 0.3 *M* sucrose, 14 m*M* 2-mercaptoethanol, 1 m*M* EDTA, 0.5 m*M* DTT (pH 7.25). The homogenate is centrifuged for 30 min at 9000 rpm and the pellet discarded. After removing the floating lipids by passing the homogenate through glass wool, ammonium sulfate is added to 30% saturation (176 g/liter). After stirring for 10 min, the solution is centrifuged as before and the pellet discarded. To the supernatant ammonium sulfate is added to 55% of saturation (162 g/liter) and after stirring for 20 min the solution is centrifuged for 30 min at 4°. The pellet is resuspended in 300 ml of 10 m*M* KP_i, 14 m*M* 2-mercaptoethanol, 1 m*M* EDTA, 0.5 m*M* DTT, pH 6.8 (buffer A). The high salt content is lowered by repeated passage through a Pelicon apparatus (Amicon, Danvers, MA), with a 100,000-kDa cutoff membrane, until the conductivity reaches 5000 μmho. Passage through a Bio-Gel DG-6 column (4 × 30 cm) can also be used at this step to remove salt. The desalted solution is applied to a 5 × 7 cm CM-Sephadex column equilibrated with buffer A. Hemoglobin sticks to CM-Sephadex under the conditions used, but 10-formyltetrahydrofolate dehydrogenase passes through. The flow-through protein solution is loaded on a 5 × 7 cm heparin–agarose column equilibrated with buffer A. The column is washed extensively with 1 liter of buffer A until the A_{280} is below 0.2. The enzyme activity is eluted by a linear gradient of 500 ml of buffer A and 500 ml of 0.5 *M* KCl in 50 m*M* KP_i, 14 m*M* mercaptoethanol, 1 m*M* DDT, 1 m*M* EDTA, pH 7.25. Fractions (15 ml) are collected and FTD activity is found near the beginning of the gradient. Ammonium sulfate is added to the pooled fractions, about 350 ml, to 60% of saturation and the suspension kept at 4° overnight. This part of the preparation can usually be completed in 1 day. If the procedure needs to be interrupted before this last step it is best to store the enzyme as an ammonium sulfate slurry in buffer containing 10% (v/v) glycerol at 4°.

The solution is centrifuged at 9000 rpm for 25 min and the pellet dissolved in a minimal amount of 50 m*M* KP_i, pH 7.25, containing 1 m*M* EDTA, 0.5 m*M* DTT, and 14 m*M* 2-mercaptoethanol. The solution is desalted by passing it through a 4 × 20-cm Sephadex G-25 column equilibrated with the 50 m*M* KP_i buffer. The desalted sample is loaded onto a 3 × 9 cm 5,8-dideazafolate Sepharose 4B affinity column equilibrated with the same buffer. After washing with 300 ml of equilibration buffer, or until the A_{280} is less than 0.1, it is washed with 300 ml of 1 *M* KCl in equilibration

TABLE III
PURIFICATION OF RABBIT LIVER 10-FORMYLTETRAHYDROFOLATE DEHYDROGENASE[a]

Purification step	Volume (ml)	Total protein (mg)	Specific activity (units/mg)	Total units[b]	Yield (%)
Homogenate	460	27,600	0.007	193	100
$(NH_4)_2SO_4$	375	15,000	0.013	195	101
Heparin–agarose	210	450	0.19	86	44
5,8-Dideazafolate affinity	80	75	0.57	43	22

[a] From two livers using 10-formyltetrahydrofolate dehydrogenase activity.
[b] One enzyme unit is defined as the formation of 1 μmol of product per minute under standard assay conditions.

buffer. The FTD is then eluted with 200 ml of 100 μM 5,8-dideazafolic acid and 1 M KCl in equilibration buffer. Fractions (5 ml) are collected and those containing FTD activity are pooled. A yield of 70 to 90 mg of enzyme is usually obtained (Table III). Ammonium sulfate is added to 60% of saturation and the solution centrifuged at 9000 g for 25 min at 4°. The pellet is suspended in a small amount of 20 mM KP_i (pH 7.2), 3 mM DDT, 14 mM 2-mercaptoethanol, and 20% (v/v) glycerol and dialyzed against this buffer overnight. The concentration of FTD is determined from its absorbance at 278 nm, using the relationship of $0.9A_{278}$ equals 1 mg of FTD per milliliter.[12] The enzyme is stored at −20°.

Properties

This tetrameric enzyme is composed of 99-kDa subunits.[8,9] Each subunit is composed of two independently folded domains connected by a polypeptide linker.[12,13] The domains can be separated by digestion with subtilisin and chromatography on HPLC size-exclusion chromatography. The large 72-kDa domain elutes as a tetramer and catalyzes only the propanal dehydrogenase activity [Eq. (7)]. The small, 32-kDa domain elutes as a monomer and catalyzes only the hydrolase reaction [Eq. (6)].[12] Neither domain catalyzes the physiological reaction (5). Both domains are less stable than the native enzyme. The rat cDNA has been cloned and the large domain expressed.[14] Cysteine at position 707 has been shown to be a critical residue at the active site of the enzyme.[15]

[12] D. Schirch, E. Villar, B. Maras, D. Barra, and V. Schirch, *J. Biol. Chem.* **269,** 24728 (1994).
[13] R. J. Cook, R. S. Lloyd, and C. Wagner, *Biochem. J.* **177,** 833 (1991).
[14] R. J. Cook, R. S. Lloyd, and C. Wagner, *J. Biol. Chem.* **266,** 4965 (1991).
[15] S. A. Krupenko, C. Wagner, and R. J. Cook, *J. Biol. Chem.* **270,** 519 (1995).

5,10-Methenyltetrahydrofolate Synthetase

The activity of 5,10-methenyltetrahydrofolate synthetase (CH^+-THF synthetase) is the only activity identified in mammalian cells to use 5-CHO-H_4PteGlu as a substrate. This small 23-kDa monomeric enzyme (EC 6.3.3.2) has been purified from bacteria and human liver as well as rabbit liver.[16–18] All three enzymes show considerable sequence identity and similar physical and catalytic properties. The primary structure for the rabbit enzyme has been determined and the cDNA for the human enzyme has been cloned, but not expressed.[18,19] The product 5,10-CH^+-H_4PteGlu$_n$ exhibits an absorption maximum at 360 nm with an ε of 25.1 mM^{-1} cm^{-1}. This is the only folate derivative to absorb above 320 nm and is used in the assay of this reaction.

$$\text{5-CHO-H}_4\text{PteGlu}_n + \text{MgATP} \rightarrow \text{5,10-CH}^+\text{-H}_4\text{PteGlu}_n + \text{MgADP} + \text{P}_i \quad (8)$$

Materials

Tween 20, polyethylene glycol 3350 (PEG 3350), DE-52 cellulose, potassium *N,N*-bis(2-hydroxyethyl)-2-aminoethane sulfonate (KBES) and potassium 2-(*N*-morpholino]ethanesulfonate (KMES), MgATP, and (6*R,S*)-5-CHO-H_4PteGlu can be purchased from most biochemical supply companies. Hydroxylapatite is obtained from Bio-Rad. The 5-CHO-H_4PteGlu affinity column is synthesized according to the procedure of Grimshaw *et al.*[11]

Assay

Except where stated, all steps are performed at room temperature. Activity is determined at 30° in 1 ml of KMES (pH 6.0), 25 m*M* MgATP, 25 m*M* (6*R,S*)-5-CHO-H_4PteGlu, and 2 to 20 μg of CH^+-THF synthetase. The reaction is followed at A_{360}, which is the absorption maximum of 5,10-CH^+-H_4PteGlu (ε_{360} of 25.1 mM^{-1} cm^{-1}).

Purification

Fifteen frozen rabbit livers are placed in a plastic freezer bag and broken into small pieces with a hammer. After partial thawing for about 10 min

[16] R. Bertrand and J. Jolivet, *J. Biol. Chem.* **264,** 8843 (1989).
[17] S. Hopkins and V. Schirch, *J. Biol. Chem.* **259,** 5618 (1984).
[18] A. Dayan, R. Bertrand, M. Beauchemin, D. Chahla, A. Mamo, M. Filion, D. Skup, B. Massie, and J. Jolivet, *Gene* **165,** 307 (1995).
[19] B. Maras, P. Stover, S. Valiante, D. Barra, and V. Schirch, *J. Biol. Chem.* **269,** 18429 (1994).

they are homogenized in a Waring blender with 1.5 liters of 20 m*M* K_2P_i, 20 m*M* 2-mercaptoethanol, and 12% (v/w) PEG 3350. After homogenization for 45 sec the slurry is centrifuged at 12,000 *g* for 25 min. The pellet is discarded and the supernatant adjusted to 40% (w/v) polyethylene glycol (0.388 × volume) and stirred for 15 min. The slurry is then centrifuged for 25 min as before. The pellet contains the activity for C_1-THF synthase and the supernatant contains the CH^+-THF synthetase activity. If only the CH^+-THF synthetase activity is desired the polyethylene glycol can be added to 40% (w/v) during the homogenization in one step. The volume of the supernatant is about 1.8 liters. If lipid debris is floating in this solution it can be removed by passing it through glass wool in a funnel.

The thick supernatant is added to a 10 × 7.5 cm DE-52 column, with a coarse glass frit, equilibrated with 50 m*M* KP_i, pH 7.6, and 20 m*M* 2-mercaptoethanol. The column is kept at room temperature and the 1.5 liters of supernatant allowed to drip slowly through the column overnight. If the DE-52 column is higher than 7.5 cm the time required to pass the homogenate through the column is greatly extended. The enzyme activity is absorbed to the column and is stable. The next morning the activity is eluted by the addition of 2 liters of 10 m*M* KP_i, pH 7.2, containing 20 m*M* 2-mercaptoethanol, and 0.5% (v/v) Tween 20. This elution buffer is added gently in 50-ml aliquots to the top of the column. Fractions of 50 ml are collected and assayed for synthetase activity, which starts to elute after about 200 ml of buffer. Fractions containing activity are pooled (about 1.8 liters).

The 1.8 liters of solution is applied directly to a 2 × 8 cm 5-CHO-THF affinity column equilibrated with the DE-52 elution buffer. This column is placed in a 4° refrigerator and the 1.5 liters of eluate allowed to pass through the column in the next 10 to 12 hr. The column can run dry without problems and is usually done during the second night of the preparation. On the

TABLE IV
PURIFICATION OF RABBIT LIVER 5,10-METHENYLTETRAHYDROFOLATE SYNTHASE[a]

Purification step	Volume (ml)	Total protein (mg)	Specific activity (units/mg)	Total units[b]
Homogenate	1880	—	—	146
DEAE-cellulose	1870	—	—	115
5-Formyl-THF affinity	210	—	—	138
Hydroxylapatite	440	16	13	211

[a] From 15 livers.

[b] One enzyme unit is defined as the formation of 1 μmol of product per minute under standard assay conditions.

morning of the third day the affinity column is washed with 300 ml of 10 m*M* KP_i, pH 7.2, containing 20 m*M* 2-mercaptoethanol, 0.5% (v/v) Tween 20, and 500 m*M* NaCl. This is followed by washing with another 300 ml of this buffer, but lacking the 500 m*M* NaCl. The CH^+-THF synthetase activity is eluted by adding to the wash buffer 100 m*M* (6*R,S*)-5-$CHO-H_4PteGlu$. Fractions (20 ml) are collected and aliquots assayed for synthetase activity. The active fractions are pooled, resulting in about 200 ml of solution.

Final purification is achieved by adding the pooled fractions to a hydroxylapatite column (2 × 10 cm) equilibrated with 5 m*M* KBES, pH 7.2. The CH^+-THF activity passes through the column and is collected in 50-ml fractions. After all the enzyme solution has been eluted the column is washed with equilibration buffer until no activity is being eluted. The final volume is about 500 ml of a dilute enzyme solution containing both Tween 20 and $5-CHO-H_4PteGlu$. The simplest method of concentration is to lyophilize the solution to a paste, which is extracted with small volumes of KBES buffer. The enzyme can be stored in this form but still contains some $5-CHO-H_4PteGlu$ and Tween 20. If required, these are removed by passing the enzyme down a small size-exclusion chromatography, such as a 3 × 20 cm Sephadex G-25 column, equilibrated with 20 m*M* KP_i, pH 7.2, 5 m*M* 2-mercaptoethanol, and 0.1% (v/v) Tween 20. Protein concentration (in milligrams per milliliter) is determined by dividing the A_{278} by 0.66. The yield is about 15 mg of enzyme. The final total activity is always greater than the activity of the homogenate. This is most likely due to inhibition of the enzyme by the high concentration of PEG. The PEG also makes protein determination difficult and is usually not performed on solutions during the first few purification steps (Table IV).

Properties

This monomeric enzyme has a broad substrate specificity with respect to the nucleotide triphosphate and divalent metal ion.[17] CTP is as effective as ATP and Mg(II) can be replaced by Co(II), Mn(II), and Ca(II). The k_{cat} is 300 min^{-1} at pH 6.0 and 30°. 5-Formyltetrahydrohomofolate is a poor substrate but a good competitive inhibitor of the natural substrate, with a K_i of 0.1 μM.[20] The K_m for $5-CHO-H_4PteGlu$ is 8 μM and 0.3 m*M* for MgATP.[17] The K_m for $5-CHO-H_4PteGlu_5$ is 0.2 μM.[19]

[20] T. Huang and V. Schirch, *J. Biol. Chem.* **270,** 22296 (1995).

[20] Human 5,10-Methenyltetrahydrofolate Synthetase

By JACQUES JOLIVET

Introduction

5,10-Methenyltetrahydrofolate synthetase (5-formyltetrafolate cyclo-ligase, EC 6.3.3.2; MTHFS) catalyzes the unidirectional transformation of 5-formyltetrahydrofolate to 5,10-methenyltetrahydrofolate and requires ATP and Mg^{2+}. MTHFS activity has been purified from sheep liver,[1] *Lactobacillus casei*,[2] rabbit liver,[3] and human liver.[4]

The substrate for MTHFS is endogenously produced by the irreversible hydrolysis of 5,10-methenyltetrahydrofolate to 5-formyltetrahydrofolate via a second catalytic activity of serine hydroxymethyltransferase.[5] The 5-formyltetrahydrofolate polyglutamates thus formed are tightly binding inhibitors of serine hydroxymethyltransferase.[6] Because endogenously formed 5-formyltetrahydrofolate polyglutamates also inhibit other folate-dependent enzymes, MTHFS and serine hydroxymethyltransferase possibly constitute together a futile cycle that buffers the intracellular 5-formyltetrahydrofolate concentrations.[7] Mammalian cell mitochondria contain one-carbon-substituted tetrahydrofolates and their metabolism is highly compartmentalized between the cytosol and mitochondria.[8] Because 5-formyltetrahydrofolate accounts for approximately 10% of total folates in rat liver mitochondria,[9] a mitochondrial MTHFS activity is required for the reentry of mitochondrial 5-formyltetrahydrofolate into the mitochondrial folate pools.

[1] D. M. Greenberg, L. K. Wynston, and A. Nagabhushan, *Biochemistry* **4,** 1872 (1965).

[2] C. E. Grimshaw, G. B. Henderson, G. G. Soppe, G. Hansen, E. J. Mathur, and F. M. Huennekens, *J. Biol. Chem.* **259,** 2728 (1984).

[3] S. Hopkins and V. Schirch, *J. Biol. Chem.* **259,** 5618 (1984).

[4] R. Bertrand, R. E. MacKenzie, and J. Jolivet, *Biochim. Biophys. Acta* **911,** 154 (1987).

[5] P. Stover and V. Schirch, *J. Biol. Chem.* **265,** 14227 (1990).

[6] P. Stover and V. Schirch, *J. Biol. Chem.* **266,** 1543 (1991).

[7] P. Stover and V. Schirch, *Trends Biochem. Sci.* **18,** 102 (1993).

[8] D. R. Appling, *FASEB J.* **5,** 2645 (1991).

[9] D. W. Horne, D. Patterson, and R. J. Cook, *Arch. Biochem. Biophys.* **270,** 729 (1989).

0076-6879/97 $25

Assay Method

Principle

MTHFS activity is measured spectrophotometrically by monitoring the formation of 5,10-methenyltetrahydrofolate (5,10-CH^+-H_4PteGlu).

$$5\text{-HCO-}H_4\text{PteGlu}_n + \text{ATP} \rightarrow 5,10\text{-CH}^+\text{-}H_4\text{PteGlu}_n + \text{ADP} + P_i \quad (1)$$

Reagents

The following reagents are required (final concentrations given):

- 4-Morpholineethanesulfonic acid (MES) (pH 6.0), 50 m*M*
- Mercaptoethanol, 10 m*M*
- Magnesium acetate, 10 m*M*
- ATP, 0.5 m*M*
- (6*R*,*S*)-5-Formyltetrahydrofolate (calcium salt), 0.2 m*M*

Procedure

Reactions are carried out in quartz cuvettes of either 1- or 10-cm optical path length and the temperature maintained at 30° in a water-jacketed sample compartment. The 1-cm cuvette is used to detect enzyme activity during purification. The kinetic measurements are initially performed using the 1-cm cuvette. Measurements at low folate concentrations are taken with the 10-cm cuvette. The increase in absorbance at 360 nm owing to the formation of 5,10-methenyltetrahydrofolate (E_{360}, $25.1 \times 10^3\ M^{-1} \times \text{cm}^{-1}$) is monitored by spectrophotometry. One unit of activity represents 1 μmol of 5,10-methenyltetrahydrofolate formed per minute. When enzyme activity under varying pH conditions is determined using different buffers {citrate buffer (pH 4.0–5.5), MES buffer (pH 5.5–6.5), PIPES buffer [piperazine-*N*,*N*′-bis(α-ethanesulfonic acid); pH 6.5–7.5], HEPES buffer (*N*-2-hydroxyethylpiperazine-*N*′-2-ethanesulfonic acid; pH 7.0–8.0), and borate buffer (pH 8.0–10.0}, a discontinuous enzyme assay is used and the reaction is allowed to proceed for 1 hr in 0.1 *M* buffer containing 10 m*M* 2-mercaptoethanol. Substrate concentrations used are 2 m*M* ATP, 20 m*M* magnesium acetate, and 1 m*M* (6*R*,*S*)-5-formyltetrahydrofolate to ensure that saturated levels of substrates are maintained over the entire pH range. The reaction is stopped by adding citric acid saturated with ammonium sulfate to a final concentration of 0.2 *M* at pH 3.5. Assay mixtures are then treated in a boiling water bath for 30 sec and chilled on ice. 5,10-Methenyltetrahydrofolate concentrations are calculated from absorbance at 360 nm after centrifugation. Blanks containing all reagents except enzyme are used to correct for spontaneous formation of 10-formyltetrahydrofolate at higher pH values.

Purification of Cytosolic 5,10-Methenyltetrahydrofolate Synthetase

MTHFS[4] can be purified ≈40,000-fold in four steps consisting of (1) homogenization and centrifugation followed by (2) $(NH_4)_2SO_4$ precipitation and (3) chromatography on Blue Sepharose and (4) 5-formyltetrahydrofolate–aminohexyl-Sepharose.

Preparation of Crude Extracts

Human liver is obtained at autopsy and kept at −80°. Approximately 40 g is homogenized in a blender, using 100 ml of 20 m*M* PIPES, 20 m*M* 2-mercaptoethanol, 10 m*M* magnesium acetate at pH 7.0 (buffer 1). The homogenate is then centrifuged at 12,000 *g* for 30 min at 4° to remove cellular debris and organelles and yield a cytosolic extract.

$(NH_4)_2SO_4$ Precipitation

The supernatant is saturated to 70% (w/v) with solid $(NH_4)_2SO_4$ (43.6 g/100 ml) and the mixture stirred for 30 min. After centrifugation (12,000 *g*, 30 min), the precipitate is resuspended in a minimal volume (20 ml) of buffer 1 (pH 7.0) and desalted by centrifugation (5000 *g*, 10 min) through a 3 × 8 cm column of Sephadex G-25. All steps are performed at 4°.

Blue Sepharose Chromatography

The fraction containing protein is applied to a Blue Sepharose column (1.6 × 100 cm; Pharmacia, Piscataway, NJ) that has been equilibrated at 4° with buffer 1 at pH 7.0. The column is then washed with 4 liters of buffer 1 containing 0.1% (v/v) Tween 20, 10% (v/v) glycerol (buffer 2), and 0.35 *M* KCl at pH 7.0 (flow rate, 120 ml/hr). MTHFS is eluted with 500 ml of buffer 2 in 0.35 *M* KCl containing 0.1 m*M* (6*R*,*S*)-5-formyltetrahydrofolate (calcium salt). Fractions containing activity are pooled and dialyzed twice against 24 liters of buffer 2 containing 24 g of acid-washed activated charcoal to remove 5-formyltetrahydrofolate prior to the next step.

5-Formyltetrahydrofolate–Aminohexyl-Sepharose Chromatography

To the dialyzed solution, 0.15 *M* KCl is added and the pH adjusted to pH 6.5. The solution is applied to a column (1 × 30 cm) of 5-formyltetrahydrofolate–aminohexyl-Sepharose equilibrated at 4° with buffer 2 containing 0.15 *M* KCl at pH 6.5. The affinity column is prepared by dissolving 5-

formyltetrahydrofolate (50 mg) in 20 ml of distilled deionized water and adding it to a suspension of activated (pH 4.5) AH-Sepharose 4B (5 g). The mixture is stirred for 30 min at 25° before adding 1-ethyl-3-(3-dimethylaminopropyl)carbodiimide (250 mg) and the pH is adjusted to pH 6.0. The final mixture is stirred for an additional 24 hr. The gel is washed and stored at 4° in 50 m*M* PIPES, 50 m*M* 2-mercaptoethanol (pH 6.0). The column is washed with 1 liter of buffer 2 and elution of the enzyme is accomplished by applying 1 m*M* 5-formyltetrahydrofolate potassium salt in buffer 2. The enzyme is eluted as a sharp band and dialysis is performed against charcoal as described above, prior to kinetic analysis.

Purification of Mitochondrial 5,10-Methenyltetrahydrofolate Synthetase

Mitochondria are purified[10] either from normal human liver obtained in the operating room from patients undergoing open liver biopsies or from human Burkitt lymphoma CA46 cells (American Type Culture Collection, Rockville, MD). Liver fragments are immersed in an ice-cold solution prepared from 0.25 *M* sucrose, 8.6 m*M* NaCl, 15 m*M* Tris-HCl, and 5 m*M* EDTA (pH 7.5). For CA46 cells, approximately 10^{11} cells are first swelled in ice-cold deionized water for 30 min and then disrupted with a Potter homogenizer. Mitochondria are isolated using standard protocols.[11] The swollen cells are homogenized in 5 vol of ice-cold deionized water while the human liver biopsies are homogenized in 5 vol of ice-cold 0.25 *M* sucrose, 8.6 m*M* NaCl, 15 m*M* Tris-HCl, and 5 m*M* EDTA (pH 7.5). The homogenate is centrifuged at 5000 *g* for 10 min at 4° and the supernatant centrifuged again at 12,000 *g* for 20 min at 4°. The supernatant solution (cytosolic fraction) is removed from the sediment (mitochondria-containing fraction) and kept at −80° in the presence of 0.1% (v/v) Tween 20 for further analysis. The mitochondria-containing fraction is resuspended in 1 vol of the homogenization buffer and applied to a sucrose gradient consisting of two layers of solutions containing (from top to bottom) 2 ml of 1.0 *M* sucrose, 10 m*M* Tris, 5 m*M* EDTA (pH 7.5) and 2 ml of 1.5 *M* sucrose, 10 m*M* Tris, 5 m*M* EDTA (pH 7.5). Following centrifugation (24,000 *g* for 30 min at 4°) the purified mitochondria are collected at the interface of the 1.0 and 1.5 *M* sucrose solutions. Tween 20 (0.1%, v/v) is immediately added to the purified mitochondria and the suspension is

[10] R. Bertrand, M. Beauchemin, A. Dayan, M. Ouimet, and J. Jolivet, *Biochim. Biophys. Acta* **1266,** 245 (1995).

[11] B. Storrie and E. A. Madden, *Methods Enzymol.* **182,** 203 (1990).

frozen at −80°. On thawing, the suspension is sonicated (5 sec, four times) prior to further analysis.

The ratios between the cytoplasmic and mitochondrial MTHFS specific activities are 1 : 1.8 and 1 : 1.09, respectively, in the two liver biopsies tested, with 85% of the total cellular activity in the cytoplasm. Mitochondria preparations are free of cytoplasmic contaminants as determined by lactate dehydrogenase activity (≤1%).

MTHFS is purified from CA46 cell mitochondria using sequential Blue Sepharose and 5-formyltetrahydrofolate–aminohexyl-Sepharose chromatographies, as described above, with yields of 17 and 6% following each step, respectively. Purification is estimated at >10,000-fold because the protein concentration is too low to be determined after the second chromatography.

Human liver mitochondria are further subfractionated in one of the human liver biopsies by low-concentration digitonin treatment, as previously described.[12] Briefly, freshly prepared mitochondria are pelleted by centrifugation and resuspended in 0.3 ml of the homogenization buffer. An equal volume of digitonin solution (1.2%, w/v) is added and the solution gently stirred in an ice bath for 15 min. The lysate is then centrifuged at 10,000 *g* for 10 min at 4°. The supernatant consists of outer mitochondrial membranes and intermembrane components, whereas the mitoplasts are pelleted. The pellet is washed twice and then resuspended in 0.1 ml of the homogenization buffer prior to sonication. The extract is centrifuged again to obtain the mitochondrial matrix components in the supernatant. Most (66%) of the mitochondrial MTHFS activity is found in the matrix components. Its distribution between the matrix and the outer membrane fraction is similar to that of the matrix marker fumarase (72% of total activity in the matrix fraction) and reflects some disruption of the mitochondrial inner membranes during digitonin treatment. The matrix preparation is relatively free of outer membrane as determined by monoamine oxidase activity (12% of total activity in the matrix fraction).

Properties of Cytosolic 5,10-Methenyltetrahydrofolate Synthetase

Stability

The activity of purified MTHFS[4,13] is labile at 4° but can be stabilized with Tween 20 (0.1%, v/v) and glycerol (10%, v/v). Under these conditions,

[12] P. L. Perdersen, J. W. Greenawalt, B. Reynafarje, J. Hullihen, G. L. Decker, J. W. Soper, and E. Bustamante, *Methods Cell Biol.* **20,** 411 (1978).

[13] R. Bertrand and J. Jolivet, *J. Biol. Chem.* **264,** 8843 (1989).

at pH 7.0, preparation of purified enzyme can be kept at −80° for several months with no substantial loss of activity.

Physical Properties

The enzyme is a single peptide with a molecular mass of 27 kDa by polyacrylamide gel electrophoresis. The isoelectric point is estimated to be at pH 7.0 by isoelectric focusing under denaturing conditions. The enzyme shows a broad pH activity profile, with an apparent maximum at pH 6.5. The Arrhenius plot is biphasic with a linear increase as the temperature is raised from 5 to 40°, with a slower increase in activity above 40°, indicating a rate-limiting step in the reaction. A Q_{10} value of 1.94 is estimated between 30 and 40° with an activation energy of 11.5 kcal/mol.

Kinetic Properties

At the final purification step, MTHFS activity is 37 μmol min^{-1} mg^{-1}, which corresponds to an approximate turnover number of 1000 min^{-1}. The K_m values of (6*R*,*S*)-5-formyltetrahydrofolate monoglutamate [purchased from B. Schircks Laboratories (Jona, Switzerland)], (6*S*)-5-formyltetrahydrofolate monoglutamate, (6*S*)-5-formyltetrahydrofolate pentaglutamate, and ATP are 4.4, 2.0, 0.6, and 20 μM, respectively. The 6*S*-stereoisomers are prepared from folic acid monoglutamate and pentaglutamate (purchased from B. Schircks Laboratories) as previously described.[4] Initial velocity plots intersect, indicating that MTHFS binds substrates in a sequential (as opposed to a Ping–Pong) mechanism. The reaction seems to involve phophorylation of the formyl group at the N-5 position, followed by displacement of the phosphate by the 10-nitrogen.[14,15]

Inhibition Studies

5-Methyltetrahydrofolate is a competitive inhibitor of MTHFS with respect to 5-formyltetrahydrofolate, with a K_i value of 18 μM, while folic acid (K_i of 55 μM) and dihydrofolate (K_i of 50 μM) are less potent and tetrahydrofolate has no inhibitory effect up to 400 μM. The pentaglutamate derivatives of folic acid and dihydrofolate are 15-fold more potent inhibitors of the enzyme than their monoglutamate forms, with K_i values of 3.5 and 3.8 μM, respectively. Similarly, methotrexate has no inhibitory capacity up to 350 μM, while methotrexate pentaglutamate is a competitive inhibitor with a K_i value of 15 μM. The most potent inhibitor of MTHFS is the

[14] T. Huang and V. Schirch, *J. Biol. Chem.* **270,** 22296 (1995).
[15] K. Kounga, D. G. Vandervelde, and R. H. Himes, *FEBS Lett.* **364,** 215 (1995).

homofolate analog 5-formyltetrahydrohomofolate, a competitive inhibitor with a K_i of 1.4 μM. 5-Formyltetrahydrohomofolate inhibits MCF-7 human breast cancer cell growth with a 50% inhibitory concentration (IC_{50}) of 2.0 μM during 72-hr exposures, and this effect is fully reversible by hypoxanthine but not thymidine, indicating specific inhibition of *de novo* purine synthesis. Furthermore, a correlation is observed between increases in intracellular 5-formyltetrahydrofolate polyglutamate levels, the accumulating substrate following MTHFS inhibition, and cell growth arrest. *De novo* purine synthesis is inhibited in the MCF-7 cells at the level of aminoimidazole carboxamide ribonucleotide formyltransferase, the second folate-dependent enzyme in this pathway, and 5-formyltetrahydrofolate pentaglutamate is a competitive inhibitor of this enzyme, with a K_i of 3.0 μM.[13]

Properties of Mitochondrial 5,10-Methenyltetrahydrofolate Synthetase

Human liver mitochondria contain MTHFS[10] with a specific activity similar to that of cytosolic extracts. The molecular masses of native human liver cytosolic and mitochondrial MTHFS are both approximately 25 kDa, as determined by gel filtration. The K_m values of (6*R*,*S*)-5-formyltetrahydrofolate monoglutamate, (6*S*)-5-formyltetrahydrofolate pentaglutamate, and ATP are almost identical to those of the cytosolic enzyme: 4.7, 0.8, and 22 μM, respectively.

Cloning and Characterization of Cytosolic 5,10-Methenyltetrahydrofolate Synthetase-Encoding cDNA

Protein Sequencing and cDNA Cloning

Human liver cytosolic MTHFS[16] is purified to homogeneity and chemically cleaved with cyanogen bromide. Following polyacrylamide gel electrophoresis, fragments are transferred to a polyvinylidene membrane prior to sequencing and a sequence of six amino acids (PGLGFD) can be firmly established from one of the fragments. A 3′RACE-PCR protocol[17] is then used to amplify a 389-bp cDNA. The first-strand cDNA for the polymerase chain reaction (PCR) is synthesized from human liver poly(A)$^+$ RNA. The 50-μl reaction contains 1 μg of poly(A)$^+$ RNA, 10 pmol of an hybrid oligo(dT_{17}) adapter primer containing an *Xho*I sequence (5′ CCCTCGAG-

[16] A. Dayan, R. Bertrand, M. Beauchemin, D. Chahla, A. Mamo, M. Filion, D. Skup, B. Massie, and J. Jolivet, *Gene* **165,** 307 (1995).

[17] M. A. Frohman, M. K. Dush, and G. R. Martin, *Proc. Natl. Acad. Sci U.S.A.* **85,** 8998 (1988).

$GTCGACGGTATCGT_{17}$ 3′), 25 units of Moloney murine leukemia virus reverse transcriptase, and a 250 μM concentration of each deoxynucleotide. Following incubation at 37° for 1 hr, an aliquot of the reaction mixture is amplified in 100 μl containing 50 mM KCl, 10 mM Tris-HCl (pH 8.3), 1.5 mM $MgCl_2$, a 200 μM concentration of each deoxynucleotide, 10 units of *Taq* I DNA polymerase, 10 pmol of the adapter primer, and 300 pmol of degenerated oligodeoxynucleotides corresponding to a heptapeptide in which methionine is assumed to precede the sequenced hexapeptide and containing an added *Eco*RI sequence [5′ GGGAATTCATGCCIGGI (C,T)TIGGITT(T,C)GAC 3′]. The PCR product is inserted in the Bluescript SK^+ vector (Stratagene, La Jolla, CA) between the *Eco*RI and *Xho*I sites for sequence analysis and found to contain nucleotides corresponding to the sequenced hexapeptide and to a 97-amino acid open reading frame. This partial cDNA clone is then used to screen a λDR2 human liver cDNA library (Clontech, Palo Alto, CA).

Characterization of Human 5,10-Methenyltetrahydrofolate Synthetase-Encoding cDNA

An 872-bp clone is isolated with an open reading frame for a 203-amino acid protein of 23,229 Da. The Genome Sequence Data Base accession number is L38928. There is a 77% amino acid identity with the rabbit liver MTHFS sequence determined directly from purified protein.[18] Sequence identity analyses reveal only one match (22% amino acid identity), with a bacterial protein of unknown function.[19] There is a consensus ATP-binding site at amino acid 147–151 and a potential Mg^{2+}-binding site at amino acids 135–140.

Synthesis and Catalytic Activity of Recombinant 5,10-Methenyltetrahydrofolate Synthetase in Escherichia coli

The MTHFS open reading frame sequence is first amplified by PCR, using specific primers containing an *Nde*I sequence at the ATG start codon and a *Bam*HI sequence at the TAA stop codon. The PCR product is inserted in the pCRII vector, sequenced, and then subcloned in the bacterial expression vector pET11c between the *Nde*I and *Bam*HI sites. *Escherichia coli* BL21(DE3) is transfected with purified plasmids. Recombinant protein expression is induced for up to 6 hr by adding 1 mM isopropyl-β-D-thiogalactopyranoside to the exponentially growing bacteria at 27°. At different time points, bacteria are pelleted by centrifugation, washed in 1 mM dithiothrei-

[18] B. Maras, P. Stover, S. Valiante, D. Barra, and V. Schirch, *J. Biol. Chem.* **269,** 18429 (1994).
[19] L. M. Hsu, J. Zagorski, and M. Fournier, *J. Bacteriol.* **161,** 1162 (1985).

tol (DTT)–50 mM Tris–2 mM EDTA prior to being pelleted again and resuspended in 20 mM PIPES–20 mM 2-mercaptoethanol–10 mM magnesium acetate–0.1% (v/v) Tween 20 at pH 7.0 and incubated with lysozyme (1 mg/ml) for 15 min at room temperature. The bacteria are lysed by two 10-sec cycles of sonication on ice. Following centrifugation at 12,000 g for 10 min at 4°, protein expression is detected in the supernatants by monitoring MTHFS enzymatic activity. MTHFS specific activity increases from 2.9×10^{-4} to a plateau of 4×10^{-2} μmol/min/mg after a 2-hr induction with 1 mM isopropyl-β-D-thiogalactopyranoside at 27°. Polyacrylamide gel electrophoresis analysis reveals the appearance of a 27-kDa protein following induction. Differences observed between the calculated (23-kDa) and observed sizes on polyacrylamide gel electrophoresis probably relate to conformational structure changes during electrophoresis as reported for rabbit liver 5,10-methenyltetrahydrofolate synthetase.[18]

mRNA Tissue Expression

A Northern blot of poly(A)$^+$ RNA from a variety of normal adult human tissues is performed using a probe generated by random priming of the partial 389-bp cDNA clone originally obtained. A transcript of approximately 0.9 kb is detected in liver with little expression in the heart, brain, placenta, lung skeletal muscle, kidney, and pancreas.

Southern Blot Analysis

Initial gene analysis studies have been performed with an MTHFS probe prepared from a restricted cDNA fragment [527-base pair (bp), *Sph*I at nucleotide 188 and *Xmn*I at nucleotide 715]. There is a single band in human placenta genomic DNA restricted with *Hin*dIII (9 kb) and *Bam*HI (25 kb) digests while two bands are seen following *Eco*RI (4.5 and 1.8 kb), *Pst*I (4.4 and 2 kb) and *Bgl*II (5 and 1 kb). These results suggest the existence of a single MTHFS gene.

Acknowledgment

Supported by the National Cancer Institute of Canada with funds from the Canadian Cancer Society.

[21] Mitochondrial NAD-Dependent Methylenetetrahydrofolate Dehydrogenase–Methenyltetrahydrofolate Cyclohydrolase

By ROBERT E. MACKENZIE

Introduction

Eukaryotes contain methylenetetrahydrofolate dehydrogenase enzymes in both cytoplasmic and mitochondrial compartments. *Saccharomyces cerevisiae* has two distinct genes that encode different trifunctional NADP-dependent methylenetetrahydrofolate dehydrogenase (D) (EC1.5.1.5)–methenyltetrahydrofolate cyclohydrolase (C) (EC3.5.4.9)–10-formyltetrahydrofolate synthetase (S) [formate–tetrahydrofolate ligase (EC6.3.4.3)] enzymes.[1] In contrast, higher eukaryotes express a cytoplasmic NADP-dependent D/C/S enzyme and a nuclear-encoded mitochondrial bifunctional NAD-dependent D/C that is unusual in that it requires both Mg^{2+} and inorganic phosphate for activity.[2] The NAD-dependent dehydrogenase activity, originally detected in Ehrlich ascites tumor cells,[3] was later found in all immortalized mammalian cell lines tested, but not in differentiated adult tissues.[4] Both NADP- and NAD-dependent activities are present in the tissues of various insects.[5] The role of the mitochondrial enzyme is not resolved, but has been proposed to contribute 10-formyltetrahydrofolate for synthesis of formylmethionyl-$tRNA^{fMet}$ to support mitochondrial protein synthesis.

Assay Method

Principle

The assays are based on the measurement of methenyltetrahydrofolate, which absorbs maximally at 350–355 nm (ε_M of 24,900), and are similar to assays described earlier.[6] Final concentrations of the components for the

[1] K. W. Shannon and J. C. Rabinowitz, *J. Biol. Chem.* **261,** 12266 (1986).
[2] X.-M. Yang and R. E. MacKenzie, *Biochemistry* **32,** 11118 (1993).
[3] K. G. Scrimgeour and F. M. Huennekens, *Biochem. Biophys. Res. Commun.* **2,** 230 (1960).
[4] N. R. Mejia and R. E. MacKenzie, *J. Biol. Chem.* **260,** 14616 (1985).
[5] G. B. Tremblay, S. S. Sohi, A. Retnakaran, and R. E. MacKenzie, *FEBS Lett.* **368,** 177 (1995).
[6] R. E. MacKenzie and L. U. L. Tan, *Methods Enzymol.* **66,** 609 (1980).

0076-6879/97 $25

dehydrogenase and cyclohydrolase spectrophotometric assays are given in Table I. Units of activity are expressed as micromoles of product per minute.

Dehydrogenase

One-milliliter aliquots of the assay mixture given in Table I are added to 13 × 100 mm test tubes and equilibrated at 30° for 3–5 min. The reaction is initiated by addition of enzyme at zero time, and terminated by acidification with 1 ml of 0.36 *M* HCl or 7% (w/v) trichloroacetic acid (TCA) at a fixed time, generally in the range of 3 to 10 min. The acidified mixtures are retained at room temperature for 10 min, centrifuged (2000 rpm for 20 min at room temperature) to remove precipitated protein if appropriate, and the absorbance at 350 nm measured against appropriate blanks.

For assay of crude extracts the reactions are stopped with 7% (w/v) TCA and centrifuged. Appropriate blanks consist of assay mixture minus inorganic phosphate and magnesium chloride treated in the same way as samples. If large amounts of extract are used, blanks can contain 100 μM

TABLE I
COMPOSITION OF SPECTROPHOTOMETRIC ASSAY MIXTURES

Assay and component	Concentration
Dehydrogenase	
Morpholinepropanesulfonic acid (MOPS), pH 7.3	25 m*M*
Potassium phosphate, pH 7.3	5 m*M*
2-Mercaptoethanol	30 m*M*
(6*R*,*S*)-Tetrahydrofolate[a,b]	0.25 m*M*
Formaldehyde	2.5 m*M*
Magnesium chloride	5 m*M*
NAD	1.0 m*M*
Enzyme[c]	0.5–5.0 IU
Cyclohydrolase	
MOPS, pH 7.3	25 m*M*
Potassium phosphate, pH 7.3	5 m*M*
2-Mercaptoethanol	30 m*M*
(6*R*,*S*)-Methenyltetrahydrofolate[a,d]	60 μM
Enzyme	1–10 IU

[a] The reduced folate substrates are now available commercially from Schircks Laboratories (Jona, Switzerland).

[b] E. J. Drury, L. S. Bazar, and R. E. MacKenzie, *Arch. Biochem. Biophys.* **169,** 662 (1975).

[c] The amount of enzyme for the dehydrogenase assay will depend on the time of incubation, generally 3 to 10 min.

[d] P. B. Rowe, *Methods Enzymol.* **XVIII B,** 733 (1971).

EDTA to ensure that any residual magnesium from the extract is not available. This is generally not required. In the case of purified protein, acidification with 0.36 *M* HCl is preferred and blanks do not contain enzyme.

The use of reagents can be minimized and the sensitivity of the assay can be increased by using 0.5-ml aliquots of assay mixture and acidifying with 60 μl of 6 *M* HCl.

Cyclohydrolase

To a 0.49-ml aliquot of assay mixture (minus methenyltetrahydrofolate), described in Table I, is added 10 μl of a solution of 300 μ*M* 5,10-methenyltetrahydrofolate dissolved in water, acidified with a drop of 0.36 *M* HCl. The decrease in absorbance at 355 nm is recorded on an appropriate spectrophotometer for up to 1 min. The blank value is the chemical rate of hydrolysis of methenyltetrahydrofolate under the same conditions, in the absence of enzyme. The difference in initial rates is the rate of the enzyme-catalyzed reaction. It is important to note that the K_m for methenyltetrahydrofolate is almost 100 μ*M*, so that the observed rates are highly dependent on the initial concentration of methenyltetrahydrofolate used in the assay.

Channeling

The methenyltetrahydrofolate product of the dehydrogenase has been observed to be hydrolyzed preferentially by the cyclohydrolase activity rather than to accumulate in the reaction medium. To determine the efficiency of this process, two time course reactions are carried out using the dehydrogenase assay mixture and initiating with enzyme. One such reaction is observed by time drive in a spectrophotometer at 355 nm. This reaction follows the production of methenyltetrahydrofolate and NADPH. The ε_M value of 28,900 (24,900 + 4000) accounts for equal concentrations of methenyltetrahydrofolate and NADPH. The second time course is achieved by the standard assay technique, stopping reactions at 20, 30, 40 and 60 sec. The acidification destroys NADPH and converts any formyltetrahydrofolate to methenyltetrahydrofolate. Using the standard ε_M of 24,900 the total concentration of formyl- plus methenyltetrahydrofolate can be obtained. Efficiency of channeling is given by the following calculation: ([methenyl + formyl] − [methenyl])/[methenyl + formyl]. The enzyme displays about 50% efficiency in channeling.

Purification

Two basic buffer systems are used in the purification of the enzyme from Ehrlich ascites tumor cells, Sf9 (*Spodoptera frugiperda*, fall armyworm ovary) cells, or the recombinant human enzyme expressed in *Escherichia coli*:

Buffer A:

Triethanolamine hydrochloride, pH 7.3	20 m*M*
Potassium phosphate, pH 7.3	5 m*M*
2-Mercaptoethanol	35 m*M*
Benzamidine	1 m*M*
Glycerol	20% (v/v)

Buffer B:

Buffer A with glycerol at 30% (v/v)

Ehrlich Ascites Tumor Cells

Purification from Ehrlich ascites tumor cells has been described earlier.[7] Well-washed Ehrlich ascites tumor cells are homogenized on ice for 2 min in 30-ml batches with 3 vol of 100 m*M* potassium phosphate (pH 7.3), 35 m*M* 2-mercaptoethanol, 1 m*M* benzamidine, and 1 m*M* phenylmethylsulfonyl fluoride (PMSF), using a Brinkmann (Westbury, NY) Polytron. Cell breakage is monitored microscopically. All further procedures are carried out at 4°. To the combined homogenates of 10 batches are added 0.24 vol of glycerol, and the suspension is mixed and then centrifuged at 25,000 *g* for 20 min at 4°. (The supernatant solution can be frozen until required for subsequent steps.) One-tenth volume of protamine sulfate (10 mg ml^{-1}, pH 7.3) is added and the solution is stirred gently for 20 min and then centrifuged at 25,000 *g* for 20 min at 4°.

The solution is concentrated to about one-third its original volume, using dry Sephadex G-25 (13.5 g of dry Sephadex per 100 ml of extract) and centrifugation. This solution is made 55% saturated in ammonium sulfate by the slow addition with stirring of 32.6 g of $(NH_4)_2SO_4$/100 ml, maintaining the pH at pH 7.3 with NH_4OH. After centrifugation the supernatant solution is discarded and the pellet is dissolved in 50 ml of buffer A and dialyzed overnight against 2 liters of buffer B. The solution is diluted to about 250 ml with buffer A and applied to a 4 × 4 cm column of Matrex Gel Blue A (Amicon, Danvers, MA) preequilibrated in buffer A. The column is washed sequentially with three column volumes each of buffer A containing 0, 100, and 600 m*M* KCl. The enzyme activity is eluted with buffer A containing 1.0 *M* KCl and pooled fractions are dialyzed overnight against 4 liters of buffer B with two changes.

[7] N. R. Mejia, E. M. Rios-Orlandi, and R. E. MacKenzie, *J. Biol. Chem.* **261,** 9509 (1986).

Dialyzed enzyme is applied to a 4 × 2.5 cm column of Matrex Gel Orange A and washed with three column volumes of buffer A, and with three column volumes of buffer A containing 100 mM KCl. The enzyme is eluted with buffer A containing 300 mM KCl and pooled fractions are again dialyzed against 2 liters of buffer B overnight.

The sample from Matrex Gel Orange A is applied to a 2.5 × 0.5 cm column of heparin-Sepharose CL-6B (Pharmacia, Piscataway, NJ) preequilibrated in buffer B, washed with 50 ml of the same buffer, and eluted with 150 ml of buffer B containing 100 mM KCl. After dialysis overnight against 2 liters of buffer B, the sample is applied to a 4 × 5.6 cm column of DEAE-Sephadex A-50, washed with two to three column volumes each of buffer B and buffer B containing 100 mM KCl, and the enzyme eluted with 200 mM KCl in the same buffer. Pooled fractions are stored in aliquots at −80°. The purified enzyme has a dehydrogenase specific activity of 30–38 U/mg. A typical purification summary showing about 5000-fold purification with a yield of about 1 mg of protein (24%) is shown in Table II.

Insect Cells: Spodoptera frugiperda Sf9

The bifunctional enzyme can be purified from Sf9 cells following the identical protocol as for its purification from ascites tumor cells. Cultured insect cells are unusual in that they do not express the normally ubiquitous NADP-dependent trifunctional D/C/S and the NAD activity resides predominantly in the cytoplasm. From 65 ml of packed cells a 3400-fold purification yields about 95 μg of enzyme (18% yield), with a specific activity of 43 units/mg.

TABLE II
PURIFICATION OF METHYLENETETRAHYDROFOLATE DEHYDROGENASE–CYCLOHYDROLASE FROM EHRLICH ASCITES TUMOR CELLS[a]

Step	Volume (ml)	Units (μmol min^{-1})	Protein (mg)	Specific activity (μmol min^{-1} mg^{-1})	Yield (%)
Extract	1,050	81	12,915	0.0063	100
Protamine sulfate	1,010	78	8,130	0.0096	96
$(NH_4)SO_4$	76	77	2,630	0.0293	95
Matrex Gel Blue A	190	55	169	0.328	68
Matrex Gel Orange A	112	50	11	4.49	62
Heparin–Sepharose CL-6B	200	53	1.36	39.4	66
DEAE-Sephadex	66	20	1.24	31.9	24

[a] Purification from 280 ml of packed cells.

TABLE III
KINETIC CONSTANTS OF NAD-DEPENDENT DEHYDROGENASE–CYCLOHYDROLASES

Substrate or effector	K_m (μM)		
	Ehrlich	Human	Sf9
NAD	111	132	13
Methylenetetrahydrofolate	19	5	<2
Mg^{2+}	250	170	90
Phosphate	170	190	—[a]
Methenyltetrahydrofolate	155	96	30

[a] —, Not reported.

Human Enzyme Expressed in Escherichia coli

Even transformed mammalian cells express low levels of this enzyme, and the cloned cDNA for the human enzyme can be used to achieve higher levels of expression in heterologous systems. The recombinant human dehydrogenase–cyclohydrolase has been expressed in *E. coli* K-38 cells, using the construct pBK-HB1 and employing the T7 promoter.[8] The enzyme is expressed at a level of 1% of soluble protein in these cells and can be purified using an abbreviated version of the procedure described above, omitting three steps. Frozen bacterial cells (45 g) are disrupted in 3 vol of sonication buffer [100 m*M* potassium phosphate (pH 7.3), 30 m*M* mercaptoethanol, 1 m*M* benzamidine, and 1 m*M* phenylmethylsulfonyl fluoride]. The 3:1 extract is centrifuged, the supernatant solution is treated with protamine sulfate but is not concentrated or fractionated with ammonium sulfate. Instead, it is applied directly to the Matrex Gel Blue A column and the purification proceeds as described above. The final step, chromatography on DEAE-Sepharose, is not required to obtain purified enzyme and is also omitted. The purified enzyme from the heparin–Sepharose column is concentrated using an Amicon Centricon and is stored at −80°. The procedure yields about 2.5 mg of purified enzyme (about 100-fold purification) with a specific activity of 40–44 U/mg.

Enzyme Properties

The enzyme is synthesized in the cytoplasm as a 37-kDa precursor protein that is imported into mitochondria. The mature enzyme is a homodimer of 34-kDa polypeptides; sequences for the human, mouse, and *Dro-*

[8] X.-M. Yang and R. E. MacKenzie, *Protein Expr. Purif.* **3,** 256 (1992).

sophila enzymes are known. Kinetic properties of the enzyme from three sources are compiled in Table III. The enzymes from all sources require both phosphate and Mg^{2+} for activity. These ions affect the binding of NAD but not methylenetetrahydrofolate and are not required for the cyclohydrolase activity. The enzyme is stable in 20–30% (v/v) glycerol-containing buffers and can be frozen at −80° for months without loss of activity.

Specificity

Although the dehydrogenase uses NAD preferentially, the activity of the human enzyme with NADP (K_m of 400 μM) is about 20% that with NAD, requires Mg^{2+}, but is weakly inhibited by phosphate. Mn^{2+} can substitute for Mg^{2+}, and phosphate can be replaced by arsenate, but not by sulfate, citrate, nitrate, or chloride.

Other Properties

The bifunctional dehydrogenase–cyclohydrolase enzymes kinetically channel the product of the dehydrogenase, methenyltetrahydrofolate, through the cyclohydrolase activity. The efficiency of this process with the human enzyme is about 45–50%. The mature enzyme has glutamate as its N-terminal residue, but additions of one, three, or seven amino acids to the amino terminus by recombinant techniques do not affect the activity. The K value for pyridine nucleotide-linked equilibrium between methylenetetrahydrofolate and formyltetrahydrofolate at pH 7.3 is 16.[9]

Distribution

The dehydrogenase activity cannot be detected in differentiated adult mammalian tissues although its mRNA is present at low levels. However, the activity is present in fetal liver and in cells obtained from bone marrow. It is readily detected in any immortalized or transformed cell line in culture. It is found at low levels in adult insect tissues as well as in larval stages of the spruce bud worm. As with mammalian cells, it is expressed at higher levels when the insect cells are maintained in culture. It has been proposed to be the higher eukaryote homolog of the yeast mitochondrial D/C/S enzyme.

Acknowledgments

The author thanks the American Society for Biochemistry and Molecular Biology for permission to adapt this chapter from N. R. Mejia, E. M. Rios-Orlandi, and R. E. MacKenzie, *J. Biol. Chem.* **261,** 9509 (1986).

[9] J. N. Pelletier and R. E. MacKenzie, *Biochemistry* **34,** 12673 (1995).

[22] Monofunctional NAD-Dependent 5,10-Methylenetetrahydrofolate Dehydrogenase from *Saccharomyces cerevisiae*

By DEAN R. APPLING and MARY G. WEST

Introduction

5,10-Methylenetetrahydrofolate (methylene-THF) dehydrogenases (EC 1.5.1.5), catalyzing the reversible oxidation of 5,10-methylene-THF to 5,10-methenyl-THF, exist in nature in several different forms. In eukaryotes, methylene-THF dehydrogenase activity is typically contained within a multifunctional enzyme. For example, mammals, birds, and yeast all possess an NADP-dependent methylene-THF dehydrogenase as part of a trifunctional enzyme, termed C_1-THF synthase, that also contains 5,10-methenyl-THF cyclohydrolase (EC 3.5.4.9) and 10-formyl-THF synthetase (EC 6.3.4.3, formate–tetrahydrofolate ligase) activities. C_1-THF synthase exits as a homodimer (subunit M_r 100,000) in which the dehydrogenase and cyclohydrolase activities are contained in a 30 to 35-kDa N-terminal domain, with the synthetase activity located in a 70-kDa C-terminal domain. Certain eukaryotic cells also contain a 34-kDa homodimeric bifunctional NAD-dependent methylene-THF dehydrogenase/methenyl-THF cyclohydrolase. In bacteria, monofunctional NAD- and NADP-dependent methylene-THF dehydrogenases, as well as bifunctional NADP-dependent methylene-THF/methenyl-THF cyclohydrolase enzymes, are found. All of these prokaryotic enzymes exist as homodimers of subunits with a molecular weight of 30,000–35,000.

A novel methylene-THF dehydrogenase isozyme has been discovered in the yeast *Saccharomyces cerevisiae.*[1] Unlike all other eukaryotic forms of methylene-THF dehydrogenases characterized to date, the yeast enzyme is monofunctional. This cytoplasmic enzyme is specific for NAD^+ and is not dependent on Mg^{2+} or phosphate for activity. The yeast gene (*MTD1*) encoding this enzyme has been cloned and sequenced[2] and mutant yeast strains lacking the enzyme have been constructed. Studies using these mutant strains reveal a metabolic role for the NAD-dependent 5,10-methylene-THF dehydrogenase in the oxidation of cytoplasmic one-carbon units in growing yeast.[3]

[1] C. K. Barlowe and D. R. Appling, *Biochemistry* **29,** 7089 (1990).
[2] M. G. West, C. K. Barlowe, and D. R. Appling, *J. Biol. Chem.* **268,** 153 (1993).
[3] M. G. West, D. W. Horne, and D. R. Appling, *Biochemistry* **35,** 3122 (1996).

0076-6879/97 $25

This chapter describes purification of the native enzyme from yeast, as well as purification of recombinant yeast enzyme from an *Escherichia coli* expression system. A summary of kinetic, structural, and metabolic properties of the enzyme, and its evolutionary relationship to other methylene-THF dehydrogenases, is presented.

Assay Method

Principle

NAD-dependent methylene-THF dehydrogenase is assayed by following the production of 5,10-methenyl-THF, as shown in Eq. (1).

$$\text{5,10-Methylene-THF} + \text{NAD}^+ \rightarrow \text{5,10-methenyl-THF} + \text{NADH} \quad (1)$$

The assay is modified from an NADP-dependent methylene-THF dehydrogenase assay[4] by optimizing pH, buffer, and temperature conditions.[1] Enzyme activity is determined by end-point assay of the production of 5,10-methenyl-THF by measuring absorbance at 350 nm after acidification of the reaction. Acidification reverses any nonenzymatic formation of 10-formyl-THF, which equilibrates with 5,10-methenyl-THF at neutral pH.[5]

Reagents

Formaldehyde [1:10 (v/v) dilution of reagent-grade 38% (w/w) stock]
(6*R*,*S*)-Tetrahydrofolate (10 m*M*)
Stock assay buffer (129 m*M* KCl–64 m*M* potassium HEPES, pH 8.0)
NAD^+ (80 m*M* in water; store at −20°)
HCl (0.36 *N*)
2-Mercaptoethanol

Procedure

(6*R*,*S*)-Tetrahydrofolate is prepared by the hydrogenation of folic acid over platinum oxide in neutral aqueous solution[6] and purified by chromatography on DEAE cellulose.[7] The stock solution contains 10 m*M* (6*R*,*S*)-THF in 0.2 *M* Tris-HCl (pH 7.0), 0.5 *M* 2-mercaptoethanol, and is stored in the dark under vacuum at 4°. The actual substrate for the reaction, 5,10-methylene-THF, is produced by the nonenzymatic condensation of formaldehyde and THF at a molar ratio of 1.5 : 1. Twelve microliters of a

[4] K. G. Scrimgeour and F. M. Huennekens, *Methods Enzymol.* **6,** 368 (1963).
[5] L. D. Kay, M. J. Osborn, Y. Hatefi, and F. M. Huennekens, *J. Biol. Chem.* **235,** 195 (1960).
[6] R. L. Blakley, *Biochem. J.* **65,** 331 (1957).
[7] N. P. Curthoys and J. C. Rabinowitz, *J. Biol. Chem.* **246,** 6942 (1971).

1 : 10 (v/v) dilution of formaldehyde is added to 1 ml of 10 m*M* THF stock solution and incubated at 37° for 5 min. The equilibrium concentration of (6*R*,*S*)-5,10-methylene-THF in this stock solution, calculated by solving the simultaneous equations for the equilibria between THF, formaldehyde, and 2-mercaptoethanol,[8] is 5.1 m*M*. This solution is stable for several hours on ice in tubes protected from light.

An assay cocktail (enough for 20 reactions) is prepared fresh by mixing 7.75 ml of stock assay buffer, 10 μl of 2-mercaptoethanol, and 250 μl of 80 m*M* NAD. Each reaction contains 400 μl of assay cocktail, 50 μl of 5,10-methylene-THF, and 50 μl of enzyme or extract. The final concentrations of the various components are 100 m*M* KCl, 64 m*M* 2-mercaptoethanol, 50 m*M* potassium HEPES (pH 8.0), 0.5 m*M* 5,10-methylene-THF, 2 m*M* NAD.

Reaction mixtures are incubated at 37° for 5 or 10 min and stopped by the addition of 1 ml of 0.36 *N* HCl. Assay tubes are vortexed and remain at room temperature for 5 min to allow for quantitative conversion any 10-formyl-THF to 5,10-methenyl-THF prior to measuring absorbance at 350 nm. For samples of high protein concentration the reaction can alternatively be stopped with 1 ml of 2% (v/v) perchloric acid to precipitate protein. The precipitate is pelleted by low-speed centrifugation after being allowed to stand 5 min at room temperature. For accurate quantitation of low amounts of activity in crude cell extracts, a background value (minus NAD) is substracted. This assay is linear for at least 10 min and is linear with respect to protein concentration. Activity is calculated using an extinction coefficient for acidified 5,10-methenyl-THF at 350 nm of 24,900 M^{-1} cm^{-1}. One unit of activity is defined as the amount of enzyme that catalyzes the formation of 1 μmol of product in 1 min.

Enzyme Purification

Native Enzyme from Yeast

The purification scheme is summarized in Table I. Expression of the *MTD1* gene varies from strain to strain, but is generally highest in cells grown in rich medium. Cells are grown at 30° to late log phase in 4 liters of YPD [1% (w/v) yeast extract–2% (w/v) peptone–2% (w/v) glucose]. The following steps are performed at 4°.

1. Collect cells by centrifugation (4000 *g*, 5 min) and wash with deionized water.

[8] R. G. Kallen and W. P. Jencks, *J. Biol. Chem.* **241,** 5851 (1966).

TABLE I
PURIFICATION OF NAD-DEPENDENT 5,10-METHYLENETETRAHYDROFOLATE DEHYDROGENASE FROM YEAST[a]

Step	Protein (mg)	Units (μmol/min)	Specific activity (units/mg)	Purification (-fold)	Yield (%)
Crude extract[b]	2056	56	0.027	1	100
pH 5.2 supernatant	1004	56	0.056	2	100
TSK-DEAE	291	47	0.162	6	84
Matrix Gel Orange A	13	12	0.932	35	21
Phenyl-Sepharose	2.2	6.3	2.25	83	11
AX500 HPLC	0.95	5.1	5.38	200	9

[a] Reprinted with permission from Ref. 1. Copyright 1990 American Chemical Society.
[b] Prepared from 40 g of cells.

2. Resuspend the cell pellet (~40 g wet weight) in 2 ml (per gram of cells) of buffer A [25 m*M* Tris-HCl (pH 7.6), 2 m*M* EDTA, 10 m*M* 2-mercaptoethanol, 1 m*M* phenylmethylsulfonyl fluoride (PMSF)] in a plastic 30-ml capped centrifuge tube and disrupt with glass beads (acid etched, 0.5-mm diameter; Sigma, St. Louis, MO) by vortexing for three 1-min periods, with 1 min of cooling on ice in between. Glass bead disruption allows a greater recovery of enzyme activity than is achieved with disruption by passage through a French pressure cell.

3. Centrifuge the crude lysate at 25,000 *g* for 30 min, and filter the supernatant fraction through cheesecloth.

4. Adjust the filtrate to pH 5.2 by dropwise addition of dilute acetic acid. Centrifuge at 20,000 *g* for 30 min to remove precipitated protein.

5. Adjust the soluble fraction to pH 7.6 by dropwise addition of 1 *N* KOH.

6. Dilute the extract with 10 m*M* 2-mercaptoethanol, 1 m*M* PMSF until the conductivity is below 1 mmho and apply to a 2.5 × 50 cm TSK-DEAE 650 (EM Science, Gibbstown, NJ) column equilibrated with buffer A. Wash the column with 1 liter of buffer A. Elute the enzyme with 400 ml of buffer A containing 0.075 m*M* KCl.

7. Pool the fractions containing activity (~85 ml) and load directly onto a 2.5 × 20 cm Matrex Gel Orange A(Amicon, Danvers, MA) affinity column equilibrated with buffer A. Desalting is not necessary to obtain binding of the enzyme to the Matrex Gel Orange A column. Wash the column with 200 ml of buffer A, then with 400 ml of 0.025 *M* KCl in buffer A. Elute activity with a 600-ml gradient of 0.025–0.80 *M* KCl in buffer A. Pool the peak activity fractions (~100 ml) and slowly add solid KCl with mixing to achieve a final concentration of 1.5 *M*. The remainder of the purification is performed at room temperature.

8. Load the pooled matrex Gel Orange A eluate onto a 1 × 10 cm phenyl-Sepharose CL-4B (Sigma) column equilibrated with 0.025 *M* Tris-HCl (pH 7.6), 1.5 *M* KC1, 1 m*M* EDTA. Wash the column with 100 ml of the equilibration buffer, followed by 100 ml of buffer B [0.015 *M* Tris-HCl (pH 7.6), 1 m*M* EDTA]. Elute activity with 40% (v/v) ethylene glycol in buffer B. Pool the peak activity fractions (~6 ml).

9. Load the pooled phenyl-Sepharose eluate at a flow rate of 1 ml/min directly onto a 0.46 × 25 cm Synchropak AX500 HPLC column (anion exchanger, SynChrome, Lafayette, IN) equilibrated with buffer B. Wash with buffer B until the absorbance at 260 nm returns to baseline (~40 ml). Elute activity with a 20-ml linear gradient of 0–0.2 *M* KCl in buffer B. Collect 1-ml fractions.

10. Analyze peak activity fractions by sodium dodecyl sulfate–polyacrylamide gel electrophoresis (SDS–PAGE) and pool fractions containing pure enzyme (apparent molecular weight of 35,000). As can be seen in Table I, the enzyme is purified 200-fold with a 9% yield by this procedure. The activity is stable for several weeks after storage in the final elution buffer at 4°.

Recombinant Enzyme Expressed in Escherichia coli

Principle. The pET system (Novagen, Madison, WI) offers a variety of host strains and vectors for the cloning and expression of recombinant proteins in *E. coli.* Target genes cloned into pET plasmids are under control of the phage T7 transcriptional promoter and expression is induced in host bacteria containing a chromosomal copy of the T7 RNA polymerase gene. The host is a lysogen of bacteriophage λDE3, which contains the polymerase gene under the control of the inducible *lacUV5* promoter. Thus, addition of isopropyl-β-D-thiogalactopyranoside (IPTG) to the growing culture induces the T7 RNA polymerase, which aggressively transcribes the target gene in the plasmid. In many cases, more than 50% of the total cell protein is the desired expression product.

Construct. The pET-11d expression vector is used to express the *MTD1* open reading frame (ORF). pET-11d has a unique *Nco*I site downstream of the T7 promoter sequence to allow cloning directly into the ATG start codon. It also has a *Bam*HI site upstream of a translational stop codon and transcriptional stop sequence. The *MTD1* gene was amplified by polymerase chain reaction (PCR) from an *MTD1*-containing plasmid using designed primers (Fig. 1). 5′ MTD1 primer incorporates an *Nco*I site into the 5′ end of the *MTD1* ORF, changing the first nucleotide of the second codon from T to G. This changes the second codon of the protein from serine (TCG) to alanine (GCG). 3′ MTD1 primer has a *Bam*HI site flanking sequence

5' MTD1 Primer

NcoI

5'-GTACTAGTCCCATGGCGAAGCCTGGTC-3'

Met**Ala**LysProGly

***MTD1* Gene**

5'-......GAGAAATGTCGAAGCCTGGTCGT... ...AAATAGAACATTTGTGGCTGTTC...-3'

Met**Ser**LysProGlyArg... ...Lys***

3' MTD1 Primer

3'-CTTGTAAACACCGACCCTAGGATAC-5'

BamHI

FIG. 1. Primers used in PCR to construct the pET11d-*MTD1* vector for expression of the yeast monofunctional 5,10-methylene-THF dehydrogenase in *E. coli.* The amino acids shown in boldface indicate the sequence change introduced by the 5′ MTD1 primer.

complementary to the *MTD1* stop codon and 3′ untranslated sequence. Thus, the 3′ end of the expressed protein is identical to that of the native protein. The amplified fragment was cloned into pET-11d and introduced into *E. coli* strains HMS174 (for long-term storage of the plasmid) and BL21(DE3) for regulated expression of the *MTD1* gene. The pET-11d and host *E. coli* strains are available from Novagen.

Procedure

1. Grow a culture of BL21(DE3)/pET11d-*MTD1* overnight at 37° in 5 ml of 2YT [1% (w/v) yeast extract–1.6% (w/v) tryptone–0.5% (w/v) NaCl]) containing ampicillin (200 μg/ml).

2. Inoculate 250 ml of fresh 2YT–ampicillin medium with 1 ml of the overnight culture and grow the cells to mid- to late log phase [A_{600} of 0.5 on a Perkin-Elmer (Norwalk, CT) Lambda 3A spectrophotometer].

3. Add IPTG to 1.0 m*M* final concentration from a 100 m*M* sterile-filtered stock to induce production of the *MTD1* gene product (NAD-dependent 5,10-methylene-THF dehydrogenase).

4. At approximately 8 hr after induction, harvest the cells by centrifugation at 7000 *g*. Wash the cell pellet once with deionized water and store at −70° until needed. Approximately 1.5 g (wet weight) of cells is typically obtained per 250 ml of culture.

Purification

1. Resuspend the cell pellet in 5–10 ml (per gram wet weight) of buffer A [25 m*M* Tris-HCl (pH 7.6), 2 m*M* EDTA, 10 m*M* 2-mercaptoethanol, 1 m*M* PMSF].

2. Transfer the suspension to two 15-ml plastic tubes and sonicate for 2 min (30-sec pulses with cooling on ice in between) to lyse the cells.

3. Centrifuge the lysate at 20,000 *g* for 30 min to pellet cellular debris.

4. Slowly (0.5 ml/min) apply the supernatant to a Matrix Gel Orange A (Amicon) affinity column equilibrated with buffer A. Wash the column with 200 ml of buffer A, followed by 200 ml of 0.025 *M* KCl in buffer A.

5. Elute the enzyme with 200 ml of 0.2 *M* KCl in buffer A. Analyze peak activity fractions by SDS–PAGE and pool fractions containing pure enzyme (apparent M_r 35,000). The yield of pure enzyme by this procedure is typically 8–10 mg from 1.5 g of cells, with a specific activity of 3.0 U/mg, similar to the value obtained for the native enzyme purified directly from yeast. The purified enzyme is stable for at least 1 year when stored at 4°.

Properties of NAD-Dependent 5,10-Methylene-THF Dehydrogenase from Yeast

Kinetic Characteristics

The enzyme is specific for NAD^+ and is not dependent on Mg^{2+} for activity. The forward reaction initial velocity kinetics are consistent with a sequential reaction mechanism. With this model, K_m values for NAD^+ and (6*R*,*S*)-5,10-methylene-THF are 1.6 and 0.06 m*M*, respectively.[1] The Michaelis constants for NAD^+ and 5,10-methylene-THF are similar to K_m values for NAD^+-dependent methylene-THF dehydrogenases from *Acetobacterium woodii*[9] and *Clostridium formicoaceticum*.[10] However, the specific activities of the purified prokaryotic enzymes are approximately 100-fold greater than for the yeast enzyme. In contrast to all other previously described eukaryotic 5,10-methylene-THF dehydrogenases, the yeast enzyme is monofunctional, lacking any detectable 5,10-methenyl-THF cyclohydrolase activity. Subcellular fractionation of yeast indicates the enzyme is cytoplasmic, with no NAD-dependent 5,10-methylene-THF dehydrogenase detectable in mitochondria.[1] The activity can be found in all *S. cerevisiae* strains examined, at all stages of growth from the lag phase through the stationary phase, although the level of expression is both strain and growth medium dependent.

Physical Characteristics

The purified enzyme is represented by a single band with an apparent molecular weight of 33,000–38,000 as determined by SDS–PAGE. A native molecular weight of 64,000 was determined by gel filtration, suggesting

[9] M. R. Moore, W. E. O'Brien, and L. G. Ljungdahl, *J. Biol. Chem.* **249,** 5250 (1974).

[10] S. W. Ragsdale and L. G. Ljungdahl, *J. Biol. Chem.* **259,** 3499 (1984).

a homodimer subunit structure. Cross-linking experiments with dimethyl suberimidate confirmed the dimeric structure.[1] Nucleotide sequencing of the yeast *MTD1* gene encoding the enzyme revealed a second in-frame AUG codon downstream from the putative translational start AUG codon.[2] Translational initiation at either of these codons would result in expressed proteins within the 33 to 38 kDa range estimated from gel electrophoresis. N-Terminal amino acid sequencing was unsuccessful, apparently due to N-terminal blockage of the enzyme purified from yeast. To determine unambiguously the molecular weight of the native enzyme, we analyzed the purified protein by electrospray ionization–mass spectrometry (ESI–MS).[11]

Electrospray ionization is a soft ionization technique that is finding increasing prominence in the biological sciences for generating molecular weight information from large biomolecules. The technique involves delivery of an analyte solution through a syringe needle to which a potential of several kilovolts is applied. Microdroplets exiting the spray region are desolvated through collisions with neutral particles or with the walls of a heated capillary. The solvent-free, intact, highly charged biomolecular ions that result are introduced to a mass analyzer for detection.

In the ESI–MS experiment performed here, a 10 μM solution of purified native yeast NAD-dependent 5,10-methylene-THF dehydrogenase in a solution of 60% methanol–38% water–2% acetic acid (v/v/v) was electrosprayed into a Fourier transform ion cyclotron resonance mass spectrometer. The mass spectrum that was generated is shown in Fig. 2. A charge state distribution corresponding to the incorporation of between 24 and 36 protons by the purified enzyme is observed, from which a calculated molecular mass of 36,125 $\pm$ 4 Da is obtained. The molecular mass calculated from the nucleotide sequence of the longest open reading frame is 36,236 Da (Ref. 2), indicating that the protein initiates from the first upstream AUG codon. The difference can be attributed to posttranslational processing of the N terminus of the mature protein.

Metabolic Role

Saccharomyces cerevisiae possesses two cytosolic methylene-THF dehydrogenases that differ in their redox cofactor specificity: the NAD-dependent enzyme (*MTD1* gene) and the NADP-dependent activity of the trifunctional enzyme (*ADE3* gene). Thus, the role of the NAD-dependent monofunctional enzyme in one-carbon unit metabolism and *de novo* purine biosynthesis must be considered in that context. Growth studies with various *mtd1* and *ade3* mutant strains show that the NAD-dependent methylene-

[11] S. A. Hofstadler and D. A. Laude, Jr., *Anal. Chem.* **64,** 572 (1992).

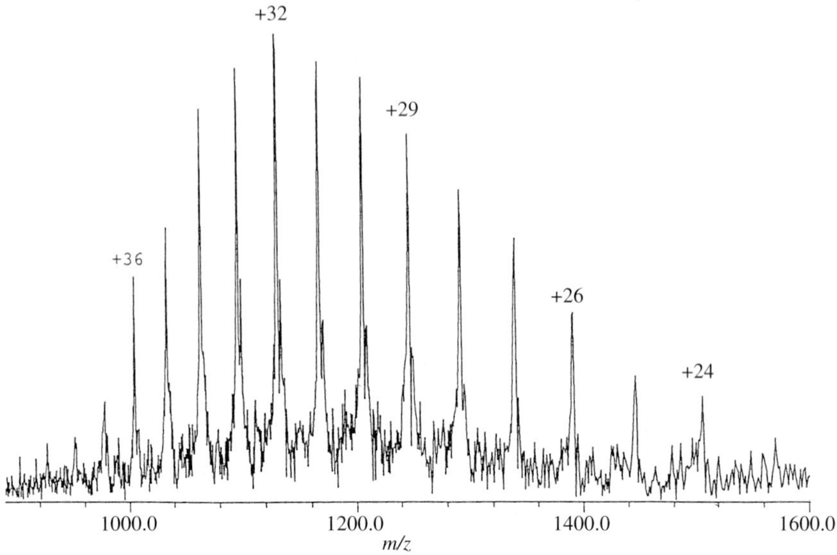

FIG. 2. ESI–MS spectrum of yeast monofunctional 5,10-methylene-THF dehydrogenase. Ion peaks representing charge states of +24 to +36 are easily resolved in this spectrum.

THF dehydrogenase is completely interchangeable with the NADP-dependent methylene-THF dehydrogenase when the flow of one-carbon units is in the oxidative direction, but the NAD-dependent enzyme does not participate significantly when one-carbon flux is in the reductive direction. ^{13}C nuclear magnetic resonance (NMR) experiments with [2-^{13}C]glycine and unlabeled formate confirmed the latter conclusion.[3] Furthermore, a yeast strain expressing only the NAD-dependent methylene-THF dehydrogenase contains substantial levels of 10-formyl-THF, the one-carbon donor used in purine synthesis. Disruption of the *MTD1* gene in this strain results in undetectable 10-formyl-THF, revealing a catalytic role for the NAD-dependent methylene-THF dehydrogenase in the oxidation of cytoplasmic one-carbon units *in vivo*.[3]

Evolutionary Relationship of Yeast NAD-Dependent 5,10-Methylenetetrahydrofolate Dehydrogenase with Other 5,10-Methylenetetrahydrofolate Dehydrogenases

Nucleotide sequence analysis of the yeast *MTD1* gene encoding the monofunctional methylene-THF dehydrogenase reveals a 320-amino acid open reading frame with only 18–25% identity to other 5,10-methylene-

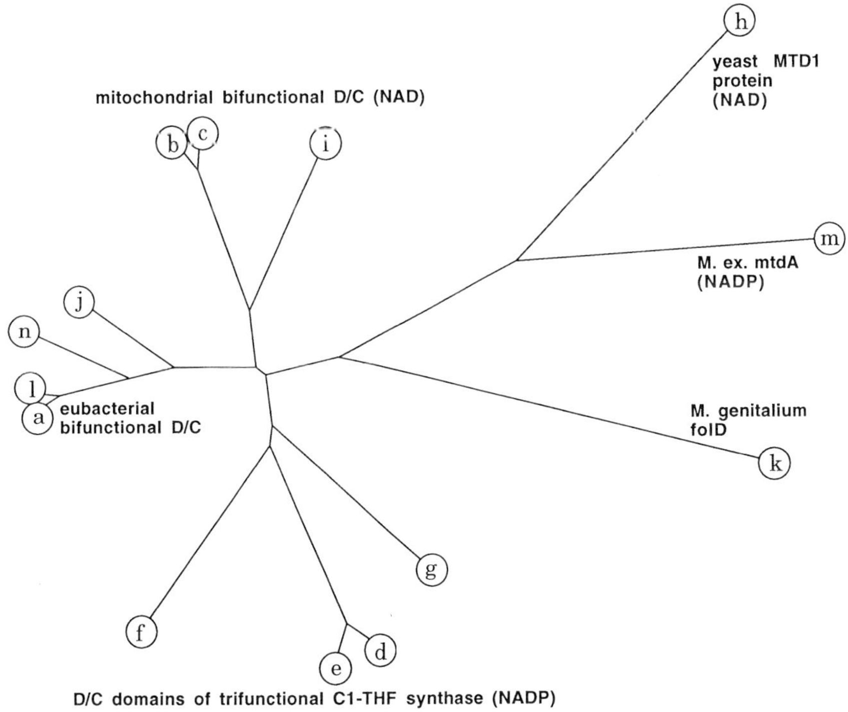

FIG. 3. Unrooted phylogenetic tree for fourteen 5,10-methylene-THF dehydrogenase sequences. (Swiss-Prot or GenBank accession numbers are in parentheses). (a) *Escherichia coli* NADP-dependent dehydrogenase/cyclohydrolase (*folD* gene) (P24186); (b) human mitochondrial NAD-dependent dehydrogenase/cyclohydrolase (P13995); (c) mouse mitochondrial NAD-dependent dehydrogenase/cyclohydrolase (P18155); (d) NADP-dependent dehydrogenase/cyclohydrolase domain of human C_1-THF synthase (P11586); (e) NADP-dependent dehydrogenase/cyclohydrolase domain of rat C_1-THF synthase (P11586); (f) NADP-dependent dehydrogenase/cyclohydrolase domain of yeast cytoplasmic C_1-THF synthase (*ADE3* gene) (P07245); (g) NADP-dependent dehydrogenase/cyclohydrolase domain of yeast mitochondrial C_1-THF synthase (*MIS1* gene) (P09440); (h) yeast NAD-dependent monofunctional dehydrogenase (*MTD1* gene) (Q02046); (i) *Drosophila* NAD-dependent dehydrogenase/cyclohydrolase (Q04448); (j) *Haemophilus influenzae folD* gene product (P44313); (k) *Mycoplasma genitalium folD* gene product (P47259); (l) *Salmonella typhimurium folD* gene product (X74603); (m) *Methylobacterium extorquens* NADP-dependent dehydrogenase (*mtdA* gene) (L27235); (n) *Photobacterium phosphoreum* dehydrogenase/cyclohydrolase (U34207).

THF dehydrogenase enzymes or dehydrogenase/cyclohydrolase domains of known sequence.[2] Although overall the yeast NAD-dependent dehydrogenase is quite divergent from other methylene-THF dehydrogenases, certain regions that are highly conserved in the other sequences are also conserved in the yeast NAD-dependent dehydrogenase.[2] This may prove useful in defining critical functional residues of methylene-THF dehydrogenases in the future.

A phylogenetic tree was created for all known 5,10-methylene-THF dehydrogenases using the PhyloTree command of the Darwin server at the Computational Biochemistry Research Group (ETH, Zurich, Switzerland). (Fig. 3). The tree is based on the relatedness of amino acid sequence between pairs of proteins. In the tree, the length of each branch is proportional to the estimated evolutionary distance between the nodes. Most of the known methylene-THF dehydrogenase sequences can be grouped on three main branches. Thus, one branch contains the mitochondrial bifunctional dehydrogenase/cyclohydrolase enzymes of human, mouse, and *Drosophila* (Fig. 3, b, c, i); a second branch carries the eubacterial bifunctional dehydrogenase/cyclohydrolase enzymes (Fig. 3, a, j, l, n); and a third branch consists of the dehydrogenase/cyclohydrolase domains of the human, rat, and yeast trifunctional C_1-THF synthase enzymes (Fig. 3, d, e, f, g). The proteins in these three branches are clearly related to each other, with amino acid sequence identities ranging from 36 to 88%.[2] The yeast monofunctional NAD-dependent dehydrogenase (*MTD1* gene product) shows less than 25% identity to any of these other proteins. In fact, it appears to be more closely related to the *mtdA* gene product of *Methylobacterium extorquens* (Fig. 3, m) and the *folD* open reading frame of *Mycoplasma genitalium* (Fig. 3, k). The *M. extorquens mtdA* gene product is an NADP-linked 5,10-methylene-THF dehydrogenase involved in the serine cycle of this methylotroph.[12] It is not known whether this enzyme is monofunctional or bifunctional. The *M. genitalium folD* ORF encodes a putative 5,10-methylene-THF dehydrogenase (determined on the basis of its homology to other bacterial *folD* genes)[13]; the protein has not been identified or characterized. Thus, the yeast monofunctional NAD-dependent enzyme represents the first characterized member of a highly divergent family of 5,10-methylene-THF dehydrogenases. It will be interesting to see whether other fungi express related enzymes.

Acknowledgments

The ESI/FTICR mass spectrometric analysis was performed by Victoria Campbell in the laboratory of Dr. Dave Laude, Jr., at The University of Texas at Austin.

This work was supported by NIH Grant RR09276.

[12] L. V. Chistoserdova and M. E. Lidstrom, *J. Bacteriol.* **176,** 1957 (1994).

[13] C. M. Fraser *et al., Science* **270,** 397 (1995).

[23] Methionine Synthase from Pig Liver

By RUMA BANERJEE, ZHIQIANG CHEN, and SUMEDHA GULATI

Mammalian methionine synthase (5-methyltetrahydrofolate–homocysteine *S*-methyltransferase, E.C. 2.1.1.13) serves two major metabolic functions [Eq. (1)]. It recycles homocysteine, a toxic catabolite of *S*-adenosylmethionine (AdoMet), and it converts the circulating form of folic acid, 5-CH_3-H_4folate, to H_4folate, which can then support a variety of cellular reactions. The activity of the enzyme is dependent on the presence of the cofactor, cobalamin. Under *in vitro* assay conditions, the activity is additionally dependent on the presence of an electron source and AdoMet, which serves as a methylating agent.

$$\text{5-Methyltetrahydrofolate} + \text{homocysteine} \rightarrow \text{tetrahydrofolate} + \text{methionine} \quad (1)$$

Assay Methods

Principle

Several methods are available for assaying methionine synthase.[1-5] Two methods that are employed for measuring methionine synthase activity from mammalian tissues or cell cultures are described here. In both, transfer of the radiolabeled methyl group from the substrate, 5-$^{14}CH_3$-H_4folate, to the product, methionine, is monitored. They differ in their source of reducing agents. The first employs dithiothreitol (DTT) and hydroxocobalamin (OH-B_{12}) and is semianaerobic. The second employs titanium(III) citrate and is anaerobic.

Reagents

Potassium phosphate (pH 7.2), 1 *M*
DTT, 500 m*M*

[1] H. Weissbach, A. Peterkovsky, B. Redfield, and H. Dickerman, *Biochem. Biophys. Res. Commun.* **17,** 17 (1964).
[2] R. E. Loughlin, H. L. Elfore, and J. M. Buchanan, *J. Biol. Chem.* **239,** 2888 (1964).
[3] V. Frasca, R. V. Banerjee, W. R. Dunham, R. H. Sands, and R. G. Matthews, *Biochemistry* **27,** 8458 (1988).
[4] Z. Chen, S. Chakraborty and R. V. Banerjee, *J. Biol. Chem.* 19246 (1995).
[5] J. T. Drummond, J. Jarrett, J. C. Gonzalez, S. Huang, and R. G. Matthews, *Anal. Biochem.* **288,** 323 (1995).

0076-6879/97 $25

Homocysteine, 100 m*M*
AdoMet (iodide salt), 3.8 m*M*
OH-B_{12}, 5 m*M*
(6*R*,*S*)-5-$^{14}CH_3$-H_4folate (specific activity, ~2000 dpm/nmol; Amersham, Arlington Heights, IL), 4.17 m*M* in 10 m*M* potassium phosphate, pH 7.2, containing 8 m*M* DTT
Titanium citrate, 84 m*M*

Procedure

Preparation of all reagents except for titanium citrate has been described by Drummond *et al.*[5] Radiolabeled CH_3-H_4folate is dissolved in potassium phosphate buffer containing 8 m*M* DTT rather than in unbuffered ascorbate.[5] Titanium citrate is prepared as follows, as has been described previously by Zehnder and Wuhrman.[6] An anaerobic 50-ml solution of 0.2 *M* sodium citrate is added to 2.6 ml of 1.9 *M* titanium(III) chloride (Aldrich, Milwaukee, WI) in 2 *M* HCl. The solution is neutralized with 6 ml of anaerobic saturated sodium carbonate. The concentration of titanium in the final solution is 84 m*M*.

I. Semianaerobic Assay

The assay mixture contains the following components [with the numbers in parentheses indicating the volumes of the stock solutions (described under Reagents) that are employed]: 100 m*M* potassium phosphate buffer, pH 7.2 (50 μl), 50 μM OH-B_{12} (5 μl), 500 μM homocysteine (2.5 μl), 152 μM AdoMet (20 μl), 25 m*M* DTT (25 μl), 250 μM (6*R*,*S*)-5-$^{14}CH_3$-H_4folate (30 μl), and enzyme in a final volume of 0.5 ml. The reaction mixture lacking CH_3-H_4folate is preincubated at 37° for 5 min. The reaction is initiated with CH_3-H_4folate, incubated for 10 min (for enzyme from porcine liver tissue) or 60 min (for enzyme from cell cultures) at 37°, and terminated by heating at 98° for 2 min, followed by cooling on ice for 2 min. Radiolabeled methionine is separated from unreacted substrate by chromatography on a Dowex 1-X8 column (chloride form, 200 mesh; Sigma, St. Louis, MO) poured in a 0.5 × 6 cm Pasteur pipette. The column is rinsed with 2 ml of water. The eluate containing methionine is collected in a scintillation vial, and 10 ml of scintillation fluid is added. Radioactivity is measured by liquid scintillation counting. Control assays in which enzyme has been omitted are run in parallel and serve as blanks. The assay is linear between 2000 and 40,000 dpm (disintegrations per minute) above background (minus enzyme controls).

[6] A. J. B. Zehnder and K. Wuhrman, *Science* **194,** 1165 (1976).

II. Anaerobic Assay

Assays are performed in 10-ml serum vials in which the atmosphere has been exchanged for an N_2:H_2 (19:1, v/v) mixture in an anaerobic chamber (Coy Products, Ann Arbor, MI). All stock reagent solutions are made anaerobic by bubbling nitrogen through them for at least 30 min and are stored in stoppered serum vials. This procedure is repeated periodically with solutions that are stored for long periods of time. The enzyme (placed in a stoppered serum vial on ice) is made anaerobic by gently passing nitrogen over the solution for 30 min. The reaction mixture contains the following [with the numbers in parentheses representing volumes of the stock solutions (described under Reagents) that are employed]: 100 m*M* potassium phosphate buffer, pH 7.2 (100 μl), 500 μ*M* homocysteine (5 μl), 19 μ*M* AdoMet (5 μl), 2 m*M* titanium citrate (24 μl), 250 μ*M* (6*R*,*S*)-5-$^{14}CH_3$-H_4folate (60 μl), and enzyme in a final volume of 1 ml. The assay can be performed either in the presence or absence of 50 μ*M* hydroxocobalamin (10 μl). The reaction mixture lacking CH_3-H_4folate is preincubated at 37° for 5 min. The reaction is initiated with CH_3-H_4folate, incubated for 10 min at 37°, and terminated by heating at 98° for 2 min. Beyond this point, the reaction vials are uncapped and the solutions are handled aerobically exactly as described for the semianaerobic assay. The assay is linear between 4000 and 80,000 dpm above background. The anaerobic assay can also be performed on a 500-μl scale for 60 min, which is more convenient when measuring activity in mammalian cell cultures. The AdoMet concentration is then increased to 152 μ*M* as described for the semianaerobic assay.

Units

One unit of enzyme activity catalyzes the formation of 1 μmol of methionine per minute at 37°. Specific activity is expressed as units per milligram of protein unless specified otherwise. Protein is determined colorimetrically with the Bradford assay (Bio-Rad, Richmond, CA), using bovine serum albumin as a standard.

Purification Procedure

All operations are carried out at 4° unless otherwise stated. The buffers used for purification contain potassium phosphate, pH 7.2, unless specified otherwise. The enzyme is purified continuously through the phenyl-Sepharose chromatography step without freezing or overnight storage. Protein solutions are concentrated under nitrogen in an Amicon (Danvers, MA) ultrafiltration cell fitted with a YM30 membrane. Small volumes of protein are concentrated in Centricon-30 microconcentrators (Amicon). The purification procedure is summarized in Table I.

Step 1. Homogenization. Two pig livers (1–1.5 kg each) that have been previously collected fresh from a slaughterhouse, cubed, and stored at −80° are thawed rapidly in cold running water. The liver cubes are placed in 4 liters of 100 m*M* potassium phosphate, pH 5.9, containing 12 mg of tosyllysine chloromethyl ketone, 100 mg of phenylmethylsulfonyl fluoride, 50 mg of trypsin inhibitor, and 5 ml of aprotinin. The livers are homogenized in a Waring blender in two batches, each of which is blended three times (1 min each) at high speed. The combined homogenate is centrifuged at 12,500 *g* for 60 min, and the supernatant is collected and filtered through three layers of Miracloth (Calbiochem, La Jolla, CA) to remove lipid and other particulate debris.

Step 2. Batch Chromatography on DEAE-Cellulose. Approximately 600 ml of a DEAE-cellulose slurry previously equilibrated with 50 m*M* potassium phosphate is added to the homogenate from step 1. The mixture is stirred for 60 min, and the resin is filtered on a Büchner funnel to remove unbound protein. Batch chromatography is performed by successive resuspension of the resin in potassium phosphate buffers of the following concentrations: 4 liters of 50 m*M*, 1 liter of 180 m*M*, and two 1-liter volumes of

TABLE I
PURIFICATION PROCEDURE FOR PORCINE LIVER METHIONINE SYNTHASE[a]

Procedure	Total units[b]	Total protein	Specific activity[c]	Yield (%)	Purification (-fold)
Homogenate	116,767	614,562	0.19	100	1
DEAE-cellulose	45,857	12,845	3.57	39	19
Q-Sepharose 1	25,278	1,442	17.53	22	92
Phenyl-Sepharose	14,893	75.6	197	13	1,037
HTP-1					
Major	2,424	5.51	440	2	2,316
Minor	3,128	23	136	3	718
HTP-2[d]					
Major	996	0.59	1,634	0.9	8,600
Minor	2,967	0.55	1,274	2.5	6,705

[a] Summary of the purification profile starting with four pig livers. [Reproduced with permission from Z. Chen, K. Crippen, S. Gulati, and R. V. Banerjee, *J. Biol. Chem.* **269,** 27193 (1994).]

[b] A unit corresponds to 1 nmol of methionine formed per minute.

[c] Specific activity is reported in units of nanomoles of methionine formed per minute per milligram protein.

[d] Only the major fraction from the previous step was loaded onto the second HTP column (see text).

400 m*M*. Between each of the washes, the slurry is stirred for 30 min before being filtered. Methionine synthase elutes in the two 400 m*M* buffer fractions, which are pooled.

Step 3. Anion-Exchange Chromatography on Q-Sepharose (Fast Flow, Sigma). The 400 m*M* washes are diluted twofold with cold water and loaded onto a 5 × 15 cm column previously equilibrated with 200 m*M* potassium phosphate. The column is washed with 500 ml of 200 m*M* potassium phosphate containing 10% (v/v) glycerol and eluted with two linear gradients. The first, with a total volume of 2 liters, ranges from 200 to 400 m*M* potassium phosphate containing 10% (v/v) glycerol. The second, with a total volume of 1 liter, ranges from 400 to 650 m*M* potassium phosphate containing 10% (v/v) glycerol. The flow rate is approximately 10 ml/min and 18-ml fractions are collected. A prominent pinkish brown band corresponding to a cytochrome b_5 impurity coelutes with methionine synthase from this column and serves as a convenient marker. Enzyme activity elutes in two peaks at ~350 and ~420 m*M* KP_i. Fractions from both peaks are pooled together.

Step 4. Hydrophobic Chromatography on Phenyl-Sepharose. The enzyme solution from the previous step is brought to 0.4 *M* ammonium sulfate and loaded immediately onto a 2.5 × 20 cm phenyl-Sepharose column (Fast Flow; Pharmacia, Piscataway, NJ) previously equilibrated with 50 m*M* potassium phosphate containing 0.4 *M* ammonium sulfate and 5% (v/v) glycerol. The column is washed with 100 ml of the equilibration buffer and eluted with a linear gradient (500-ml total volume) using the following two buffers: Buffer A [50 m*M* potassium phosphate–0.4 *M* ammonium sulfate–5% (v/v) glycerol] and buffer B [50 m*M* potassium phosphate–20% (v/v) glycerol]. The flow rate is ~2 ml/min and 5-ml fractions are collected. The cytochrome-like impurity does not bind tightly to this column and is present in the eluate before the start of the gradient. Methionine synthase activity elutes in a sharp peak at ~100 m*M* ammonium sulfate. The pooled fractions are concentrated and brought to 20% (v/v) glycerol before being frozen at −80°.

Step 5. First Hydroxyapatite Column. When enzyme from two livers is processed through step 4, the eluate is combined and diluted 20-fold with cold water and brought to 0.2 *M* KCl with solid potassium chloride. The enzyme solution is loaded onto a 2.5 × 10 cm hydroxyapatite column (Calbiochem) equilibrated with 5 m*M* potassium phosphate–0.2 *M* KCl. The column is eluted with a 500-ml linear gradient ranging from 5–150 m*M* potassium phosphate at a flow rate of 1.6 ml/min. The enzyme elutes in three partially resolved (Fig. 1) peaks. Two pools corresponding to protein eluting around ~80 m*M* (major) and ~60 m*M* (minor) potassium phosphate are made, and the protein is concentrated and frozen at −80°.

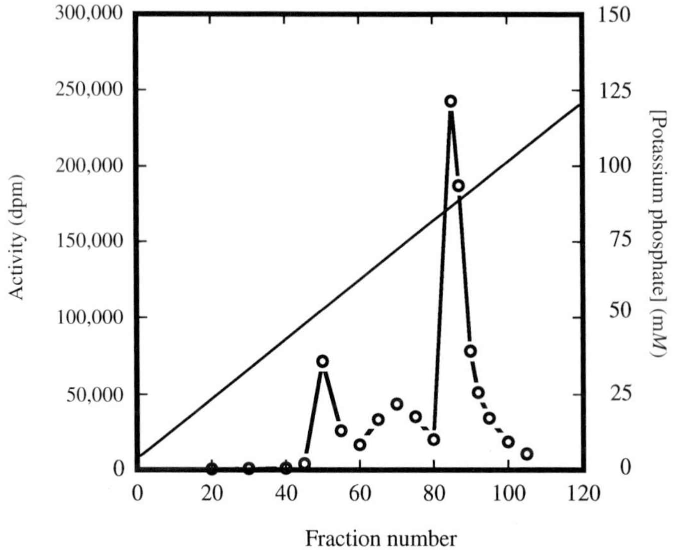

FIG. 1. Elution profile of methionine synthase from first HTP column, indicating partial resolution of enzyme activity into three fractions. This column was eluted with a linear gradient ranging from 5 to 120 m*M* potassium phosphate, pH 7.2.

Step 6. Second Hydroxyapatite Column. The pooled enzyme from the major fraction in the previous step is brought to ~60 m*M* potassium phosphate and 200 m*M* KCl and loaded onto a 1 × 5 cm column equilibrated with 60 m*M* potassium phosphate containing 0.2 *M* KCl. The protein is eluted with a 100-ml linear gradient ranging from 60 to 150 m*M* potassium phosphate containing 200 m*M* KCl. The flow rate is 0.5 ml/min, and 2-ml fractions are collected. Enzyme activity elutes at ~85 m*M* potassium phosphate.

Properties

pH Optimum. The activity-versus-pH dependence for the porcine methionine synthase has a bell-shaped profile with a maximum at pH 7.0.

Size and Subunit Composition. The native enzyme has an apparent molecular mass of 155 kDa as determined by gel filtration on a calibrated Superose 12 FPLC column (Pharmacia), assuming that it is a globular protein (Fig. 2). The subunit molecular weight of methionine synthase determined by electrophoresis on a denaturing sodium dodecyl sulfate (SDS)–polyacrylamide gel is 151 kDa. The virtually identical molecular

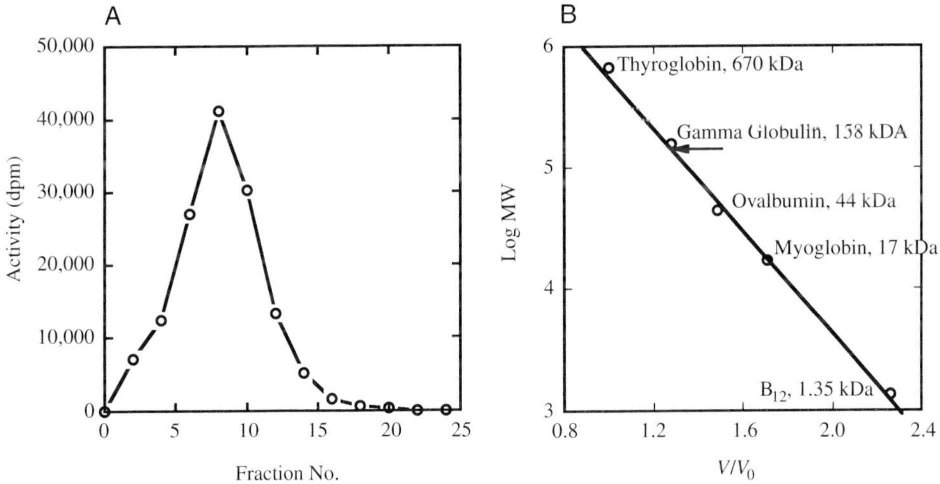

FIG. 2. (A) Elution profile of methionine synthase from a Superose 12 (Pharmacia) gel filtration column. The column was eluted with 50 m*M* potassium phosphate, pH 7.2, containing 150 m*M* KCl at a flow rate of 0.5 ml/min. (B) Calibration curve with molecular mass markers. The arrow indicates the position of methionine synthase.

mass estimates under denaturing and nondenaturing conditions indicate that methionine synthase from pig liver is a large monomeric protein.

Metal Content. The purified porcine methionine synthase contains 1 mol of cobalt per mole of enzyme. No other metal ions are detected by plasma emission spectroscopy, which detects 20 metals, including the common transition metal ions.

Ultraviolet–Visible Absorption Spectrum. The absorption spectrum of methionine synthase at pH 7.2 has maxima at 358, 506, and 536 nm and resembles that of hydroxocobalamin (OH-B_{12}) rather than aquocobalamin (H_2O-B_{12}).[7] Quantitative analysis of the absorption spectrum reveals the presence of stoichiometric cobalamin, indicating that the enzyme purified under these conditions is predominantly in the holoenzyme form.

Michaelis Constants. The porcine enzyme employs an ordered sequential mechanism in which CH_3-H_4folate binds before homocysteine, and methionine is released before H_4folate. The K_m for (6*S*)-CH_3-H_4folate is between 13 to 17 μM.[8] The K_m for homocysteine is low, and could not be determined accurately by Chen *et al.*[8] However, an earlier study reported the following

[7] C. Giannotti, *in* "B_{12}" (D. Dolphin, ed.), Vol. 1, pp. 394–430. John Wiley & Sons, New York, 1982.

[8] Z. Chen, K. Crippen, S. Gulati, and R. V. Banerjee, *J. Biol. Chem.* **269,** 27193 (1994).

Michaelis constants for the hog kidney enzyme: AdoMet, 200 n*M*; (6*S*)-CH_3-H_4folate, 25 μM; homocysteine, 100 n*M*.[9]

Acknowledgment

This work was supported by a grant from the National Institute of Health (NIDDK 45776).

[9] G. T. Burke, J. H. Mangum, and J. D. Brodie, *Biochemistry* **10,** 3079 (1971).

[24] Purification and Assay of Cobalamin-Dependent Methionine Synthase from *Escherichia coli*

By JOSEPH T. JARRETT, CELIA W. GOULDING, KERRY FLUHR, SHA HUANG, and ROWENA G. MATTHEWS

Introduction

Cobalamin-dependent methionine synthase, or 5-methyltetrahydrofolate–homocysteine *S*-methyltransferase (EC 2.1.1.13), from *Escherichia coli* is a 136,902-Da peptide containing one molecule of bound cobalamin. As shown in Fig. 1, the enzyme catalyzes the transfer of a methyl group from methyltetrahydrofolate to the cob(I)alamin form of the cofactor, to form methylcobalamin and tetrahydrofolate, and then transfers the methyl group from methylcobalamin to homocysteine, forming methionine and regenerating the cob(I)alamin form of the enzyme. Cob(I)alamin is not only a powerful nucleophile, but also a potent reductant, capable of reducing the protons of solvent to hydrogen gas.[1] This form of the enzyme occasionally becomes oxidized to an inactive cob(II)alamin species. Return of this species to the catalytic cycle requires a reductive methylation, in which the methyl group is supplied by adenosylmethionine rather than methyltetrahydrofolate. In *E. coli,* the electrons for reductive activation are supplied by reduced flavodoxin.[2,3] For *in vitro* assays, other reducing systems are often used; most frequently electrons for reductive activation are supplied by using dithiothreitol or 2-mercaptoethanol and catalytic amounts of hydroxocobalamin.[2,4]

[1] S. L. Tackett, J. W. Collat, and J. C. Abbott, *Biochemistry* **2,** 919 (1963).
[2] K. Fujii and F. M. Huennekens, *J. Biol. Chem.* **249,** 6745 (1974).
[3] C. Osborne, L.-M. Chen, and R. G. Matthews, *J. Bacteriol.* **173,** 1729 (1991).
[4] R. T. Taylor and H. Weissbach, *J. Biol. Chem.* **242,** 1502 (1967).

0076-6879/97 $25

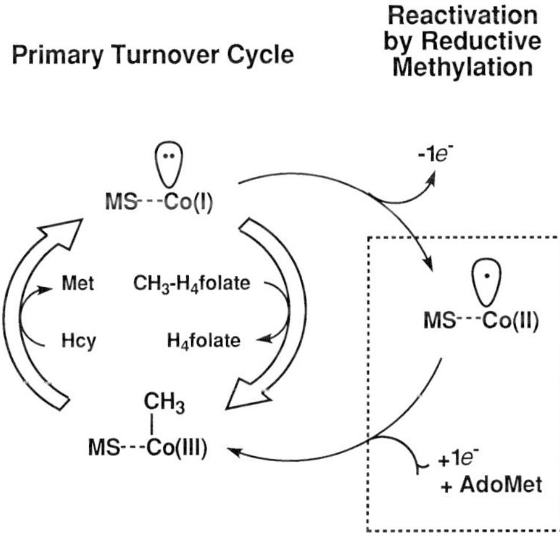

FIG. 1. Schematic summary of the reactions catalyzed by methionine synthase. The enzyme cycles in catalysis between methylcobalamin (CH_3-Co) and cob(I)alamin [Co(I)]. Homocysteine (Hcy) demethylates methylcobalamin to form methionine and enzyme in the cob(I)alamin form. Methyltetrahydrofolate supplies the methyl group to reform methylcobalamin and produce tetrahydrofolate. The broad arrows indicate that turnover is largely described by these events. However, occasionally the cob(I)alamin cofactor is oxidized to cob(II)alamin. Return of this form of the enzyme to the catalytic cycle requires a reductive methylation in which the methyl group is derived from adenosylmethionine (AdoMet). In *E. coli* the electron required for reductive methylation is supplied by reduced flavodoxin.

The enzyme has a modular construction, with distinct regions responsible for homocysteine binding and activation, for methyltetrahydrofolate binding and activation, for cobalamin binding, and for adenosylmethionine binding and reductive methylation.[5–7] Although the overall methyl transfer from methyltetrahydrofolate to homocysteine requires all the modules, other assays can be used to ascertain the catalytic capabilities of individual modules. For example, the enzyme catalyzes methyl transfer from exogenous methylcobalamin to homocysteine, in a reaction that is first order in methylcobalamin.[1,7,8] This activity can also be demonstrated in a truncated 71-kDa protein that contains only the homocysteine and methyltetrahydrofolate-binding modules at the N terminus of methionine synthase[7] and

[5] R. V. Banerjee, N. L. Johnston, J. K. Sobeski, P. Datta, and R. G. Matthews, *J. Biol. Chem.* **264,** 13888 (1989).

[6] J. T. Drummond, S. Huang, R. M. Blumenthal, and R. G. Matthews, *Biochemistry* **32,** 9290 (1993).

[7] C. W. Goulding and R. G. Matthews, *FASEB J.* **10,** A973 (1996).

[8] J. R. Guest, S. Friedman, D. D. Woods, and E. L. Smith, *Nature* (*London*) **195,** 340 (1962).

in an even smaller N-terminal 38-kDa fragment that contains only the homocysteine-binding module.[8a] Similarly, the enzyme catalyzes methyl transfer from methyltetrahydrofolate to exogenous cob(I)alamin, and this activity can also be demonstrated in the 71-kDa protein.[7]

Sequences homologous to cobalamin-dependent methionine synthase from *E. coli* have been identified in *Salmonella typhimurium*,[9] in *Haemophilus influenzae*,[10] in *Mycobacterium leprae*,[11] in the cyanobacterium *Synechococcus*,[12] and in the roundworm *Caenorhabditis elegans*.[13] The enzyme activity has been identified in mammals,[14–16] the porcine enzyme has been purified to homogeneity,[17] and more recently a sequence homologous to the extreme C terminus of methionine synthase from *E. coli* has been identified in the mouse.[18] Cobalamin-dependent methionine synthase has also been purified from *Plasmodium falciparum*.[19] Cobalamin-dependent methionine synthase is not found in plants or fungi, where a similar reaction is catalyzed by a cobalamin-independent methionine synthase.

Assay Methods

Two methods for measuring methyltetrahydrofolate–homocysteine methyltransferase activity have been developed. The assay using radioactively labeled methyltetrahydrofolate, modified from that described by Tay-

[8a] C. W. Goulding, D. Postigo, and R. G. Matthews, *Biochemistry* in press (1997).

[9] M. L. Urbanowski and G. V. Stauffer, *Gene* **73,** 193 (1988).

[10] R. D. Fleischmann, M. D. Adams, O. White, R. A. Clayton, E. F. Kirkness, A. R. Kerlavage, C. J. Bult, J.-F. Tomb, B. A. Dougherty, J. M. Merrick, K. McKenney, G. Sutton, W. FitzHugh, C. Fields, J. D. Gocayne, J. Scott, R. Shirley, L.-I. Liu, A. Glodek, J. M. Kelley, J. F. Weidman, C. A. Phillips, T. Spriggs, E. Hedblom, M. D. Cotton, T. R. Utterback, M. C. Hanna, D. T. Nguyen, D. M. Saudek, R. C. Brandon, L. D. Fine, J. L. Fritchman, J. L. Fuhrmann, N. S. M. Geoghagen, C. L. Gnehm, L. A. McDonald, K. V. Small, C. M. Fraser, H. O. Smith, and J. C. Venter, *Science* **269,** 496 (1995).

[11] D. R. Smith, GenBank Accession No. U00017 (1994).

[12] T. Kaneko, A. Tanaka, S. Sato, H. Kotani, T. Sazuka, N. Miyajima, M. Sugiura, and S. Tabata, GenBank Accession No. D64002 (1995).

[13] J. Swinburne, *Nature* (*London*) **368,** 32 (1994).

[14] R. E. Loughlin, H. L. Elford, and J. M. Buchanan, *J.Biol. Chem.* **239,** 2888 (1964).

[15] J. H. Mangum and J. A. North, *Biochemistry* **10,** 3765 (1971).

[16] C. S. Utley, P. D. Marcell, R. H. Allen, A. C. Antony, and J. F. Kolhouse, *J. Biol. Chem.* **260,** 13656 (1985).

[17] Z. Chen, K. Crippen, S. Gulati, and R. Banerjee, *J. Biol. Chem.* **269,** 27193 (1994).

[18] M. Marra, L. Hillier, M. Allen, M. Bowles, N. Dietrich, T. Dubuque, S. Geisel, T. Kucaba, M. Lacy, M. Le, J. Martin, M. Morris, K. Schellenberg, M. Steptoe, F. Tan, K. Underwood, B. Moore, B. Theising, T. Wylie, G. Lennon, B. Soares, R. Wilson, and R. Waterston, GenBank Accession No. W33307 (1996).

[19] J. Krungkrai, H. K. Webster, and Y. Yuthavong, *Parasitol. Res.* **75,** 512 (1989).

lor and Weissbach,[4] is most suitable for measuring activity in crude extracts, while the nonradioactive assay is less expensive and more convenient for use with purified enzyme.[20]

5-^{14}CH$_3$-H$_4$folate–Homocysteine Methyltransferase Assay

Principle. During steady-state turnover of methionine synthase, the N^5-methyl group of (6*S*)-5-$^{14}CH_3$-H_4folate is transferred to homocysteine, generating [^{14}C]methionine [Eq. (1)].

$$(6S)\text{-}5\text{-}^{14}CH_3\text{-}H_4\text{folate} + \text{homocysteine} \rightarrow (6S)\text{-}H_4\text{folate} + [^{14}C]\text{methionine} \quad (1)$$

At pH 7.2, the unreacted radiolabeled CH_3-H_4folate is anionic and is bound to a strong anion-exchange resin, AG 1-X8 (Bio-Rad, Richmond, CA), while the labeled methionine is a zwitterion and is not bound to the resin. This separation allows for an accurate assay of steady-state activity when the amount of methionine produced is 3–30% of the initial (6*S*)-5-$^{14}CH_3$-H_4folate. An advantage of the radioactive assay method is that by increasing the specific radioactivity of the labeled substrate, one can increase the sensitivity of the assay to detect low levels of activity. The nonphysiological isomer, (6*R*)-$^{14}CH_3$-H_4folate, is also present in the assay but does not react with methionine synthase and is completely bound to the anion-exchange resin.

Reagents

L-Homocysteine, 100 m*M* in 0.1 *M* sodium acetate (pH 5)
(6*R,S*)-5-$^{14}CH_3$-H_4folate, 4.17 m*M* in 8 m*M* sodium ascorbate
S-Adenosyl-L-methionine, iodide salt, 3.8 m*M* in 1 m*M* HCl
Dithiothreitol (DTT), 500 m*M* in 5 m*M* sodium acetate (pH 4.5)
Hydroxocobalamin, 5 m*M* in 50% (v/v) ethanol
Potassium phosphate buffer (pH 7.2), 1 *M*
Bio-Rad AG 1-X8 strong anion-exchange resin, chloride form, in deionized water

Procedure. L-Homocysteine is prepared by hydrolysis of the L-homocysteine thiolactone (Sigma, St. Louis, MO). L-Homocysteine thiolactone (150 mg in 5 ml of H_2O) is mixed with 0.8 *M* NaOH (2.5 ml), bubbled with argon for 5 min, and incubated at 45° for 6 min. The solution is acidified with 0.5 *M* acetic acid to pH 5, diluted to 10 ml with H_2O, and stored in 1-ml portions at −80°. (6*R,S*)-5-$^{14}CH_3$-H_4folate is prepared by adding 50

[20] J. T. Drummond, J. Jarrett, J. C. González, S. Huang, and R. G. Matthews, *Anal. Biochem.* **228,** 323 (1995).

μCi of (6*R,S*)-5-$^{14}CH_3$-H_4folate (Amersham, Arlington Heights, IL), barium salt, to 100 μmol of unlabeled (6*R,S*)-CH_3-H_4folate (Schirks Laboratory, Jona, Switzerland), calcium salt, in 8 m*M* sodium ascorbate buffer (24-ml final volume), yielding a 4.17 m*M* solution with a specific activity of ~1200 dpm/nmol. This stock solution is stored in 1-ml portions at −80°. The initial reaction mixture is prepared by adding in order: 760 μl of H_2O, 100 μl of potassium phosphate buffer, 50 μl of dithiothreitol, 5 μl of homocysteine, 5 μl of AdoMet, and 10 μl of enzyme containing ~1–5 × 10^{-5} units of activity. The enzyme is preactivated by adding 10 μl of hydroxocobalamin (Sigma) and incubating at 37° for 5 min. Steady-state turnover is initiated by adding 60 μl of $^{14}CH_3$-H_4folate and the reaction mixture is incubated at 37° for 10 min. The reaction is stopped by denaturing the enzyme at 90° for 2 min, after which the reaction mixture is cooled on ice. A small AG 1-X8 column (0.5 × 2 cm) is prepared by plugging a Pasteur pipette with glass wool, followed by the addition of ~2 ml of resin slurried in water. After the resin bed has settled, a 20-ml scintillation vial is placed below the column to collect the eluant. The reaction mixture is applied to the top of the column and the [^{14}C]methionine is eluted with two 1-ml volumes of H_2O. Scintillation fluid (10 ml) is added to the pooled washes and the sample is counted to determine the total methionine produced. One unit of activity is defined as 1 μmol of product formed per minute. A blank is usually included that contains all of the reaction components except enzyme; the blank value is subtracted from the values of samples containing enzyme.

Spectrophotometric Methyltetrahydrofolate–Homocysteine Methyltransferase Assay

Principle. The differences in the spectra of the substrate CH_3-H_4folate and the product H_4folate are small, and are not easily observed in the presence of an excess of CH_3-H_4folate.[20] However, at low pH, the product H_4folate is quantitatively derivatized with formic acid to yield CH^+=H_4folate [Eq. (2)], which absorbs strongly at 350 nm.

$$H_4\text{folate} + \text{formic acid} + H^+ \rightarrow CH^+{=}H_4\text{folate} + 2H_2O \qquad (2)$$

After heating in strong acid, the other components of the reaction mixture have little absorbance at 350 nm. This assay method is preferred over the radioactive assay for routine assays using overexpressed or purified enzyme preparations. In particular, it avoids the precautions necessary for handling radioactive samples and also eliminates the time-consuming column separation step.

Reagents

L-Homocysteine, 100 mM in 0.1 M sodium acetate buffer, pH 5
(6R,S)-5-CH_3-H_4folate, calcium salt, 4.17 mM in 8 mM sodium ascorbate
S-Adenosyl-L-methionine, iodide salt, 3.8 mM in 1 mM HCl
Dithiothreitol, 500 mM in 5 mM sodium acetate buffer, pH 4.5
Hydroxocobalamin, 1 mM in deionized water
Potassium phosphate buffer, pH 7.2, 1M
Formic acid, 13 M in 5 M HCl

Procedure. L-Homocysteine is prepared as described above. The initial reaction mixture is prepared by adding in order: 576 μl of deionized water, 80 μl of potassium phosphate buffer, 40 μl of dithiothreitol, 48 μl of CH_3-H_4folate (Schircks Laboratory), 4 μl of AdoMet, and 8 μl of enzyme containing ~1–4 $\times 10^{-5}$ units of activity. The enzyme is preactivated by adding 40 μl of hydroxocobalamin (Sigma) and incubating at 37° for 5 min. Steady-state turnover is initiated by adding 4 μl of homocysteine and the reaction mixture is incubated at 37° for 10 min. The reaction is quenched by adding 200 μl of formic acid/HCl and heating to 90° for 10 min. The samples are cooled to room temperature and the absorbance at 350 nm is measured. The derivatized product, CH^+=H_4folate, has an extinction coefficient at 350 nm of 26,500 M^{-1} cm^{-1}.[21] One unit of activity is defined as 1 μmol of product formed per minute. A blank is usually run that contains all of the reaction components except homocysteine, and this blank typically has an A_{350} of 0.3–0.4 vs deionized water. If crude protein samples are assayed, the acid derivatization step will result in a protein precipitate that will interfere with absorbance readings. The precipitate can be removed by centrifugation at 14,000 g for 5 min at room temperature prior to measuring the absorbance.

Methyltetrahydrofolate–cob(I)alamin Methyltransferase Assay

Principle The methyltetrahydrofolate–cob(I)alamin methyltransferase is followed using a radioactive assay similar to that described for the methyltetrahydrofolate–homocysteine methyl transferase assay, except that cob(I)alamin rather than homocysteine serves as the methyl acceptor.

$$CH_3\text{-}H_4\text{folate} + \text{cob(I)alamin} \rightarrow H_4\text{folate} + CH_3\text{-cobalamin} \quad (3)$$

Cob(I)alamin is produced by the reduction of exogenous cob(III)alamin with an excess of titanium(III) citrate [Eq. (4)].[22] Titanium(III) citrate is a one-electron reductant with a redox potential of −500 mV calculated for

[21] F. M. Huennekens, P. P. K. Ho, and K. G. Scrimgeour, *Methods Enzymol.* **6,** 806 (1963).
[22] A. J. Zehnder and K. Wuhrmann, *Science* **194,** 1165 (1976).

the titanium(IV) citrate/titanium(III) citrate couple at pH 7.3 and 25°,[23] and is known to reduce free corrinoids to the cob(I)alamin form.

$$2\text{Ti(III) citrate} + \text{cob(III)alamin} \rightarrow 2\text{Ti(IV) citrate} + \text{cob(I)alamin} \quad (4)$$

The reactive nucleophile cob(I)alamin will remain in the +1 oxidation state in the presence of a 10-fold excess of titanium(III) citrate in an anaerobic environment. As in the radioactive assay to determine the turnover of methionine synthase, the 6*R*-isomer of the reactant (6*R*,*S*)-5-$^{14}CH_3$-H_4folate will not react with exogenous cob(I)alamin, which should be taken into account when calculating the concentration of 5-$^{14}CH_3$-H_4folate utilized in the reaction.

Reagents

(6*R*,*S*)-5-$^{14}CH_3$-H_4folate, 4.17 m*M* in 8 m*M* sodium ascorbate (specific activity of ~1200 dpm/nmol)
Titanium(III) citrate solution, pH 7.3, 95 m*M* in titanium in 35 m*M* Tris-chloride buffer containing 0.1 *M* NaCl
Hydroxycobalamin, 5 m*M* in deionized water
Tris-chloride buffer, pH 7.3, 500 m*M*
KCl, 1 *M*
AG 1-X8 strong anion-exchange resin, chloride form, in deionized water

Procedure. The titanium(III) citrate solution [~82 m*M* in Ti(III)] is prepared anaerobically from $TiCl_3$ as described by Zehnder and Wuhrmann.[22] The synthesis is carried out under anaerobic conditions, in a vessel equipped with a syringe port. Nine milliliters of 0.5 *M* sodium citrate is added to 1 ml of 1.9 *M* titanium(III) chloride in 2 *M* HCl (available as a prepared solution from Aldrich); this mixture is stirred at room temperature for 30 min. The solution is neutralized by the slow addition of 3 ml of saturated sodium bicarbonate solution, followed by addition of 7 ml of 100 m*M* Tris chloride buffer, pH 7.3. The titanium(III) citrate solution is left to stir overnight under argon and is stored in the vessel used for synthesis. The concentration of titanium(III) citrate is determined by titration of benzyl viologen ($E^{0\prime} = -350$ mV vs the standard hydrogen electrode, SHE) and is typically found to be ~85% of the total titanium concentration.[23]

The assay reaction is carried out anaerobically in the dark at 37° in a modified Schlenk tube with a side chamber and syringe port. The assay mixture consists of 100 μl of hydroxocobalamin, 120 μl of $^{14}CH_3$-H_4folate, 100 μl of Tris-chloride buffer, 100 μl of KCl, and 450 μl of deionized water, 10 μl of enzyme (~1.6 units in the methyltetrahydrofolate–homocysteine methyltransferase assay) is placed in the side chamber of the Schlenk tube.

[23] P. Gartner, D. S. Weiss, U. Harms, and R. K. Thauer, *Eur. J. Biochem.* **226,** 465 (1994).

The reaction mixture is taken through seven cycles of partial evacuation followed by replacement with argon over a 30-min period. One hundred and twenty microliters of titanium(III) citrate solution is then added and the reaction mixture is incubated at 37° for 10 min; at this point one should observe a change in color of the cob(III)alamin from pink to greenish-black. The reaction is initiated by mixing the enzyme with the assay mixture. At intervals during the reaction, 100-μl portions of the reaction mixture are removed with a syringe inserted into the side-arm port while the reaction mixture remains under a positive pressure of argon. The samples are cooled on ice to terminate the reaction. A series of AG 1-X8 columns in Pasteur pipettes are prepared, one for each time point, as described for the methyltetrahydrofolate–homocysteine methyltransferase assay. A 50-μl aliquot of the reaction mixture is applied to the column, and the [^{14}C]methylcobalamin and cob(III)alamin are eluted with two 1-ml volumes of deionized water. Any unreacted $^{14}CH_3$-H_4folate, which is anionic, remains bound to the column. Scintillation fluid (10 ml) is added to the pooled washes and the sample is counted to determine the total amount of methylcobalamin produced. A blank that contains all the reaction components except the enzyme is also chromatographed and counted. To determine the total number of counts if the reaction had gone to completion, 50 μl of the blank reaction mixture is added to 2 ml of deionized water and 10 ml of scintillation fluid and counted. Because only the 6*S*-isomer of methyltetrahydrofolate is a substrate for the enzyme, the number of counts is halved due to the presence of the unreactive 6*R*-isomer of 5-$^{14}CH_3$-H_4folate in the racemic substrate. The standard reaction mixture contains 0.5 m*M* cob(I)alamin and 0.25 m*M* (6*S*)-5-$^{14}CH_3$-H_4folate. This reaction shows first-order dependence on the concentration of cob(I)alamin, but at these initial substrate concentrations the reaction rate is approximately constant for the first 20 min. The amount of methylcobalamin produced at a given time *t* can be calculated using Eq. (5) from the ratio of counts per minute (cpm) of methionine produced at time *t* to total cpm of (6*S*)-CH_3-H_4folate initially present in the assay.

$$[\text{Methylcobalamin produced}]_t = [(6S)\text{-}5\text{-}^{14}\text{CH}_3\text{-H}_4\text{folate}_{t=0}] \left[\frac{(\text{cpm}_t - \text{cpm}_{\text{blank}})}{\text{cpm}_{\text{total}}/2}\right] \quad (5)$$

Methylcobalamin–homocysteine Methyltransferase Assay

Principle. The methylcobalamin–homocysteine methyltransferase assay is followed by ultraviolet–visible (UV–Vis) spectrophotometry using a procedure modified from that described by Taylor and Weissbach.[4] This reaction may be carried out under anaerobic or aerobic conditions. Under anaerobic conditions the product formed is cob(II)alamin [Eq. (6)], presumably because the highly reactive cob(I)alamin product reduces the protons of solvent to hydrogen gas. The reaction is followed by monitoring the consumption of methylcobalamin at 525 nm. Under aerobic conditions the product formed is cob(III)alamin (Eq. (7)] and the rate of reaction is followed by monitoring the production of cob(III)alamin at 352 nm.

$$\text{Methylcobalamin} + \text{L-homocysteine} \rightarrow \text{L-methionine} + \text{cob(II)alamin} + 1/2\,H_2 \quad (6)$$

$$\text{Methylcobalamin} + \text{L-homocysteine} + H^+ + O_2 \rightarrow \text{L-methionine} + \text{cob(III)alamin} + H_2O_2 \quad (7)$$

Reagents

Homocysteine, 100 m*M* in 0.1 *M* sodium acetate buffer, pH 5.0
Methylcobalamin, 5 m*M* in deionized water
Tris-chloride buffer, pH 7.3, 500 m*M*
KCl, 1 *M*

Procedure. The reaction is carried out in the dark at 37° in an anaerobic curvette with a side arm. The initial reaction mixture consists of 100 μl of methylcobalamin (Sigma), 100 μl of Tris-chloride buffer, 100 μl of KCl, 685 μl of H_2O, and 10 μl of enzyme (~1.6 units using the methyltetrahydrofolate–homocysteine methyltransferase assay); 5 μl of homocysteine is added to the side arm of the cuvette. The reaction mixture is made anaerobic by the method described in the previous section [Eq. (3)]. The cuvette is incubated at 37° for 10 min and the reaction is then initiated by the addition of homocysteine from the side arm. The reaction is followed by observing the disappearance of the methylcobalamin peak at 525 nm; prior to addition of homocysteine, the A_{525} is monitored to obtain a blank reading, and the initial rate of reaction is then determined by monitoring the absorbance changes at 525 nm for 7 min after the addition of homocysteine. To determine the change in A_{525} associated with consumption of all the methylcobalamin, once the reaction has finished the cuvette is immersed in an ice bath and exposed to a tungsten–halogen lamp (650 W) at a distance of 10 cm. Two exposures of 30 sec are usually sufficient for complete photolytic conversion of the exogenous methylcobalamin to cob(II)alamin.

Conditions for Growth of Strains That Overproduce Cobalamin-Dependent Methionine Synthase and Purification of Enzyme

Principle

Escherichia coli does not synthesize cobalamin, and production of the methionine synthase holoenzyme requires that cobalamin be added to the medium. If cobalamin is lacking in the medium, the methionine synthase apoenzyme is unstable and rapidly undergoes proteolytic degradation (J. D. Drummond and R. G. Matthews, unpublished data, 1993). When growing a strain that overexpresses methionine synthase, the rate of cobalamin transport from the medium into the cells must be sufficiently rapid to keep pace with the rate of methionine synthase synthesis. It was found that growth of overexpressing strains in Luria broth resulted in production of enzyme of low specific activity,[24] suggesting that the rate of protein synthesis was faster than that of cobalamin transport in this medium, so that most of the enzyme was expressed as apoenzyme and degraded before the cells were harvested. To obtain enzyme of high specific activity, the overexpressing strain XL1-Blue/pKF5A is grown in glucose M9 medium supplemented with 20 amino acids, thiamin, ampicillin, micronutrients, and cobalamin. Under these growth conditions, the specific activity of the supernatant from sonicated cells is 0.8–1.3 units/mg.

The overproducing strain XL-1-Blue/pKF5A[24] contains the *metH* gene incorporated into the expression vector pKK223-3 (Pharmacía, Piscataway, NJ), in which expression is controlled by the *tac* promoter. Although expression from the *tac* promoter is normally induced by addition of isopropyl-β-D-thiogalactopyranoside (IPTG) to the medium, the promoter is "leaky" and considerable amounts of methionine synthase are synthesized in the absence of IPTG. Although the level of expression of methionine synthase is much higher when cells are grown in medium containing 50 μM IPTG than in medium lacking IPTG, the specific activity of the enzyme is not increased. These findings suggest that addition of IPTG again leads to a rate of protein synthesis that exceeds the rate of cobalamin transport into the cell, resulting in the production and then degradation of apoprotein.

The preparation[24] involves 2 days of cell growth, and then two column steps that take up most of the following 2 days. The protein is susceptible to proteolysis until it is pure, so it is important to keep solutions cold, and to minimize the time between sonication of the cells and the first chromatographic step. Enzyme activity is reduced by successive cycles of freezing and thawing, so do not freeze fractions between steps, but keep them on ice in the cold room.

Reagents

M9 minimal medium,[25] 10×
Glucose

[24] M. Amaratunga, K. Fluhr, J. T. Jarrett, C. L. Drennan, M. L. Ludwig, R. G. Matthews, and J. D. Scholten, *Biochemistry* **35,** 2453 (1996).

[25] J. H. Miller, "A Short Course in Bacterial Genetics." Cold Spring Harbor Laboratory Press, Cold Spring Harbor, New York, 1992.

Micronutrient stock,[26] 10,000×
Ampicillin, 2.5 g in 50 ml of deionized water
$MgSO_4 \cdot 7H_2O$
$CaCl_2 \cdot 2H_2O$
Hydroxocobalamin
Thiamin
Amino acid supplement
$ZnSO_4 \cdot 7H_2O$, 1 *M*, in 1 m*M* H_2SO_4
Tosyl-L-lysine chloromethyl ketone (TLCK), 1 mg/ml in deionized water
Phenylmethylsulfonyl fluoride (PMSF), 20 mg/ml in 2-propanol
Potassium phosphate buffer, pH 7.2, 1 *M* stock in deionized water

Procedure

Preparation of Growth Medium. The 10× M9 concentrate is prepared by a modification of the procedure of Miller.[25] Dissolve 256 g of $Na_2HPO_4 \cdot 7H_2O$, 60 g of anhydrous KH_2PO_4, 10 g of NaCl, 8.92 g of $MgSO_4$, and 20 g NH_4Cl in 1.89 liters of deionized water and adjust to pH 7.4 with solid NaOH. Filter to remove undissolved particulate matter. To prepare the 10,000× micronutrient stock, 37.1 mg of $(NH_4)_6(Mo)_7O_{24} \cdot 4H_2O$, 71.4 mg of $CoCl_2 \cdot 6H_2O$, 25 mg of $CuSO_4 \cdot 5H_2O$, 247 mg of H_3BO_3, 197.9 mg of $MnCl_2 \cdot 4H_2O$, and 28.8 mg of $ZnSO_4 \cdot 7H_2O$ are dissolved in 1 liter deionized water. To prepare the cobalamin supplement, combine 4 ml of 10,000× micronutrient stock, 9.13 g of $MgSO_4 \cdot 7H_2O$, 0.544 g of $CaCl_2 \cdot 2H_2O$, 37 mg of thiamin, 126 mg of hydroxocobalamin, and 148 g of glucose and bring to 500 ml with deionized water. Filter sterilize and store at 4°. The amino acid supplement contains sufficient concentrations of all 20 amino acids found in proteins to allow growth of cultures to an A_{420} of 10.[26,27] Add 1.43 g of alanine, 1.69 g of arginine hydrochloride, 1.20 g of asparagine, 1.37 g of potassium aspartate, 0.35 g of cysteine hydrochloride monohydrate, 2.22 g of potassium glutamate, 1.75 g of glutamine, 1.20 g of glycine, 0.84 g of histidine hydrochloride dihydrate, 1.05 g of isoleucine, 2.10 g of leucine, 1.46 g of lysine hydrochloride, 0.60 g of methionine, 1.32 g of phenylalanine, 0.92 g of proline, 21 g of serine, 0.95 g of threonine, 0.41 g of tryptophan, 0.74 g of tyrosine, and 1.41 g of valine to deionized water to yield a total volume of 2 liters, and autoclave for 20 min at 120°. To prepare 1 liter of the bacterial growth medium, combine in order: 100 ml of 10× M9 stock, 100 ml of amino acid supplement, 50 μl

[26] F. C. Neidhardt, P. L. Bloch, and D. F. Smith, *J. Bacteriol.* **119,** 736 (1974).
[27] B. Wanner, R. Kodaira, and F. C. Neidhardt, *J. Bacteriol.* **130,** 212 (1977).

$ZnSO_4$, 786.5 ml of deionized water, 13.5 ml of cobalamin supplement, and 1 ml of ampicillin and filter sterilize.

Growth of Cell Cultures. Cells from strain XL1-Blue/pKF5A are plated on a Luria broth–agar plate containing ampicillin (40 μl/plate). The plate is incubated at 37° until colonies ~2 mm in diameter are formed. In the late afternoon, 2 days before the desired preparation date, a 5-ml culture is started from a single colony by innoculating the colony into a filter-sterilized solution containing 0.5 ml of 10× M9 stock, 0.25 ml of amino acid supplement, 67.5 μl of cobalamin supplement, 5 μl of ampicillin, and 4.25 ml of distillized water. On the morning of the next day, 10 ml of 10× M9 stock, 5 ml of amino acid supplement, 83.5 ml of deionized water, 1.35 ml of cobalamin supplement, 5 μl $ZnSO_4$, and 100 μl of ampicillin are mixed and filter sterilized. The 5-ml overnight culture is added to the medium and incubated for 8 hr with shaking (300 rpm) at 37°. On the day before the desired preparation date, two to six 3.5-liter flasks are prepared for culture. In each flask, 100 ml of 10× M9 stock, 50 ml of amino acid supplement, 50 μl $ZnSO_4$, and 850 ml of deionized water are added, and the flasks are sterilized for 20 min at 120° in the autoclave. The flasks are cooled, and 13.5 ml of cobalamin supplement and 1 ml of ampicillin are added; the flasks are then preheated to 37° before innoculation. Approximately 15 ml of the 100-ml culture is added to each flask and the flasks are incubated overnight, at 37°, with shaking at 300 rpm. Cells are harvested at OD_{420} = 5.0–5.5, typically 10–12 hr after inoculation.

Preparation of Cell Extract. The cells are collected by centrifugation in 500-ml tubes at 12,000 *g* for 10 min at 4°. The cell pellets are resuspended in 50 ml of 10 m*M* potassium phosphate buffer, pH 7.2, in a single tube and collected by centrifugation. The wet weight of the cell pellet is determined. At this point, the cell pellet may be stored at −80° overnight. The cells are suspended in 100 ml of 10 m*M* potassium phosphate buffer, pH 7.2 and placed in a beaker just large enough to hold the cells. TLCK (200 μl) and PMSF (100 μl) are added. (These volumes are for a 6-liter preparation, and should be scaled down accordingly for smaller preparations.) The beaker is submerged in ice. The cell suspension is sonicated four times for 1-min intervals at 70% power with 4-min breaks between sonications to allow the heat to dissipate, and then centrifuged in a titanium fixed-angle rotor in an ultracentrifuge at 64,000 *g* for 1 hr at 4°. The supernatant fractions are pooled.

Chromatography on Fast-Flow DEAE-Sepharose. The pooled supernatant is loaded onto a 2 × 10 cm column of fast-flow DEAE-Sepharose (Sigma) that has been preequilibrated with 10 m*M* potassium phosphate buffer, pH 7.2, and the column is washed with 100 ml of the same buffer. The enzyme is eluted with a 400-ml linear gradient from 10 to 500 m*M*

TABLE I
PURIFICATION OF COBALAMIN-DEPENDENT METHIONINE SYNTHASE FROM *Escherichia coli*

Purification step	Total activity[a] (units)	Total protein[b] (mg)	Specific activity (units/mg)	Purification (-fold)	Recovery (%)
Sonication[c]	698	810	0.86	1	100
Chromatography on DEAE-Sepharose	526	259	2.03	2.4	75
FPLC on Mono Q	358	31.8	11.3	13.1	51

[a] Activity (in micromoles per minute) determined using the methyltetrahydrofolate–homocysteine methyltransferase assay.
[b] Protein concentration determined using the Bradford protein assay (Bio-Rad).
[c] Preparation from 2 liters of cell culture.

potassium phosphate buffer, pH 7.2. Fractions are collected every 2–3 min. The red-colored fractions are pooled together and concentrated to 30 ml in an Amicon (Danvers, MA) concentrator with a PM30 membrane. The sample is transferred to a dialysis bag and dialyzed overnight against 1 liter of 50 m*M* potassium phosphate buffer, pH 7.2.

Fast Protein Liquid Chromatography on Mono Q. A Mono Q 16/10 column (Pharmacia) is equilibrated with 50 m*M* potassium phosphate buffer, pH 7.2, at 20° and the enzyme is loaded on the column. The column is washed with 100 ml of 50 m*M* potassium phosphate buffer, then with 120 ml of 118 m*M* potassium phosphate buffer, and finally with a 150-ml linear gradient between 118 and 320 m*M* potassium phosphate buffer. Methionine synthase elutes at ~200 m*M* buffer concentration. The reddish-colored fractions are pooled, concentrated to ~10 ml in an Amicon concentrator, exchanged into 10 m*M* potassium phosphate buffer, pH 7.2, and concentrated to 1–2 ml in a Centricon-30 concentrator prior to storage at −80°. At this point the enzyme is 90–95% pure as assessed by polyacrylamide gel electrophoresis in the presence of sodium dodecyl sulfate. A representative purification is shown in Table I. If homogeneous enzyme is required (as for crystallization), the procedure described below is followed.

Chromatography on Bio-Gel HTP. This step was described by Frasca *et al.*[28] Enzyme from the previous step is equilibrated with 5 m*M* potassium phosphate buffer, pH 7.2, containing 0.2 *M* KCl, and is applied to a 1.5 × 20 cm column of Bio-Gel HTP (Bio-Rad) that has previously been

[28] V. Frasca, R. V. Banerjee, W. R. Dunham, R. H. Sands, and R. G. Matthews, *Biochemistry* **27,** 8458 (1988).

equilibrated with the same buffer. The column is rinsed with 50 ml of the same buffer, and then eluted with a 300-ml linear gradient from 5 to 60 m*M* potassium phosphate buffer in 0.2 *M* KCl. Fractions of 3 ml are collected. The fractions with the highest specific activity are pooled and equilibrated with 10 m*M* potassium phosphate buffer, pH 7.2, and then concentrated prior to storage at −80°.

Preparation of Enzyme in Methylcob(III)alamin, Hydroxocob(III)alamin, and Cob(II)alamin Forms

Principle

Methionine synthase is inactivated during turnover by oxidation of the cob(I)alamin intermediate to the cob(II)alamin or hydroxocob(III)alamin forms of the enzyme. These oxidized forms of the enzyme can be returned to the methylcobalamin form by supplying a methyl group from AdoMet and reducing equivalents from electrochemical, enzymatic, or chemical reducing systems [Eq. (8)]. The fully methylated enzyme can be converted quantitatively to the cob(II)alamin form by anaerobic demethylation with homocysteine [Eqs. (9) and (10)] or by photolysis [Eq. (11)], and can be converted to the hydroxocob(III)alamin form by aerobic demethylation with homocysteine [Eqs. (9) and (12)]. The oxidation state of the cobalamin cofactor in methionine synthase is readily determined by examining the absorbance spectrum of the protein from 300 to 600 nm, as illustrated in Fig. 2.[29]

$$\text{Cob(II)alamin} + e^- + \text{AdoMet} \rightarrow \text{CH}_3\text{-cob(III)alamin} + \text{AdoHcy} \quad (8)$$

$$\text{CH}_3\text{-cob(III)alamin} + \text{Hcy} \rightarrow \text{Met} + \text{cob(I)alamin} \quad (9)$$

$$\text{Cob(I)alamin} \rightarrow \text{cob(II)alamin} + e^- \quad (10)$$

$$\text{CH}_3\text{-cob(III)alamin} + h\nu \rightarrow \text{cob(II)alamin} + \text{CH}_3\cdot \quad (11)$$

$$\text{Cob(I)alamin} + \text{O}_2 \rightarrow \text{hydroxocob(III)alamin} + \text{H}_2\text{O}_2 \quad (12)$$

Electrochemical Formation of Methylcobalamin [Eq. (8)]

Reagents

Adenosyl-L-methionine, iodide salt, 38 m*M* in 1 m*M* HCl
Methyl viologen, 50 m*M* in deionized water
KCl, 2 *M*

[29] J. T. Jarrett, M. Amaratunga, C. L. Drennan, J. D. Scholten, R. H. Sands, M. L. Ludwig, and R. G. Matthews, *Biochemistry* **35,** 2464 (1996).

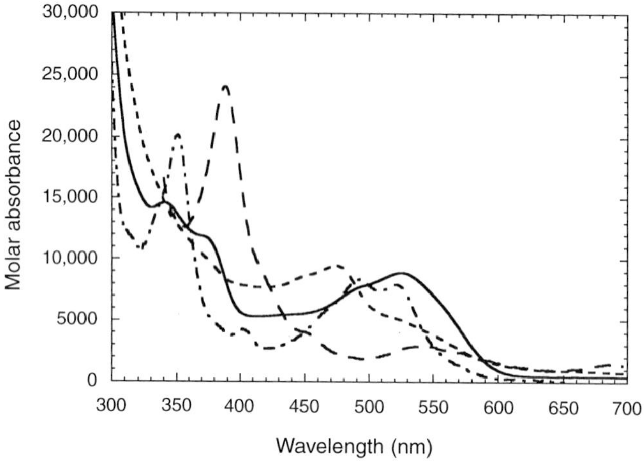

FIG. 2. UV–Vis spectra of the different forms of enzyme-bound cobalamin in methionine synthase. (—) Methylcobalamin; (---) cob(II)alamin; (–·–·–) hydroxocob(III)alamin; (– – –) cob(I)alamin. Methylcobalamin enzyme was prepared by reductive methylation in an electrochemical cell as described in text; the determination of the molar absorbance of the enzyme-bound cofactor has been described previously.[29] Hydroxycob(III)alamin enzyme was prepared by addition of homocysteine to methylcobalamin enzyme in air-saturated buffer and enzyme in the cob(II)alamin form was prepared by anaerobic photolysis of methylcobalamin enzyme by procedures described in text. Enzyme in the cob(I)alamin form was prepared in a strictly anaerobic stopped-flow spectrophotometer by mixing methylcobalamin enzyme (10 μM) with excess homocysteine (500 μM) in the presence of limiting methyltetrahydrofolate (100 μM). All spectra were recorded in 0.1 M potassium phosphate buffer at pH 7.2.

Procedure. Methionine synthase, containing a mixture of cob(II)alamin and cob(III)alamin forms, can be reductively methylated in a small anaerobic electrochemical cell similar to that described by Harder *et al.*[30] This cell has a gold foil working electrode immersed in the sample and Ag|AgCl reference and counter electrodes that are separated from the sample by porous Vycor frits. The enzyme (10–100 nmol in 1 ml of 0.1 M potassium phosphate buffer, pH 7.2) is placed in the cell with a small stir bar and 15 μl of AdoMet, 12 μl of methyl viologen, and 200 μl of KCl are added. The sample is gently stirred and the electrochemical cell is made anaerobic by evacuating and filling with oxygen-free argon gas five times every few minutes for 1 hr. Using a potentiostat, the cell is poised at −450 mV vs SHE for 45 min at room temperature. After 2–5 min, the cell becomes purple-violet as the concentration of reduced methyl viologen increases. The reaction mixture is removed from the cell and the excess reagents are

[30] S. R. Harder, B. A. Feinberg, and S. W. Ragsdale, *Anal. Biochem.* **181,** 283 (1989).

removed by one of two possible methods. The sample can be repeatedly washed by concentrating in a Centricon-30 concentrator (Bio-Rad) followed by a 10-fold dilution in 0.01 *M* potassium phosphate buffer (four or five washes are usually necessary). Alternatively, the protein can be loaded on a gel-filtration column (1 × 30 cm, Superose 12; Pharmacia-LKB) and eluted with 0.01 *M* potassium phosphate buffer, giving baseline separation between protein and reagents.

Enzymatic Method for Formation of Methylcobalamin Enzyme [Eq. (8)]

Reagents

Adenosyl-L-methionine, iodide salt, 38 m*M* in 1 m*M* HCl
NADPH, 25 m*M* in 10 m*M* Tris base.
Escherichia coli flavodoxin, 500 μ*M* in 10 m*M* potassium phosphate buffer, pH 7.2
E. coli ferrodoxin (flavodoxin) : $NADP^+$ oxidoreductase, 500 μ*M* in 10 m*M* potassium phosphate buffer, pH 7.2

Procedure. Methionine synthase in the cob(II)alamin form can be reductively methylated by its *in vivo* methylating system: flavodoxin, ferrodoxin (flavodoxin) : $NADP^+$ oxidoreductase, and NADPH. The same reaction is much slower with the hydroxocob(III)alamin enzyme. This system is highly species specific; flavodoxins from other species will not serve as reductants. *Escherichia coli* flavodoxin is purified from the overexpressing strain XL1-Blue/pDH2 as previously described.[31] *Escherichia coli* ferrodoxin (flavodoxin) oxidoreductase is purified from the overexpressing strain C600/pEE1010 as also previously described.[31] The methylation reaction is run in an anaerobic cuvette that has a rubber septum, allowing addition of reagents while simultaneously purging with argon. Methionine synthase (10–100 nmol in 1 ml of 0.02 *M* potassium phosphate buffer, pH 7.2) is placed in the cuvette with a small stir bar and AdoMet (15 μl), flavodoxin (5 μl), and NADPH (40 μl) are added. The sample is gently stirred and made anaerobic by evacuating and filling with oxygen-free argon gas five times every few minutes for 1 hr. While the cell is under argon pressure, ferrodoxin (flavodoxin) oxidoreductase (5 μl) is added with a syringe through the septum. The methylation reaction is followed spectrally (see Fig. 2) and is typically complete in 30–60 min. The protein is removed from the cuvette and is immediately applied to an FPLC Mono Q ion-exchange column equilibrated with 25 m*M* potassium phosphate buffer. Methionine

[31] V. Bianchi, P. Reichard, R. Eliasson, E. Pontis, M. Krook, H. Jornvall, and E. Haggard-Ljungquist, *J. Bacteriol.* **175,** 1590 (1993).

synthase is eluted with a 150-ml linear gradient to 500 m*M* potassium phosphate.

Chemical Method for Formation of Methylcobalamin Enzyme [Eq. (8)]

Reagents

Adenosyl-L-methionine, iodide salt, 38 m*M* in 1 m*M* HCl
Titanium(III) citrate in 35 m*M* Tris-chloride buffer, pH 7.3

Procedure. The cob(II)alamin form of methionine synthase can be reductively methylated by the chemical reductant titanium(III) citrate. The methylation reaction is run in an anaerobic cuvette that has a rubber septum allowing addition of reagents while simultaneously purging with argon. Methionine synthase (10–100 nmol in 1 ml of 0.1 *M* Tris-chloride buffer, pH 7.2) is placed in the cuvette with a small stir bar and 15 μl of AdoMet. The sample is gently stirred and made anaerobic by evacuating and filling with oxygen-free argon gas five times every few minutes for 1 hr. While the cell is under argon pressure, titanium(III) citrate (50 μl) is added with a syringe through the septum. The methylation reaction is followed spectrally (see Fig. 2) and is complete in 5–10 min. The protein is loaded on a Superose 12 (Pharmacia-LKB) gel-filtration column (1 $\times$ 30 cm) and eluted with 0.1 *M* Tris chloride buffer, giving baseline separation between protein and reagents. If the methylation reaction has not gone to completion, enzyme in the methylcobalamin form can be separated from cob(II)alamin enzyme by anion-exchange chromatography on a Mono Q FPLC column as described in the preceding section.

Formation of Cob(II)alamin Enzyme by Homocysteine Demethylation [Eqs. (9) and (10)]

Reagents

L-Homocysteine, 100 m*M* in 0.1 *M* sodium acetate (pH 5).
Dithiothreitol, 500 m*M* in 5 m*M* sodium acetate (pH 4.5).

Procedure. Methionine synthase in the methylcobalamin form (10–100 nmol in 1 ml of 0.1 *M* potassium phosphate buffer, pH 7.2) is placed in an anaerobic cuvette with a small stir bar and 50 μl of dithiothreitol is added. The cuvette is covered with a rubber septum and made anaerobic by evacuating and filling with oxygen-free argon gas five times every few minutes for 1 hr. Homocysteine (5 μl) is added with a syringe through the septum and the reaction is followed spectrally (see Fig. 2). After 1–2 hr, the protein is removed from the cuvette and the excess reagents are removed by gel filtration or concentration in a Centricon-30 concentrator as previously described.

Formation of Cob(II)alamin Enzyme by Photolysis [Eq. (11)]

Reagents

2,2,6,6-Tetramethyl-1-piperidinyloxy (TEMPO), 8 m*M* in deionized water

Procedure. Methionine synthase in the methylcobalamin form (10–100 nmol in 1 ml of 0.1 *M* potassium phosphate buffer, pH 7.2) is placed in an anaerobic cuvette with a small stir bar and 50 μl of freshly prepared TEMPO is added. The cuvette is covered with a rubber septum and made anaerobic by evacuating and filling with oxygen-free argon gas five times every few minutes for 1 hr. The cuvette is immersed in a beaker filled with ice–water and is exposed to a tungsten–halogen lamp (650 W) at a distance of 10 cm. Two exposures of 10 sec are usually sufficient for complete photolytic demethylation of the enzyme. Excess reagents are removed by gel filtration or Centricon-30 concentration as previously described.

Formation of Hydroxocob(III)alamin Enzyme [Eqs. (9) and (12)]

Reagents

L-Homocysteine, 100 m*M* in 0.1 *M* sodium acetate (pH 5)

Procedure. Methionine synthase in the methylcobalamin form (10–100 nmol in 1 ml of 0.1 *M* potassium phosphate buffer, pH 7.2) is placed in a cuvette with a small stir bar and 5 μl of homocysteine is added. The air-equilibrated protein solution contains enough oxygen to oxidize completely the resulting cob(I)alamin protein. The reaction is followed spectrally and is typically complete within 1–10 min. Longer incubation times will result in damage to the protein due to the H_2O_2 by-product. Excess reagents are removed by gel filtration or Centricon-30 concentration as previously described.

Acknowledgments

Work in this laboratory has been supported in part by NIH Grant R37 GM24908 and by an NIH postdoctoral fellowship (GM17455) to Joseph T. Jarrett.

[25] Formyltetrahydrofolate Hydrolase from *Escherichia coli*

By HOWARD ZALKIN

Formyltetrahydrofolate hydrolase, encoded by the *Escherichia coli purU* gene, is a regulatory enzyme that catalyzes the hydrolysis of N^{10}-formyltetrahydrofolate (formyl-FH_4)[1]:

$$\text{Formyl-FH}_4 + \text{H}_2\text{O} \rightarrow \text{FH}_4 + \text{formate}$$

Formyl-FH_4 hydrolase has two important roles in *E. coli.* First, the enzyme monitors the relative pools of FH_4 and FH_4 one-carbon adducts (C_1-FH_4) and functions to replenish FH_4 when other routes of FH_4 regeneration are shut down due to growth conditions that repress biosynthetic pathways for histidine, methionine, and purine nucleotides. Tetrahydrofolate is needed by serine hydroxymethyltransferase for glycine synthesis. A second role for the enzyme is to generate formate during aerobic growth. Formate is utilized by 5′-phosphoribosylglycinamide (GAR) transformylase-T for synthesis of 5′-phosphoribosyl-*N*-formylglycinamide (FGAR)[2]:

$$\text{GAR} + \text{formate} + \text{ATP} \rightarrow \text{FGAR} + \text{ADP} + \text{P}_i$$

Although *E. coli* contains a second GAR transformylase-N that utilizes formyl-FH_4 as the one-carbon donor for synthesis of FGAR, formate can contribute up to 50% of the carbon for position 8 of the purine ring in wild-type cells.[3] Mutant analysis has indicated that formyl-FH_4 hydrolase supplies the formate that is utilized by the GAR transformylase-T reaction during aerobic growth.[4]

Assay Method

A polyglutamate derivative of formyl-FH_4 is assumed to be the natural substrate for formyl-FH_4 hydrolase. Although formyl-FH_4 serves as a substrate, difficulty of preparation and oxygen lability limit its routine use.

[1] P. L. Nagy, A. Marolewski, S. J. Benkovic, and H. Zalkin, *J. Bacteriol.* **177,** 1292 (1995).
[2] A. Marolewski, J. M. Smith, and S. J. Benkovic, *Biochemistry* **33,** 2531 (1994).
[3] I. K. Dev and R. J. Harvey, *J. Biol. Chem.* **257,** 1980 (1982).
[4] P. L. Nagy, G. M. McCorkle, and H. Zalkin, *J. Bacteriol.* **175,** 7066 (1993).

METHODS IN ENZYMOLOGY, VOL. 281
0076-6879/97 $25

10-Formyl-5,8-dideazafolate (fDDF),[5] which is not sensitive to aerobic oxidation, is the preferred substrate for routine enzyme assays. 10-Formyl-5,8-dideazafolate is presently available from J. B. Hynes (Department of Pharmaceutical Chemistry, Medical University of South Carolina, Charleston, SC).

An assay mixture of 100 μl contains 50 m*M* Tris-HCl (pH 7.5), 60 μM fDDF, 2 m*M* methionine, and enzyme. The reaction is carried out at room temperature (~23°) and is started by the addition of enzyme. The initial rate of hydrolysis is recorded at 295 nm ($\Delta\varepsilon$ of 18.9 mM^{-1} cm^{-1}).

Enzyme Overproduction

Escherichia coli purU has been placed under the control of the phage T7 promoter ϕ10 in plasmid pT7-7 to yield plasmid pT7-PU1.[1] The enzyme is overproduced in *E. coli* BL21(DE3). A single colony of BL21(DE3)/pT7-PU1 is used to inoculate 10 ml of LB medium (10 g of tryptone, 5 g of yeast extract, 10 g of NaCl dissolved in 1 liter and adjusted to pH 7.0 with NaOH) supplemented with 100 μg of ampicillin per milliliter. After overnight growth six 2-liter flasks, each containing 0.5 liter of LB medium with ampicillin, are inoculated with 1 ml of the overnight culture and grown with shaking at 37° to a turbidity of 180 as measured with a Klett colorimeter using a 660-nm filter (equivalent to 0.84 optical density units at 600 nm as measured in a spectrophotometer). At this point lactose is added to a final concentration of 1% for induction of T7 polymerase and the flasks are incubated at 30° with shaking for 24 hr. Cells are harvested by centrifugation and stored frozen at −20°. Approximately 30–40% of the soluble protein is formyl-FH_4 hydrolase.

Enzyme Purification

The enzyme is purified approximately threefold to electrophoretic homogeneity in four steps.[1] A typical purification is summarized in Table I. After ammonium sulfate precipitation the enzyme is approximately 85–90% pure, on the basis of sodium dodecyl sulfate-polyacrylamide gel electrophoresis. The residual contaminating proteins can be removed, if required, by ion-exchange chromatography to give an essentially homogeneous enzyme.

Step 1. Crude Extract. All steps are carried out at 4° in a buffer containing 50 m*M* Tris-HCl (pH 7.5), 1 m*M* EDTA, plus other additions specifically noted. Frozen cells are suspended in buffer containing 1 m*M* phenylmethylsulfonyl fluoride (PMSF) in the ratio of 1 g of cells per 4 ml of buffer. The

[5] J. B. Hynes, D. E. Eason, C. M. Garrett, and P. L. Colvin, Jr., *J. Med. Chem.* **20,** 588 (1977).

TABLE I
SUMMARY OF PURIFICATION

Fraction	Volume (ml)	Protein (mg)	Specific activity (nmol/min/mg)	Total activity (nmol/min)
Disrupted cells	135	2,160	30	65,800
Streptomycin sulfate	135	850	55	46,800
Ammonium sulfate	10	280	72	20,200
DEAE-Sepharose	22	205	85	17,400

cell suspension is broken by two passages through a French press at 20,000 lb/in^2.

Step 2. Precipitation with Streptomycin Sulfate. To the broken cell suspension, 0.1 vol of 10% (w/v) streptomycin sulfate is added slowly with stirring. After the last addition, stirring is continued for 15 min, followed by centrifugation at 18,000 *g* for 30 min at 4°. The supernatant containing formyl-FH_4 hydrolase is used for the next step.

Step 3. Precipitation with Ammonium Sulfate. Ammonium sulfate is added slowly to 30% saturation (0.176 g/ml) with stirring. Stirring is continued for 15 min following the last addition, after which the enzyme is collected by centrifugation for 30 min at 18,000 *g* at 4°. The pellet is dissolved in buffer and dialyzed overnight against 100 vol of the same buffer. Any insoluble protein is removed by centrifugation at 18,000 *g* for 30 min.

Step 4. Chromatography on DEAE-Sepharose. A 1.5 × 5 cm column of DEAE-Sepharose is used for 200–400 mg of protein. The dialyzed enzyme is loaded onto the column, the column washed with 50 ml of buffer, and the enzyme is eluted with a 300-ml linear gradient of 0 to 0.5 *M* NaCl in the buffer. Fractions in the main protein peak can be checked by sodium dodecyl sulfate-polyacrylamide gel electrophoresis and those fractions lacking contaminating proteins are pooled. Ordinarily, the entire protein peak can be pooled without prior analysis. The enzyme is concentrated by precipitation with ammonium sulfate, dissolved in buffer, dialyzed, and adjusted to approximately 10 mg/ml. The enzyme is stable to storage at −20°.

Enzyme Properties

Formyltetrahydrofolate hydrolase is a hexamer of 32-kDa subunits.[1] Enzyme activity is stimulated 25-fold by methionine (K_a 200 μM) and is inhibited uncompetitively by glycine. The inhibition constant for glycine is dependent on the methionine concentration and varies between less than

20 μM in the absence of methionine to 40 μM with saturating methionine. Cooperativity is observed for activation by methionine and inhibition by glycine, with Hill coefficients of 1.3 to 2.7 for methionine and 0.83 to 1.7 for glycine. The Hill coefficient for activator and inhibitor is maximal in the presence of a saturating concentration of the opposing ligand. These properties are consistent with regulatory roles for methionine and glycine.

The native protein of 280 amino acids appears to be composed of three domains. A C-terminal segment, residues 85–280, is identical in 53 positions with the two domains of *E. coli* GAR transformylase-N.[4] 5′-Phosphoribosylglycinamide transformylase-N catalyzes the reaction GAR + formyl-FH_4 $\rightarrow$ FGAR + FH_4.[6] Formyltetrahydrofolate hydrolase corresponds to an N-terminal 84-amino acid domain fused to the 196-residue GAR transformylase-like domains.[4] Comparison of conserved residues in the two sequences and the X-ray structure of GAR transformylase[7] suggest that a binding site for formyl-FH_4 but not for GAR has been retained in the C domains of the enzyme. According to this model the 84-amino acid N domain would thus function to bind methionine and glycine and regulate catalysis by the transformylase-related C domains.

The k_{cat} value with formyl-FH_4 is fourfold higher than with fDDF. However, as a result of a sevenfold lower K_m for fDDF the k_{cat}/K_m is nearly twofold higher using the substrate analog compared to formyl-FH_4.

Role of Enzyme

One role of the enzyme is to balance pools of FH_4 and C_1-FH_4 to ensure that synthesis of glycine can be maintained when cells have excess purines, methionine, and histidine and biosynthetic pathways for these molecules are shut down. Purine biosynthesis regenerates FH_4 from formyl-FH_4 in two transformylation reactions. Methionine biosynthesis regenerates FH_4 as a product of the methionine synthase reactions and the histidine biosynthetic pathway produces 5′-phosphoribosyl-4-carboxamide-5-aminoimidazole (AICAR), which is transformylated in a reaction yielding FH_4. In a *purU* mutant lacking formyl-FH_4 hydrolase, repression of purine synthesis by adenine and repression of either the methionine or histidine pathways lead to a growth requirement for glycine as a result of a deficiency in regenerating FH_4 from formyl-FH_4 and methyl-FH_4.[4] The methionine/glycine ratio thus reflects the relative pools of C_1-FH_4 and FH_4 and regulates

[6] J. Inglese, D. L. Johnson, A. Shiau, J. M. Smith, and S. J. Benkovic, *Biochemistry* **29,** 1436 (1990).

[7] R. J. Almassy, C. A. Janson, C. C. Kan, and Z. Hostomska, *Proc. Natl. Acad. Sci. U.S.A.* **89,** 6114 (1992).

the hydrolase to generate FH_4 or to preserve formyl-FH_4 as required.[1] A high ratio of methionine to glycine reflects a high C_1-FH_4/FH_4 ratio and is a signal to activate formyl-FH_4 hydrolase to regenerate FH_4 for glycine synthesis.

A second role of the hydrolase is to generate formate for use by GAR transformylase-T for purine synthesis. A *purN* mutant lacking GAR transformylase-N requires formyl-FH_4 hydrolase or an external source of formate for aerobic growth.[4]

In animals formyl-FH_4 dehydrogenase may function to regenerate FH_4.[8] Tetrahydrofolate dehydrogenase is a bifunctional enzyme that catalyzes the reaction formyl-FH_4 + $NADP^+ \rightarrow FH_4 + CO_2 + NADPH + H^+$ as well as the $NADP^+$-independent hydrolysis of formyl-FH_4 to FH_4 and formate.[9] An N-terminal 200-amino acid domain of the dehydrogenase is related in sequence to *E. coli* GAR transformylase-N but does not contain the putative regulatory domain corresponding to residues 1–84 of the *E. coli* formyl-FH_4 hydrolase and is not regulated by methionine or glycine in response to fluctuations in FH_4 and C_1-FH_4 pools.

Acknowledgment

The author's research was supported by Public Health Service Grant GM24658 from the National Institutes of Health.

[8] H. Min, B. Shane, and E. L. R. Stokstad, *Biochim. Biophys. Acta* **967,** 348 (1988).
[9] R. J. Cook, R. S. Lloyd, and C. Wagner, *J. Biol. Chem.* **266,** 4965 (1991).

[26] Use of ^{13}C Nuclear Magnetic Resonance to Evaluate Metabolic Flux through Folate One-Carbon Pools in *Saccharomyces cerevisiae*

By DEAN R. APPLING, EVDOKIA KASTANOS, LAURA B. PASTERNACK, and YAKOV Y. WOLDMAN

Introduction

Figure 1 summarizes the major pathways of folate-mediated one-carbon metabolism. In most organisms, the major source of one-carbon units is carbon 3 of serine, derived from glycolytic intermediates. The one-carbon unit is transferred to tetrahydrofolate (THF) in a reaction catalyzed by serine hydroxymethyltransferase (reaction 4, Fig. 1), generating methylene-

0076-6879/97 $25

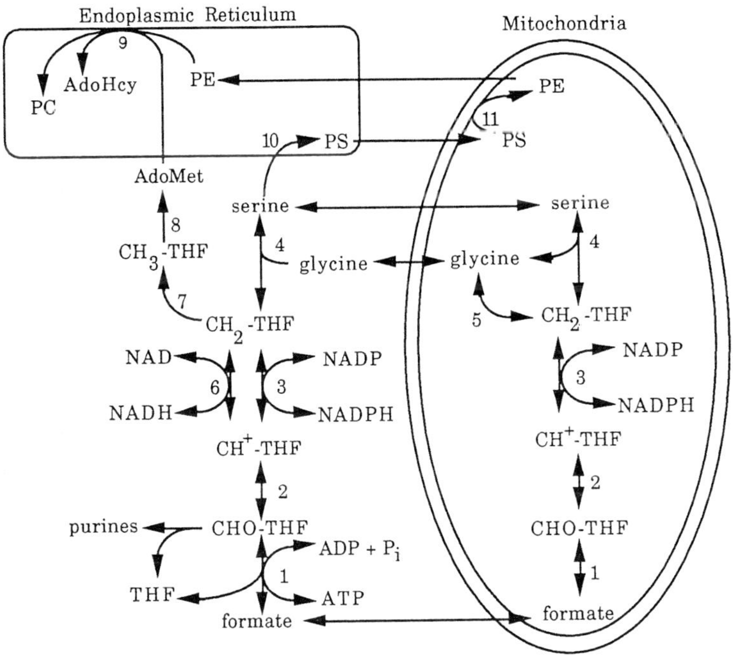

FIG. 1. Proposed organization of the enzymes of one-carbon metabolism in *Saccharomyces cerevisiae.* Reactions 1, 2, and 3, CHO-THF synthetase (EC 6.3.4.3, formate–tetrahydrofolate ligase), CH^+-THF cyclohydrolase (EC 3.5.4.9, methenyltetrahydrofolate cyclohydrolase), and NADP-dependent CH_2-THF dehydrogenase (EC 1.5.1.5, methylenetetrahydrofolate dehydrogenase), respectively, are catalyzed by cytoplasmic or mitochondrial C_1-THF synthase. Reaction 4 is cytoplasmic or mitochondrial serine hydroxymethyltransferase (EC 2.1.2.1). Reaction 5 is the glycine cleavage system (EC 2.1.2.10, aminomethyltransferase). Reaction 6 is the monofunctional NAD-dependent CH_2-THF dehydrogenase. Reaction 7 is CH_2-THF reductase (EC 1.5.1.2.20, methylenetetrahydrofolate reductase); reaction 8 is homocysteine methyltransferase (EC 2.1.1.14, 5-methyltetrahydropteryltriglutamate–homocysteine-methyltransferase) and ATP:L-methionine *S*-adenosyltransferase (EC 2.5.1.6, methionine adenosyltransferase); reaction 9 is the phosphatidylethanolamine methyltransferases; reaction 10 is phosphatidylserine synthase (EC 2.7.8.8, CDP diacylglycerol–serine *O*-phosphatidyltransferase); and reaction 11 is phosphatidylserine decarboxylase (EC 4.1.1.65). (Reprinted with permission from Ref. 5. Copyright 1994 American Chemical Society.)

THF and glycine. This form of the coenzyme is required for *de novo* thymidylate synthesis by thymidylate synthase. Methylene-THF may also be reduced to 5-methyl-THF (for methyl group biogenesis) (reaction 7, Fig. 1) or oxidized to 10-formyl-THF depending on the needs of the cell. In rapidly growing cells, the synthesis of purines is a critical folate-dependent pathway, requiring 2 mol of 10-formyl-THF per mole of purine ring. Methylene-THF is converted to 10-formyl-THF via the sequential enzymes methylene-THF dehydrogenase and methenyl-THF cyclohydrolase (reactions 3 and 2, Fig. 1). In eukaryotes, the mitochondrial and cytosolic compartments each contain a parallel set of these one-carbon unit interconverting enzymes.[1]

These pathways have been difficult to study by traditional ^{14}C metabolic labeling strategies owing to the potential for tracer carbons to become incorporated at multiple positions in several end products and the presence of parallel pathways in the two compartments. ^{13}C Nuclear magnetic resonance (NMR) spectroscopy is ideally suited for the study of metabolism. The major advantage of ^{13}C NMR over conventional radioisotope-labeling studies is its ability to distinguish which carbons in a particular metabolite are labeled without complex extraction and metabolite isolation procedures. In some cases, detection and identification can be achieved directly within the living cell. In addition, the unique properties of spin–spin interaction allow quantitation of ^{13}C-labeled metabolites by isotopomer analysis.

These properties have made ^{13}C NMR particularly useful in the study of folate-mediated one-carbon metabolism. We have chosen the yeast *Saccharomyces cerevisiae* as a eukaryotic model to study the compartmentation of folate-mediated one-carbon metabolism. *Saccharomyces cerevisiae* provides a tremendously powerful system in which to study these pathways. The genes for all of the reactions listed in Fig. 1 have been cloned from *Saccharomyces*, facilitating the construction of defined mutations in these genes. Our basic approach has been to create blocks at specific steps in the one-carbon pathway and to examine the resulting metabolic effects of the blocks. The analytical power of ^{13}C NMR, used in combination with yeast strains harboring specific metabolic blocks, enable the enzymatic route and compartment of biosynthesis of a given metabolite to be determined.

The principle of this method is to label metabolically the one-carbon pools with ^{13}C-labeled one-carbon donors such as serine, glycine, or formate, followed by NMR analysis of metabolic end products. To control the point of entry of one-carbon units into the pool, yeast strains harboring a *ser1* mutation are used. In wild-type cells the C-3 of serine, synthesized from glycolytic intermediates, is the major source of one-carbon units. However, *ser1*$^{-}$ strains are blocked at phosphoserine aminotransferase and require supplemental serine for growth.[2] Glycine and formate can substitute

[1] D. R. Appling, *FASEB J.* **5,** 2645 (1991).

[2] E. W. Jones and G. R. Fink, Regulation of amino acid and nucleotide biosynthesis in yeast. *In* "The Molecular Biology of the Yeast Saccharomyces: Metabolism and Gene Expression" (J. N. Strathern, E. W. Jones, and J. R. Broach, eds.), pp. 181–299. Cold Spring Harbor Laboratory Press, Cold Spring Harbor, New York.

for serine. Thus, use of *ser1*$^-$ strains allows us to introduce a one-carbon unit at the level of the 10-formyl-THF synthetase reaction of C_1-THF synthase by providing formate, serine hydroxymethyltransferase by providing serine, or the glycine cleavage system by providing glycine.

[2-^{13}C]Glycine has proved to be the most useful exogenous one-carbon source in yeast. Because glycine catabolism via the glycine cleavage system is strictly a mitochondrial process, [2-^{13}C]glycine provides a solely mitochondrial source of $^{13}CH_2$-THF. Incorporation of ^{13}C-labeled one-carbon units can be detected in choline and purines, two metabolites whose synthesis depends on the one-carbon pools of compartments outside the mitochondria. Choline synthesis from glycine involves the incorporation of mitochondrial CH_2-THF during serine synthesis, conversion to phosphatidylserine (PS), decarboxylation to phosphatidylethanolamine (PE), and subsequent methylation by *S*-adenosylmethionine (AdoMet) to produce phosphatidylcholine (PC)[3] (Fig. 1). The active methyl carbon of AdoMet is derived from CH_3-THF pools and donated for choline synthesis in the endoplasmic reticulum. Purine synthesis involves 10 enzymatic steps, 2 of which require the incorporation of cytosolic CHO-THF (Fig. 1). Thus comparison of the labeling pattern of these two metabolites allows direct comparison of the one-carbon pools in three structural compartments of the cell.

Methods

Growth Media and Yeast Strains

Rich media (YPD) contain 1% (w/v) yeast extract, 2% (w/v) Bacto Peptone, and 2% (w/v) glucose. Synthetic minimal media consist of 0.7% (w/v) yeast nitrogen base without amino acids (Difco, Detroit, MI), 2% (w/v) glucose, and the following supplemental nutrients when indicated (final concentration): L-serine (375 mg/liter), L-leucine (30 mg/liter), L-histidine (20 mg/liter), L-tryptophan (20 mg/liter), uracil (20 mg/liter), adenine (20 mg/liter), glycine (20 mg/liter), formate (1000 mg/liter). Yeast are grown at 30° with ample aeration in a 250-rpm rotary shaker.

Any laboratory strain of *S. cerevisiae* can be used. We typically use haploid strains carrying the *ser1* allele to block production of serine from glycolytic intermediates.

[3] F. Paltauf, S. D. Kohlwein, and S. A. Henry, Regulation and compartmentalization of lipid synthesis in yeast. *In* "The Molecular and Cellular Biology of the Yeast *Saccharomyces*: Gene Expression" (E. W. Jones, J. R. Pringle, and J. R. Broach, eds.), Vol II, pp. 415–500. Cold Spring Harbor Laboratory Press, Cold Spring Harbor, New York, 1992.

^{13}C-Labeled One-Carbon Donors

[2-^{13}C]Glycine, [^{13}C]formate, and L-[3-^{13}C] serine are available commercially at 99% enrichment from Cambridge Isotope Laboratories (Woburn, MA) or ICN (Cambridge, MA). Other potential one-carbon donors, such as glucose, pyruvate, acetate, or histidine, can also be used.

Metabolic Labeling

Metabolic labeling of yeast can be accomplished by either a short-term exposure to the ^{13}C substrate[4] or a longer term growth on the substrate.[5,6] The short-term experiment has the advantage that kinetic experiments can be performed,[7] but incorporation levels are not high for many metabolites. The long-term experiments, in which a steady-state enrichment of the one-carbon pools is achieved, give good incorporation into several useful end products of one-carbon metabolism.

Long-Term Labeling. Yeast cultures of 1 liter are grown aerobically (250 rpm on a rotary shaker) at 30° in yeast minimal media [Difco; 0.7% (w/v) yeast nitrogen base without amino acids] supplemented with amino acids and other nutrients (as required by a given strain) and the ^{13}C-labeled precursor. Cultures are inoculated with 10-ml overnight cultures of the appropriate strain and allowed to grow to late log phase in the presence of the ^{13}C-labeled precursor. Appropriate supplement concentrations for optimal growth of yeast and other yeast tips can be found in Sherman.[8] [2-^{13}C]Glycine is typically used at 100 mg/liter and [^{13}C]formate at 250 mg/liter in order to satisfy the serine requirement of *ser1*$^{-}$ strains. If both glycine and formate are used, the [2-^{13}C]glycine can be lowered to 20 mg/liter. L-[3-^{13}C]Serine should be used at 375 mg/liter, although its cost can be prohibitive. Glucose (2%, w/v) is typically used as carbon source.

Short-Term Labeling. Yeast cultures, grown aerobically at 30° in rich media [1% (w/v) yeast extract, 2% (w/v) Bacto Peptone, 2% (w/v) glucose] are harvested at midlog phase by centrifugation at 4000 rpm for 4 min at room temperature and resuspended to 37% (w/v) in yeast minimal media (described above). Cultures are incubated aerobically at 30° and provided with ^{13}C-labeled precursor for up to 6 hr.

Extract Preparation

Three different extraction procedures are used, depending on the desired metabolic end product.

[4] L. B. Pasternack, D. A. Laude, Jr., and D. R. Appling, *Biochemistry* **31,** 8713 (1992).
[5] L. B. Pasternack, D. A. Laude, Jr., and D. R. Appling, *Biochemistry* **33,** 74 (1994).
[6] L. B. Pasternack, L. E. Littlepage, D. A. Laude, Jr., and D. R. Appling, *Arch. Biochem. Biophys.* **326,** 158 (1996).
[7] L. B. Pasternack, D. A. Laude, Jr., and D. R. Appling, *Biochemistry* **33,** 7166 (1994).
[8] F. Sherman, *Methods Enzymol.* **194,** 3 (1991).

Purines and Choline. Yeast are harvested by centrifugation (4000 rpm for 4 min at 4°) and washed once with deionized water. The washed pellet is resuspended in 30 ml of 0.3 *N* HCl and steamed over a boiling water bath until the cell suspension is reduced to approximately 15 ml. This cleaves intracellular phosphatidylcholine, phosphocholine, and purine nucleotides to yield acid-soluble choline and purine bases. The acid suspension is centrifuged at 10,000 *g* for 30 min at 4° and the supernatant is transferred to a round-bottomed flask for drying of the sample by rotary evaporation. The resulting residue is resuspended in 1.0 ml of deuterated dimethyl sulfoxide (DMSO) and transferred to a 5-mm NMR sample tube. If the sample is to be analyzed by ^{1}H NMR, additional purification is required to remove paramagnetic ions that cause line broadening. After evaporation, the residue is resuspended in 5–10 ml of H_2O (instead of DMSO) and adjusted to pH 4.0 with NaOH. A 3 × 1 cm Chelex 100 column (Sigma, St. Louis, MO) is prepared by washing first with 0.1 *N* HCl (pH 1.0), then with 0.1 *M* NaOH (pH 12.5). The column is then equilibrated with 0.2 *M* sodium acetate buffer (pH 4), followed by 50 m*M* sodium acetate buffer (pH 4). Both conductivity and pH of the effluent are monitored to ensure complete equilibration. The sample is applied to the column, and the effluent is monitored at A_{260}. A flow rate of 2–3 drops per minute is used for sample loading and elution. The highest absorbing fractions (typically 30–40 ml) are pooled and dried completely in a rotary evaporator. The residue is resuspended in 1–2 ml of D_2O and reevaporated. The final residue is resuspended in 1.0 ml of D_2O and adjusted to pH 0.5 with deuterated sulfuric acid.

Free Amino Acids. Cultures are centrifuged at 4000 rpm for 4 min at 4°, washed with 25 ml of 1.8 m*M* KH_2PO_4–118 m*M* NaCl (pH 7.4), and then resuspended in 1.8 ml of the same buffer containing 4 *M* 2-mercaptoethanol. Cells are disrupted by vortexing for 4 min with glass beads, boiled for 5 min, and centrifuged at 25,000 *g* for 25 min at 4°. A portion of the resulting supernatant is transferred to a 5-mm NMR tube and overlaid with argon prior to capping.

Proteins and Nucleic Acids. Cells are harvested by centrifugation and washed twice with cold water. The washed cell pellet is resuspended in 90% (v/v) methanol (10 ml/g wet weight of cells) and vortexed. The precipitate (protein and nucleic acid) is collected by centrifugation, washed once with the same volume of 90% (v/v) methanol, resuspended in 5% (v/v) perchloric acid (5 ml/g wet weight of cells), heated at 70° for 1 hr to release the purine bases, cooled to 4°, and centrifuged. The supernatant is adjusted to pH 4–6 with KOH, centrifuged, and precipitated salt is discarded. The supernatant, containing the purine bases, is evaporated to dryness and the residue is dissolved in 1–2 ml of deuterated DMSO (for ^{13}C NMR) or D_2O (for ^{1}H NMR).

The remaining perchloric acid precipitate (protein fraction) is washed twice with 90% methanol, dried *in vacuo*, resuspended in 6 *M* HCl (300 ml/g protein dry weight), evacuated (or bubbled with argon for 0.5 hr), sealed, and placed in oven for 18–20 hr at 110°. After hydrolysis, the mixture is filtered or centrifuged, the clear supernatant is evaporated to dryness on a rotary evaporator, and the final residue resuspended in 1–2 ml of D_2O for NMR analysis.

Nuclear Magnetic Resonance Analysis

^{13}C Spectra. ^{13}C NMR spectra are obtained with a 5-mm probe on either a Nicolet Analytical Instrument NT 360 (Bruker Instruments, Billerica, MA) or on a Bruker AMX 500 (Bruker Instruments, Billerica, MA). ^{13}C data are collected with a 5-sec delay at 90 MHz (NT 360) or 125 MHz (Bruker AMX 500) with a 45° pulse and proton broad-band decoupling. A total of 1200–2400 scans of 32K data points are acquired over a sweep width of 11,904 Hz (NT 360) or 31,250 Hz (Bruker AMX 500). Data processing includes line broadening of 3.0 Hz and 32K zero filling prior to Fourier transformation to yield the frequency domain spectrum.

Proton Spectra. ^{1}H NMR spectra are obtained on a Bruker 250 MHz or a Varian (Palo Alto, CA) Unity Plus 500 MHz, both equipped with a 5-mm probe. A pulse width corresponding to a 90° flip angle is used, and data are collected during a 3-sec acquisition time with a 7-sec delay. The residual water signal is presaturated during the delay. Typically, 200–500 scans of 54K data points are acquired over a spectral width of 5000 Hz (10 ppm). A total of 260K points is used for Fourier transform. An exponential line broadening of 0.2 Hz is used to improve the signal-to-noise ratio.

Data Analysis and Quantification

Metabolites are identified by characteristic chemical shift values established from natural abundance spectra using extracts spiked with unlabeled compounds. Quantitation of adenine and choline resonances is determined by a combination of integration and isotopomer analysis. Peak integration rather than peak height is used because line widths sometimes vary from sample to sample. The integrated area for each resonance is multiplied by a normalization factor (NC_{xx}) to correct for different signal intensities for carbons of the same compound, which depend on the nuclear Overhauser effect (NOE) and relaxation time of the particular carbon under the conditions employed. This normalization factor for the carbons within choline and adenine is determined by signal integration of a natural abundance sample of the metabolite added to an extract of strain DAY4[4] grown in unlabeled glycine and recorded under the same conditions used for the experimental samples.

Incorporation of ^{13}C-labeled one-carbon units into C-2 and C-8 of adenine and C-1 and C-4 of choline is compared to the direct (non-folate-dependent) incorporation of glycine (C-5 in adenine; C-2 in choline) to calculate a relative enrichment (RE) for each one-carbon oxidation state pool:

$$\mathrm{RE_{CHO}} = \frac{(C_{2\mathrm{A}})(\mathrm{NC}_{2\mathrm{A}}) + (C_{8\mathrm{A}})(\mathrm{NC}_{8\mathrm{A}})}{2(C_{5\mathrm{A}})}$$

$$\mathrm{RE_{CH_2}} = (F_{\mathrm{C1}})/(F_{\mathrm{C2}})$$

$$\mathrm{RE_{CH_3}} = \frac{(C_{4\mathrm{C}})(\mathrm{NC}_{4\mathrm{C}})}{(C_{2\mathrm{C}})}$$

where $\mathrm{RE_{CHO}}$ is the relative enrichment of the 10-formyl-THF pool; $\mathrm{RE_{CH_2}}$ is the relative enrichment of the 5,10-methylene-THF pool; $\mathrm{RE_{CH_3}}$ is the relative enrichment of the 5-methyl-THF pool; F_{C1} is the fractional enrichment of C-1 of choline, determined by isotopomer analysis of the C-2 of choline as described below; F_{C2} is the fractional enrichment of C-2 of choline, determined by isotopomer analysis of the C-1 of choline as described below; $C_{2\mathrm{A}}$ is the integral of the C-2 peak of adenine; $C_{8\mathrm{A}}$ is the integral of the C-8 peak of adenine; $C_{5\mathrm{A}}$ is the integral of the C-5 peak of adenine; $C_{4\mathrm{C}}$ is the integral of the C-4 peak of choline; $C_{2\mathrm{C}}$ is the integral of C-2 of choline, entire peak; $\mathrm{NC}_{8\mathrm{A}}$ is the integral of C-5 divided by the integral of C-8 in natural abundance spectra (normalization factor for adenine C-8); $\mathrm{NC}_{2\mathrm{A}}$ is the integral of C-5 divided by the integral of C-2 in natural abundance spectra (normalization factor for adenine (C-2); and $\mathrm{NC}_{4\mathrm{C}}$ is the integral of C-2 divided by the integral of C-4 in natural abundance spectra (normalization factor for choline C-4).

Fractional enrichment (F_{C1}, F_{C2}) for choline C-1 and C-2 resonances is determined by deconvolution and integration. The resonances for C-1 and C-2 of choline are a combined singlet and doublet. This indicates the presence of three labeled species with respect to the C-1 and C-2 position: label at the C-1 position only (singlet at 56.4 ppm), label at the C-2 position only (singlet at 68.2 ppm), and label at both the C-1 and C-2 positions (doublets surrounding each singlet). Thus each resonance is composed of a singlet indicating the absence of a ^{13}C-labeled neighbor, and a doublet indicating presence of ^{13}C label at the neighboring carbon. Fractional enrichment of the neighboring carbon can be determined by dividing the integral of the doublet by the integral of the whole resonance.[9] The following equations apply:

$$F_{C1} = \frac{D_{C2}}{S_{C2} + D_{C2}}$$

$$F_{C2} = \frac{D_{C1}}{S_{C1} + D_{C1}}$$

where S_{C2} is the integral of the singlet at C-2; D_{C2} is the integral of the doublet at C-2; S_{C1} is the integral of the singlet at C-1; and D_{C1} is the integral of the doublet at C-1.

^{1}H NMR spectra are quantitated by Lorenzian line fitting of the corresponding adenine and guanine proton resonances and calculation of the area under the peaks. Depending on the carbon isotope in positions 8, 2, and 5 of the purine ring, the proton signal will occur as singlet, doublet, or quartet, and the spectrum of a partially labeled compound will be represented by a characteristic multiplet composed of individual isotopomers. The area of the whole multiplet for each position is set as 100%, and the relative contribution of the different isotopomers is calculated according to multiplet structure.

Whole-Cell Experiments

Yeast cultures are exposed to [^{13}C]formate as described above for short-term labeling. Aliquots of 25 ml are removed at the appropriate time, frozen in methanol–dry ice, and stored at −70°. Just prior to NMR analysis, samples are thawed at 4°, centrifuged cold for 4 min at 4000 rpm, resuspended in 1 ml of D_2O, and transferred to 10-mm NMR tubes. This provides a total volume of approximately 7 ml of cell suspension. Cells are allowed to settle for 7 hr at 4° prior to data aquisition. This results in a cell volume that exceeds the active probe volume and ensures that each spectrum is acquired from the same volume of cells. Incubation media can be saved for analysis if necessary. ^{13}C NMR spectra are obtained as described above with the following changes. ^{13}C NMR spectra are acquired using a 10-mm probe. A total of 1800 scans of 16K data points is acquired over an 11,904-Hz bandwidth. Pulse sequence parameters include a 45° flip angle and 1.5-sec delay time. For each experiment, all samples and a standard curve of serine in D_2O (external standard) are run consecutively with the same pulse conditions and receiver gain. To accommodate day-to-day variation in NMR performance, integrated values for serine resonances are extrapolated from the standard curve run with each data set. No adjustments

[9] C. R. Malloy, A. D. Sherry, and F. M. H. Jeffrey, *Am. J. Physiol.* **259,** H987 (1990).

are made for differences such as T_1 typical of external standards and thus values obtained in this manner, while accurately reflecting relative changes in concentration, may not reflect actual concentrations.

Discussion

This section illustrates the kinds of experiments that can be done using these methods.

Monitoring the incorporation of [2-^{13}C]glycine into the purine ring and the choline moiety of phosphatidylcholine allows one to follow the flow of one-carbon units from donor (glycine) through three different levels of oxidation: CHO-THF at C-2 and C-8 of purines; CH_2-THF at C-1 of choline; and CH_3-THF at C-4 of choline (Fig. 2). Both choline and purine synthesis involve the direct incorporation of a glycine unit. This acts as an internal control to normalize relative enrichments between choline and purines and to compare directly the state of enrichment of the different oxidation states of the folate derivatives.[5] This methodology can be used to determine how one-carbon units are distributed in response to varying growth and nutritional conditions. It can be used in strains harboring defined mutations to decipher metabolic pathways. It can also be used to study how yeast cells handle alternative, competing one-carbon sources.[6] Cells are grown with [2-^{13}C]glycine and an unlabeled one-carbon donor such as formate or folinic acid (leucovorin, 5-formyltetrahydrofolate). Dilution of the glycine-derived ^{13}C-labeled one-carbon units into choline and purines indicates competition by the unlabeled one-carbon source.

Whole-cell experiments can be used to study the kinetics of metabolic pathways *in vivo*. For example, [^{13}C]formate has been used to follow serine synthesis by the C_1-THF synthase/serine hydroxymethyltransferase pathway[7] (reactions 1–4; Fig. 1). The intracellular pool of [3-^{13}C]serine produced by this pathway can be detected directly in whole cells (Fig. 3), allowing measurement of rates, substrate dependence, and inhibitor effects *in vivo*.

^{1}H NMR spectroscopy of ^{13}C-labeled metabolites provides additional quantitative information.[10] Isotopomer analysis of ^{13}C spectra allows determination of relative enrichments only at specific positions. Absolute enrichments cannot be determined because the size of the unlabeled (^{12}C) pool of metabolite, which is invisible in the ^{13}C spectrum, is not known. ^{1}H NMR, on the other hand, detects all molecules of a metabolite because ^{1}H is the most abundant hydrogen isotope. Furthermore, the proton spectrum of a molecule containing ^{13}C differs from the spectrum of the same molecule containing only ^{12}C; proton resonances are split to doublets owing to spin–

[10] R. E. London, *Prog. NMR Spectrosc.* **20,** 337 (1988).

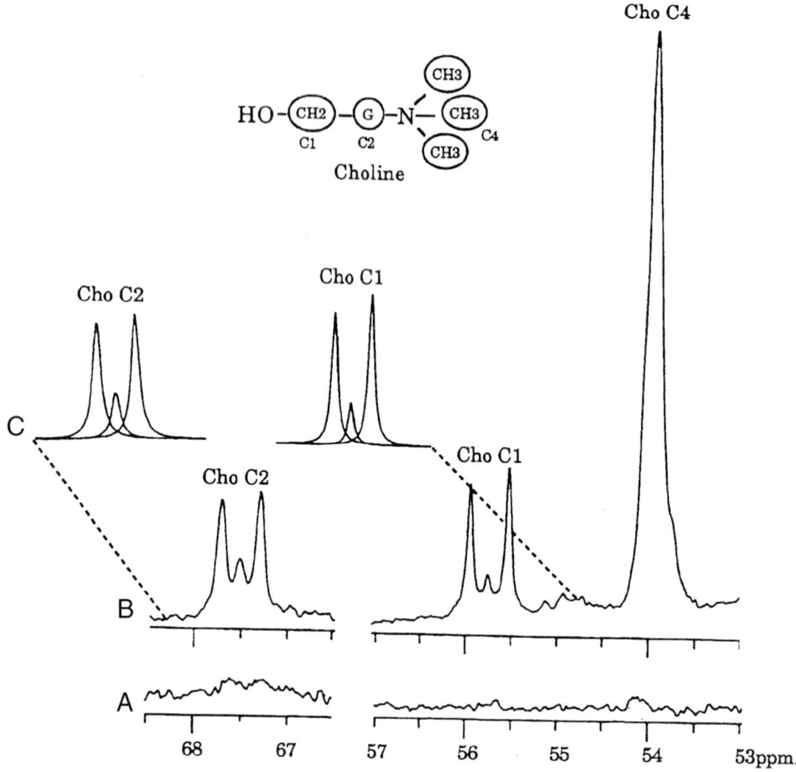
Cho C4
HO
CH2
G
N
CH3
CH3
CH3
C1
C2
C4
Choline
Cho C2
Cho C1
C
Cho C2
Cho C1
B
A
68
67
57
56
55
54
53ppm.

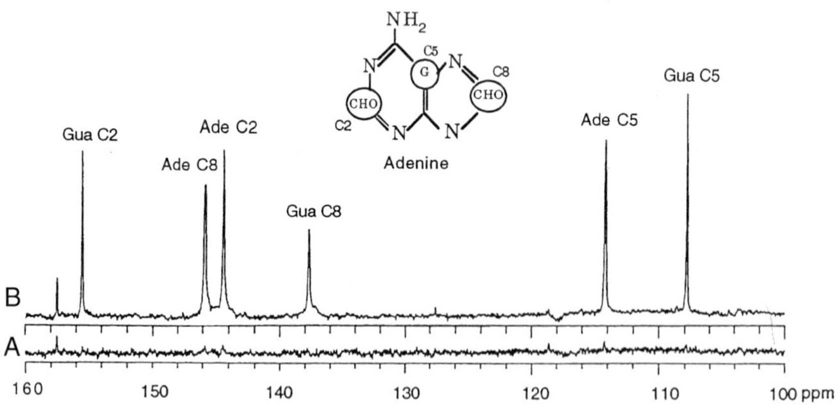
NH2
N
C5
G
N
C8
CHO
CHO
C2
N
N
Adenine
Gua C5
Ade C5
Gua C2
Ade C2
Ade C8
Gua C8
B
A
160
150
140
130
120
110
100 ppm

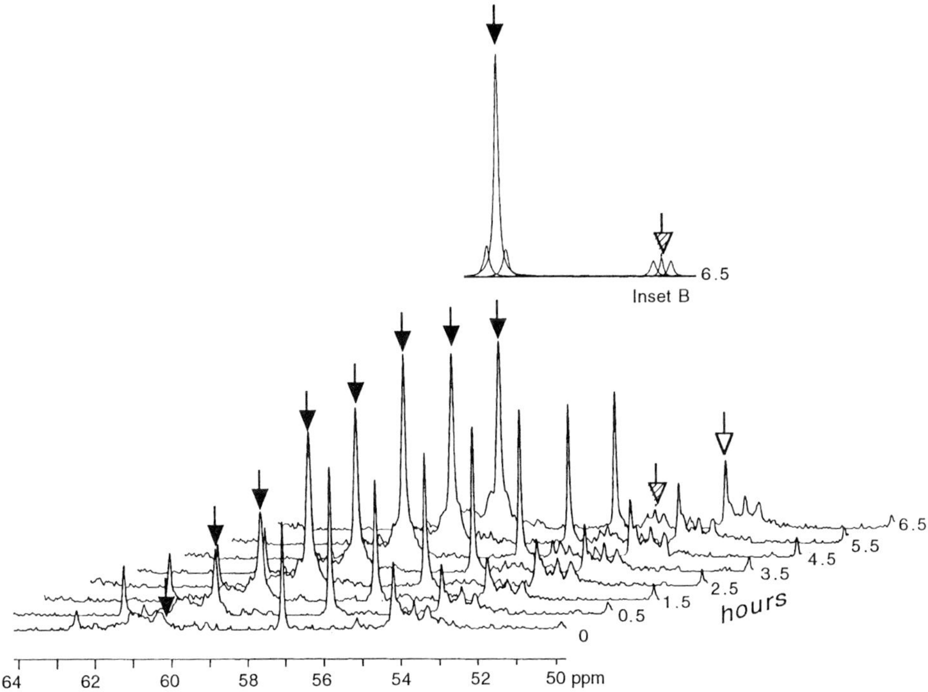

FIG. 3. Whole-cell ^{13}C NMR spectra of [3-^{13}C]serine accumulation in strain DAY4 exposed to [^{13}C]formate. Each spectrum is identified by the number of hours of [^{13}C]formate exposure. *Inset B:* Deconvolution of the C-3 and C-2 serine resonances after 6.5 hr of exposure. Filled arrow indicates C-3 serine resonance. Hatched arrow indicates C-2 serine resonances. Open arrow indicates C-2 glutamate resonance. (Reprinted with permission from Ref. 7. Copyright 1994 American Chemical Society.)

spin interaction with ^{13}C (there is no such interaction with ^{12}C, because its spin is zero). The split resonances have characteristic coupling constants (measured in hertz) that depend on the number of bonds between the coupled atoms, with one-bond coupling exhibiting the greatest constant. Two-, three-, and four-bond couplings are visible as long as the coupling

---

FIG. 2. ^{13}C NMR spectra of choline resonances (*top*) and purine resonances (*bottom*) in cell extract of strain DAY4 grown in (A) unlabeled glycine or (B) [2-^{13}C]glycine (100 mg/liter). (C) Deconvolution and Lorenzian line fit of spectrum for integration. The choline structure indicates the source of each carbon; C-1, CH_2-THF; C-2, direct incorporation of glycine; C-4, CH_3-THF via AdoMet. The purine structure indicates the source of each carbon; C-2 and C-8, CHO-THF; C-5, direct incorporation of glycine. (Reprinted with permission from Ref. 5. Copyright 1994 American Chemical Society.)

constant is greater than the line width. The signal of a specific proton in the spectrum of a partially ^{13}C-labeled metabolite is a superposition of the individual resonances (singlet plus doublets) of different ^{13}C isotopomers. This phenomenon is particularly useful for purines, in which the protons attached to C-2 (A-H2) and C-8 (A-H8) of adenine and C-8 of guanine (G-H8) can be visualized (Fig. 4). Thus, the resonance of a proton attached to the C-8 of adenine (A-H8) will be split both by a ^{13}C at C-8 (211-Hz constant) and a ^{13}C at C-5 (9-Hz constant). By integrating the individual multiplet components of the proton resonances, absolute enrichments at adenine C-5, C-2, and C-8 can be quantitated.

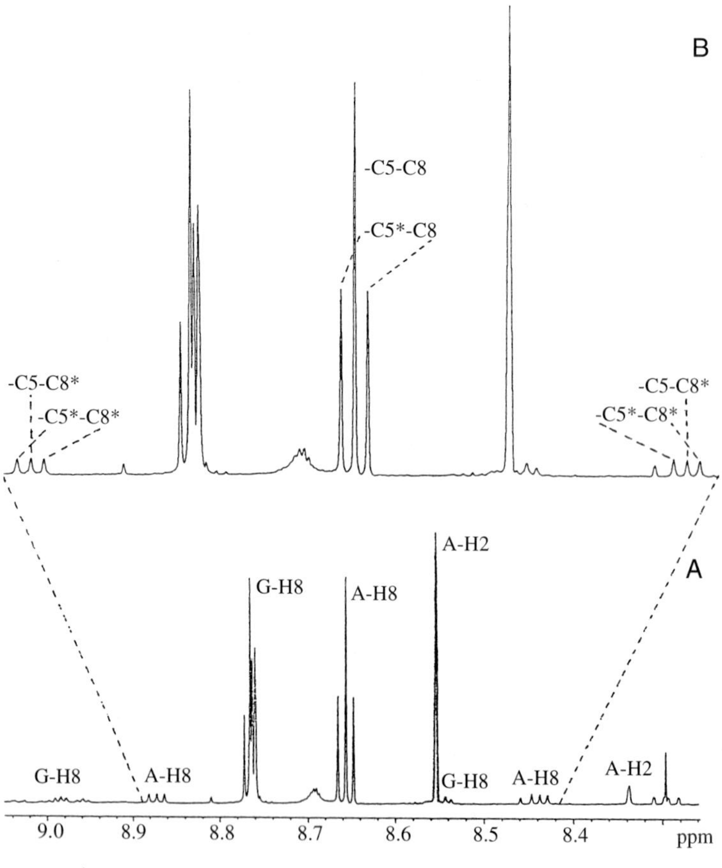

FIG. 4. (A.) ^{1}H NMR spectrum of purine extract from yeast grown on [2-^{13}C]glycine plus unlabeled serine. (B.) Expansion of part of the spectrum in (A), indicating the ^{13}C isotopomers responsible for each split proton resonance. An asterisk indicates a ^{13}C label.

Finally, combining these ^{13}C NMR methods with the use of yeast mutants blocked at specific steps of one-carbon metabolism allows a dissection of the compartmentation and the roles of the various isozymes of these pathways in a model eukaryote. For example, comparing the ability of unlabeled formate to compete with one-carbon units derived from [2-^{13}C]glycine in the labeling of choline in strains lacking either the NADP- or NAD-dependent 5,10-methylene-THF dehydrogenase (reaction 3 vs reaction 6; Fig. 1) reveals that only the NADP-dependent enzyme contributes to the flow of one-carbon units in the reductive direction.[11]

Acknowledgment

This work was supported by NIH Grant RR09276.

[11] M. G. West, D. W. Horne, and D. R. Appling, *Biochemistry* **35,** 3122 (1996).

Section II

Vitamin B_{12} and Cobalamins

[27] Use of Magnetic Field Effects to Study Coenzyme B_{12}-Dependent Reactions

By CHARLES B. GRISSOM and ETTAYA NATARAJAN

Introduction

The rate of thermochemical, photochemical, and biological reactions with radical pair (RP) intermediates can be altered by the application of an external magnetic field.[1–5] The effect of the magnetic field is to change the rate of intersystem crossing (ISC) between the singlet and triplet electron spin states. Because only reactions with at least two unpaired electrons show this behavior, the observation of a magnetic field effect is unambiguous proof of the existence of a kinetically significant RP intermediate.

This technique has been used to probe the RP that is generated by homolysis of the carbon–cobalt (C–Co) bond in coenzyme B_{12}-dependent photochemical and enzymatic reactions.[1,6–8] This chapter summarizes the theory of magnetic field effects (MFEs) on chemical and enzymatic reactions in a form that is useful to the biochemist, and outlines the experimental methods necessary to use this powerful technique to study a variety of enzymatic and chemical reactions with RP intermediates.

Theory of Magnetic Field Effects on Radical Pair Reactions

A radical pair consists of two radical species that are spatially close (typically <10–20 Å). If the electron spin vectors of the two radicals are close enough to interact through space or through bonds, the RP is said to be "spin correlated," and either the singlet state (antiparallel orientation) or the triplet state (parallel orientation) will be the lower energy orientation.

[1] C. B. Grissom, *Chem. Rev.* **95,** 3 (1995).

[2] U. E. Steiner and T. Ulrich, *Chem. Rev.* **89,** 51 (1989).

[3] H. Hayashi, *in* "Photochemistry and Photophysics" (J. F. Rabek, ed.), Vol. I, Ch. 2. CRC Press, Boca Raton, Florida, 1990.

[4] U. E. Steiner and H.-J. Wolff, *in* "Photochemistry and Photophysics" (J. F. Rabek, ed.), Vol. IV, Ch. 1. CRC Press, Boca Raton, Florida, 1991.

[5] K. M. Salikhov, Y. N. Molin, R. Z. Sagdeev, and A. L. Buchachenko, "Spin Polarization and Magnetic Effects in Radical Reactions." Elsevier, Amsterdam, 1984.

[6] A. M. Chagovetz and C. B. Grissom, *J. Am. Chem. Soc.* **115,** 12152 (1993).

[7] C. B. Grissom and A. M. Chagovetz, *Z. Phys. Chem.* **182,** 181 (1993).

[8] E. Natarajan and C. B. Grissom, *Photochem. Photobiol.* **64,** 286 (1996).

0076-6879/97 $25

When an RP is formed by homolytic cleavage of a covalent bond, the initial RP will be in the singlet or triplet electron spin state, depending on the multiplicity of the precursor (Fig. 1). Photochemically induced homolysis can occur from the excited singlet state to generate the singlet RP or, more commonly, photohomolysis can occur from the excited triplet state to produce the triplet RP. Thermally induced homolysis, of the type that might occur by steric deformation or geometric approximation of a molecule that is bound in an enzyme active site, is expected to generate the singlet RP, because the electrons are spin paired in the precursor molecule by virtue of the Pauli exclusion principle.

A diffusible RP can undergo one of several fates, including *recombination* (*reaction*) to form a covalent bond, or *separation* (*diffusion*) to produce free radicals that are noninteracting and whose reactivity is not altered by a magnetic field. In the typical radical pair, only the singlet RP has the proper electron spin orientation to form a covalent bond. In contrast, singlet or triplet radical pairs will undergo diffusive separation with equal efficiency (Fig. 1).

If the RP intermediate maintains close contact for 10^{-10}–10^{-6} sec, a transition between singlet and triplet electron spin state is possible and this process is called *intersystem crossing* (ISC). Because only the singlet RP can undergo recombination in most instances, a triplet RP must typically undergo ISC to the singlet state before reaction to form a covalent bond (Fig. 1).

An externally applied magnetic field can either increase or decrease the rate of ISC, and thereby affect the competition between RP recombination and diffusion. In this way, a magnetic field can be used to alter the distribution of products that arise from either recombination or diffusion; or alter the net rate of a multistep reaction if a kinetically significant RP

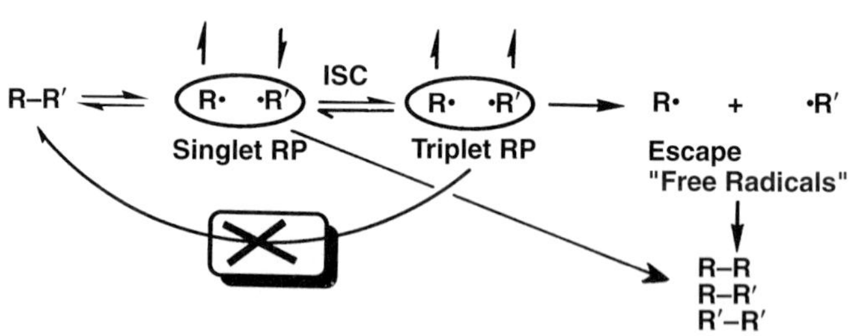

FIG. 1. Homolysis of a covalent bond reversibly produces a singlet radical pair with antiparallel electron spins. Intersystem crossing produces the triplet electron spin state with the electron spin vectors in a parallel orientation. The triplet radical pair cannot undergo direct recombination. In this scheme, both singlet and triplet radical pairs can undergo nonrecombination processes, including escape, with equal efficiency. ISC, Intersystem crossing.

intermediate is involved. In the latter case, a magnetic field effect on the rate occurs by altering the forward partitioning to product, relative to reverse partitioning to regenerate the starting substrate.

There are at least three mechanisms by which a magnetic field can change the rate of intersystem crossing in radical pair intermediates: the hyperfine interaction mechanism, the Δg mechanism, and the level crossing mechanism.[2,3,5]

Hyperfine Interaction Mechanism

The hyperfine interaction (HFI) mechanism is operative in weakly coupled radical pairs, in which the singlet and triplet electronic spin states are nearly equal in energy.[5] This occurs when the exchange integral between the radical centers is nearly zero (i.e., $J = 0$ and $A \neq 0$, where J is the exchange integral between two radicals and A is the isotropic hyperfine coupling constant of the individual radical species). The value $2J$ defines the difference in energy between the singlet and triplet spin state. In the absence of an external magnetic field, $J = 0$ and S–T conversion occurs through the mixing between the S_0 and the T_{+1}, T_0, and T_{-1} states. As the magnetic field increases, Zeeman splitting removes the energy degeneracy among the three triplet spin states and now S–T conversion can occur only through mixing of the S_0 and T_0 spin states. Thus, the overall S–T conversion rate decreases with increasing magnetic field (B). This mechanism is most important at magnetic field strengths below 1000 G (Fig. 2A).[9]

Δg Mechanism

The Δg mechanism can be operative in radical pairs where $J = 0$, and the difference in Larmor precession frequency (Δg) for the two radical species does not equal zero.[5] In contrast to the HFI mechanism, the Δg mechanism increases the rate of ISC as the magnetic field increases (Fig. 2B). This occurs through rephasing of the electron spin vectors in the magnetic field. This mechanism is most important at high magnetic fields (typically greater than 1000 G). For many organic radical pairs, $\Delta g \approx 0$ and this mechanism becomes significant only at high magnetic field strengths (typically greater than 10,000–50,000 G). The HFI and Δg mechanisms can operate in the same system and produce a biphasic response of ISC vs magnetic field (Fig. 2C).

[9] 1 tesla = 10,000 gauss (G).

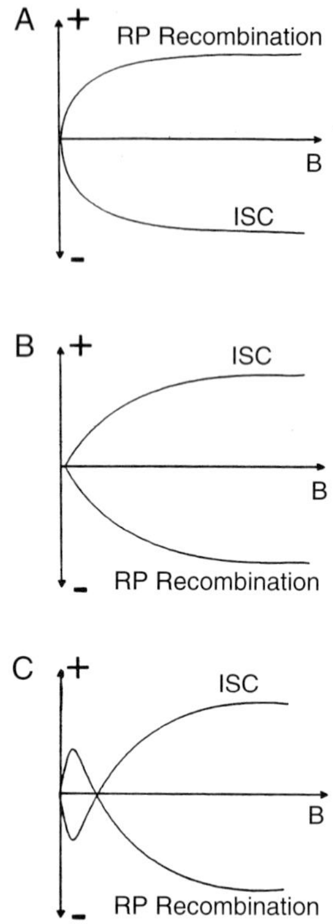

FIG. 2. (A) Nuclear hyperfine interactions (HFI mechanism) promote intersystem crossing. If the radical pair is formed in the singlet spin state, intersystem crossing to the triplet spin state will decrease the overall amount of radical pair recombination because of the lower singlet radical pair concentration. The HFI mechanism decreases the amount of intersystem crossing as the magnetic field is increased. (B) Different Larmor precession frequencies (g value) will promote electron spin rephasing. The Δg mechanism increases the amount of intersystem crossing as the magnetic field is increased. (C) If the HFI mechanism and the Δg mechanism operate in the same system, the result can be a biphasic dependence of radical pair recombination on applied magnetic field.

Level Crossing Mechanism

The level crossing (LC) mechanism can operate in radical pairs when $J \neq 0$ and $A \neq 0$. At $B = 0$, ISC does not occur because of the large difference in energy between the singlet and triplet spin states ($2J > 0$).[5] However, in a narrow region of magnetic field, S–T conversion occurs through the mixing of the S_0 and T_{-1} spin states where these energy surfaces overlap. This produces a sharp increase in ISC, followed by a sharp decrease in ISC at higher magnetic field strengths.

Predicted Magnetic Field Dependence of Coenzyme B_{12}

Consider the case of coenzyme B_{12} in the active site of an enzyme. Homolysis of the C–Co bond should produce a singlet radical pair.[10] Diffusive separation of the 5′-deoxyadenosyl radical and cob(II)alamin (Cbl^{II}) will be prevented by multiple hydrogen-bonding interactions in the enzyme active site.[11] These binding interactions will allow sufficient time for ISC to mix the singlet and triplet spin states via the HFI mechanism, as previously described. Hyperfine coupling is especially large for Co(II) because of the nonzero nuclear spin ($I = 7/2$) associated with the metal center. This yields an effective hyperfine coupling constant of $A = 95$ G for Cbl^{II}.[8]

The 5′-deoxyadenosyl radical will typically have an electronic g value near the free electron ($g \approx 2.025$). Cob(II)alamin that is generated in the absence of enzyme[12] has a g value near 2.16, whereas the g value of Cbl^{II} bound in the active site of an enzyme may approach 2.3.[13,14] A large g value for Cbl^{II} ($g > 2.025$) will increase the effectiveness of the Δg mechanism and increase the net rate of ISC at high magnetic field strengths (see below).

Magnetic Field Effects in Enzymatic Reactions

Note: Not all enzymes with radical intermediates will exhibit a magnetic field effect.[1] Recall that a chemical or enzymatic reaction must have at least two unpaired electrons for a magnetic field effect to be possible.[1,15,16] This excludes all but the radical–radical termination steps in a free radical chain process.[1] In an enzymatic reaction where a protein radical (E·) abstracts a hydrogen atom from substrate to generate a substrate radical (S·), only one unpaired electron spin will exist at any given time and a singlet or triplet spin state cannot be defined[1,17] [see Eqs. (1)–(3); P· is the product radical and PH is the final product].

$$E\cdot + SH \rightarrow EH + S\cdot \tag{1}$$
$$S\cdot \rightarrow P\cdot \tag{2}$$
$$P\cdot + EH \rightarrow PH + E\cdot \tag{3}$$

[10] D. Dolphin, "B_{12}," Vol. 1, and 2. John Wiley & Sons, New York, 1982.
[11] W. B. Lott, A. M. Chagovetz, and C. B. Grissom, *J. Am. Chem. Soc.* **117,** 12194 (1995); see Ref. 36 therein.
[12] G. N. Schrauzer and L.-P. Lee, *J. Am. Chem. Soc.* **90,** 6541 (1968).
[13] J. A. Hamilton, R. Yamada, R. L. Blakley, H. P. C. Hogenkamp, F. D. Looney, and M. E. Winfield, *Biochemistry* **10,** 347 (1971).
[14] G. R. Buettner and R. E. Coffman, *Biochim. Biophys. Acta* **480,** 495 (1977); K. N. Joblin, A. W. Johnson, M. F. Lappert, M. R. Hollaway, and W. A. White, *FEBS Lett.* **53,** 193 (1975).
[15] V. K. Vanag and A. N. Kuznetsov, *Izv. Akad. Nauk. SSSR Ser. Biol.* **2,** 215 (1988).
[16] V. K. Vanag and A. N. Kuznetsov, *Biofizika* **29,** 23 (1984).
[17] J. Stubbe, *Biochemistry* **27,** 3893 (1988).

In contrast, if the radical is generated through homolysis of a covalent bond, a radical pair will be formed [see Eqs. (4)–(6)]. If bond homolysis occurs without atomic motion (separation of E· and S·), then recombination of E· and S· to regenerate E + S will proceed with a nearly zero activation energy barrier [Eq. (4)]. This will provide an opportunity for magnetic field-dependent recombination. In this simple kinetic scheme, a second opportunity exists to observe a magnetic field effect. Regeneration of the resting state of the enzyme, E, also requires RP recombination and this will also be spin selective, as only the singlet RP, consisting of E· and P·, can combine to form the resting state of E and P.

$$E + S \rightleftharpoons E\cdot + S\cdot \tag{4}$$
$$S\cdot \rightarrow P\cdot \tag{5}$$
$$P\cdot + E\cdot \rightleftharpoons P + E \tag{6}$$

Methods

The use of magnetic field effects to probe enzymatic reactions is usually limited to static (DC) magnetic fields that do not vary with time. Furthermore, the magnetic field range of interest is typically 0–10,000 G, where 0 G is approximated by the geographically constant geomagnetic field of 0.2–0.5 G that depends on latitude.[9] According to accepted and tested theories of magnetic field-dependent recombination in radical pair reactions, alternating magnetic fields near the powerline frequencies of 50 or 60 Hz exert an effect that is described by the absolute value of the instantaneous magnetic field produced by the alternating magnetic field vector, because the rate of field oscillation is orders of magnitude slower than the rate of RP recombination reactions of interest.[18]

Measuring Magnetic Fields

Magnetic flux density within the range of interest is commonly measured with a gaussmeter (teslameter) with a Hall effect transducer. This device will precisely measure magnetic flux densities to less than 0.01 G, although the absolute accuracy often depends on the zero-field reference. A transverse-measuring Hall probe tends to be better than an axial probe for most applications with laboratory electromagnets. Because the type of magnetic field effect experiment described herein is not a narrow resonance phenomenon [as would be encountered in nuclear magnetic resonance (NMR) or

[18] J. C. Scaiano, N. Mohtat, F. L. Cozens, J. McLean, and A. Thansandote, *Bioelectromagnetics* **15,** 549 (1994).

electron spin resonance (ESR) spectrometry], magnetic field measurement to within 1–2% is usually adequate.

Laboratory Electromagnets and Instrumentation

Electromagnets with metal cores of high magnetic permeability are suitable for producing static magnetic fields up to 1.8 T. Magnetic flux densities in the range of 1.8–16.5 T are obtained by superconducting magnets, although these fields are mostly of interest to evaluate the importance of the Δg mechanism in a given system. In practice, it is difficult to carry out absolute rate measurements with different electromagnets, superconducting magnets, or permanent magnets, as small temperature differences or cell configuration differences between magnets can be a confounding factor. For this reason, it is preferable to vary the applied magnetic field within the same magnet.

Magnetic field homogeneity of 2–3% is sufficient for most magnetic field effect experiments, because the process being studied is not a narrow resonance phenomenon. Most laboratory electromagnets provide a sufficiently homogeneous field. A Hall probe mounted on an x–y–z translator can be used to map the variation in magnetic flux density within the magnet to define the region that is suitable for experimentation. Temporal magnetic field stability is often a more critical parameter than overall homogeneity. Short-term instability (over the course of one assay) might only produce noisy data, whereas long-term instability (over the course of a series of assays) can introduce a systematic error that further complicates data analysis. For this reason, it is important to operate laboratory electromagnets with a constant-current power supply and, if possible, add a field-sensitive feedback circuit from the gaussmeter to the programmable power supply.

A simple system might use an analog voltage signal that is proportional to magnetic field, and use a differential op-amp circuit to compare this measured value to a reference value, and use any difference to increase or decrease the current supplied to the electromagnet. In practice, this simple system does not work as well as a full-featured PC-based digital-to-analog controller to set and maintain the desired magnetic field value, in part because of the hysteresis of the electromagnet. Similarly, hysteresis of the electromagnet often necessitates a bipolar power supply to achieve "zero" G. Commercial systems that integrate digital gaussmeters and electromagnet controllers are available.

Small commercial electromagnets that cover the desired field range are available, or satisfactory "home-built" electromagnets can be constructed with only modest expertise in metal fabrication techniques. High-impedance electromagnets can be constructed from relatively thin [22–32 American wire gauge (AWG)] varnish-coated magnet wire and operated at high voltage (50–1000 VDC) and low current (up to 1 A), or low-impedance electromagnets can be constructed from thicker (10–18 AWG) varnish-coated magnet wire and operated at low voltage (1–10 VDC) and high current (up to 50 A). When constructing an electromagnet, the choice between a high- and low-impedance system is often dictated by the available power supply. A basic physics textbook should be consulted for help in calculating the field produced by a wire-wound solenoid where the length-to-width aspect ratio exceeds 10. The available magnetic field density can be increased by "concentrating" the lines of magnetic flux by increasing the magnetic permeability of the core of the electromagnet. A satisfactory material for home-built electromagnets is low-carbon high-iron steel that is machined into a cylinder of the desired configuration. Alternatively, a toroid-shaped core can be constructed with a gap in the ring to create the sample compartment. In the latter design, the wire is wound around the toroid in the same fashion as when constructing a toroidal ferrite transformer.

Magnetic Shielding

Care must be taken to shield nearby photomultiplier tubes from the magnetic field. This can be accomplished by commercial mu-metal shields that fit over the photomultiplier tube. This material is brittle in its final form and it is difficult to fabricate shields on-site. A lower-cost alternative is to use several sheets of thin low-carbon high-iron steel that is laminated with a thin dielectric to create the desired shield.

Steady State Enzyme Kinetics as Function of Magnetic Field

Figure 3 shows the experimental setup for carrying out steady-state enzyme kinetic studies as a function of magnetic field. This is useful for any enzyme with a spectrophotometrically observable chromophore in the substrate or product. Any ultraviolet–visible (UV–Vis) spectrophotometer (diode array or PMT detection) with a sufficiently large cell compartment can be retrofitted with an electromagnet. Diode array spectrophotometers, with their open cell compartments, are especially well suited to this application. As in any isothermal experiment, care must be taken to thermostat the cuvette holder independently of any thermostatting applied to the electromagnet itself.

Stopped-Flow Kinetics as Function of Magnetic Field

Any stopped-flow spectrophotometer or rapid-quench mixing device can be outfitted with an electromagnet surrounding the cell compartment.

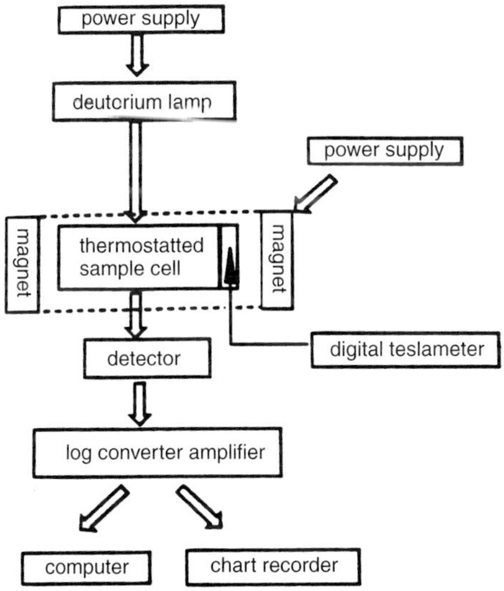

FIG. 3. UV–Vis spectrophotometer retrofitted with an electromagnet.

Figure 4 illustrates an OLIS, Inc. (Bogart, GA) RSM-1000 rapid-scanning stopped-flow spectrophotometer that has been retrofitted with a GMW Associates, Inc. (Redwood City, CA), electromagnet (7.5-cm cylindrical poles; model 5403). This system is especially well suited to study the coenzyme B_{12} cofactor, as true absorbance spectra can be acquired at a rate of 1 kHz, while maintaining the sample in the dark 97% of the time, on a time-averaged basis.

Magnetic Field Effect Results with Alkylcobalamin

Magnetic Field Dependence of Alkylcobalamin Photolysis

Nonenzymatic carbon–cobalt (C–Co) bond cleavage of methylcob(III)-alamin can be initiated by photolysis in solution to generate the singlet RP

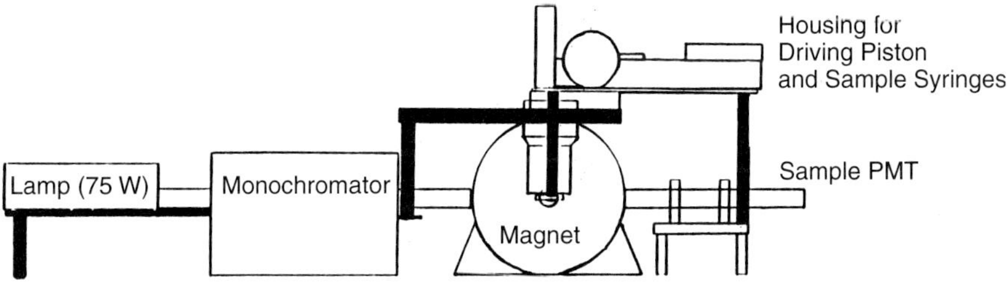

FIG. 4. Olis, Inc., stopped-flow spectrophotometer retrofitted with an electromagnet surrounding the cell compartment.

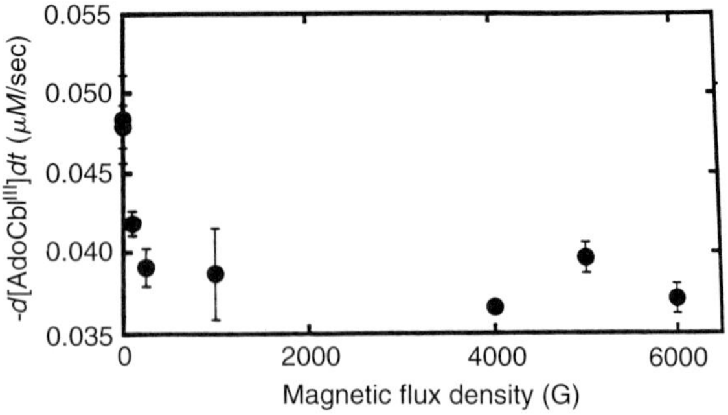

FIG. 5. Magnetic field dependence of the relative rate of CH_3Cbl^{III} photolysis at 520 nm under anaerobic conditions. Photolysis was monitored by the appearance of cob(II)alamin at 460 nm. Conditions: 0.091 m*M* CH_3Cbl^{III}, 50 m*M* HEPES, pH 7.0, 22°, 75% (w/v) glycerol.

that consists of the methyl radical and cob(II)alamin $\{CH_3\cdot\ Cbl^{II}\}$.[6–8] In a buffered solution of 75% (w/v) glycerol with a viscosity of ~30 cP, the photochemical quantum yield of CH_3Cbl^{III} decreases by 40% in the range 1000–6000 G (Fig. 5), as mixing of the singlet and triplet spin states is decreased by less efficient hyperfine coupling from Co(II).[8] The rate of RP recombination following picosecond laser flash photolysis of adenosylcob(III)alamin increases by up to fourfold at 500 G.[6,19] The magnetic field dependence of RP recombination following the photolysis of alkylcob(III)-alamins shows the utility of this technique to study B_{12}-dependent enzymatic reactions. A quantitative model of magnetic field-dependent recombination has been applied to continuous-wave photolysis of the B_{12} cofactor.[8]

Magnetic Field Dependence of Coenzyme B_{12}-Dependent Ethanolamine Ammonia Lyase

Ethanolamine ammonia lyase from *Salmonella typhimurium* is a prototypical B_{12}-dependent enzyme for which radical intermediates have been observed by ESR spectroscopy.[19,20] The proposed reaction mechanism is shown in Fig. 6.[21] The C–Co bond of adenosylcob(III)alamin undergoes homolysis to produce the 5′-deoxyadenosyl radical (Ado·) and cob(II)-alamin radical pair $\{Ado\cdot\ Cbl^{II}\}$. This activation step is reversible and will be reflected in the kinetic parameter V_{max}/K_m, but not V_{max}. Accordingly, V_{max}/K_m decreases with applied magnetic field, but V_{max} is insensitive to magnetic field (Fig. 7).[22] If the substrate ethanolamine is deuterated, hydro-

[19] W. B. Lott, A. M. Chagovetz, and C. B. Grissom, *J. Am. Chem. Soc.* **117,** 12194 (1995).
[20] B. M. Babior, *in* "B_{12}" (D. Dolphin, ed.), Vol. 2, p. 263. John Wiley & Sons, New York, 1982.
[21] T. T. Harkins and C. B. Grissom, *J. Am. Chem. Soc.* **117,** 566 (1995).
[22] T. T. Harkins and C. B. Grissom, *Science* **263,** 958 (1994).

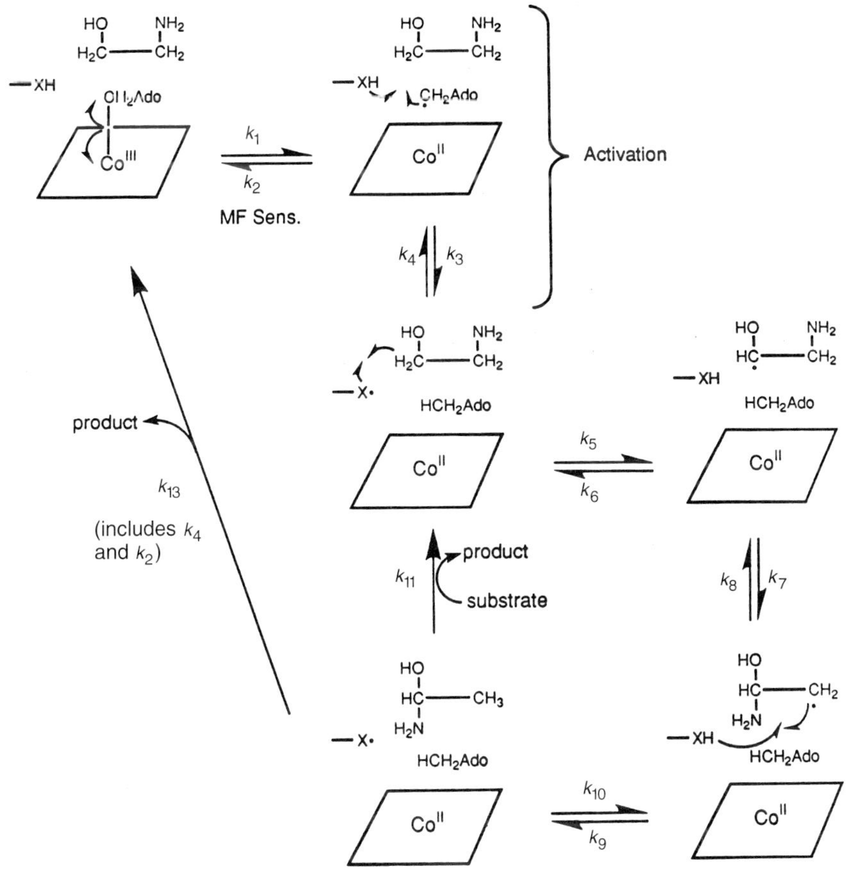

FIG. 6. Reaction mechanism for ethanolamine ammonia lyase. Enzyme-induced homolysis of the C–Co bond produces the 5′-deoxyadenosyl radical and cob(II)alamin in the singlet spin state. The 5′-deoxyadenosyl radical abstracts H· from C-2 of ethanolamine to generate the initial substrate radical. The amine group migrates to form the carbinolamine radical that abstracts H· from 5′-CH_3-adenosine to produce the hydrolytically unstable carbinolamine product and regenerate the 5′-deoxyadenosyl radical. Under V_{max} conditions, the enzyme always has ethanolamine bound and the $\{\cdot CH_2Ado:Cbl^{II}\}$ radical pair does not have to recombine between turnover (k_{11} includes product dissociation) and substrate binding before $\{\cdot CH_2Ado:Cbl^{II}\}$ recombination occurs. Under V_{max}/K_m conditions, recombination of the $\{\cdot CH_2Ado:Cbl^{II}\}$ radical pair (k_9) is more likely. This would begin the catalytic cycle with the transient $\{\cdot CH_2Ado:Cbl^{II}\}$ radical pair in the singlet spin state. [Reprinted from T. T. Harkins and C. B. Grissom, *J. Am. Chem. Soc.* **117,** 566 (1995). Copyright 1995 American Chemical Society.]

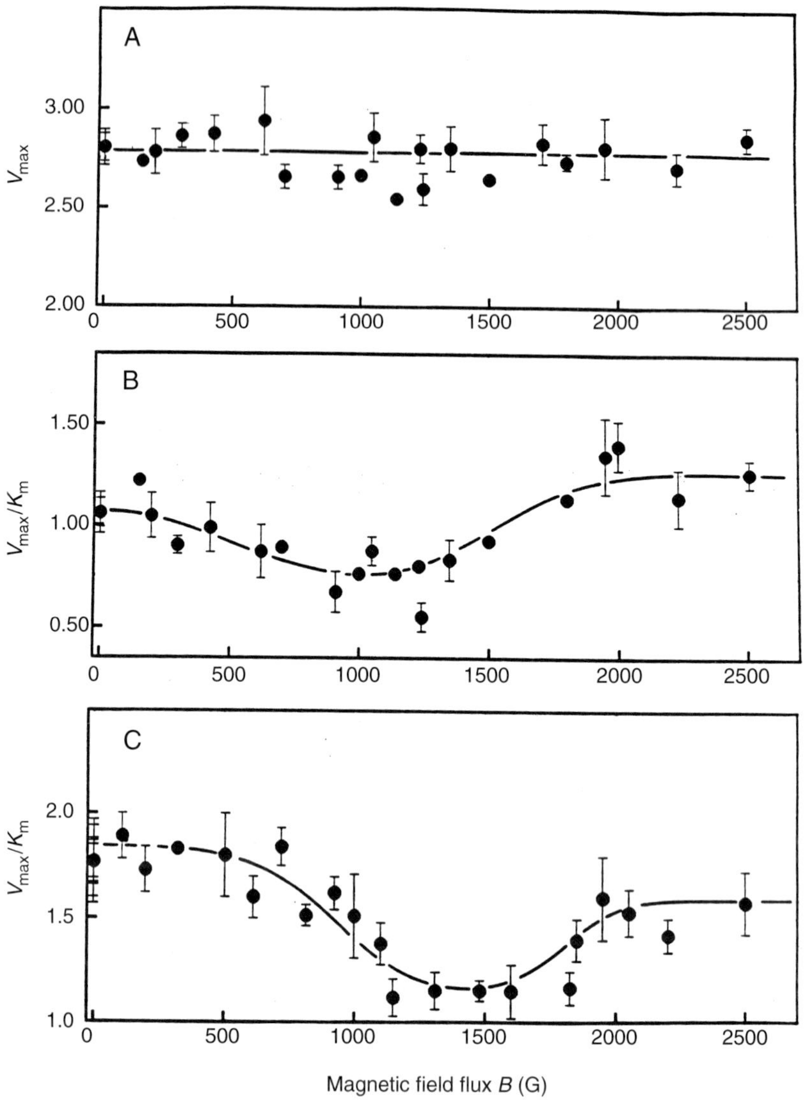

FIG. 7. Magnetic field dependence of ethanolamine ammonia lyase with: (A) V_{max} with unlabeled ethanolamine; (B) V_{max}/K_m with unlabeled ethanolamine; (C) V_{max}/K_m with [1,1,2,2-^{2}H]ethanolamine. Each assay contained 100 mM N-2-hydroxyethylpiperazine-N'-2-ethanesulfonic acid (HEPES, pH 7.48), 5 μM adenosylcob(III)alamin, and ethanolamine ammonia lyase at 25°. Each data point represents the kinetic parameter derived by fitting observed $d[P]/dt$ vs [ethanolamine] data to $d[P]/dt = V_{max}[S]^n/K_m + [S]^n$ by nonlinear methods. The Hill number, n, varied only slightly between 0.75 and 0.85. To keep the measured rates with deuterated and unlabeled substrates similar, 8.59-fold more EAL enzyme was used in assays with deuterated ethanolamine than in assays with unlabeled ethanolamine. This yields an observed kinetic isotope effect of ${}^{D}V_{max} = 6.8 \pm 0.2$ and ${}^{D}V_{max}/K_m = 5.4 \pm 0.4$ at 0 G. [Reprinted from T. T. Harkins and C. B. Grissom, *Science* **263,** 958 (1994). Copyright 1994 American Association for the Advancement of Science.]

gen atom abstraction from substrate (k_5) is slowed, and an even larger dependence of V_{max}/K_m on magnetic field is observed. The C–Co bond homolysis step can be monitored directly by rapid-scanning stopped-flow spectrophotometry.[21] As expected, the net rate of C–Co bond homolysis decreases as the magnetic field is increased, owing to a decrease in ISC from the singlet state to the triplet spin state and an increase in RP recombination. This results in an overall decrease in the catalytic efficiency (V_{max}/K_m) of the reaction.[22]

Conclusions

Magnetic field effects provide a new method with which to study biological reactions with radical pair intermediates. Coenzyme B_{12}-dependent enzymes with radical pair intermediates are well suited for study by this technique, as the RP is formed reversibly and the hyperfine interactions from the paramagnetic Co(II) provide a mechanism by which magnetic field-dependent intersystem crossing can occur. Not all enzymes with radical intermediates will exhibit magnetic field-dependent chemistry—only the subset of enzymes with a pair of radicals.

[28] Analysis of Cobalamin and Cobalamin Analogs by Gas Chromatography–Mass Spectrometry

By DAVID P. SUNDIN and ROBERT H. ALLEN

Introduction

The natural production of cobalamin (Cbl, vitamin B_{12}) occurs exclusively in microorganisms. In addition, naturally occurring analogs of Cbl are common in nature as a result of microorganismal synthesis.[1,2] As shown in Fig. 1, the only difference between native Cbl and these analogs lies in the lower axial ligand or nucleotide loop portion of the molecule. Cobalamin contains 5,6-dimethylbenzimidazole in the nucleotide of the lower axial ligand while a number of other bases are substituted in the naturally occurring Cbl analogs.

In the past, studies involving Cbl and Cbl analogs have used the intact molecules for analyses. High-performance liquid chromatography (HPLC)

[1] H. P. C. Hogenkamp, *Annu. Rev. Biochem.* **37,** 225 (1968).
[2] H. A. Barker, *Annu. Rev. Biochem.* **41,** 55 (1972).

0076-6879/97 $25

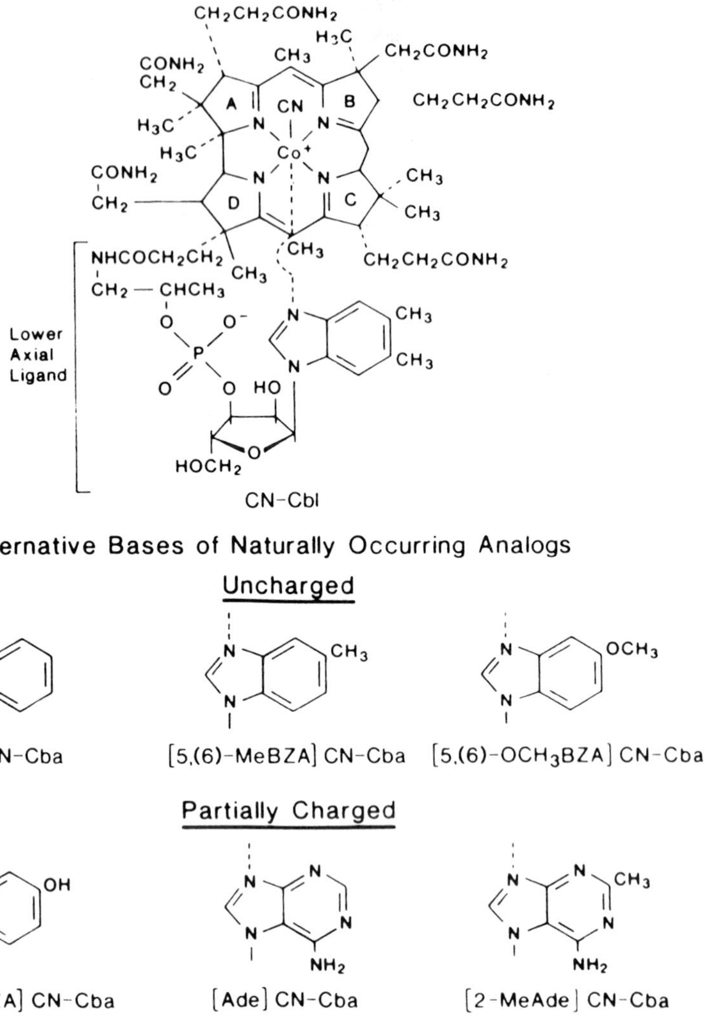

FIG. 1. Structure of CN-Cbl (*top*) and partial structures of six naturally occurring analogs (*bottom*). Only the portions of the analogs that differ from native Cbl are shown (i.e., the base of the nucleoside in the lower axial ligand). The top three bases are uncharged at neutral pH, while the bottom three are partially negatively charged in the case of 5(6)-OHBZA, and partially positively charged in the case of the two adenine-containing bases. (Reprinted from Ref. 4, with permission.)

is useful in this regard, but marked similarities in absorption spectra and overlapping of retention times have resulted in ambiguities of analog identification.[3] This chapter describes a methodology[4] for analyzing the nucleoside of the lower axial ligand separate from the corrin ring portion of the Cbl and Cbl analog molecules.

Assay Method

Principle

In this assay method,[4] cerous hydroxide is utilized to cleave the phosphodiester bonds on either side of the phosphate group in Cbl and Cbl analogs.[5] This disrupts the cobalt/nitrogen coordination producing cobinimide, inorganic phosphate, and the nucleoside. The nucleosides are then isolated using mixed-bed columns containing C_{18}/CM-Sephadex. Samples are applied to the mixed-bed columns at low pH to ensure that the hydroxyl group of the nucleoside from [5(6)hydroxybenzimidazole]cyanocobamide ([5(6)-OHBZA]CN-Cba) is not ionized and is therefore retained by the C_{18} portion of the column, together with the benzimidazole-containing nucleosides, and that the amino groups of the nucleosides from [adenine] cyanocobamide ([Ade]CN-Cba) and [2-methyladenine]cyanocobamide ([2-MeAde]CN-Cba) are fully protonated and therefore retained by the cationic CM-Sephadex portion of the column.

Attempts to form the *tert*-butyldimethylsilyl derivatives of the intact nucleosides were unsuccessful because of steric hindrance resulting from hydroxyl groups on the adjacent 2′ and 3′ carbon atoms of the ribose moiety. This problem is overcome by the periodate oxidation–cleavage/borohydride reduction scheme[6,7] that is illustrated in Fig. 2. The treated nucleosides are then reisolated using the same mixed-bed column technique mentioned above and are now readily converted to their *tert*-butyldimethylsilyl derivatives, which can now be analyzed using gas chromatography–mass spectrometry (GC–MS).

Reagents and Materials

C_{18} (120-Å pore size, 63-210 mesh, YMC, Inc., Morris Plains, NJ) CM-Sephadex (C-25, 40–120 μm; Pharmacia, Piscataway, NJ)

Polypropylene 0.9 × 6.3 cm (4 ml) disposable columns with a 20-μm frit

[3] M. J. Binder, J. F. Kolhouse, K. C. Van Horne, and R. H. Allen, *Anal. Biochem.* **125,** 253 (1982).

[4] D. P. Sundin and R. H. Allen, *Arch. Biochem. Biophys.* **298,** 658 (1992).

[5] W. Fredrich and K. Bernauer, *Chem.* (*Berlin*) **89,** 2507 (1956).

[6] G. Schmidt, *Methods Enzymol.* **XII B,** 230 (1968).

[7] S. H. Leppla, B. Bjoraker, and R. M. Bock, *Methods Enzymol.* **XII B,** 236 (1968).

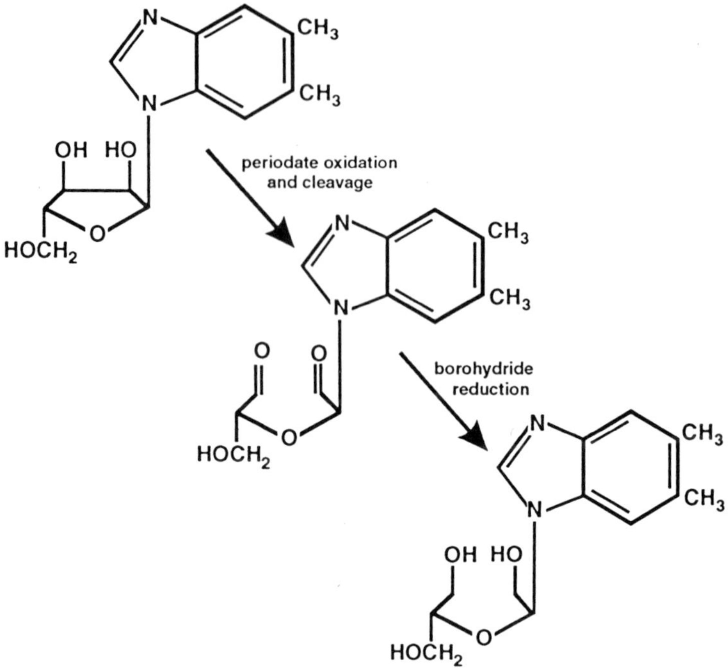

FIG. 2. Scheme for periodate oxidation and cleavage and borohydride reduction of the nucleoside in the lower axial ligand of CN-Cbl. It is the same for Cbl analogs except that different bases are present in the nucleoside.

HCN, 1% (w/v)
HCN, 0.1% (w/v)
HCl, 0.17 *N*
NH_4OH, 6.67% (w/v) in H_2O
NH_4OH, 0.3% (w/v) in H_2O
NH_4OH, 4 *N* in CH_3OH
Acetic acid, 2% (v/v)
$CeNO_3$, 64 mg
Sodium periodate, 8 m*M*, in 0.015% (w/v) NH_4OH
Sodium borohydride, 40 m*M*
N-Methyl-(*tert*-butyldimethylsilyl)trifluoroacetamide (MTBSTFA), mixed 1:2 (v/v) with acetonitrile

Preparations

CM-Sephadex in Hydrogen Ion Form. Approximately 35 g of the sodium form is washed with 1.2 liters of 0.17 *N* HCl, followed by a 500-ml H_2O wash. The yield is approximately 200 ml of fully hydrated material.

C_{18}/CM-Sephadex Columns. C_{18}, 0.5 cm in height, is measured into a 0.9 × 6.3 cm disposable column and an additional 20-μm frit is placed on top of the C_{18} to keep it in place. The C_{18} is activated by washing with 4 ml of methanol, followed by a 4 ml H_2O wash. A 0.6-ml aliquot of a 50 : 50 (v/v) mix of CM-Sephadex (in the hydrogen ion form) : H_2O is pipetted on top of the C_{18} and washed with 4 ml of H_2O.

Cerous Hydroxide Suspension. Cerous nitrate, 64 mg, is weighed into a 1.5-ml centrifuge tube. NH_4OH [750 μl of a 6.67% (w/v) solution in H_2O] is added and the solution is mixed well and then centrifuged at 1000 *g* for 2 min in a microcentrifuge. The supernatant is discarded and the pellet resuspended in 1 ml of 0.3% (w/v) NH_4OH followed by centrifugation as described above. This washing step is repeated three more times. After the last wash, the pellet is brought up in 1 ml of H_2O and 20 μl of a 1% (w/v) HCN solution is added. This suspension should be prepared fresh each time it is used.

HCN (1%, w/v). Prepare 1% (w/v) HCN in a fume hood by passing 4 ml of a 300-mg/ml KCN solution over a 1.0 × 30 cm column containing approximately 20 ml of Dowex 50W-X8 (BioRad Laboratories, Richmond, CA) in the hydrogen ion form. HCN is eluted by washing with H_2O. The first 12.5 ml of the wash is discarded and the next 50 ml is collected as 1% (w/v) HCN. After collection, the HCN is kept at −20° until used.

Procedure

Cerous Hydroxide Hydrolysis. Cobalamin or Cbl analog (5 μg) is dried by vacuum centrifugation in 1.5-ml lock-top microcentrifuge tubes. Cerous hydroxide suspension (50 μl) is then aliquotted onto the dried Cbl and Cbl analog samples and they are incubated for 45 min at 90°. Samples are mixed vigorously every 15 min. After cooling to room temperature, 1 ml of 2% (v/v) acetic acid is added followed by mixing and centrifugation for 2 min in a microcentrifuge. Supernatants are then removed and applied to the individual C_{18}/CM-Sephadex columns prepared as previously described. Each column is washed with 4 ml of 0.1% (w/v) HCN, followed by a 4-ml H_2O wash. Nucleosides are eluted into 4 ml polypropylene conical tubes with two 0.55-ml aliquots of 4 *N* NH_4OH in methanol (prepared by diluting stock 14.8 *N* NH_4OH in methanol), and dried by vacuum centrifugation.

Oxidation/Reduction. For oxidation of the hydrolyzed nucleosides, 50 μl of 8 m*M* sodium periodate in 0.015% (w/v) NH_4OH is added to the dried samples from above. They are incubated for 30 min at 25° in the dark, with mixing every 10 min. Reduction is then performed by adding 50 μl of 40 mM sodium borohydride followed by a 1-hr incubation at 25°. Samples are mixed every 20 min and after 1 hr, 1 ml of 2% (v/v) acetic

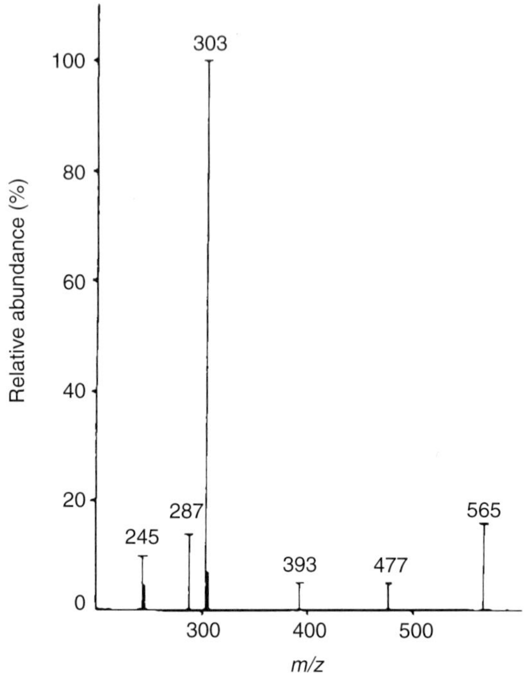

FIG. 3. Spectrum of the nucleoside from CN-Cbl after it was oxidized, reduced, and derivatized as described in text. (Reprinted from Ref. 4, with permission.)

acid is added. The samples are mixed and spun in a microcentrifuge as previously described. The supernatants are applied to C_{18}/CM-Sephadex columns and eluted (into 1-ml tapered glass autosampler vials) and dried as previously described.

Derivatization. The *tert*-butyldimethylsilyl derivatives of the oxidized/reduced nucleosides are prepared by adding 30 μl of a 1:2 mixture of MTBSTFA and acetonitrile to the dried samples, capping the vials, and incubating for 2 hr at 90°.

Gas Chromatography–Mass Spectrometry. Analysis of derivatized samples is performed on a Hewlett-Packard (Palo Alto, CA) 5890 GC/5970 MS equipped with a 7673A autosampler system. A Supelco (Bellefonte, PA) SPB-1 (nonpolar, methyl silicone) GC column is used for all experiments. The column is 10 m in length with an internal diameter of 0.25 mm and a film thickness of 0.25 μm. All experiments are carried out under standard autotune conditions with an injection port temperature of 300° and a column head pressure of 7.5 psi. Initial temperature of the gas chromatograph oven is 80°, which is held for 1 min after sample injection. The

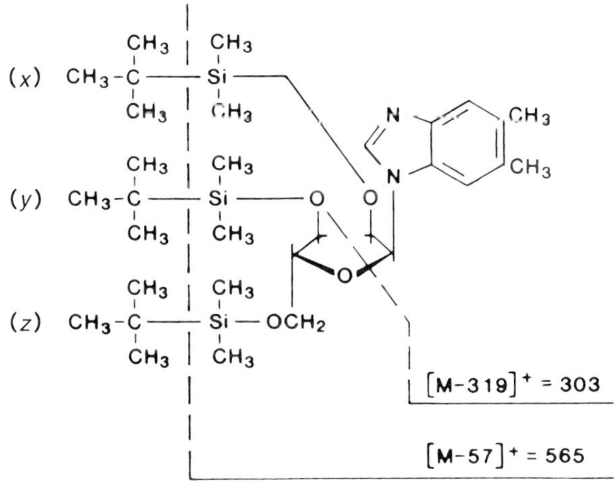

FIG. 4. Structure of the fully derivatized, oxidized, and reduced 5,6-dimethylbenzimidazole nucleoside, showing the fragmentation points and *m/z* values of the major distinguishing ions. The M−57 ion may be formed by loss of the *tert*-butyl group at any one of the positions labeled *x*, *y*, and *z*. (Reprinted from Ref. 4, with permission.)

temperature is then increased at a rate of 30°/min until 300° is reached, where it is again held for 1 min. Data are collected from 6 to 10 min, using a single-ion monitor (SIM) program designed to monitor the ions with *m/z* values equal to those of the molecular ions minus 57 (M−57) and minus 319 (M−319), which are specific for Cbl and each Cbl analog (see the next section).

Analysis

Figure 3 shows the observed mass spectrum, and Fig. 4 shows the proposed structure and fragmentation points of the two most abundant ions of the fully derivatized 5,6-dimethylbenzimidazole nucleoside of native Cbl. The large ion 565 represents the fully derivatized, M−57 form of the nucleoside. This form represents a species in which a *tert*-butyl fragment (atomic mass of 57) has been lost at any one of the positions *x*, *y*, or *z*. The smaller ion 303 represents a form of the nucleoside in which the ribose ring has been cleaved, with a fragment of the ribose ring remaining attached to the base portion of the nucleoside. It was this ribose fragment/base form (M−319) that was consistent and was monitored with the M−57 form. The M−57 and M−319 ions proved to be descriptive of each nucleoside, the only difference in the atomic mass being attributed to variation in the base of the nucleoside.

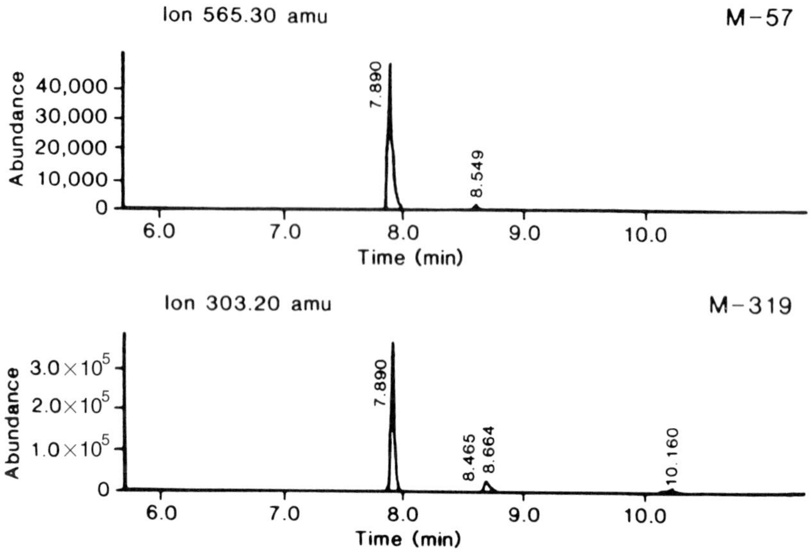

FIG. 5. Chromatograms of the oxidized, reduced, and derivatized nucleoside derived from CN-Cbl. (Reprinted from Ref. 4, with permission.)

Figure 5 shows actual chromatograms of the oxidized, reduced, and derivatized nucleoside from CN-Cbl in which ions 565 and 303 were monitored with the corresponding retention times of 7.890 min. These data and those obtained for the nucleosides present in the six naturally occurring

TABLE I
MAJOR IONS AND RETENTION TIMES OF NUCLEOSIDES ISOLATED FROM COBALAMIN AND SIX NATURALLY OCCURRING ANALOGS[a,b]

Base present in nucleoside	Retention time on gas chromatography (min)	Mass spectrometer fragmentation ions	
		M−57	M−319
Benzimidazole	7.59	537	275
5(6)-Methylbenzimidazole	7.72	551	289
5,6-Dimethylbenzimidazole	7.89	565	303
5(6)-Methoxybenzimidazole	8.03	567	305
5(6)-Hydroxybenzimidazole	8.68	667	405
α-Adenosine	9.16	668	406
α-2-Methyladenosine	8.96	682	420

[a] Reprinted from Ref. 4, with permission.

[b] Nucleosides were isolated, oxidized, reduced, and derivatized as described in text.

Cbl analogs shown in Fig. 1, are presented in Table I. Each of the seven nucleosides give distinct retention times and distinct values for the M−57 and M−319 ions.

The most common sources of naturally occurring adenosine, such as ATP and RNA, contain β-adenosine rather than the α-adenosine present in [Ade]CN-Cba. Although both forms of adenosine give the same values for the fragmentation ions 668 for M−57 and 406 for M−319, they are readily distinguishable with this method of analysis because the β-adenosine derivative elutes at 8.61 min, which is much earlier than the 9.16 min required for α-adenosine.

It is possible to incorporate stable isotope label into the lower axial ligand of Cbl and Cbl analogs by guided biosynthesis[8,9] of *Propionibacterium shermanii* (ATCC 9615; American Type Culture Collection, Rockville, MD) using deuterated bases. Using [D_2]CN-Cbl, for example, it is possible to quantitate the amount of Cbl from various sources using the stable isotope dilution technique. This procedure has the potential to serve as a gold standard for radioisotope dilution methods that are used to assay Cbl in human plasma and other sources.

[8] D. Perlman and J. M. Barrett, *Can. J. Microbiol.* **4,** 9 (1958).
[9] J. F. Kolhouse and R. H. Allen, *Anal. Biochem.* **84,** 486 (1978).

[29] Expression of Functional Intrinsic Factor Using Recombinant Baculovirus

By MARILYN M. GORDON, GREG RUSSELL-JONES, and DAVID H. ALPERS

Introduction

Intrinsic factor (IF) is produced in foregut tissues of mammals and mediates the uptake of cobalamin (Cbl, vitamin B_{12}). It is a protein with unusual properties, being soluble and functional at very acidic pH (pH <3.0). Intrinsic factor content in gastric mucosa is low for two reasons: first, IF is produced in only a subpopulation of cells (parietal or chief, depending on the species), and second, the majority of IF produced is rapidly secreted. When gastric mucosa is the source for IF purification, large amounts of tissue are needed. When gastric secretions are used, large-volume collections are needed after pharmacological stimulation of secretion, coupled with continuous neutralization of the fluid to delay inacti-

0076-6879/97 $25

vation of the protein. Thus, to obtain sufficient IF for structural and metabolic studies an *in vitro* expression system would be desirable.

A number of systems have been tried and found inadequate for production of even microgram amounts of IF. No cell line derived from gastric carcinoma or normal mucosa has been found to produce IF. Gastric mucosal explants secrete only nanogram quantities, similar to the yield from COS-1 cells transfected with the IF cDNA.[1] Cell factories could be used to convert the limited production per milliliter in culture media of transfected mammalian cells, but these are not convenient for testing mutated versions of IF. Many different prokaryotic expression systems have been tested by the authors, and in none of them was any detectable IF produced, as assayed either by incorporation of labeled amino acids or by Western blot. Recombinant baculovirus was chosen as a system that provided flexibility and convenience of use.

Insertion into Vectors

Since the introduction of baculovirus for *in vivo* protein production there have been numerous improvements in transfer vectors and expression systems. These improvements have included reducing the size of the transfer vector, providing more restriction sites for the insertion of foreign DNA, and creating systems for increasing the percentage of recombinants. Some of the newest vectors simplify the recombinant protein purification by inserting a tag sequence and a protease cleavage site upstream from the multiple cloning site, which will produce a fusion protein that can be purified on an affinity column.[2] Such a tag sequence can be removed proteolytically from the fusion protein, but the final product will contain an amino acid extension corresponding to the length of the unused multiple cloning site. This capability may have limited value for recombinant IF purification, because affinity chromatography using Cbl–agarose is readily available. However, use of a tag sequence would allow IF purification without denaturation with guanidine.

The initial rat and human IF cDNA constructs were made using the transfer vector pBlueBac (Invitrogen, San Diego, CA).[3] This transfer vector contains twin promoters derived from the *Autographa californica* nuclear polyhedrosis virus (AcNPV).[3] The ETL (early-to-late) promoter directs

[1] M. Gordon, C. Hu, H. Chokshi, J. E. Hewitt, and D. H. Alpers, *Am. J. Physiol.* **260** (*Gastrointest. Liver Physiol.* **23**), G736 (1991).

[2] D. Polayes, R. Harris, D. Anderson, and V. Ciccarone, *Focus* **18,** 10 (1996).

[3] J. Vialard, M. Lalumiere, T. Vernet, D. Briedis, G. Alkhatib, D. Henning, D. Levin, and C. Richardson, *J. Virol.* **64,** 37 (1990).

the synthesis of β-galactosidase and the polyhedrin promoter controls the synthesis of the cloned gene. Recombinant virus produces blue plaques when X-Gal (5-bromo-4-chloro-3-indoyl-β-D-galactopyranoside) is included in the agarose overlay. These blue plaques are easy to distinguish, but two or three rounds of plaque purification may be required to separate plaques containing recombinant virus from those with wild-type virus. There are a number of manuals available for using the baculovirus system, and the reader is referred to these for an introduction to the use of these systems.[4,5] Companies that supply the vectors also provide specific information for using their proprietary material.

All of our more recent IF baculovirus constructs have been made using a system developed by Monsanto Corp. and licensed to Life Technologies (Gaithersburg, MD) under the name Bac-to-Bac. This system is based on the site-specific transposition of an expression cassette from a donor plasmid (pFastBac) into a baculovirus shuttle vector grown in *Escherichia coli.* The expression cassette contains the polyhedrin promoter from AcNPV and a multiple cloning site downstream from the promoter. The plasmid with the inserted gene is used to transform DH10Bac (GIBCO-BRL) containing the parent baculovirus shuttle vector (bacmid) and a helper plasmid supplying transposition proteins required for recombination. The useful aspect of this site-specific transposition is that it takes place in *E. coli* and provides *lacZ* complementation, so that recombinant bacmids produce white colonies and nonrecombinants produce blue colonies in the presence of X-Gal. High molecular weight recombinant bacmid DNA is then isolated from the *E. coli* and used to transfect Sf9 (*Spodoptera frugiperda* fall armyworm ovary) cells. Because the recombinant selection is performed in *E. coli* instead of by plaque development, it can be accomplished more rapidly, making the system responsive to expression of mutations for structure–function studies.

Infection

Cell lines derived from the fall armyworm *Spodoptera frugiperda* (Sf9) have been grown in Grace's medium containing lactalbumin hydrolysate and yeastolate (each 3.3 g/liter) and supplemented with 10% heat-inactivated fetal calf serum and penicillin and streptomycin (each 100 units/ml). The use of serum-containing medium does not pose a problem, because IF is purified by affinity column chromatography. Sf9 cells are grown at

[4] M. D. Summers and G. E. Smith, *Texas Agric. Exp. Station Bull. No. 1555* (1987).

[5] C. I. Murphy and H. Piwnica-Worms, *in* "Current Protocols in Molecular Biology" (F. M. Ausubel *et al.*, eds.), Ch. 16, Section II. John Wiley & Sons, New York, 1994.

27° and do not require CO_2. Sf9 cells are not anchorage dependent, and can be transferred between monolayer and suspension cultures. Further information on the growth of these cells and the production and titering of viral stocks is available in manuals[4,5] and in the instructional material provided with expression systems sold by biotechnology companies.

The first evaluation of newly selected recombinant clones for their ability to produce IF is performed by infecting cells in monolayers. All subsequent IF production is done in suspension cultures. Sf9 cells at a density of 1–2 $\times$ 10^6 cells/ml are infected with recombinant baculovirus stock at a multiplicity of infection (MOI) of 1 plaque-forming unit (PFU)/cell. After infection IF production is monitored on aliquots of culture medium, using a scaled-down modification of the [^{57}Co]Cbl charcoal-binding assay.[6] Intrinsic factor can be detected in the medium at 48 hr postinfection, and continues until 4–5 days postinfection. Little increase in binding activity can be detected thereafter.

Modified Charcoal-Binding Assay

Charcoal-Binding Assay Materials

Albumin-coated charcoal: Prepare by adding 2.5 g of activated, acid-washed charcoal (BDH Chemicals Ltd., Poole, England) to 100 ml of water in which has been dissolved 500 mg of bovine serum albumin (BSA, fraction V; Sigma Chemical Co., St. Louis, MO). The solution is stirred on a magnetic stir platform for at least 1 hr before use

Cyano[^{57}Co]cobalamin: Intermediate specific activity; purchase from Johnson & Johnson Clinical Diagnostics (Markham, Ontario, Canada). A working solution is prepared by making a 1 : 700 (v/v) dilution in 10 m*M* Tris-HCl, pH 7.3, containing 1 mg of BSA/ml. This solution contains 1 μg of cyanocobalamin/ml

Diluent: 10 m*M* Tris-HCl, pH 7.3, containing 1 mg of BSA/ml

Binding Assay. All operations are performed in a 1.5-ml microcentrifuge tube. The final assay volume is 1.0 ml. The unknown sample is diluted to 0.4 ml with diluent. To the sample add 0.2 ml of the diluted cyano[^{57}Co]cobalamin. Mix vigorously on a vortex mixer, and incubate at room temperature for 5 min. Add 0.4 ml of the albumin-coated charcoal solution and mix vigorously. The uptake of free cyano[^{57}Co]cobalamin occurs rapidly, and thus after just a few minutes at room temperature the sample can be centrifuged at 16,000 *g* for 10 min. An aliquot of the supernatant fraction containing the IF-bound cyano[^{57}Co]cobalamin is counted in a γ counter.

[6] C. Gottlieb, K. S. Lau, L. R. Wasserman, and V. Herbert, *Blood* **25,** 875 (1965).

The molar ratio of IF:cobalamin is about 30:1, and the IF:cobalamin binding ratio is 1:1. Therefore, estimates of IF concentration can be made as follows: nanograms of IF in the sample equals nanograms of bound cyano[^{57}Co]cobalamin × 30.

Glycosylation of IF appears to be maintained by Sf9 cells during the first 24 hr of infection (M_r 46,000), but the bulk of IF production occurs from days 2 to 5 postinfection. During that time period the relative molecular weight of IF is 43,000, the size of the nonglycosylated protein.[6] Thus, the majority of recombinant IF produced in the baculovirus system is nonglycosylated. Insect cells do not often add the larger complex oligosaccharide side chains seen in glycoproteins of higher eukaryotes, but they do provide trimmed high-mannose units. Glycosylation of IF only during the first 24 hr could be due either to efficient glycosylation only when expression of recombinant protein is at a low level, or to direct inhibition of host cell glycosylation as infection proceeds. The composition of the carbohydrate side chains of baculovirus-derived recombinant IF is not known, but it is probably different from that of native gastric IF. Despite this difference, the functional characteristics of the purified recombinant IF are identical to those of native IF (see Yield and Properties of Recombinant Intrinsic Factor).

Purification

At 4–5 days postinfection the suspension culture producing recombinant IF is ready for purification. The amount of IF remaining in intact Sf9 cells is not sufficient to make recovery from the cells worth the effort.

Buffers

Wash I: 100 mM Tris HCl, pH 7.3, containing 2 mM sodium azide
Wash II: 100 mM glucose, 100 mM glycine, 1 M NaCl, pH 10.0
Wash III: 1 M NaCl, 2% (v/v) Triton X-100, 100 mM Tris-HCl, pH 7.3

The culture medium is centrifuged at 17,000 g for 20 min at 4°. The clarified medium from 2–3 liters of virus-infected cells is pooled and applied to a Cbl (B_{12})–agarose column (25 ml of packed gel; Sigma), packed in a 2.5 × 20 cm Econo-column (Bio-Rad, Hercules, CA). All column operations are carried out at 4° and the column is wrapped in aluminum foil to protect the light-sensitive Cbl. Typically, a flow rate of 1 ml/min has been used when loading, washing, and eluting the column. After loading the column with medium containing the recombinant IF, it is washed with a series of buffers described at the beginning of this section. We use 200 ml of wash

I, 200 ml of wash II, 200 ml of wash I, 200 ml of wash III, and finally 200 ml of wash I. Initially wash III containing Triton X-100 was not included in the procedure, but a 65-kDa baculovirus-derived protein copurified with IF under these conditions.[7] Addition of the step using wash III removes the contaminating protein. The column can be used numerous times for purifying the identical protein, but should be washed extensively with wash I between runs. When a new column is prepared, it is washed with the same sequence of buffers, although in somewhat smaller volume, before addition of the recombinant proteins.

Intrinsic factor is removed from the Cbl–agarose column by adding 20 ml of 4 *M* guanidine hydrochloride in 100 m*M* Tris, pH 7.0, and maintaining contact with the adsorbed protein for 3–4 hr, allowing it to remove and denature the IF. To renature the IF, the column eluate is dialyzed against a large volume (8 liters) of water at 4° for at least 48 hr. The process of dialysis continues as the sample is concentrated approximately 20-fold by vacuum dialysis in a collodion membrane (Schleicher & Schuell, Keene, NH). Because there are minor contaminating proteins that elute in variable amounts from the affinity column, the concentrated solution is then applied to either a Superose 12 HR 10/30 or a Superdex 75 HR 10/30 column (the latter is somewhat better) in an FPLC system (Pharmacia-LKB, Piscataway, NJ), using 50 m*M* sodium phosphate, pH 7.4, with 150 m*M* NaCl as the eluant solution. Such a purification results in a single major protein peak eluting from the FPLC column, and this peak contains all of the Cbl-binding activity. Multiple column runs are required to accommodate the 1–2 ml of concentrated eluant from the affinity column. After pooling, the column fractions containing IF are concentrated by vacuum dialysis. Because the degree of Cbl binding to IF in the charcoal assay is not linearly related to the amount of IF added when large dilutions of concentrated IF are used, we have found it necessary to use amino acid analysis or densitometric scans of stained sodium dodecyl sulfate-polyacrylamide gel electrophoresis (SDS–PAGE) gels to determine the final yield of IF.

Yield and Properties of Recombinant Intrinsic Factor

The yield of recombinant human IF is modest, but reproducible. The protein is expressed at concentrations varying from 1 to 3 mg/liter of medium. Recovery of human IF during purification is quite good, varying from 25 to 35%, and comparable to the yield of native IF recovered from rat stomach. For reasons that are unclear, the production of recombinant rat IF is much lower and more variable than that of human IF, although

[7] M. Gordon, H. Chokshi, and D. H. Alpers, *Biochim. Biophys. Acta* **1132,** 276 (1992).

80% of the amino acid residues are identical in those proteins. Recovery of recombinant and functional IF might be improved by the use of a tag sequence and nondenaturing elution from an affinity column,[2] but this approach has not yet been used by the authors.

The functional properties of the purified recombinant IF are indistinguishable from those of native IF. This result is consistent with the observation that nonglycosylated IF is functionally identical with native IF, except for its increased sensitivity to proteolysis.[1] The K_a of [^{57}Co]Cbl binding to recombinant human IF is $2\text{–}3 \times 10^{10}\ M^{-1}$. The K_a of the [^{57}Co]Cbl–IF complex binding to guinea pig brush borders was $2\text{–}4 \times 10^{10}\ M^{-1}$. Despite the fact that much of the recombinant IF was either not glycosylated or was incompletely glycosylated, the sensitivity to proteases and pH was the same as for native IF.[6] Moreover, as for native IF, saturation with Cbl stabilizes IF even more against the action of proteases. Recombinant IF is relatively stable when kept frozen at $-70°$.

[30] Quantitative Methods for Measurement of Transcobalamin II

By SHELDON P. ROTHENBERG and EDWARD V. QUADROS

Transcobalamin II (TCII) is a plasma protein that binds and transports cobalamin (Cbl, vitamin B_{12}) to tissues where cellular uptake of the TCII–Cbl complex occurs by receptor-mediated endocytosis. Mutations of the TCII gene result in either a complete deficiency of TCII or expression of a functionally abnormal protein and infants affected by such genetic mutations develop clinical evidence of Cbl deficiency within several months following birth.

Studies to characterize the structure–function relationship of TCII have been difficult because the plasma concentration of this protein is low (~0.4–1.1 n*M*). Even with affinity chromatography using Cbl as the binding ligand, only ~3 mg of TCII can be purified from 10 kg of Cohn fraction III of human plasma.

In this chapter analytical methods are described to quantify TCII in plasma and other biological fluids. Molecular methods are provided in [31] in this volume.[1] In addition, a recombinant baculovirus containing the TCII complementary DNA (cDNA) has been prepared to generate recombinant

[1] L. Qian, E. V. Quadros, and S. P. Rothenberg, *Methods Enzymol.* **281,** [31], 1997 (this volume).

0076-6879/97 $25

TCII in *Spodoptera frugiperda* fall armyworm ovary cells (Sf9). The recombinant TCII has the functional and immunoreactive properties of native TCII.

Quantitative Analysis of Transcobalamin II in Biological Fluids

Measurement of Apotranscobalamin II by Saturation with [^{57}Co]Cobalamin

Principle. There are two major plasma proteins that bind Cbl: TCII and transcobalamin I (TCI). A third Cbl-binding protein, transcobalamin III (TCIII), is a minor component of plasma that comprises a small fraction of protein-bound Cbl because it is rapidly cleared from plasma by the hepatic asialoglycoprotein receptor pathway. Although TCIII and TCI cross-react with an antiserum to TCI, they differ in carbohydrate composition.

Most of the unsaturated Cbl-binding protein in normal plasma is apo-TCII. To quantify apo-TCII in plasma, sufficient [^{57}Co]Cbl is added to an aliquot of plasma to saturate both TCII and TCI (this equals total Cbl-binding capacity of plasma). The unbound [^{57}Co]Cbl is removed by adsorption to protein-coated charcoal. The bound [^{57}Co]Cbl remaining in the plasma is the total Cbl-binding capacity [apo-TCII + apo-TCI). The plasma is then treated with QUSO G-32 (a silica preparation; Philadelphia Quartz Co., Philadelphia, PA), which adsorbs only the TCII–[^{57}Co]Cbl. The remaining [^{57}Co]Cbl is bound to TCI (and any TCIII) in the sample. Therefore,

Total [^{57}Co]Cbl bound − bound [^{57}Co]Cbl (in supernatant) after QUSO treatment = [^{57}Co]Cbl-TCII.

Because the binding of Cbl to TCII is equimolar, the molar concentration of apo-TCII in plasma can be computed from the moles of bound [^{57}Co]Cbl. The procedure described has been adapted from the method of Jacob and Herbert[2] and is used routinely in our laboratory.

Reagents

Sodium phosphate buffer (pH 7.4), 0.10 *M*
[^{57}Co]Cbl, specific activity greater than 200 μCi/μg, diluted to 25 pg/100 μl in water

[2] E. Jacob and V. Herbert, *J. Lab. Clin. Med.* **86,** 505 (1975).

Crystalline cyanocobalamin, prepared as stock solutions of 400 pg/100 μl and 25 ng/100 μl

Activated charcoal, acid washed and neutralized (Sigma, St. Louis, MO)

Bovine hemoglobin solution, 0.125% (w/v) in water

QUSO G-32 (Philadelphia Quartz Co.): Prepare a fresh suspension of 30 mg/ml of water

Blood collection tubes: Serum or plasma can be used to determine TCII in blood because blood clotting does not affect the concentration of this protein. However, if the method is also to be used to measure transcobalamin I in plasma, the blood should be collected in tubes containing EDTA and sodium fluoride (Becton Dickinson, Mountain View, CA) and the plasma quickly separated to minimize release of this protein from granulocytes

Procedure

1. Prepare a set of three tubes in duplicate:

   100 μl [^{57}Co]Cbl (25 pg) + 100 μl Cbl (400 pg) + 750 μl buffer + 50 μl sample

   100 μl [^{57}Co]Cbl (25 pg) + 100 μl Cbl (400 pg) + 700 μl buffer + 100 μl sample

   100 μl[^{57}Co]Cbl (25 pg) + 100 μl Cbl (400 pg) + 650 μl buffer + 150 μl sample

   Note: The sample to be assayed is always added last.
2. Prepare a charcoal blank as follows: 700 μl of buffer plus 100 μl of Cbl (25 ng) plus 100 μl of serum (or plasma). Incubate for 15 min and then add 100 μl of [^{57}Co]Cbl.
3. Prepare the tracer standard: 900 μl of buffer plus 100 μl of [^{57}Co]Cbl (25 pg).
4. Incubate the tubes for 15–30 min at room temperature.
5. Add 0.5 ml of the hemoglobin solution containing 2.5% (w/v) of the charcoal to all tubes, including the charcoal blank, but not the tracer standard. Vortex briefly and pellet the charcoal by centrifugation (2000 *g* for 10 min at 4°).
6. Determine the radioactivity (counts per minute) contained in 750 μl of the supernatant fraction. This is half the incubation volume. Determine the total radioactivity in the tracer standard by scintillation γ counting.
7. Compute the total protein-bound Cbl (Cbl bound to TCII and TCI) (plus TCIII) as follows:

$$\text{Total Cbl/ml} = \left[\frac{(\text{cpm in charcoal supernatant}) - (\text{charcoal blank})}{(\text{tracer standard cpm}) - (\text{counter background})}\right] \times 2 \times 425 \text{ pg} \times \left(\frac{1000\ \mu\text{l}}{\mu\text{l serum}}\right) \quad (1)$$

Notes: (a) The supernatant in Eq. (1) is one-half the assay reaction and, therefore, the net counts per minute is multiplied by 2. (b) The charcoal blank corrects for nonspecific binding and counter background. (c) The standard counts per minute is the total [^{57}Co]Cbl in the assay reaction. (d) The total Cbl in the assay reaction is 425 pg (25 pg of [^{57}Co]Cbl plus 400 pg of Cbl). (e) The calculation (1000 μl/μl serum) normalizes the total bound Cbl to 1 ml of serum.

8. Assay for TCII and TCI (plus TCIII): To the same 750 μl of incubation sample that was used to determine the total bound Cbl (i.e., charcoal supernatant), add 100 μl of the QUSO G-32 suspension. QUSO settles rapidly and it is necessary to maintain constant mixing with a magnetic stirring bar while dispensing the suspension. Vortex the samples and pellet the QUSO by centrifugation at 2000 *g* for 10 min at 4°.
9. Remove 425 μl of the supernatant fraction and determine the radioactivity. (This is one-half the new volume of 850 μl following the addition of the QUSO suspension.)
10. Compute the Cbl bound to TCI (plus TCIII) and TCII as follows:

$$\text{pg TCI (plus TCIII)-bound Cbl/ml} = \left[\frac{(\text{cpm in QUSO supernatant} \times 2) - (\text{counter background})}{\text{net cpm in charcoal supernatant}}\right] \times (\text{total bound Cbl/ml}) \quad (2)$$

Notes: (a) The counts per minute in the supernatant fraction of the QUSO-treated sample is multiplied by 2 to normalize to the total sample volume of 850 μl. (b) Net counts per minute in the charcoal supernatant is the corrected numerator from Eq. (1). (c) Total bound cobalamin per milliliter is the value determined in Eq. (1),

$$\text{pg TCII-bound Cbl/ml} = (\text{total pg bound Cbl}) - [\text{pg TCI (+ TCIII)-bound Cbl}] \quad (3)$$

where the total picograms of bound cobalamin per milliliter of serum is the value obtained from Eq. (1), and the picograms of TCI

TABLE I
ANALYTICAL PROCEDURES TO MEASURE APOCOBALAMIN-BINDING PROTEINS IN PLASMA

Step	Procedure	Result[a]
1. Total apo-Cbl-binding protein capacity	a. Saturate sample with [^{57}Co]Cbl + Cbl b. Adsorb free Cbl with charcoal suspension c. Determine bound Cbl in supernatant fraction	Total bound Cbl = total apo Cbl-binding proteins
2. Apo-TCI (+ TCIII)-binding capacity	Treat supernatant of step 1 with QUSO G-32	Remaining bound Cbl in supernatant = apo-TCI (+ TCIII)
3. Apo-TCII-binding capacity	Step 1 minus step 2	Total Cbl binding − apo-TCI (+ TCIII) = apo-TCII

[a] Apocobalamin-binding protein(s) is the molar equivalent of bound Cbl.

(plus TCIII)-bound cobalamin per milliliter of serum is the value obtained from Eq. (2). Separate values for the unsaturated binding capacity of TCIII and TCI may be obtained by a modification of the above assay and is based on the difference in ionic charge of the two proteins. Accordingly, a mixture of TCI and TCIII can be separated by chromatography through a small DEAE-cellulose column using a low-molarity neutral buffer. Alternatively, Jacob and Herbert[2] describe the addition of a suspension of DEAE-cellulose prepared in 0.02 *M* phosphate buffer, pH 6.3, to the mixture of holo-TCI (plus holo-TCIII). The DEAE-cellulose binds TCI and is pelleted by centrifugation, leaving the TCIII–[^{57}Co]Cbl in the supernatant fraction. A summary of the steps for measuring the Cbl-binding capacity of proteins in serum or other biological samples is presented in Table I.

11. For direct measurement of TCII when apo-TCII is the only Cbl-binding protein of interest (i.e., for studies measuring the secretion of TCII by cultured cells or the expression of recombinant TCII), it is only necessary to incubate the sample with sufficient Cbl (total of radiolabeled and unlabeled) to saturate the total Cbl-binding capacity as in step 1, and then to add the QUSO suspension to adsorb the TCII–Cbl. The QUSO is pelleted by centrifugation and all the supernatant fluid is carefully removed. The radioactivity in the pellet is determined and the apo-TCII is calculated as follows:

pg TCII-bound Cbl

$$= \left[\frac{(\text{cpm in QUSO pellet}) - (\text{counter background})}{(\text{total cpm in the assay}) - (\text{counter background})} \right] \times 425 \text{ pg} \quad (4)$$

Notes: (a) QUSO removes only TCII-bound Cbl. Less than 1% of free [^{57}Co]Cbl binds to QUSO. (b) Equation (4) is the fraction of total TCII–Cbl bound to the QUSO and in this example the total Cbl in the assay is 425 pg (25 pg of [^{57}Co]Cbl plus 400 pg of Cbl).

Measurement of Total Transcobalamin II (Apo- Plus Holotranscobalamin II)

Principle. A polyclonal antiserum has been raised in rabbits to holo-TCII purified from Cohn fraction III of human plasma.[3] The antiserum has high affinity for both apo- and holo-TCII and this provides a sensitive competitive radioimmunoassay (RIA); i.e., it can measure as little as 5–10 pg of TCII. The protocol for the assay is based on the principle described by Yalow and Berson for measuring insulin in plasma.[4] To measure TCII, recombinant TCII–[^{57}Co]Cbl serves as the radiolabeled tracer and unlabeled human TCII is the competing ligand with which to generate the standard curve.[5] An aliquot of human serum is assayed under identical conditions and the concentration of TCII is obtained by referring the value for antibody-bound tracer in the assay to the corresponding value of TCII on the standard curve.

Reagents

Assay buffer: Phosphate buffered saline (PBS, pH 7.4), 0.15 *M*, containing 0.025% (w/v) human serum albumin and 0.5% (v/v) Triton X-100

Staphylococcus aureus, 10% (w/v) membrane suspension (Omnisorb; Calbiochem, La Jolla, CA)

[^{57}Co]Cobalamin, specific activity greater than 200 μCi/μg

TCII for the standard curve[5]

[3] E. V. Quadros, S. P. Rothenberg, Y.-Ch. E. Pan, and S. Stein, *J. Biol Chem.* **261,** 15455 (1986).

[4] R. S. Yalow and S. A. Berson, *J. Clin. Invest.* **39,** 1157 (1960).

[5] Three sources of TCII can be used for the standard curve: purified TCII from human plasma, recombinant human TCII (rhTCII), or TCII in normal human serum. The concentration of the apo-TCII in serum is determined as described in the first part of this chapter and represents approximately 90% of total TCII in serum. Whereas purified human TCII and rTCII in solution have a tendency to aggregate because of their hydrophobicity, TCII in serum remains monomeric and stable and is preferable for generating the standard curve.

Rabbit anti-serum to human TCII (anti-TCII) (not commercially available but may be provided by the authors on request)
Normal rabbit serum (NRS)

Note: The endogenous TCII in the rabbit antiserum and NRS is removed by mixing 2 vol of the QUSO G-32 suspension (60 mg/ml) with each volume of serum.

Procedure

1. Prepare the [^{57}Co]Cbl–TCII tracer by diluting TCII in the standard PBS buffer to bind ~100% of 5000 cpm of [^{57}Co]Cbl. This can be prepared as a stock reagent for subsequent assays.
2. Prepare the TCII standard at a concentration of 1000 pg of protein/100 μl of PBS.
3. Prepare a 1:100 dilution of normal rabbit serum in the PBS buffer.
4. Using the assay method described below, prepare dilutions of the anti-TCII antiserum in the diluted normal rabbit serum to determine the dilution that will bind ~50% of the [^{57}Co]Cbl–TCII tracer.

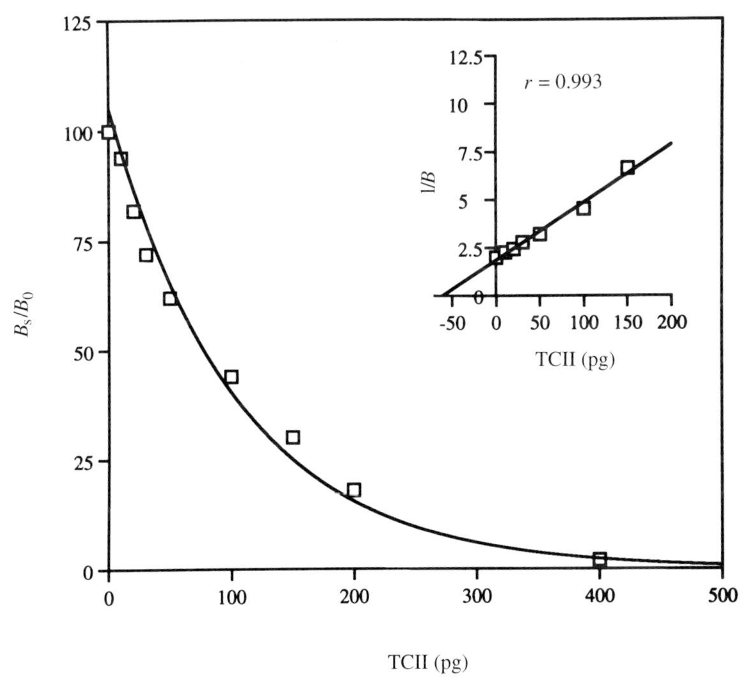

FIG. 1. Radioimmunoassay standard curve for measuring immunoreactive TCII. *Inset:* Linearization of the standard curve by plotting $1/B$ on the y axis against the TCII protein standards on the x axis. B, Fraction of [^{57}Co]TCII bound to the anti-TCII antibody.

5. Prepare the standard assay reactions as follows: PBS buffer plus standard TCII (0–500 pg) plus [^{57}Co]Cbl–TCII (100 μl) plus anti-TCII (100 μl). *Note:* The anti-TCII is always added last.
6. A control blank is prepared using a volume of the diluted normal rabbit serum equal to the volume of the anti-TCII, which, in the above method, is 100 μl.
7. A volume of serum (or any source of TCII) to be assayed is substituted for the TCII standard in the reaction mixture.
8. The final volume of the assay reaction mixture is 500 μl. The components can be dispensed in any volume prepared in the PBS buffer. For our assay we prepare the [^{57}Co]Cbl–TCII in PBS such that 100 μl contains 5000 cpm; the standard is prepared as a stock solution of 1000 pg of protein/100 μl. The volume of the TCII standard preparation dispensed to obtain a reference curve from 0 to 500 pg determines the volume of PBS buffer required to bring the final volume of the assay mixture to 500 μl.
9. Incubate the assay mixture for 24 hr at 4°.
10. Add 50 μl of the protein A membrane suspension, vortex the mixture, and incubate for 15 min at 4° with constant mixing. Centrifuge at 14,000 rpm for 15 min at 4°. Wash the pellet once with assay buffer and count in a scintillation γ counter.
11. The standard reference curve can be plotted in a number of ways:
    a. Bound/free (B/F) [^{57}Co]Cbl–TCII on the y axis against the standard TCII on the x axis (B equals the percentage of [^{57}Co]Cbl–TCII antibody bound; $F = 100 - \%B$).
    b. B_s/B_0 on the y axis against the standard TCII on the x axis; where B_s is the fraction of total [^{57}Co]Cbl–TCII bound at each corresponding concentration of the standard TCII and B_0 is the fraction of total [^{57}Co]Cbl–TCII bound in the absence of competing TCII.
    c. $1/B$ on the y axis against the standard TCII on the x axis. B for this plot is the fraction bound and not the percentage bound. This will linearize the reference curve for values of bound tracer above 10%. An example of a standard reference curve is shown in Fig. 1. The range of the dose–response curve is 6.25–500 pg (0.15–11.6 fmol) of competing TCII. The inset shows the linear reference curve obtained by plotting $1/B$ against the concentration of the TCII standard.

[31] Molecular Methods for Analysis and Expression of Transcobalamin II

By LIAN QIAN, EDWARD V. QUADROS, and SHELDON P. ROTHENBERG

Transcobalamin II (TCII) has been reported to be expressed in several mammalian tissues, including liver, kidney, intestinal epithelium, cultured fibroblasts, and human umbilical vein endothelial cells. The expression of TCII has been determined by the direct assay for apo-TCII and holo-TCII as described in [30] in this volume,[1] and by analysis for the TCII transcript in cellular RNA.

Three methods have been used in our laboratory to identify TCII expression at the molecular level. These include (1) Northern analysis for the TCII transcript [total RNA, or poly(A)$^+$ RNA] in tissues or cells with a moderate to substantial level of expression, (2) an RNase protection assay for low level of expression, and (3) reverse transcription followed by the polymerase chain reaction (RT-PCR) for amplification of the cDNA generated from tissues with a low level of TCII mRNA. Because Northern analysis for messenger RNA (mRNA) has been described in protocol manuals as well as in previous volumes of this series, we provide the methods used in our laboratory to identify the TCII transcript by ribonuclease (RNase) protection and RT-PCR.

RNase Protection

Principle

The RNase protection requires the synthesis of a radiolabeled TCII riboprobe, incubation of this probe with the mRNA prepared from the tissues or cells to be analyzed, followed by RNase A and RNase T digestion. The RNase(s) will digest regions of the riboprobe that do not form hybrids with the complementary regions of the TCII mRNA. Only the RNA–RNA hybrids will remain intact and can be identified by polyacrylamide gel electrophoresis and autoradiography.

Reagents

Transcription buffer, 10× (Ambion, Austin, TX)
Diethyl pyrocarbonate (DEPC)-treated water

[1] S. P. Rothenberg and E. V. Quadros, *Methods Enzymol.* **281,** [30], 1997 (this volume).

0076-6879/97 $25

Gel loading buffer: 95% (v/v) formamide, 0.025% (w/v) xylene cyanol, 0.025% (w/v) bromphenol blue, 0.5 m*M* EDTA, 0.025% (w/v) sodium dodecyl sulfate (SDS)

Additional reagents:

Dithiothreitol (DTT), 0.1 *M*
ATP, GTP, and CTP ribonucleotides, 10 m*M*
UTP ribonucleotide, 0.05 m*M*
[α-^{32}P]UTP (400–800 Ci/mmol, 10 mCi/ml)
RNA polymerase (Ambion)
DNase I, free of RNase (Ambion)
tRNA, 10 mg/ml
Ribonucleases A (10 mg/ml) and T_1 (2 mg/ml)
Proteinase K, 20 mg/ml
Ethanol (100%)
2-Propanol (100%)
Sodium acetate buffer (pH 5.5), 3 *M*
Sodium dodecyl sulfate (SDS), 10% (w/v) solution in water
Ammonium acetate, 7.5 *M*
TE buffer: 10 m*M* Tris-HCl, 1 m*M* EDTA, pH 8.0
TBE buffer: 90 m*M* Tris base, 90 m*M* boric acid, 1 m*M* EDTA
Phenol–chloroform–isoamyl alcohol (25 : 24 : 1, by volume)
Chloroform–isoamyl alcohol (24 : 1, v/v)
Cesium chloride
RNase inhibitor (Promega, Madison, WI)
Annealing buffer: 80% (v/v) formamide, 40 m*M* piperazine-*N*,*N*′-bis[2-ethanesulfonic acid] (PIPES, pH 6.4), 0.4 *M* NaCl, 1 m*M* EDTA
Bovine serum albumin (BSA)
RNase buffer: 10 m*M* Tris-HCl (pH 7.5), 5 m*M* EDTA, 0.3 *M* NaCl
Guanidinium lysis buffer: 4 *M* guanidinium isothiocyanate, 30 m*M* sodium acetate, 0.1 *M* 2-mercaptoethanol

Procedure

1. Preparation of Transcription Plasmid. To identify the TCII transcript, a region of exon II was selected for preparation of the riboprobe because this domain of the TCII gene has no significant sequence homology with the genes encoding intrinsic factor and transcobalamin I (haptocorrin). Accordingly, a 94-nucleotide (nt) sequence is obtained from the TCII cDNA[1a] (nt 173–267) by restriction digestion with *Hin*dIII and *Fok*I and

[1a] O. Platica, R. Janeczko, E. V. Quadros, A. Regec, R. Romain, and S. P. Rothenberg, *J. Biol. Chem.* **266,** 7860 (1991).

the fragment is purified and subcloned into pGEM-9Zf(−) plasmid (Promega). The orientation of the insert is determined by restriction analysis of several clones, using *Bbs*I and *Sac*I, which will generate a 166-bp antisense fragment when the 3′ terminus is in proximity to the T7 promoter [pGEM-9Zf(−)TCII 94$^-$]. A 78-bp sense fragment will be obtained if the insert is in the opposite orientation [PGEM-9Zf(−)TCII 94$^+$]. Alternatively, the orientation of the inserted fragment can be established by sequencing from either polylinker site. These plasmids are linearized by digestion with *Eco*RI, which cuts in the polylinker region. These constructs generate runoff sense and antisense riboprobes under the control of the T7 polymerase promoter. (*Note:* Use only a restriction enzyme that generates a 5′ protruding or blunt end on the plasmid in order to obtain high-efficiency transcription. An enzyme that linearizes the plasmid and leaves a 3′ overhang may reduce transcription efficiency.)

The plasmids are linearized with *Eco*RI and an aliquot of the reaction is analyzed by agarose gel electrophoresis to confirm complete digestion. Proteinase K (to a final concentration of 200 μg/ml) is then added to this reaction mixture and incubated at 37° for 30 min. The volume is adjusted to 100 μl with TE buffer and 0.1 vol of 3 *M* sodium acetate is added. The sample is extracted with an equal volume of phenol–chloroform–isoamyl alcohol (25:24:1, by volume) followed by an equal volume of chloroform–isoamyl alcohol (24:1, v/v). The linear plasmid is precipitated at −20° for 60 min following the addition of 2.5 vol of 100% ethanol. The precipitate is pelleted at 13,000 rpm for 10 min, washed twice with 80% (v/v) ethanol, and dried. The linearized template DNA is dissolved in TE buffer at a concentration of 0.5 mg/ml.

2. Preparation of Total and Poly(A)$^+$ RNA from Cultured Cells. The tissue (or cultured cells) is homogenized [preferably using a Polytron (Brinkmann, Westbury, NY)] in guanidinium lysis buffer and the RNA pelleted by ultracentrifugation over a CsCl–EDTA cushion (5.7 *M* CsCl, 5 m*M* EDTA), overnight at 32,000 *g* in an SW41 rotor (Beckman, Palo Alto, CA) at 18°.[2] If the level of expression of the TCII transcript is low, the sensitivity of the assay can be increased by using poly(A)$^+$ RNA, which can be prepared from total RNA using a BioMag mRNA purification kit (PerSpective Biosystems, Cambridge, MA). Alternatively, poly(A)$^+$ RNA can also be prepared using the standard oligo(dT)-cellulose method.[3]

3. Preparation of Radiolabeled Riboprobes. ^{32}P-labeled TCII riboprobes are generated from pGEM-9Zf(−)TCII 94$^+$ and 94$^-$ with T7 RNA polymer-

[2] J. M. Chirgwin, A. E. Przybyla, R. J. MacDonald, and W. J. Rutter, *Biochemistry* **18,** 5294 (1979).

[3] H. Aviv and P. Leder, *Proc. Natl. Acad. Sci. U.S.A.* **69,** 1408 (1972).

ase according to the manufacturer instructions (Ambion). The labeling reaction is carried out in a 20-μl volume containing 0.5 μg of linearized plasmid, 2 μl of 10× transcription buffer, 1 μl of 200 m*M* DTT, 1 μl of 10 m*M* ATP, 1 μl of 10 m*M* CTP, 1 μl of 10 m*M* GTP, 1 to 3 μl of 0.05 m*M* UTP, 1 μl of RNase inhibitor, 5 μl of [α-^{32}P]UTP (50 μCi), and 1 μl of T7 RNA polymerase (10 U/μl). The reaction mixture is incubated at 37° for 1 hr and then treated with 1 μl of RNase-free DNase I for 15 min at 37° to destroy the DNA template. The reaction is stopped by addition of 1 μl of 0.5 *M* EDTA. The total volume is adjusted to 100 μl with TE buffer, followed by the addition of 0.1 vol of 3 *M* sodium acetate and then extracted twice with phenol–chloroform and once with chloroform. The riboprobes are precipitated with 2.5 vol of ethanol. The precipitate is dissolved in 5 μl of gel loading buffer, heated for 5 min at 70°, chilled on ice for 3–5 min, and electrophoresed in a 8% (w/v) polyacrylamide gel containing 7 *M* urea and 1× TBE.

The riboprobe is recovered from the gel by the following method. After the electrophoresis is completed, the gel is covered with plastic wrap and exposed to X-ray film (Cronex 4, Du Pont, DE) for ~30 sec. A light band should be obtained to determine the position of the riboprobe. The orientation of the film on the gel should be ascertained by cutting one edge of the film in order to localize the exact area of the gel that contains the full-length labeled transcripts. A helpful method to locate the exact position of the riboprobes is to align the film accurately over the gel and then pass a straight pin through the film at each corner of the band into the gel. After removing the film, the pin holes will correspond to the four corners of the gel containing radiolabeled riboprobe. The gel is then excised, crushed, incubated at room temperature for at least 2 hr in 400 μl of RNase-free TE buffer, and then centrifuged at 13,000 *g* for 30 min at 4°. The supernatant solution is filtered through a 0.2- to 0.4-μm syringe filter. The filter can be washed with an additional 400 μl of buffer to recover more of the labeled probe. The sample is then extracted twice with phenol–chloroform, once with chloroform, and then precipitated by the addition of 0.6 vol of 100% 2-propanol. The riboprobes are resuspended in 50 μl of RNase-free water and the radioactivity contained in 1 μl is determined.

4. RNase Protection Assay. The radiolabeled riboprobe (10^6 cpm) is first incubated with 1–3 μg of poly(A)$^+$ RNA (or 30–40 μg of total RNA) in 30 μl of annealing buffer at 85° for 10 min and then at 45° for a minimum of 8 hr. An RNase cocktail consisting of 0.35 ml of RNase buffer and RNase A and T_1 at a final concentration of 40 and 2 μg/ml, respectively, is then added to the annealing mixture. Following incubation for 30 min at 30°, SDS (20%, w/v) to a final concentration of 0.5% (w/v), and proteinase K to a final concentration of 200 μg/ml, are added to destroy the RNase. The sample is incubated at 37° for 30 min and the volume of the reaction mixture is increased to 400 μl with DEPC-treated water. The protected riboprobe is extracted twice with an equal volume of phenol–chloroform, once with chloroform, and then concentrated by ethanol precipitation as described above. The pellet is dissolved in the gel loading buffer and electrophoresed in an 8% (w/v) polyacrylamide gel containing 7 *M* urea and 1× TBE. The gel is then dried and exposed to X-ray film (Cronex 4). For better resolution of the protected fragment(s), a sequencing gel is recommended. Figure 1 shows an autoradiograph of an RNase protection assay using total RNA from normal human placenta and from skin fibroblasts cultured from a normal subject and from patients with congenital TCII deficiency.

Reverse Transcription–Polymerase Chain Reaction

Principle

The size of the TCII transcript in the RNA preparation can be determined by RT-PCR. A cDNA copy is generated from total or poly(A)$^+$ RNA by reverse transcription using an oligo(dT) primer (GIBCO-BRL, Gaithersburg, MD). The TCII cDNA is then amplified by PCR, using specific primers to the extremities of the coding region of the TCII cDNA.[1a]

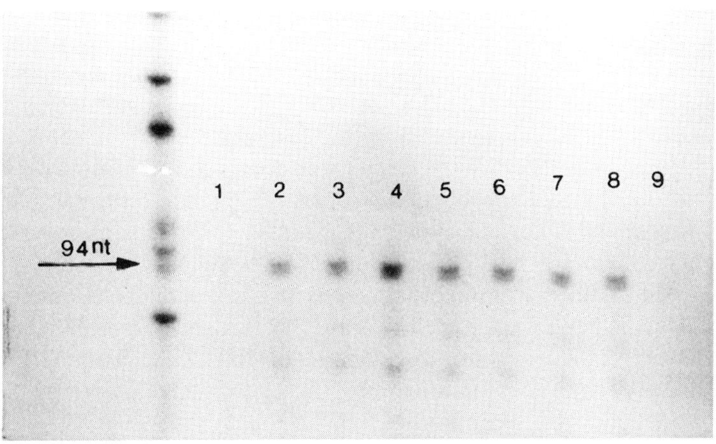

FIG. 1. RNase protection assay to identify the TCII transcript in total RNA. Lane 1, human placenta RNA; lanes 2 and 4–8, fibroblast RNA from patients with congenital TCII deficiency; lane 3, fibroblast RNA from a normal subject; lane 9, negative control containing tRNA. The signal for the human placenta is low and more visible on the autoradiography than here.

The amplified cDNA is purified by gel electrophoresis and cloned into a plasmid for propagation and nucleotide sequencing.[4]

Reagents

Reverse transcription (RT) buffer, 10×: 200 m*M* Tris-HCl (pH 8.4), 500 m*M* KCl, 25 m*M* $MgCl_2$, BSA (1 μg/μl)
dNTP(s), 10 m*M*
PCR buffer, 10×: 100 m*M* Tris-HCl (pH 8.8), 100 m*M* KCl, 0.02% (v/v) Tween 20 (Perkin-Elmer, Norwalk, CT)

Procedure

The reverse transcription reaction is performed as follows: 0.5 μg of oligo(dT) is first mixed with 1 μg of poly(A)$^+$ RNA (or 5–10 μg of total RNA) in 14 μl of DEPC-treated water, denatured at 70° for 10 min, and then chilled on ice for 1 min. The contents of the microcentrifuge tube are briefly centrifuged and the following components are added: 1 μl of 10× RT buffer, 1 μl of 10 m*M* dNTPs, 2 μl of 0.1 *M* DTT, and 1 μl of SuperScript II reverse transcriptase (GIBCO-BRL, Gaithersburg, MD) (20 U/μl). The tube is gently agitated and incubated for 10 min at room temperature and then transferred to 42° for 1 hr. The reaction is terminated by heating the mixture at 70° for 15 min.

The PCR is then initiated in a 25-μl volume that includes 1 μl of template cDNA generated from the RT reaction, 2.5 μl of 10× PCR buffer, 5 μl of 1 m*M* dNTP(s), 1.5 μl of 25 m*M* $MgCl_2$, 1 μl of the sense primer (10 μ*M*), 1 μl of the antisense primer (10 μ*M*), 12.8 μl of H_2O, and 0.2 μl (1 U) of *Taq* polymerase (Perkin-Elmer). These primers are complementary to the 5′ and 3′ extremities of the TCII cDNA. The PCR is performed for 30 cycles under the following conditions for each phase: denaturation, 94° for 0.8 min; annealing, 55° for 0.8 min; extension, 72° for 3 min. Following amplification, RT-PCR products are separated by electrophoresis in a 1% (w/v) agarose gel. Figure 2 shows the agarose gel electrophoresis of the fragments generated by RT-PCR from poly(A)$^+$ RNA prepared from fibroblast cell cultures from two patients with congenital TCII deficiency.

The specific band is excised, and the DNA is recovered and then cloned into pCR-Script SK(+) vector (Stratagene, La Jolla, CA). The TCII cDNA clones are propagated in liquid culture, the plasmid DNA purified and the insert sequenced by the method of Sanger *et al.*[4]

Note: (1) A positive control is included in the PCR assay using TCII cDNA (1 ng) as the template; (2) a negative control is also included in which the PCR amplification reaction mixture contains no DNA template.

[4] F. Sanger, S. Nicklen, and A. R. Coulson, *Proc. Natl. Acad. Sci. U.S.A.* **74,** 5463 (1977).

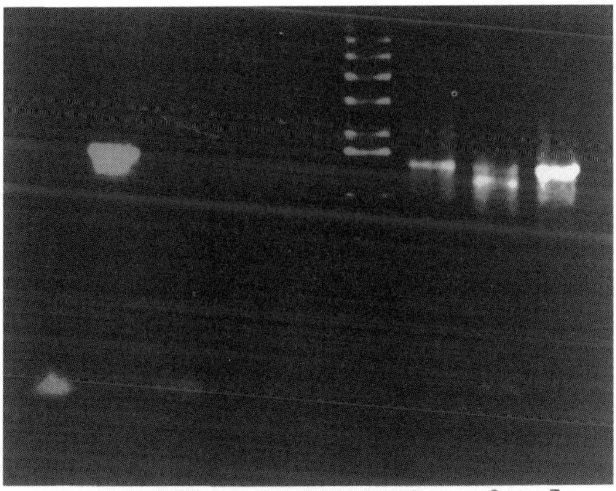

FIG. 2. RT-PCR using fibroblast poly(A)$^+$ RNA from two patients with congenital TCII deficiency and from a normal subject. Lane 1, negative control lacking the template (the primers migrate to the bottom of the lane); lane 2, positive control with plasmid TCII cDNA; lane 3, repeat negative control; lane 4, DNA size markers; lane 5, normal fibroblast RNA; lanes 6 and 7, fibroblast RNA from patients with congenital TCII deficiency. The doublet in lane 6 represents amplification of the cDNA for two allelic genes. The size of the generated fragments approximate 1.5 kb.

Production of Recombinant Human Transcobalamin II

The cloning of the full-length cDNA encoding human TCII facilitates the application of recombinant techniques to produce sufficient quantities of this protein for structural studies.[5] Baculovirus-mediated expression of recombinant TCII was selected because a prokaryotic expression system failed to generate an intact functional protein. The baculovirus system, on the other hand, has provided recombinant proteins that are structurally and functionally identical to the native counterparts.[6]

Principle

Spodoptera frugiperda (Sf9) cells derived from the ovary of the fall armyworm are highly susceptible to infection with *Autographa californica* nuclear polyhedrosis virus (AcMNPV). In the replicative stage, the viral particles are embedded in proteinaceous occlusions called *polyhedra.* The polyhedrin protein is the major structural component of occlusion bodies

[5] E. V. Quadros, P. Sai, and S. P. Rothenberg, *Blood* **81,** 1239 (1993).

[6] D. R. O'Reilly, L. K. Miller, and V. A. Luckow, "Baculovirus Expression Vectors." W. H. Freeman, New York, 1992.

and accounts for ~50% of the protein in infected cells. Foreign gene sequences inserted downstream of the polyhedrin promoter have been efficiently transcribed and translated to produce recombinant proteins. In addition, recombinant proteins produced in insect cells have been shown to be processed and glycosylated like the wild-type protein expressed in mammalian cells.

The production of recombinant proteins in insect cells involves the insertion of the cDNA into a plasmid vector that contains a segment of the baculovirus genome and the polyhedrin promoter. The nucleotide sequence encoding the protein of interest is transferred to the viral genome downstream of the polyhedrin promoter by homologous recombination with wild-type virus to produce the recombinant virus. The highly active polyhedrin promoter then drives the expression of the recombinant protein in the infected insect cells.

Reagents

cDNA for TCII
Sf9 cells (ATCC CRL-1711)
TNM-FH medium (Sigma, St. Louis, MO)[7]
TE buffer: See previous section
Grace's insect medium (Sigma)
Plasmid: pGEM3Zf(−) (Promega)
Plasmid: pVL1393 (Invitrogen, San Diego, CA)
Heat-inactivated fetal bovine serum (FBS; GIBCO-BRL)
Transfection buffer: 25 m*M* HEPES (pH 7.1), 140 m*M* NaCl, 125 m*M* $CaCl_2$
Proteinase K, 20 mg/ml (Sigma)
Sarkosyl, 10% (w/v) (Sigma)
Low melting point agarose (BRL)
Calf intestine alkaline phosphatase (U.S. Biochemical, Cleveland, OH)
Phenol reagent: Phenol–chloroform–isoamyl alcohol (25:24:1, by volume)

Procedure

1. Preparation of Transcobalamin II cDNA Insert. The full-length cDNA for human TCII was cloned from a human umbilical vein endothelial cell (HUVE) cDNA library in λgt11.[1a] The cDNA, subcloned into the plasmid pGEM3z(f−), is propagated in liquid culture in *Escherichia coli* JM109 and then isolated by alkaline lysis and cesium chloride density gradient

[7] *Trichoplusia ni* medium—Frank Hink [W. F. Hink, *Nature* (*London*) **266,** 466 (1970)].

centrifugation at 45,000 rpm for 36 hr in an SW65 rotor (Beckman). The plasmid is digested with the restriction enzyme *Eco*RI to release the full-length cDNA as a single fragment, which is isolated by electrophoresis in low melting point agarose (LMPA). The fragment is recovered by extraction with phenol reagent followed by precipitation with ethanol (final concentration 70%, v/v) and 0.3 *M* sodium acetate.

2. Insertion of cDNA into Baculovirus Plasmid Vector. The plasmid, pVL1393, is propagated in *E. coli* JM109 and purified through cesium chloride as described above. The plasmid is digested with *Eco*RI, dephosphorylated using calf intestine alkaline phosphatase, and the linear plasmid isolated by electrophoresis in LMPA followed by extraction with phenol reagent and ethanol precipitation. The isolated fragment contains the 1866-bp TCII cDNA, which includes a 54-nucleotide (nt) sequence encoding the signal peptide, a 37-nt 5′ untranslated region, a 548-nt 3′ untranslated region that includes the polyadenylation signal sequence, and the poly(A)$^+$ tail. The *Eco*RI sticky arms flank both ends of the insert. This cDNA is ligated into the linear plasmid pVL1393, using T_4 DNA ligase and propagated in *E. coli* DH5α. Several clones are grown up in 5-ml liquid cultures. The plasmid is isolated and digested with *Sma*I, which cuts once in the polylinker region and once in the 3′ untranslated region of the TCII cDNA and generates a 1.7-kb fragment if the 5′ end of the cDNA is located next to the polyhedrin promoter. This restriction pattern identifies clones with the TCII cDNA in the correct orientation for expression.

3. Culture of Sf9 Cells. Sf9 cells are maintained in TNM-FH medium supplemented with 10% (v/v) heat-inactivated FBS. The cultures are seeded at a density of 0.2×10^5 cells/ml and subcultured every 48–72 hr. The cells, which grow as a loosely attached monolayer, are detached with culture medium streamed repetitively over the cells, using a Pasteur pipette. The cells are diluted 10-fold in fresh medium and 5-ml aliquots are grown in 25-cm^2 flasks in an incubator at 27° and ambient air.

4. Isolation of Wild-Type Baculovirus. Sf9 cells are seeded at a density of 1×10^6/ml in Grace's insect medium (GIM), and 20 ml are distributed in 150-cm^2 flasks and incubated at room temperature for 1 hr to allow the cells to attach to the flask. The medium is removed and the cells are infected with the *A. californica* nuclear polyhedrosis virus (AcMNPV) in 4 ml of GIM for 1 hr at a multiplicity of infection (MOI) of 10 plaque-forming units (PFU). The cells are then cultured at 27° for 72 hr. The medium is clarified by centrifugation at 3000 *g* for 10 min at 4° and the supernatant fraction is centrifuged at 100,000 *g* for 30 min at 4° to pellet the virus. The viral pellet is then resuspended in 1 ml of 0.1× Tris–EDTA, pH 8.0 (TE), and 0.5 ml is layered onto a 25–56% sucrose gradient in 0.1× TE (approximately 11 ml/tube) and centrifuged at 100,000 *g* in an SW 41Ti rotor

(Beckman) for 90 min at 4°. The opaque region containing the virus is collected, diluted 10-fold in 0.1× TE, and pelleted at 100,000 *g* for 90 min at 4°. The pellet is resuspended in 2 ml of 1× digestion buffer containing proteinase K (100 μg) and sarkosyl (6%) and digested overnight at 50°. The viral DNA is extracted with the phenol reagent followed by ethanol precipitation, then dissolved in 0.1× TE and stored at −20°.

5. Cotransfection of pVL1393 Plasmid and Wild-Type Baculovirus into Sf9 Cells. The cells are seeded at a density of 2×10^6 cells in 25-cm^2 flasks and allowed to attach for 1 hr. The viral DNA (1 μg) and plasmid DNA (2 μg) are mixed in a 5-ml polypropylene tube containing 0.75 ml of transfection buffer.

The medium in the flask is removed and 0.75 ml of GIM containing 10% (v/v) FBS and antibiotics (gentamicin, 50 μg/ml and amphotericin B, 2.5 μg/ml) is added. The DNA, diluted in the transfection buffer, is added dropwise to the flask with gentle mixing. A fine precipitate of calcium phosphate forms from the calcium chloride in the transfection buffer and from the phosphate in GIM and can be observed with a microscope. The flasks are incubated at 27° for 4 hr, the medium is replaced with 5 ml of TNM-FH plus 10% (v/v) FBS plus antibiotics and the cells are incubated at 27° for 4–5 days. Infection of the insect cells with the virus is monitored by examining the cells with an inverted microscope. Infected cells appear refractile, increase in size, and detach from the surface. The medium is then collected, the cells removed by spinning at 3000 *g* for 10 min at 4°, and the supernatant fraction containing the virus is stored at 4°.

6. Plaque Purification of Recombinant Virus. Sf9 cells (2×10^6) are seeded into 60-mm culture dishes and allowed to attach for 1 hr. The medium from the initial transfection, which now contains recombinant as well as wild-type virus, is diluted 10^2- to 10^5-fold; 1 ml is added to each culture dish after removing the medium from the attached cells and incubated at 27° for 1 hr. Low melting point agarose (1.5%, w/v) prepared in water is autoclaved for 15 min and then equilibrated at 37–40° and mixed with an equal volume of 2× TNM-FH–FBS–antibiotics prior to use. After the cells are exposed to the virus, the medium is removed, the cell layer is gently washed with TNM-FH, and 4 ml of the 1× LMPA in TNM-FH–FBS is added to each plate. The plates are left at room temperature for ~1 hr to allow the overlay to solidify and then incubated at 27° in a humidified incubator for 4–5 days.

Plates are examined under a dissecting microscope for recombinant plaques that are polyhedra negative and appear less refractile than plaques infected with wild-type virus. A dilution (10^{-4}–10^{-6}) that shows well-separated plaques is reexamined for polyhedra-negative plaques and an agarose plug from the center of each negative plaque is collected using a Pasteur pipette and transferred to a microcentrifuge tube containing 200 μl of medium.

7. Confirmation of Recombinant Plaques by Dot-Blot Hybridization. Twenty-four-well culture plates are seeded with 0.5×10^6 cells in 0.5 ml of medium and infected with 10 μl of the medium from the tubes containing individual plaques. After 3 days, the medium is removed and the cells are lysed by adding 200 μl of 0.5 *N* NaOH, mixing for 5 min, followed by addition of 20 μl of 1 *M* ammonium acetate. Fifty microliters of the lysate is applied to the nylon membrane using a dot-blot apparatus and probed with the TCII cDNA labeled with [^{32}P]dCTP by the random primer method.[8] Plaques with the strongest signal are selected for additional purification and are plated at a dilution of 10^{-2}–10^{-4}. One or more plaques that are well separated are selected from the highest dilution and plated for a third round. Single plaques from the third plating are collected into 200 μl of medium. Such multiple plating and plaque lifts are necessary to eliminate any carryover of wild-type virus with the recombinant virus. Individual plaques from the second and third round of purification are also tested for the presence of recombinant virus by reinfecting fresh cultures with each plaque and assaying for recombinant human TCII (rhTCII) in the culture medium. For this assay, Sf9 cells (0.5×10^6) are seeded in 0.5 ml of medium in 24-well plates and infected with 10 μl of purified recombinant virus. Four days postinfection, the medium is collected, the cells removed by spinning at 3000 *g* for 10 min at 4°, and 50 μl of the supernatant fraction assayed for rhTCII by measuring the binding of [^{57}Co]Cbl as described in [30] of this volume.[1]

8. Determination of Virus Titer and Optimization of Recombinant Human Transcobalamin II Production. The virus titer is determined by the plaque assay in LMPA or by end-point dilution assay.[9] Both methods use visual identification and counting of plaque-forming units at various dilutions of the virus and are, therefore, error prone. Vectors containing the β-galactosidase gene when propagated with X-Gal provide easy visual identification because of the blue color formed and are less prone to such error.

For the TCII recombinant virus, the titer also may be easily determined by a modification of the end-point dilution method. Instead of visual assessment of virus infection, rhTCII is assayed in the culture medium as an indicator of infection with recombinant virus. For this method, the original virus stock is diluted 10^5- to 10^{10}-fold; 100 μl of each dilution is mixed with

[8] A. P. Feinberg and B. Vogelstein, *Anal. Biochem.* **137,** 266 (1984).

[9] M. D. Summers and G. E. Smith, "A Manual of Methods for Baculovirus Vectors and Insect Cell Culture Procedures." The Texas A&M University Press, 1988.

100 μl of Sf9 cells (2.5×10^5/ml) and 10-μl aliquots are distributed into each well of 96-well microtiter plates, providing 20 replicate wells for each dilution. The plates are tightly sealed and incubated at 27° for 5 days in a humidified chamber. Recombinant human TCII is assayed in 1–2 μl of the medium by the binding of 10 pg of [^{57}Co]Cbl (described in [30] of this volume).[1] The total number of positive and negative wells is recorded for each dilution and used to calculate the viral titer. A detailed description of the computation of the viral titer is provided by O'Reilly *et al.*[6] The viral stock is stored at 4° and should be protected from prolonged exposure to white light during long-term storage.

Because production of a recombinant protein is a function of synchronous infection of the Sf9 cells and cell viability, it is necessary to optimize conditions for expression of rhTCII. When Sf9 cells at a density of 1×10^6/ml are infected with recombinant baculovirus, the highest rhTCII concentration in the medium is reached between 72 and 110 hr in cells infected at an MOI of 1 PFU.[9] Approximately 4–5 μg of rhTCII per milliliter can be obtained by increasing the initial cell density of Sf9 cells to $2–3 \times 10^6$ cells/ml.

9. Purification of Recombinant Human Transcobalamin II. For large-scale production and purification of rhTCII, the culture medium from ten to twenty 150-cm^2 flasks or from 500 ml to 1 liter from spinner flasks is collected and the cells are pelleted by centrifugation at 3000 *g* for 15 min at 4°. The supernatant fluid is removed and titrated to pH 5.2 with 1 *N* HCl. CM-Sephadex C-50 (dry powder) is added to the medium (2 g/liter) and mixed for 6–12 hr using an overhead stirrer. The Sephadex binds the rhTCII and is then washed with ~1 liter of 0.02 *M* $NaPO_4$–0.1 *M* NaCl, pH 5.8, followed by 500 ml of 0.02 *M* $NaPO_4$–0.2 *M* NaCl, pH 5.8. The TCII is eluted with 100 ml of 0.1 *M* $NaPO_4$–1 *M* NaCl and purified by Cbl–Sephacryl photolabile affinity chromatography as described for the purification of TCII from human plasma.[10] Briefly, 0.5 ml of the washed affinity matrix is added to 100 ml of the 1 *M* NaCl eluate and mixed overnight by rotating the suspension end over end. All procedures involving the affinity matrix are carried out in a dark room with a red light. The matrix is collected in a sintered glass funnel and washed with 1 liter of 0.1 *M* $NaPO_4$–1 *M* NaCl, pH 5.8. The washed affinity matrix is resuspended in 5 vol of buffer and the rhTCII–Cbl is dissociated from the matrix by exposure to a 300-W tungsten light for 30 min. The photolysis is repeated two additional times and the photolysates pooled. The high salt concentration in the photolysis buffer helps to minimize the aggregation and precipitation of rhTCII. For this reason, purified rhTCII and all concentrated stock

[10] E. V. Quadros, S. P. Rothenberg, Y.-Ch. E. Pan, and S. Stein, *J. Biol. Chem.* **261,** 15455 (1986).

preparations of the protein (i.e., the supernatant fraction from Sf9 cultures) are stored in a high salt concentration (1 *M* NaCl) and freeze–thawing cycles should be avoided. We routinely store the purified protein and culture medium containing rhTCII in 1 *M* NaCl–50% (v/v) glycerol at −15 to −20°.

[32] Purification, Membrane Expression, and Interactions of Transcobalamin II Receptor

By SANTANU BOSE and BELLUR SEETHARAM

The plasma transport and tissue exchange of cobalamin (Cbl; vitamin B_{12}) occurs by the binding of Cbl to transcobalamin II (TC II), a 43-kDa nonglycoprotein,[1] followed by receptor-mediated endocytosis.[2] The tissue/cellular uptake of TC II–Cbl is mediated by a cell surface receptor (TC II-R). Transcobalamin II receptor is synthesized as a single polypeptide of 44 or 45 kDa and is both N- and O-glycosylated to yield a mature protein with a molecular mass of 62 kDa. Transcobalamin II receptor is expressed in the plasma membranes of all human tissues as a functional dimer with a molecular mass of 124 kDa.[3] This chapter outlines some of the more recent findings on the membrane expression and interactions of human TC II-R.

Purification of Human Placental Transcobalamin II Receptor

Transcobalamin II receptor activity has been detected in many human tissues.[4–7] Owing to unrestricted availability and relatively high levels of TC II-R activity, however, human placenta has been the primary source for its purification.

Principle

The major step in the purification of TC II-R is ligand (TC II) affinity chromatography. The source of TC II is rabbit plasma/serum. Commercially

[1] E. V. Quadros, S. P. Rothenberg, Y. C.-E. Pan, and S. Stein, *J. Biol. Chem.* **261,** 15455 (1986).
[2] P. Youngdhal-Turner, L. E. Rosenberg, and R. H. Allen, *J. Clin. Invest.* **61,** 133 (1978).
[3] S. Bose, S. Seetharam, and B. Seetharam, *J. Biol. Chem.* **270,** 8152 (1995).
[4] P. A. Seligman and R. H. Allen, *J. Biol. Chem.* **253,** 1766 (1978).
[5] P. A. Friedman, M. A. Shia, and J. K. Wallace, *J. Clin. Invest.* **59,** 51 (1977).
[6] E. Nexo and M. D. Hollenberg, *Biochim. Biophys. Acta* **628,** 190 (1980).
[7] J. S. D. Scott, E. P. W. Bowman, and W. G. E. Cooksley, *Clin. Sci.* **68,** 357 (1985).

0076-6879/97 $25

purchased (GIBCO-BRL, Gaithersburg, MD) rabbit plasma contains free TC II capable of binding about 0.015 μmol of Cbl per liter. When passed through a Sepharose 4B–Cbl[8] column, 1 liter of plasma provides enough TC II bound to the solid matrix to facilitate the binding of TC II-R present in a Triton X-100 extract from 250 g of human placenta. Between 50 and 60% of bound TC II-R can be eluted from the affinity column. It is important to make sure (1) that the placental extract that is passed over the affinity matrix is free of endogenous TC II and other interfering proteins that are normally present in the placental tissue, and (2) that both the extract and the affinity matrix are at pH 8.0 buffer and contain 5 mM $CaCl_2$. After extensive washing of the affinity matrix, the receptor is eluted with pH 5 buffer containing 0.1% (v/v) Triton X-100 and 5 mM EDTA.

Methods

The various steps involved in the purification of TC II-R are as follows. Freshly obtained human placenta (1 kg) is cut into small pieces and homogenized in a Waring blender (full speed for 3–5 min) with 3 liters of pH 5.5/EDTA buffer [50 mM acetate buffer, pH 5.5, containing 140 mM NaCl, 0.1 mM phenylmethylsulfonyl fluoride (PMSF), 2 mM benzamidine, and 5 mM EDTA]. Prior to centrifugation at 25,000 g for 30 min at 4°, the homogenate is incubated for 1 hr with stirring at 4°. The membranes thus obtained are rehomogenized in the same buffer and incubated as before. The repeated washing of the placental membranes with pH 5.5/EDTA buffer removes all endogenous TC II (the primary source is the blood present in this tissue) and other proteins that may interfere in the assay for TC II-R. Placental membranes treated in this way reveal much higher TC II-R activity than without washing and the yields from the affinity columns are much higher owing to higher binding of TC II-R to the ligand matrix. Further purification of TC II-R involves Triton X-100 solubilization, solid ammonium sulfate precipitation, and affinity chromatography. The details of these steps are as described by Seligman and Allen.[4] To process TC II-R present in 1 kg of human placenta it is essential to pass 4 liters of rabbit plasma through a Cbl–Sepharose column. By this procedure it is possible to recover at least 50% (0.03 μmol) of TC II-R from 1 kg of the tissue. The purified TC II-R demonstrates a single 62-kDa band under nonreducing sodium dodecyl sulfate–polyacrylamide gel electrophoresis (SDS–PAGE).[3]

[8] R. H. Allen and P. W. Majerus, *J. Biol. Chem.* **247,** 7695 (1972).

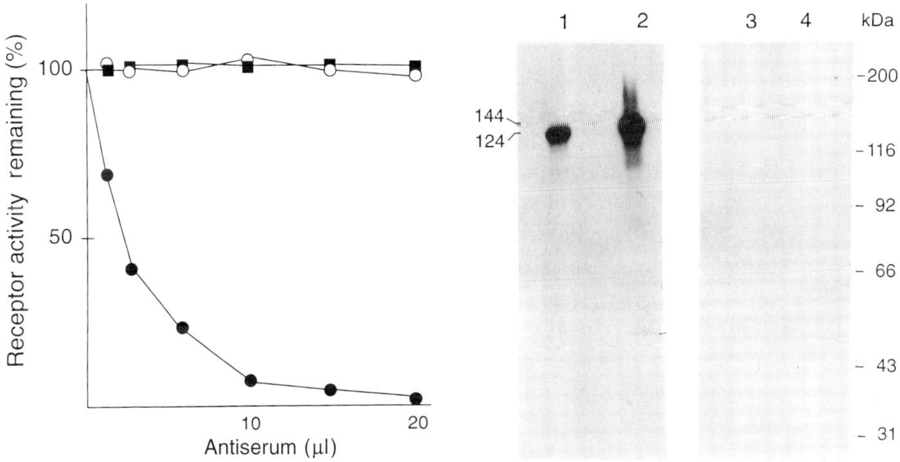

FIG. 1. *Left:* Immunoblocking of the binding of TC II–[^{57}Co]Cbl to purified TC II-R. Pure TC II-R (0.5 μg) was preincubated with the indicated amounts of antiserum to human TC II-R (●), TC II (■), and rat intrinsic factor–Cbl receptor (○) for 1 hr at 22°. TC II–[^{57}Co]Cbl (1.5 pmol) was then added and assayed for ligand binding. The activity is expressed as the percentage of TC II-R activity in a sample incubated without serum. *Right:* Immunoblotting of human placental membranes. Placental membranes (10 μg of protein) were incubated with (lanes 2 and 4) and without (lanes 1 and 3) 2-mercaptoethanol (2.5%, v/v) for 1 hr at 22° and then subjected to SDS–PAGE (7.5%, w/v). The separated proteins were transferred to a nitrocellulose membrane and probed with unabsorbed (lanes 1 and 2) or absorbed (lanes 3 and 4) antiserum. The diluted antiserum was absorbed with total human kidney membranes. (Reprinted with permission from Ref. 3.)

Preparation and Properties of Polyclonal Antibody to Human Placental Transcobalamin II Receptor

The 62-kDa TC II-R (30 μg) mixed with Freund's complete adjuvant is subcutaneously injected at multiple sites into the backs of New Zealand White rabbits. The rabbits are boosted with 15 μg of TC II-R mixed with Freund's incomplete adjuvant 10 days after the initial dose and blood is drawn from the ear vein every week after the booster dose.

The harvested TC II R antiserum inhibits the binding of TC II–[^{57}Co]Cbl to pure TC II-R in a dose-dependent manner and the inhibition of ligand binding is specific to TC II-R antiserum. Polyclonal antiserum raised to pure TC II,[9] the ligand, or to intrinsic factor–cobalamin receptor,[10] has no effect on ligand binding (Fig. 1, left). In these assays, the antiserum

[9] K. S. Ramanujam, S. Seetharam, M. Ramasamy, and B. Seetharam, *Am. J. Physiol.* **260,** G416 (1991).

[10] B. Seetharam, J. S. Levine, M. Ramasamy, and D. H. Alpers, *J. Biol. Chem.* **263,** 4443 (1988).

is first incubated with the pure receptor for 60 min prior to the addition of the ligand. The receptor-bound and free TC II–[^{57}Co]Cbl is separated by the DEAE–Sephadex method of Seligman and Allen.[4] A monoclonal antiserum raised to human TC II has been shown to inhibit the cellular uptake of holo-TC II by K 562 cells.[11] This inhibition of uptake is probably due to blocking of the receptor-binding site of TC II. These results suggest that the receptor recognition site on TC II is confined to a specific domain that is recognized by a specific monoclonal antibody, but not by the polyclonal antibody used in this study.

The antiserum to TC II-R recognizes in placental total membranes two proteins (Fig. 1, right), on SDS–PAGE: one with a molecular mass of 124 kDa (lane 1, Fig. 1) and the other with a molecular mass of 144 kDa (lane 2, Fig. 1) under nonreducing and reducing conditions, respectively. In these membranes TC II-R monomer is not seen owing to its extremely low levels and because of the transfer time of 90 min used during the electroblotting of the separated proteins onto the nitrocellulose membranes (Optitran, pore size 0.45 μm; Schleicher & Schuell, Keene, NH) (see the following sections). The specificity of the antiserum to recognize only TC II-R is evident, as preabsorbed antiserum fails to recognize these bands (Fig. 1, right, lanes 3 and 4). These initial immunoblot studies have revealed that the antiserum raised to the 62-kDa TC II-R recognizes in the placental membranes a protein of 124 kDa that has intramolecular disulfide bonds, the reduction of which produces an extended form of the receptor with reduced mobility on SDS–PAGE; as a consequence there is an apparent increase in its molecular mass. Pure 62-kDa TC II-R, when reduced and alkylated, reveals an apparent increase in its molecular mass by 10 kDa.[3] These results have suggested that intramolecular disulfide bonds of TC II-R monomer remain intact during its dimerization in the plasma membrane.

Standardization and Application of Immunoblotting Technique Using Polyclonal Antibody to Transcobalamin II Receptor

In the laboratory of the authors an immunoblotting procedure has been developed to study tissue membrane expression of TC II-R monomer and dimer forms.

Principle

Transcobalamin II receptor has an unusual property that enables it to remain a dimer even after boiling it with SDS buffer both in the

[11] E. V. Quadros, S. P. Rothenberg, and P. McLoughlin, *Biochem. Biophys. Res. Commun.* **222,** 149 (1996).

presence and absence of a reducing agent. This property has been used to study the expression and distribution of TC II-R monomer and dimer (with molecular masses of 62 and 124 kDa, respectively) in tissue membranes. In addition, this procedure can also be used to study any changes in the physical state of TC II-R in membranes following experimental manipulations that favor the existence of one or the other form of the receptor in membranes.

Methods

Initial immunoblotting is carried out with pure placental TC II-R (0.1–2 μg) first by subjecting it to nonreducing SDS–PAGE (7.5%, w/v) and then by electroblotting it onto nitrocellulose membranes at 90 V for 45 min. The filters are probed with diluted (1:2000, v/v) TC II-R antiserum and ^{125}I-labeled protein A. The bands are visualized by autoradiography and quantified by an Ambis (Image Acquisition & Analysis, San Diego, CA) radioimaging system. The band intensity is linear between 0.1 and 1.2 μg of TC II-R. This linearity corresponds to image density counts between 0.006 and 1×10^6, or arbitrary units of 0.24–40 (1 unit = 25,000 counts). The transfer time of 45 min to transfer pure TC II-R, with a molecular mass of 62 kDa, appears to be crucial, as less than 45 min of transfer results in incomplete transfer and more than 45 min of transfer results in the loss of TC II-R to the second nitrocellulose membrane.

The optimal time of transfer of TC II-R monomer and dimer forms in tissue membranes is standardized as follows: Homogenates of rat intestinal mucosal membranes (100 μg of protein) are loaded onto several lanes and subjected to nonreducing SDS–PAGE (7.5%, w/v). After separation, the proteins are transferred at 90 V to nitrocellulose for various time intervals (15–90 min) and probed with antiserum to TC II-R and ^{125}I-labeled protein A. The bands from all the lanes are visualized by autoradiography after exposure of the X-Omat plates for 18 hr. A typical immunoblot is shown in Fig. 2. The top and the bottom strips (Fig. 2A) represent the 124-kDa TC II-R dimer and 62-kDa monomer, respectively. When quantified by the Ambis radioimaging system, the ratio of the dimer to the monomer is 10:1. When transferred for 60 min, both the monomer and the dimer could be visualized on the same gel following an 80-hr exposure (Fig. 2B). When the bands are quantified, the ratio of the dimer to the monomer is 2:1. The lower ratio of the dimer to monomer noted during the 60-min transfer of proteins from the SDS–polyacrylamide gel is due not only to the loss of the 62-kDa monomer but also to the incomplete transfer of the dimer.

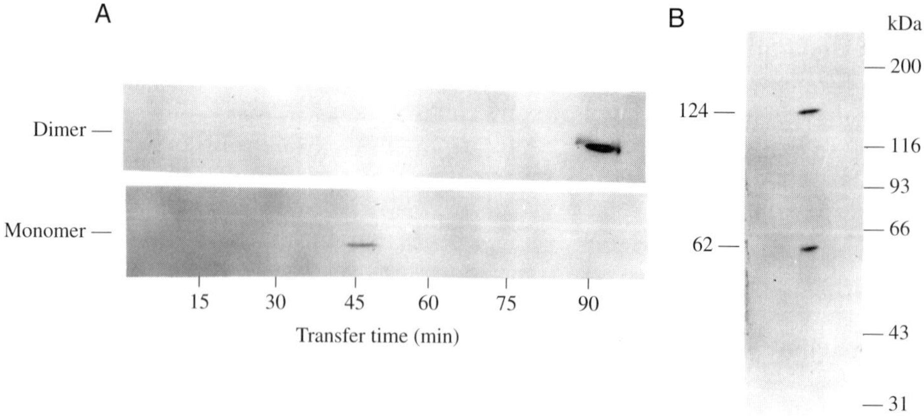

FIG. 2. Optimization of transfer time during immunoblotting of TC II-R monomer and dimer forms. (A) Rat intestinal mucosal total membranes (100 μg of protein) were separated on nonreducing SDS–PAGE (7.5%, w/v) and transferred to nitrocellulose membranes for 15–90 min. The blotted proteins were probed with diluted antiserum to human TC II-R and ^{125}I-labeled protein A. The top strip represents the 124-kDa dimer and the bottom strip represents the 62-kDa monomer. (B) The same membranes (300 μg of protein) were separated by SDS–PAGE and proteins separated were transferred for 60 min and probed as before. (Reprinted with permission from Ref. 12.)

Immunoblot studies[3,12] using pure TC II-R bound to Triton X-100 micelles (62-kDa monomer) or TC II-R bound to egg phosphatidylcholine (PC)/cholesterol vesicles (124-kDa dimer) have shown that with a 60-min transfer from the SDS–polyacrylamide gels, the 124-kDa dimer form is incompletely transferred and the 62-kDa monomer is lost to the second nitrocellulose membrane. Thus, the time chosen to detect both the monomer and dimer in the same gel will not reflect the actual amount of each species of the receptor present in any given membrane. The likely explanation for the twofold time differential for the optimal transfer of the two forms of TC II-R to the nitrocellulose membranes could be that this phenomenon is related to the receptor status (i.e., whether it is a monomer or a dimer). The immunoblotting procedure has been used to understand several aspects of membrane expression of TC II-R.

Relative Tissue Expression and Distribution of Transcobalamin II Receptor Monomer and Dimer and Their Interconversion

Using both the 45- and 90-min transfer time, the relative distribution of TC II-R monomer and dimer has been studied in rat tissues (Table I).

[12] S. Bose, J. Feix, S. Seetharam, and B. Seetharam, *J. Biol. Chem.* **271,** 11718 (1996).

TABLE I
TISSUE DISTRIBUTION OF TRANSCOBALAMIN II RECEPTOR MONOMER AND DIMER FORMS IN RAT[a]

Tissue	Arbitrary units/μg protein: Monomer (62 kDa)	Dimer (124 kDa)	Dimer/monomer
Intestine	0.365	3.01	8.24
Heart	0.055	0.54	9.82
Liver	0.037	0.29	7.83
Kidney	0.372	3.12	8.38

[a] The immunoblots were quantified by an Ambis radioimaging system. One arbitrary unit equals 25,000 image density counts. The data reported are the average of separate immunoblot analyses from two separate tissue homogenates.

The data show that in all the tissues, the dimer form was enriched between eight- and ninefold, and that, unlike in human,[3] the relative TC II-R protein distribution in rat kidney and intestine is the same. The TC II-R protein distribution in these four tissues correlates well with TC II-R activity.[13]

Immunoblotting technique using the polyclonal antiserum to human TC II-R has revealed TC II-R dimer with a molecular mass 124 kDa in human,[3] rat,[13] and rabbit[14] tissue membranes. In the rat kidney, the 62-kDa monomer was present only in the microsomes, and the 124-kDa dimer was the only species present in the apical and basolateral membranes.[12,13] These observations have suggested that dimerization of TC II-R is a postmicrosomal event and may occur following insertion of TC II-R in the plasma membrane. The immunoblotting technique was used to gain some insights into the mechanism of TC II-R dimerization. Membrane perturbations were carried out to examine their effects on the physical state of TC II-R (Table II). *In situ* conversion of TC II-R dimer to the monomer form occurred when the plasma membranes were treated with organic solvents, digitonin, and phospholipase A_2. Nonionic detergent Triton X-100 also converted the dimer form into the monomer form, but the TC II-R was solubilized. These results suggested that TC II-R dimer formation is due to its interaction with the annular lipids and that cholesterol and the fatty acyl residue in position 2 of the membrane phospholipid play a strong

[13] S. Bose, S. Seetharam, T. G. Hammond, and B. Seetharam, *Biochem. J.* **310,** 923 (1995).
[14] S. Bose, R. A. Komorowski, S. Seetharam, B. Gilfix, D. S. Rosenblatt, and B. Seetharam, *J. Biol. Chem.* **271,** 4195 (1996).

TABLE II
INTERCONVERSION OF TRANSCOBALAMIN II RECEPTOR PHYSICAL FORMS

Treatment[a]	Monomer (62 kDa)	Dimer (124 kDa)	Ref.
Tissue plasma membrane			
Untreated	−	+	
Triton X-100 (1%, v/v)[b]	+	−	3
Digitonin	+	−	12
$CHCl_3:CH_3OH$ (2:1, v/v)	+	−	3
Phospholipase A_2	+	−	12
Phospholipase C	−	+	12
Phospholipase D	−	+	12
Microsomal membrane			
Untreated	+	−	
Cholesterol (37°)	−	+	12
7-Ketocholesterol (37°)	+	−	12
Dihydrocholesterol (37°)	+	−	12

[a] Membranes were treated with either digitonin (2 μg/mg protein) or cholesterol and its analogs (25 μg/mg protein).
[b] With solubilization. +, Indicates physical state of TC II-R detected; −, indicates physical state of TC II-R not detected.

role in lipid interactions of TC II-R and the ensuing dimerization. Direct evidence for the role of cholesterol in the dimerization of TC II-R was obtained when the monomer form of TC II-R in the microsomal membranes was able to dimerize when its cholesterol content was increased from about 4 to 12 mol% by incubations with added cholesterol (Table II). The specificity of cholesterol in mediating the dimerization was evident from the observations that cholesterol analogs 7-ketocholesterol and dihydrocholesterol had no effect when incubated with the microsomes.

Evidence for the role of cholesterol in mediating the dimerization of TC II-R by affecting the fluidity of the lipid bilayer was obtained by immunoblotting pure TC II-R (bound to Triton X-100 micelles, 62-kDa monomer) reconstituted with egg phosphatidylcholine (PC) and synthetic phosphatidylcholine vesicles.[12] Transcobalamin II receptor was able to dimerize only when the cholesterol content of the egg PC vesicles was greater than 10 mol%. Transcobalamin II receptor, when reconstituted with dimyristoyl (14:0), dipalmitoyl (16:0), and disteroyl PC (18:0) vesicles, dimerized in the presence of 50 mol% of cholesterol at a temperature above the transition temperatures of 37, 45, and 65°, respectively. At 5, 22, and 22°, temperatures below that required for phase transition, cholesterol was not needed for the dimerization of TC II-R.[12] Taken together these results have shown that the cholesterol-mediated increase in order is important for the dimerization of TC II-R.

Conclusions and General Comments

Recognition of the unusual property of TC II-R dimers to remain dimers following treatment with SDS and the preparation of monospecific antiserum to TC II-R have enabled the authors to study the tissue expression of TC II-R and the mechanism of its dimerization. With the ease of preparing several hundred micrograms of pure placental TC II-R, combined with the availability of its monospecific antiserum, further studies on the structure and regulation of expression of this important vitamin B_{12} receptor are now possible.

Acknowledgments

The work presented in this chapter was supported by Research Grants NIDDK-26638 and 7816-01P from the Veteran Affairs Administration.

[33] Purification and Characterization of Aquacobalamin Reductase from *Euglena gracilis*

By FUMIO WATANABE and YOSHIHISA NAKANO

$$2 \text{ Aquacobalamin} + NADPH_2 \rightarrow 2 \text{ cob(II)alamin} + NADP^+$$

The conversion of aquacobalamin to 5′-deoxyadenosylcobalamin in bacteria[1] probably involves three enzymatic steps: the reduction of Co^{3+} in aquacobalamin to Co^{2+} by aquacobalamin reductase, the reduction of Co^{2+} to Co^+ by cob(II)alamin reductase, and the adenosylating reaction of Co^+ to 5′-deoxyadenosylcobalamin by cob(I)alamin adenosyltransferase. Activity of aquacobalamin reductase catalyzing the reduction of aquacobalamin to cob(II)alamin has been detected in some microorganisms[1,2] *Clostridium tetanomorphum*[1] and *Euglena gracilis*[2] contain NADH (EC 1.6.99.8)- and NADPH (EC 1.6.99.11)-linked aquacobalamin reductases, respectively. The *Clostridium* enzyme[1] requires FAD or FMN as a cofactor for the reduction of aquacobalamin, but the *Euglena* enzyme[2] does not. The *Eu-*

[1] G. A. Walker, S. Murphy, and F. M. Huennekens, *Arch. Biochem. Biophys.* **134,** 95 (1969).
[2] F. Watanabe, Y. Oki, Y. Nakano, and S. Kitaoka, *Agric. Biol. Chem.* **51,** 273 (1987).

0076-6879/97 $25

glena enzyme is located only in the soluble fraction of the mitochondria.[2,3] The enzyme has been purified to homogeneity only from *E. gracilis.*[3]

This chapter summarizes our work on NADPH-linked aquacobalamin reductase from *E. gracilis,* a photosynthetic protozoan.

Aquacobalamin Reductase Assay

Principle

The aquacobalamin reductase activity is assayed by spectrophotometric estimation of the amount of aquacobalamin converted to cob(II)alamin.[3]

Reagents

Tris–acetate buffer (pH 7.0), 100 m*M*
Hydroxocobalamin hydrochloride, 10 m*M* (Sigma, St. Louis, MO)
NADPH, 20 m*M*

Procedures

The reaction mixture in a total volume of 1.0 ml consists of 0.5 ml of Tris–acetate buffer, 0.01 ml of hydroxocobalamin, 0.47 ml of distilled water, and 0.01 ml of enzyme. The mixture is incubated at 40° in a 1.5-ml cuvette, and then the recation is started by the addition of 0.01 ml of NADPH. The reaction is monitored by measurement of the decrease in absorbance by aquacobalamin at 525 nm and is calculated on the basis of the differential molecular extinction coefficient of aquacobalamin to cob(II)alamin (5.57×10^3 M^{-1} cm^{-1} at 525 nm). This assay is performed with a dual-beam spectrophotometer with water-jacketed cuvette holder. One unit of the enzyme activity is defined as the amount of enzyme that catalyzes the reduction of aquacobalamin at the rate of 1 μmol/min.

Comment

Aquacobalamin is the conjugated acid of hydroxocobalamin. Aqua- and hydroxocobalamins are interconvertible in solution depending on pH. Above pH 8.0, hydroxocobalamin tends to predominate over aquacobalamin at equilibrium.[4]

Cob(II)alamin is reoxidized by oxygen in solution. The oxygen adduct of cob(III)alamin is represented as a superoxide anion coordinated to cob

[3] F. Watanabe, Y. Oki, Y. Nakano, and S. Kitaoka, *J. Biol. Chem.* **262,** 11514 (1987).
[4] D. Lexa, J. M. Saveant, and J. Zickler, *J. Am. Chem. Soc.* **99,** 2786 (1977).

(III)alamin. This oxidation reaction, however, is quite slow ($t_{1/2}$ = 21 min)[5]; it hardly affects this enzyme assay. The enzyme must be assayed under aerobic conditions because nonenzymatic reduction of aquacobalamin by NADPH occurs under anaerobic conditions.

Isolation of *Euglena* Mitochondria

Euglena gracilis Z is obtained from the Research Reactor Institute (Kyoto University, Kyoto, Japan). *Euglena gracilis* SM-ZK (a streptomycin-bleached mutant of *E. gracilis* Z), which lacks chloroplasts without any change in other cellular components, is prepared by S. Kitaoka and Y. Nakano (Department of Applied Biological Chemistry, Osaka Prefecture University, Osaka, Japan). *Euglena gracilis* SM-ZK is cultured for 5 days at 27° with illumination (2000 lux) in cobalamin-limited (0.05 μg/liter) Koren–Hutner medium.[6] A cell homogenate of *E. gracilis* SM-ZK is obtained by partial trypsin digestion of the pellicles followed by mild mechanical disruption by a modification of the method of Tokunaga *et al.*,[7] and the homogenate is fractionated by differential centrifugation. *Euglena* cells (about 6 g, wet weight) are washed twice with 10 m*M* 3-(*N*-morpholino)propanesulfonic acid (MOPS)–KOH buffer, pH 7.5, containing 0.3 *M* sucrose, and suspended in 5 ml of the same buffer. Bovine pancreas trypsin (type III, 30 mg; Sigma, St. Louis, MO) is added to the suspension. The mixture is gently stirred in an ice bath for 45 min and centrifuged at 2000 *g* for 5 min at 4° to remove trypsin and broken cells in the supernatant. The treated cells are suspended in 5 ml of 10 m*M* MOPS–KOH buffer, pH 7.5, containing 0.25 *M* sucrose and 3 mg of egg white trypsin inhibitor (type II-0; Sigma) to inhibit the action of trypsin remaining in the cell suspension; within 2 min, the mixture is centrifuged at 2000 *g* for 5 min at 4° to remove trypsin inhibitor and broken cells. The partly digested, unbroken cells that are at the bottom of the tube are suspended in 10 ml of 10 m*M* MOPS–KOH buffer, pH 7.5, containing 0.25 *M* sucrose and stirred gently for 10 min to cause the cells to burst. The suspension is centrifuged at 2000 *g* for 5 min at 4° to obtain the cell homogenate. The precipitate is resuspended in 10 ml of the same buffer, and stirred under the same conditions. The cell disruption procedures are repeated several times. The supernatants (cell homogenate), combined, are centrifuged at 2000 *g* for 5 min at 4° to remove the pellicle, paramylon, and undisrupted cells. The supernatant is centri-

[5] H. P. C. Hogenkamp, *in* "Cobalamin: Biochemistry and Pathophysiology" (B. M. Babior, ed.), p. 21. John Wiley & Sons, New York, 1975.

[6] L. Koren and S. H. Hutner, *J. Protozool.* **14** (Suppl.), 17 (1967).

[7] M. Tokunaga, Y. Nakano, and S. Kitaoka, *J. Protozool.* **26,** 471 (1979).

fuged at 10,000 *g* for 10 min at 4°. The pellet is washed twice and suspended in the same buffer and the suspension is used as the mitochondrial preparation.

Purification Procedure

Purification procedures are at 0–4° unless otherwise specified. Mitochondria (986 mg of protein) of *E. gracilis* are suspended in 33 ml of 10 m*M* Tris–acetate buffer, pH 7.0, containing 1 m*M* EDTA and 1 μM dithiothreitol (DTT), disrupted by sonic oscillation (10 kHz, four 20-sec pulses), and centrifuged at 100,000 *g* for 60 min to remove the membrane fraction. The supernatant fraction (32 ml) is put on a column (2.0 × 20.0 cm) of DEAE-BioGel A (Bio-Rad Laboratories, Richmond, CA) equilibrated with 10 m*M* Tris–acetate buffer, pH 7.0, containing 1 m*M* EDTA and 1 μM DTT, and eluted at a flow rate of 21 ml/hr. The column is washed with 100 ml of the same buffer and then eluted with 300 ml of a linear gradient (0–0.5 *M*) of potassium chloride in the same buffer. The active fractions (27 ml) are combined, dialyzed overnight against the same buffer (2 liters), and put on a column (1.5 × 5.0 cm) of DEAE-BioGel A equilibrated with the same buffer. The column is washed with 30 ml of the same buffer and eluted with 100 ml of a linear gradient (0–0.3 *M*) of potassium chloride in the same buffer at a flow rate of 13 ml/hr. The active fractions (11 ml) are combined and concentrated in a Centricon-30 microconcentrator (Amicon, Danvers, MA). The concentrated solution is put on a column (1.0 × 90.0 cm) of Sephadex G-100 (Pharmacia-LKB Biotechnology, Piscataway, NJ) equilibrated with 100 m*M* Tris–acetate buffer, pH 7.0, containing 1 m*M* EDTA and 1 μM DTT, and eluted with the same buffer at a flow rate of 3 ml/hr. The active fractions (8 ml) are combined and dialyzed against 10 m*M* Tris–acetate buffer, pH 7.0, containing 1 m*M* EDTA and 1 μM DTT (1 liter). The dialyzed solution is put on an affinity column (1.5 × 10.0 cm) of Affi-Gel Blue (Bio-Rad Laboratories) equilibrated with the same buffer at a flow rate of 0.7 ml/hr. The column is washed with 20 ml of the same buffer and eluted with 50 ml of a linear gradient (0–0.5 *M*) of potassium chloride in the same buffer. The active fractions (15 ml) are combined, concentrated to a final volume of 2 ml in the Centricon-30, and stored at −20°. No loss of activity is observed in several months when the final preparation is stored at −20 to −80°. A typical purification of the enzyme is shown in Table I.[3]

Properties of *Euglena* Aquacobalamin Reductase

The optimum pH for activity was pH 7.0.[2] The enzyme, when treated at various pH values for 10 min at 55°, was stable between pH 6.0 and 8.0

TABLE I
PURIFICATION OF AQUACOBALAMIN REDUCTASE FROM *Euglena gracilis*[a]

Step	Protein (mg)	Total activity (units)	Specific activity (milliunits/mg protein)	Yield (%)
Mitochondria	986	14.7	14.9	100
Ultracentrifugation	424	12.6	29.7	85.7
First DEAE-BioGel A	29.7	12.1	407.4	82.3
Second DEAE-BioGel A	15.9	12.2	767.3	83.0
Sephadex G-100	6.6	7.9	1197.0	53.7
Affi-Gel Blue	2.5	5.6	2240.0	38.1

[a] Adapted, with permission, from Ref. 3. Copyright 1987 American Society for Biochemistry and Molecular Biology.

but completely lost its activity below pH 5.0.[3] The optimum temperature was 40°.[2] The enzyme, when incubated at various temperatures for 10 min at pH 7.0, was stable up to 50°; activity was completely lost at 60°.[3] The enzyme reaction followed Michaelis–Menten kinetics for NADPH and aquacobalamin. The apparent K_m values were 43 μM for NADPH and 55 μM for aquacobalamin.[3]

The *Euglena* aquacobalamin reductase reduced approximately 2 mol of aquacobalamin/mol of NADPH.[8] The enzyme was specific for aquacobalamin, but not for cyanocobalamin, and NADH could not replace NADPH.[3] The enzyme did not require FAD or FMN as cofactors for the reduction of aquacobalamin.[3] The enzyme reduced 2,6-dinitrophenolindophenol, potassium ferricyanide, and cytochrome *c* as well as aquacobalamin under aerobic conditions, and under anaerobic conditions it reduced FAD, FMN, methyl viologen, and benzyl viologen.[8] Cyanocobalamin was not reduced by the purified enzyme under anaerobic conditions.[8] The results indicate that the *Euglena* enzyme has an NADPH-diaphorase-like activity (Table II).

The enzyme activity was inhibited 73, 87, and 100% by incubation for 10 min with 5,5′-dithiobis(2-nitrobenzoic acid), *N*-ethylmaleimide, or mersalyl at 1 mM, respectively.[3] The results indicate that an SH group is in the active center of the enzyme. The activity was inhibited 49 and 81% by 1 mM Zn^{2+} or Al^{3+}, respectively; other metal ions (Na^+, K^+, Ni^{2+}, Mg^{2+}, Mn^{2+}, Co^{2+}, and Ca^{2+}, all at 1 mM) and 1 mM EDTA did not cause inhibition.[3] Inhibition studies with various cobalamin analogs suggest that

[8] F. Watanabe, R. Yamaji, Y. Isagawa, T. Yamamoto, Y. Tamura, and Y. Nakano, *Arch. Biochem. Biophys.* **305,** 421 (1993).

TABLE II
SUBSTRATE SPECIFICITY OF *Euglena* AQUACOBALAMIN REDUCTASE[a]

Compound	Activity (units/mg protein)
Aerobic conditions	
Aquacobalamin	2.9
Cyanocobalamin	0
2,6-Dinitrophenolindophenol	14.2
Potassium ferricyanide	25.0
Cytochrome *c*	37.0
Anaerobic conditions	
Cyanocobalamin	0
FAD	0.69
FMN	1.08
Methyl viologen	7.69
Benzyl viologen	14.01

[a] Adapted, with permission, from Ref. 8.

multiple interactions with cobalamin can occur at the binding site of the enzyme.[7]

The purified enzyme was pale yellow. Absorption peaks of the purified native enzyme occurred around 385 and 455 nm, and absorption disappeared completely on the addition of 0.2 m*M* NADPH under anaerobic conditions. The *Euglena* enzyme contained one molecule of FAD or FMN as a prosthetic group (calculated on the basis of the molecular extinction coefficients of FAD and FMN at 375 nm).[3]

The molecular weight of the *Euglena* enzyme was estimated to be 66,000 by Sephadex G-100 gel filtration and 65,000 by sodium dodecyl sulfate-polyacrylamide gel electrophoresis.[3] From the values of the amino acid composition of the purified enzyme, the molecular weight of the enzyme was calculated to be 63,559.[8]

The NH_2 terminal of the enzyme was Ala-Ala-Ala-Pro-Ser-Gly-Asn-His-Val-Thr-Ile-Leu-Tyr-Gly-Ser-Glu-.[8] The 16 amino acid residues at the NH_2 terminal of the enzyme were identical with those of the NADPH-diaphorase domain of *Euglena* pyuvate : $NADP^+$ oxidoreductase,[8] which is involved in *Euglena* wax ester fermentation.[9,10] *Euglena* pyruvate : $NADP^+$ oxidoreductase consists of two identical subunits with a monomeric molecu-

[9] H. Inui, K. Ono, K. Miyatake, Y. Nakano, and S. Kitaoka, *J. Biol. Chem.* **262,** 9130 (1987).
[10] H. Inui, R. Yamaji, H. Saidoh, K. Miyatake, Y. Nakano, and S. Kitaoka, *Arch. Biochem. Biophys.* **286,** 270 (1991).

lar weight of 166,000, and each subunit has two functional domains (the NADPH-diaphorase domain that contains FAD and a domain that contains an iron–sulfur cluster).[10] Peptide mapping of the aquacobalamin reductase and the NADPH-diaphorase domain of the pyruvate : NADP$^+$ oxidoreductase was done with a reversed-phase high-performance liquid chromatography column to find the degree of primary structural similarity between them. The peptides from the aquacobalamin reductase showed elution behavior identical to that of the NADPH-diaphorase domain.[8] Immunoblotting indicated that the *Euglena* aquacobalamin reductase had a higher molecular weight (166,000) in the intact mitochondria than the purified enzyme (65,000), and that the molecular weights of the native and purified enzyme were identical to those of the subunit and the NADPH-diaphorase domain of the pyruvate : NADP$^+$ oxidoreductase, respectively.[8] These results showed that the aquacobalamin reductase isolated earlier was the NADPH-diaphorase domain, cleaved by trypsin during preparation of the mitochondrial homogenate from the native enzyme. Purified *Euglena* pyruvate : NADP$^+$ oxidoreductase also had the activity of aquacobalamin reductase, which indicates that the enzyme in *Euglena* mitochondria has more than one function in the synthesis of cobalamin coenzymes.

An enzyme involved in the decyanation of cyanocobalamin has been found in the cell homogenate of *E. gracilis* and has been designated as cyanocobalamin reductase (NADPH; CN-eliminating) (EC 1.6.99.12).[11] The enzyme essentially requires FAD or FMN, and NADPH as cofactors and is located in the outer membrane of mitochondria.[11] Aquacobalamin reductase and cyanocobalamin reductase have different enzymological properties.

[11] F. Watanabe, Y. Oki, Y. Nakano, and S. Kitaoka, *J. Nutr. Sci. Vitaminol.* **34,** 1 (1988).

[34] Purification and Characterization of Aquacobalamin Reductases from Mammals

By Fumio Watanabe and Yoshihisa Nakano

Activity of NADH (EC 1.6.99.8)- or NADPH (EC 1.6.99.11)-linked aquacobalamin reductase (or both) has been found in various tissues of mammals; liver contains both enzymes with higher specific activities than

0076-6879/97 $25

any other tissue.[1] Both NADH- and NADPH-linked enzymes are distributed in mitochondria and microsomes of human and rat livers.[2,3] Cobalamin deficiency increases the specific activities of rat liver NADH- and NADPH-linked enzymes.[4] Mitochondrial NADH- or NADPH-linked enzyme activity decreases significantly in human skin fibroblasts with inherited disorders in the synthesis of cobalamin coenzymes.[5] The mammalian enzymes have been purified and characterized only from rat liver.[6–9]

This chapter summarizes our work on the NADH- and NADPH-linked aquacobalamin reductases from rat liver.

Aquacobalamin Reductase Assay

Principle

Rat liver aquacobalamin reductases are assayed by the method described in [33] in this volume.[9a]

Reagents

Tris–acetate buffer (pH 7.5), 100 m*M*, for the microsomal NADPH-linked enzyme
Tris–acetate buffer (pH 6.6), 100 m*M*, for the microsomal NADH-linked enzyme
Tris–acetate buffer (pH 8.2), 100 m*M* for the mitochondrial NADPH-linked enzyme
Tris–acetate buffer (pH 7.1), 100 m*M*, for the mitochondrial NADH-linked enzyme

[1] F. Watanabe, Y. Nakano, N. Tachikake, Y. Tamura, H. Yamanaka, and S. Kitaoka, *J. Nutr. Sci. Vitaminol.* **36,** 349 (1990).
[2] F. Watanabe, Y. Nakano, N. Tachikake, S. Kitaoka, Y. Tamura, H. Yamanaka, S. Haga, S. Imai, and H. Saido, *Int. J. Biochem.* **23,** 531 (1991).
[3] F. Watanabe, Y. Nakano, S. Maruno, N. Tachikake, Y. Tamura, and S. Kitaoka, *Biochem. Biophys. Res. Commun.* **165,** 675 (1989).
[4] F. Watanabe, Y. Nakano, N. Tachikake, H. Saido, Y. Tamura, and H. Yamanaka, *J. Nutr.* **121,** 1948 (1991).
[5] F. Watanabe, H. Saido, R. Yamaji, K. Miyatake, Y. Isegawa, A. Ito, T. Yubisui, D. S. Rosenblatt, and Y. Nakano, *J. Nutr.* **126,** 2947 (1996).
[6] F. Watanabe, Y. Nakano, H. Saido, Y. Tamura, and H. Yamanaka, *Biochim. Biophys. Acta* **1119,** 175 (1992).
[7] F. Watanabe, Y. Nakano, H. Saido, Y. Tamura, and H. Yamanaka, *J. Nutr.* **122,** 940 (1992).
[8] H. Saido, F. Watanabe, Y. Tamura, Y. Fumae, S. Imaoka, and Y. Nakano, *J. Nutr.* **123,** 1868 (1993).
[9] H. Saido, F. Watanabe, Y. Tamura, K. Miyatake, A. Ito, Y. Yubisui, and Y. Nakano, *J. Nutr.* **124,** 1037 (1994).
[9a] F. Watanabe and Y. Nakano, *Methods Enzymol.* **281,** [33], 1997 (this volume).

Hydroxocobalamin hydrochloride, 10 m*M* (Sigma, St. Louis, MO)
NADPH, 20 m*M*, for the mitochondrial and microsomal NADPH-linked enzymes
NADH, 20 m*M*, for the mitochondrial and microsomal NADH-linked enzymes

Procedures

The reaction mixture in a total volume of 1.0 ml consists of 0.5 ml of Tris–acetate buffer, 0.01 ml of hydroxocobalamin, 0.47 ml of distilled water, and 0.01 ml of enzyme. The mixture is incubated at 45, 50, 50, or 40° (for the microsomal NADPH-, NADH-, mitochondrial NADPH-, or NADH-linked enzyme, respectively) in a 1.5-ml cuvette, and the reaction is then started by the addition of 0.01 ml of NAD(P)H. The reaction is monitored by measurement of the decrease in absorbance by aquacobalamin at 525 nm and calculated on the basis of the differential molecular extinction coefficient of aquacobalamin to cob(II)alamin ($5.57 \times 10^3\ M^{-1}\ cm^{-1}$ at 525 nm). This assay is performed with a dual-beam spectrophotometer with water-jacketed cuvette holder.

One unit of the enzyme activity is defined as the amount of enzyme that catalyzes the reduction of aquacobalamin at the rate of 1 μmol/min.

Purification Procedures

Preparation of Rat Liver Mitochondrial Membranes and Microsomes

All procedures are done at 4°. Livers are obtained from male Wistar strain rats starved for 24 hr; the livers are washed with chilled 0.9% (w/v) NaCl solution, cut into small pieces with a razor blade, and homogenized in about 10 vol of 10 m*M* Tris-HCl buffer, pH 7.5, containing 0.25 *M* sucrose, 50 m*M* KCl, 2 m*M* $MgCl_2$, and 0.1 m*M* EDTA by use of a glass homogenizer with a Teflon pestle. The homogenate is filtered through a double layer of gauze to remove unbroken tissues and then centrifuged at 500 *g* for 10 min to remove unbroken cells. The supernatant, after centrifugation at 12,000 *g* for 20 min, is further centrifuged at 100,000 *g* for 60 min. The precipitate is resuspended in 10 m*M* Tris-HCl buffer, pH 7.0, containing 1 m*M* EDTA and 0.1 m*M* dithiothreitol (DTT) and centrifuged at 100,000 *g* for 60 min to wash the precipitate. The washing of the precipitate is repeated twice. The final precipitate is suspended in the washing buffer and used as the microsomal fraction.

In the case of preparation of mitochondrial membranes, livers are homogenized in 10 m*M* Tris–acetate buffer, pH 7.5, containing 0.25 *M* sucrose,

1 m*M* EDTA, and 1 μM DTT. The homogenate, after having been filtered through a double layer of gauze and centrifuged at 500 *g* for 10 min, is further centrifuged at 10,000 *g* for 20 min. The precipitate (mitochondrial fraction) is resuspended in 10 m*M* Tris–acetate buffer, pH 7.5, containing 0.25 *M* sucrose, 1 m*M* EDTA, and 1 μM DTT, and centrifuged at 10,000 *g* for 20 min to wash the mitochondria. The washing procedures are repeated twice. The mitochondria are suspended in the washing buffer and disrupted by sonic treatment for 80 sec at 10 kHz. The treated mitochondria are centrifuged at 100,000 *g* for 60 min. The precipitate is resuspended in 10 m*M* Tris–acetate buffer, pH 7.5, containing 30% (v/v) ethylene glycol, 1 m*M* EDTA, and 1 μM DTT, and centrifuged at 100,000 *g* for 60 min to wash the precipitate. The washing procedures are repeated twice. The final precipitate is suspended in the same buffer and used as mitochondrial membranes.

Microsomal NADPH-Linked Aquacobalamin Reductase from Rat Liver

Solubilization mixture for the rat liver microsomal NADPH-linked aquacobalamin reductase contains 50 m*M* Tris-HCl buffer (pH 7.5), 20% (w/v) ethylene glycol, 1% (w/v) Triton X-100, 1% (w/v) sodium deoxycholate, and the rat liver microsomal fraction (4 mg of protein/ml of mixture).[6] The enzyme is solubilized with stirring at 0° for 60 min. The mixture is centrifuged at 100,000 *g* for 60 min and the supernatant is subjected to the following purification.

Purification procedures are performed at 0–4°. The solubilized fraction (250 ml) is put on a column (3.0 × 10.0 cm) of DEAE-cellulose (Wako Pure Chemical Industries, Osaka, Japan) equilibrated with 10 m*M* Tris-HCl buffer, pH 8.0, containing 20% (w/v) ethylene glycol, 0.2% (w/v) Triton X-100, 1 m*M* EDTA, and 0.1 m*M* DTT, and eluted at a flow rate of 20 ml/hr. The column is washed with 100 ml of the same buffer and then eluted with 300 ml of a linear gradient (0–0.5 *M*) of potassium chloride in the same buffer. The active fractions (34 ml) are combined, dialyzed overnight against the same buffer (2 liters), and put on a column (1.0 × 5.0 cm) of 2′,5′-ADP–Sepharose 4B (Pharmacia-LKB Biotechnology, Piscataway, NJ) equilibrated with the same buffer. The column is washed with 100 ml of the same buffer and eluted with 30 ml of 1 m*M* NADPH in the same buffer. The active fractions are combined and concentrated by dialyzing against polyethylene glycol 20,000. The concentrated solution is put on a column (1.0 × 100.0 cm) of Toyopearl HW55 (Tosoh, Tokyo, Japan) equilibrated with 100 m*M* Tris-HCl buffer, pH 8.0, containing 20% (w/v) ethylene glycol, 0.2% (w/v) Triton X-100, 1 m*M* EDTA, and 0.1 m*M* DTT and eluted with the same buffer at a flow rate of 20 ml/hr. The active fractions are combined, concentrated to a final volume of 1.0 ml with polyethylene glycol 20,000, and stored at −20°. A typical purification of the enzyme is shown in Table I.

Mitochondrial NADPH-Linked Aquacobalamin Reductase from Rat Liver

The mixture for solubilization of the mitochondrial NADPH-linked aquacobalamin reductase contains 50 m*M* Tris–acetate buffer (pH 7.5), 30% (w/v) ethylene glycol, 0.1% (w/v) Triton X-100, 0.1% (w/v) sodium deoxycholate, and the mitochondrial membranes (4 mg of protein/ml of mixture).[8] The enzyme is solubilized with stirring at 0° for 30 min, and centrifuged at 100,000 *g* for 60 min. The supernatant is subjected to the following purification.

All purification procedures are done at 4°. The solubilized fraction (110 ml) is put on a column (2.0 × 10.0 cm) of DEAE-BioGel A (Bio-Rad Laboratories, Richmond, CA) equilibrated with 10 m*M* Tris–acetate buffer, pH 7.5, containing 30% (w/v) ethylene glycol, 0.5% (w/v) Triton X-100, 1 m*M* EDTA, and 1 μM DTT at a flow rate of 15 ml/hr. The column is washed with 100 ml of the same buffer, and then eluted with 300 ml of a linear gradient (0–0.5 *M*) of potassium chloride in the same buffer at a flow rate of 20 ml/hr. The active fractions (39 ml) are combined and dialyzed overnight against the same buffer (2 liters). The dialyzed solution is put on a column (1.0 × 5.0 cm) of 2′,5′-ADP–Sepharose 4B (Pharmacia-LKB Biotechnology) equilibrated with the same buffer at a flow rate of 4 ml/hr. The column is washed with 12 ml of the same buffer and eluted with 10 m*M* NADPH in the same buffer at a flow rate of 4 ml/hr. The active fractions (4 ml) are combined and concentrated with a Centricon-10 microconcentrator (Amicon, Danvers, MA). The concentrated solution (0.5 ml) is put on a column (1.0 × 90.0 cm) of Sephadex G-150 (Pharmacia-LKB Biotechnology) equilibrated with 50 m*M* Tris–acetate buffer, pH 7.5, containing 30% (w/v) ethylene glycol, 0.5% (w/v) Triton X-100, 0.3 *M* potassium chloride, 1 m*M* EDTA, and 1 μM DTT, and eluted with the same buffer at a flow rate of 8 ml/hr. The active fractions (30 ml) are combined and dialyzed overnight against 2 liters of 10 m*M* Tris–acetate buffer, pH 7.5, containing 30% (w/v) ethylene glycol, 0.5% (w/v) Triton X-100, 1 m*M* EDTA, and 1 μM DTT. The dialyzed solution is put on a column (1.0 × 6.0 cm) of cobalamin-conjugated Affi-Gel 102 equilibrated with the same buffer at a flow rate of 5 ml/hr. The column is washed with 20 ml of the same buffer and eluted with 30 ml of a linear gradient (0–1.0 *M*) of potassium chloride in the same buffer at a flow rate of 20 ml/hr. The active fractions (5 ml) are combined and concentrated with a Centricon-10. The

TABLE I

PURIFICATION OF NADPH-LINKED AQUACOBALAMIN REDUCTASE FROM RAT LIVER MICROSOMES[a]

Step	Protein (mg)	NADPH-aquacobalamin reductase: Specific activity (units/mg protein)	NADPH-aquacobalamin reductase: Yield (%)	NADPH-cytochrome *c* reductase: Specific activity (units/mg protein)	NADPH-cytochrome *c* reductase: Yield (%)	Activity ratio (cytochrome-*c* reductase/aquacobalamin reductase)
Microsomal membrane	1000	0.021	100	0.099	100	4.71
Solubilization	978.4	0.021	97.9	0.099	97.9	4.71
DEAE-cellulose	466.4	0.031	68.8	0.146	68.8	4.71
2′,5′-ADP–Sepharose	5.8	1.345	37.1	6.673	39.1	4.96
Toyopearl HW55	0.6	11.220	32.1	54.865	33.2	4.89

[a] Adapted from *Biochim. Biophys. Acta* **1119,** F. Watanabe *et al.*, pp. 175–177, Copyright 1992 with kind permission of Elsevier Science–NL, Sara Burgerhartstraat 25, 1055 KV Amsterdam, The Netherlands.

TABLE II
PURIFICATION OF RAT LIVER MITOCHONDRIAL NADPH-LINKED AQUACOBALAMIN REDUCTASE[a]

Step	Total activity (milliunits)	Total protein (mg)	Specific activity (milliunits/mg protein)	Yield (%)
Membrane fraction	10,199.6	575.5	17.7	100
Solubilized fraction	8,046.5	332.0	24.2	78.9
DEAE-BioGel A	6,976.5	106.7	65.4	68.4
2′,5′-ADP–Sepharose 4B	4,695.5	6.6	711.4	46.0
Sephadex G-150	2,383.3	0.99	2,407.4	23.0
Cobalamin-conjugated Affi-Gel 102	1,514.8	0.25	6,059.2	14.9
Asahipak GS-320	765.0	0.12	6,375.0	7.5

[a] Adapted, with permission, from Ref. 8. Copyright *J. Nutr.* **123,** pp. 1868–1874, American Institute of Nutrition.

concentrated solution (1 ml) is applied on a high-performance liquid chromatography (HPLC) gel-filtration column (7.5 × 500.0 mm) of Asahipak GS-320 (Asahi Chemical Industry, Tokyo, Japan) equilibrated with 50 m*M* Tris–acetate buffer, pH 7.5, containing 10% (w/v) ethylene glycol, 0.5% (w/v) Triton X-100, and 0.3 *M* potassium chloride, and eluted with the same buffer at a flow rate of 1 ml/min at 10°. The active fractions (12 ml) are combined, concentrated with a Centricon-10, and stored at −60°. A typical purification of the enzyme is shown in Table II.

Preparation of Cobalamin-Conjugated Affi-Gel 102. Cobalamin-conjugated Affi-Gel 102 is prepared according to the method of Sato *et al.*[10] Affi-Gel 102 (Bio-Rad Laboratories) is used instead of aminobutyl-Sepharose 4B. The amount of cobalamin bound to Affi-Gel 102 is 35 nmol/ml of gel.

Microsomal NADH-Linked Aquacobalamin Reductase from Rat Liver

Most of the enzyme is solubilized with 1% (w/v) sodium deoxycholate from rat liver microsomal membranes.[7] The elution behavior of the solubilized enzyme is identical to that of NADH–cytochrome-*c* reductase (cytochrome b_5/cytochrome-b_5 reductase complex) during DEAE-Toyopearl 650 (Tosoh) column chromatography. By mixing both purified cytochrome b_5 and cytochrome-b_5 reductase, cob(II)alamin is immediately formed from

[10] K. Sato, E. Hiei, S. Shimizu, and R. H. Abeles, *FEBS Lett.* **85,** 73 (1978).

TABLE III

REDUCTION OF SOME ELECTRON ACCEPTORS BY NADH WITH THE PURIFIED CYTOCHROME-b_5 REDUCTASE AND/OR CYTOCHROME b_5[a]

	Reduction activity (units/mg protein)		
Cytochrome	Potassium ferricyanide (1 mM)	Cytochrome c (20 μM)	Aquacobalamin (0.1 mM)
Cytochrome-b_5 reductase	1849.3 ± 54.8	ND[b]	ND
Cytochrome b_5	ND	ND	ND
Cytochrome b_5 plus cytochrome-b_5 reductase		35.1 ± 1.7	3.3 ± 0.1

[a] Adapted, with permission, from Ref. 7. Copyright *J. Nutr.* **123,** pp. 1868–1874, American Institute of Nutrition.

[b] ND, Not detectable.

aquacobalamin with NADH (Table III). The results provide evidence that cytochrome b_5/cytochrome-b_5 reductase complex has the activity of the NADH-linked aquacobalamin reductase.

Cytochrome b_5 and cytochrome-b_5 reductase are solubilized and purified from rat liver microsomes by the methods of Stritmatter *et al.*[11] and Mihara and Sato,[12] respectively.

Mitochondrial NADH-Linked Aquacobalamin Reductase from Rat Liver

Most of the enzyme is solubilized with 1% (w/v) Triton X-100 from rat liver mitochondrial membranes.[9] The elution behavior of the solubilized enzyme is identical to that of NADH-cytochrome-c reductase (b-type cytochromes/cytochrome-b_5 reductase complex) during DEAE-Sepharose Fast Flow (Pharmacia-LKB Biotechnology) column chromatography. By mixing both purified cytochrome b_5-like hemoprotein (outer membrane cytochrome b) and cytochrome-b_5 reductase, cob(II)alamin is immediately formed from aquacobalamin with NADH (Table IV). The results provide evidence that the outer membrane cytochrome b/mitochondrial cytochrome-b_5 reductase complex has the activity of the NADH-linked aquacobalamin reductase in rat liver mitochondria.

Outer membrane cytochrome b and cytochrome-b_5 reductase are solubilized and purified from rat liver mitochondrial membranes by the methods of Ito[13] and Mihara and Sato,[12] respectively.

[11] P. Stritmatter, M. Rogers, and L. Spats, *J. Biol. Chem.* **247,** 7188 (1972).

[12] K. Mihara and R. Sato, *Methods Enzymol.* **52,** 102 (1978).

[13] A. Ito, *J. Biochem.* **87,** 63 (1980).

TABLE IV
REDUCTION OF SOME ELECTRON ACCEPTORS BY NADH WITH THE PURIFIED OUTER MEMBRANE CYTOCHROME-b AND/OR CYTOCHROME-b_5 REDUCTASE

Cytochrome	Reduction activity (units/mg protein)		
	Potassium ferricyanide (1 mM)	Cytochrome c (20 μM)	Aquacobalamin (0.1 mM)
Cytochrome-b_5 reductase	18.57 ± 0.05	ND[b]	ND
OM cytochrome b	ND	ND	ND
OM cytochrome b plus cytochrome-b_5 reductase		6.89 ± 0.03	0.34 ± 0.03

[a] OM, Outer membrane. Adapted, with permission, from Ref. 9. Copyright *J. Nutr.* **124,** pp. 1037–1040, American Institute of Nutrition.
[b] ND, Not detectable.

Properties

Microsomal NADPH-Linked Aquacobalamin Reductase

The optimum pH and temperature for activity are pH 7.5 and 45°, respectively.[6] The enzyme reaction follows Michaelis–Menten type kinetics for NADPH and aquacobalamin. The apparent K_m values were 58 μM for aquacobalamin, and 14 μM for NADPH. The enzyme was specific for aquacobalamin, but not for cyanocobalamin.

The monomeric molecular weight of the purified enzyme was estimated to be 79,000 by sodium dodecyl sulfate (SDS)–polyacrylamide gel electrophoresis.

The purified enzyme had the ability to reduce cytochrome c, potassium ferricyanide, and 2,6-dichlorophenolindophenol as well as aquacobalamin. About 33% of the NADPH-cytochrome-c reductase activity that was found in rat liver microsomal fraction was recovered in the final purified preparation. The activity ratio of NADPH-cytochrome-c reductase/NADPH-linked aquacobalamin reductase was about 5.0 through the purification step. Mammalian NADPH-cytochrome-c reductase has been reported to be a flavoprotein with a monomeric molecular weight of 78,000–79,000.[14] These results indicate that the rat liver microsomal NADPH-linked aquacobalamin reductase is the NADPH-cytochrome-c (P-450) reductase.

[14] Y. Yasukochi and B. S. S. Masters, *J. Biol. Chem.* **251,** 5337 (1976).

Mitochondrial NADPH-Linked Aquacobalamin Reductase from Rat Liver

The optimum pH and temperature for activity were pH 8.2 and 50°, respectively.[8] The enzyme reaction followed Michaelis–Menten kinetics for aquacobalamin and NADPH. The apparent K_m values were 208 μM for aquacobalamin, and 4.8 μM for NADPH. The enzyme activity was completely inhibited by sulfhydryl inhibitors, *N*-ethylmaleimide, and mersalyl at 1 mM. The enzyme activity was activated slightly by chelators, EDTA, and *o*-phenanthroline at 1 mM, and inhibited completely by 1 mM Zn^{2+}, Cu^{2+}, and Hg^{2+}. Other bivalent and trivalent cations (20–50% inhibition by Mg^{2+}, Ca^{2+}, Ba^{2+}, Mn^{2+}, Co^{2+}, Fe^{2+}, and Al^{3+}, all at 1 mM) were more effective inhibitors than monovalent cations (<10% inhibition by K^+, Na^+, and Li^+, all at 1 mM). The enzyme had the ability to reduce cytochrome *c*, potassium ferricyanide, 2,6-dichlorophenolindophenol, as well as aquacobalamin.

The molecular weight of the enzyme was estimated to be 65,000 by Toyopearl HW55 gel filtration and by SDS–polyacrylamide gel electrophoresis.

The purified enzyme was a light yellow; the enzyme contained 1 mol of FAD and FMN/mol of enzyme as prosthetic groups.

Although the enzyme immunoreacted with an antibody against NADPH-cytochrome *P*-450 reductase, which had the activity of the NADPH-linked aquacobalamin reductase in rat liver microsomes, the mitochondrial enzyme and the microsomal enzyme had different peptide maps on C_{18} reversed-phase high-performance liquid chromatography (HPLC).

Microsomal NADH-Linked Aquacobalamin Reductase from Rat Liver

The optimum pH and temperature for the reduction of aquacobalamin were pH 6.6 and 50°, respectively.[7] The specific activity (127.5 ± 11.52 milliunits/mg protein) of the enzyme was determined under physiological conditions (pH 7.0 at 37°). The apparent K_m values were 32 μM for aquacobalamin and 10 μM for NADH. The enzyme was specific for aquacobalamin, and cyanocobalamin was not reduced by the enzyme.

Mitochondrial NADH-Linked Aquacobalamin Reductase from Rat Liver

The specific activity (109.5 ± 14.3 milliunits/mg protein) of the enzyme was assayed under physiological conditions (pH 7.1 at 40°).[9] The optimum pH and temperature for activity were pH 7.1 and 40°, respectively. The apparent K_m values were 41.9 μM for aquacobalamin and 14.4 μM for

NADH. The enzyme was specific for aquacobalamin, and cyanocobalamin could not be reduced by the enzyme.

Aquacobalamin Reductase Activities in Cobalamin-Deficient Rats and Human Skin Fibroblasts with Inherited Disorders

The specific activities of NADH- and NADPH-linked enzymes[4] in homogenates of liver, kidney, or upper intestine were shown to be 3- to 20-fold greater in cobalamin-deficient rats than in control (cobalamin-sufficient) rats. In liver, cobalamin deficiency specifically elevated the specific activities of the mitochondrial NADH-linked and microsomal NADPH-linked enzymes. In human mutant fibroblasts (cblC and A cells),[5] the activities of both mitochondrial NADH-linked and NADPH-linked enzymes decreased significantly, but both microsomal enzymes did not. They are likely the isozymes involved in cobalamin–coenzyme synthesis in mammalian cells.

Section III

Heme

[35] Continuous Coupled Assay for 5-Aminolevulinate Synthase

By PETER M. SHOOLINGIN-JORDAN, JEREMY E. LELEAN, and ADRIAN J. LLOYD

The fundamental tetrapyrrole macrocycle, uroporphyrinogen III, is biosynthesized from eight molecules of 5-aminolevulinic acid by three enzyme reactions common to all organisms that synthesize tetrapyrroles.[1] 5-Aminolevulinic acid is formed via two pathways. One is found in nonphotosynthetic bacteria and plants that utilize glutamate as the carbon skeleton and is known as the C_5 pathway.[2] The second route is the C_4 or Shemin pathway, which uses glycine and succinyl-CoA and is found in animals, fungi, and photosynthetic bacteria.[1] The latter involves the key enzyme 5-aminolevulinate synthase (EC 2.3.1.37), which catalyzes the condensation between glycine and succinyl-CoA to yield 5-aminolevulinic acid, coenzyme A (CoASH), and CO_2.

$$\text{Glycine} + \text{succinyl-CoA} \xrightarrow{\text{5-aminolevulinate synthase}} \text{5-aminolevulinic acid} + \text{CoASH} + CO_2$$

The enzyme was first described in the photosynthetic bacterium *Rhodobacter spheroides*[3] and has since been isolated from several sources including fungi and animals (for a review see Ferreira and Gong[4]). 5-Aminolevulinate synthase is highly specific for glycine, with no other amino acid acting as a substrate.[4,5] However, there is evidence[5] that the specificity for the acyl-CoA substrate is less rigid, although no supporting data have yet been published.

The genes/cDNAs specifying several 5-aminolevulinate synthases have been cloned and sequenced and some have been overexpressed in bacterial or fungal systems.[4] In humans, two genes specifying 5-aminolevulinate synthase have been identified,[6] one encoding the ubiquitous (housekeeping) enzyme and the other encoding a distinct erythroid form. Mutations in the

[1] P. M. Jordan, *in* "New Comprehensive Biochemistry" (P. M. Jordan, ed.; A. Neuberger and L. L. M. van Deenen, series eds.), Vol. 19, p. 1. Elsevier, Amsterdam, 1991.

[2] C. G. Kannangara, S. P. Gough, P. Bruyant, J. K. Hoober, A. Kahn, and D. von Wettstein, *Trends Biochem. Sci.* **13,** 139 (1988).

[3] G. Kikuchi, A. Kumar, P. Talmage, and D. Shemin, *J. Biol. Chem.* **233,** 1214 (1958).

[4] G. C. Ferreira and J. Gong, *J. Bioenerg. Biomembr.* **27,** 151 (1995).

[5] P. M. Jordan and D. Shemin, "The Enzymes" (P. D. Boyer, ed.), 3rd Ed., Vol. VII, p. 339. Academic Press, New York, 1973.

[6] D. F. Bishop, *Nucleic Acids Res.* **18,** 7187 (1990).

0076-6879/97 $25

latter gene lead to some forms of X-linked sideroblastic anemia.[7] In mouse two 5-aminolevulinate synthases also exist[8] and the murine erythroid enzyme has been overexpressed and characterized.[9] Two 5-aminolevulinate synthase genes, *hemA* and *hemT*, have also been characterized in *R. spheroides.*[10]

Comparison of the nucleotide-derived amino acid sequences of all 5-aminolevulinate synthases[4] shows a considerable amount of structural conservation, particularly in the catalytic region, which is found in a C-terminal 48-kDa core.[11]

The 5-aminolevulinate synthase enzymes have been studied extensively by enzyme kinetics.[9,12] The mechanism has been shown to proceed by an ordered Bi–Bi reaction with glycine binding before succinyl-CoA and with CoA, CO_2, and 5-aminolevulinic acid being released from the enzyme in that order. Detailed stereochemical studies[13,14] using stereospecifically labeled glycine have established that the *pro-R* hydrogen atom of glycine is lost during the course of the reaction and that the mechanism proceeds overall by stereochemical inversion, with the enzyme also catalyzing the decarboxylation.[15] 5-Aminolevulinate synthases belong to the α-family of pyridoxal 5′-phosphate-requiring enzymes[16] and are closely related to several other enzymes that catalyze condensation/decarboxylation reactions.[17]

The availability of a large number of 5-aminolevulinate synthases has led to a renewed interest in the enzyme because it is the key regulatory step in tetrapyrrole biosynthesis. The implication of the erythroid enzyme in human genetic diseases and studies on the enzyme mechanism by site-directed mutagenesis have also meant that it is important to be able to analyze the activity in an accurate assay. These factors have made it necessary to have available a continuous assay method that has the capability of being used for investigating a range of enzymatic properties. The classic assay for 5-aminolevulinate synthase,[18] while being relatively sensitive, is

[7] T. C. Cox, S. S. Bottomley, J. S. Wiley, M. J. Bawden, C. S. Matthews, and B. K. May, *N. Engl. J. Med.* **330,** 675 (1994).

[8] R. D. Riddle, M. Yamamoto, and J. D. Engel, *Proc. Natl. Acad. Sci. U.S.A.* **86,** 2717 (1989).

[9] G. C. Ferreira and H. A. Dailey, *J. Biol. Chem.* **268,** 584 (1993).

[10] E. L. Niedle and S. Kaplan, *J. Bacteriol.* **175,** 2292 (1993).

[11] H. Munakata, T. Yamagani, T. Najai, M. Yamamoto, and N. Hayashi, *J. Biochem.* **114,** 103 (1993).

[12] M. Fancia-Gaigner and J. Clement-Metral, *Eur. J. Biochem.* **40,** 19 (1973).

[13] P. M. Jordan and M. Akhtar, *Tetrahedron Lett.* **11,** 875 (1969).

[14] Z. Zaman, P. M. Jordan, and M. Akhtar, *Biochem. J.* **135,** 257 (1973).

[15] Abboud, P. M. Jordan, and M. Akhtar, *J. Chem. Soc. Chem. Commun.* 643 (1974).

[16] F. W. Alexander, E. Sandmeier, P. K. Mehta, and P. Christen, *Eur. J. Biochem.* **219,** 953 (1994).

[17] O. Ploux and A. Marquet, *Eur. J. Biochem.* **236,** 301 (1996).

[18] D. Mauzerall and S. Granick, *J. Biol. Chem.* **219,** 435 (1956).

discontinuous and is not ideal for kinetic analysis. Furthermore, this assay is not suitable for investigating quantitatively substrate specificity with acyl-CoA species other than succinyl-CoA because the chemical stability and reaction of the α-aminoketone products with acetylacetone to form pyrroles are rather unpredictable. Because of these limitations an alternative direct assay method has been developed that involves measuring the glycine-dependent production of CoASH by 5-aminolevulinate synthase (ALS). The CoASH is determined by conversion into acetyl-CoA with pyruvate dehydrogenase (PDH). The concomitant oxidation of pyruvate to acetyl-CoA is accompanied by the reduction of NAD^+ to NADH, resulting in an increase of the absorbance at 340 nm. The reactions are summarized as follows in reactions (1) and (2):

$$\text{Glycine} + \text{succinyl-CoA} \xrightarrow{\text{ALS}} \text{5-AL} + \text{CoASH} + CO_2 \quad (1)$$

$$\text{CoASH} + \text{pyruvate} + NAD^+ \xrightarrow{\text{PDH}} \text{acetyl-CoA} + CO_2 + \text{NADH} + H^+ \quad (2)$$

Another direct assay has also been developed that involves the continuous regeneration of succinyl-CoA using 2-ketoglutaric acid dehydrogenase.[19] However, the assay described here has the important advantage that the acetyl-CoA generated does not interfere in the 5-aminolevulinate synthase reaction studied here and thus the assay can be used reliably for the analysis of acyl-CoA substrates other than succinyl-CoA.

Reagents

Morpholinepropanesulfonic acid (MOPS) buffer (pH 6.8), 0.5 *M*: Prepare by dissolving 104.7 g of MOPS in 500 ml of distilled water and adjusting the pH with concentrated (11 *M*) HCl
Sodium pyruvate, 10 m*M* (Sigma, St. Louis, MO)
Pyridoxal 5′-phosphate, 10 m*M* (Sigma)
Succinyl-CoA, 10 m*M*: Prepare from 10 mg of CoA, 1 mg of freshly powdered succinic anhydride, and 4.5 mg of $NaHCO_3$ at 0° according to the method of Simon and Shemin[20]
Glycine, 1 *M*: Dissolve in water and adjust to pH 6.8 with 1 *M* NaOH
Solution A: Combine the following components and adjust to pH 7 (prepare fresh before use):

NAD^+, 0.8 m*M* (Park Scientific, Northampton, UK)
Thiamin pyrophosphate, 0.8 m*M* (Sigma)

[19] G. A. Hunter and G. C. Ferreira, *Anal. Biochem.* **226,** 221 (1995).
[20] E. J. Simon and D. Shemin, *J. Am. Chem. Soc.* **75,** 2520 (1953).

MgCl$_2$, 40 m*M*
Dithiothreitol (DTT), 4 m*M*

Pyruvate dehydrogenase: Bovine heart enzyme from Sigma contains 3 μmol/min/ml of enzyme activity
Acyl-CoA derivatives, 10 m*M* (Sigma): Prepare fresh from lyophilized powder

Reagents Used for Discontinuous Assay

5-Aminolevulinate synthase
Solution B: Combine the following components:

Potassium phosphate buffer (pH 6.8), 0.25 *M*
Glycine, 0.25 *M*
Pyridoxal 5′-phosphate, 50 μ*M*

Succinyl-CoA, 10 m*M* (see above)
Modified Ehrlich's reagent: Prepare fresh by dissolving 1 g of 4-dimethylaminobenzaldehyde (Merck, Rahway, NJ) in 42 ml of glacial acetic acid and 8 ml of 70% (v/v) perchloric acid)
Trichloroacetic acid (10%, w/v)
Sodium acetate buffer (pH 4.6), 1 *M*
Acetylacetone (freshly distilled *in vacuo* if yellow in color)

Preparation of 5-Aminolevulinate Synthases

Purification of 5-Aminolevulinate Synthase from Rhodobacter spheroides

5-Aminolevulinate synthase is purified to homogeneity[21] from cells of *R. spheroides* (NCIB 8253) grown semianerobically at 28–30°, according to Lascelles.[22,23] The enzyme is purified from 100 g (wet weight) of cells in a yield of 30% with a specific activity of 120 μmol/mg protein/hr. The enzyme is stable for several weeks when stored at −20°.

Purification of Mouse Erythroid 5-Aminolevulinate Synthase

Purification of mouse erythroid ALS is carried out using an *Escherichia coli* strain overexpressing the mouse erythroid ALS gene.[9] The enzyme is purified from 5 g (wet weight) of *E. coli* cells in a yield of approximately

[21] P. M. Jordan and A. Laghai-Newton, *Methods Enzymol.* **123,** 435 (1986).
[22] J. Lascelles, *Biochem. J.* **62,** 78 (1956).
[23] J. Lascelles, *Biochem. J.* **72,** 508 (1959).

50% with a specific activity of 180 nmol/mg protein/hr. The enzyme is stable for several weeks at −20°.

Methods

Assay of 5-Aminolevulinate Synthase

5-Aminolevulinate Synthase: Pyruvate Dehydrogenase-Linked Enzyme Assay. The assay is carried out at 37° in a final volume of 1 ml containing 200 μl of MOPS buffer (pH 6.8), 20 μl of sodium pyruvate, 30 μl of glycine, 250 μl of solution A, 20 μl of succinyl-CoA, 10 μl of pyridoxal 5′-phosphate, 40 μl of pyruvate dehydrogenase, and 5-aminolevulinate synthase (~0.15 μmol/min). The change in absorbance at 340 nm is monitored by a Uvikon 930 spectrophotometer (Park Scientific, Northampton, UK) and enzyme rates are calculated assuming an extinction coefficient for NADH of 6220 M^{-1} cm^{-1}.

Mauzerall and Granick Method to Determine 5-Aminolevulinic Acid by Formation of 5-Aminolevulinic Acid Pyrrole and Its Reaction with Ehrlich's Reagent. The assay is carried out by a modification of the method of Mauzerall and Granick.[18] Incubations are carried out at 37° for 30 min in a final volume of 175 μl containing 100 μl of reagent B, 20 μl of succinyl-CoA, and 5-aminolevulinate synthase (0.15 μmol/min). Incubations are terminated by the addition of 150 μl of 10% (w/v) trichloroacetic acid and any protein precipitate is removed by centrifugation for 3 min in a bench-top centrifuge. The supernatant is decanted to a fresh tube containing 300 μl of sodium acetate buffer (pH 4.6), and 25 μl of acetylacetone is then added. The mixture is heated in covered test tubes in a boiling water bath for 10 min and, after cooling the solution, an equal volume of modified Ehrlich's reagent is added. The absorbance of the resultant solution is measured at 553 nm in a Hitachi (Tokyo, Japan) U-200 spectrophotometer. Enzyme rates are calculated assuming an extinction coefficient for the pyrrole : Ehrlich's reagent adduct of 60,400 M^{-1} cm^{-1}.

Discussion and Comments

A direct 5-aminolevulinate synthase enzyme assay has been developed on the basis of linking 5-aminolevulinate synthase to pyruvate dehydrogenase [Eqs. (1) and (2)]. Initial investigations demonstrated that pyruvate dehydrogenase was unaffected by components of the 5-aminolevulinate synthase assay system and vice versa.

When 5-aminolevulinate synthase activity was determined using this coupled assay a linear increase in absorbance was obtained. The rate was dependent on the presence of both succinyl-CoA and glycine and was linear

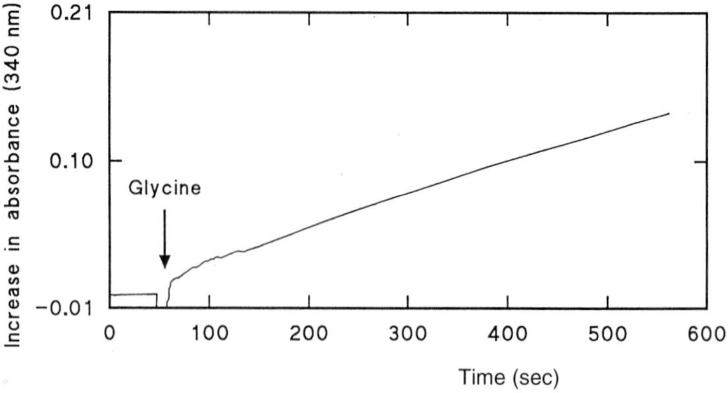

FIG. 1. Time course of *R. spheroides* 5-aminolevulinate synthase-catalyzed CoA accumulation. The reaction was initiated by the addition of glycine at the point shown (↓).

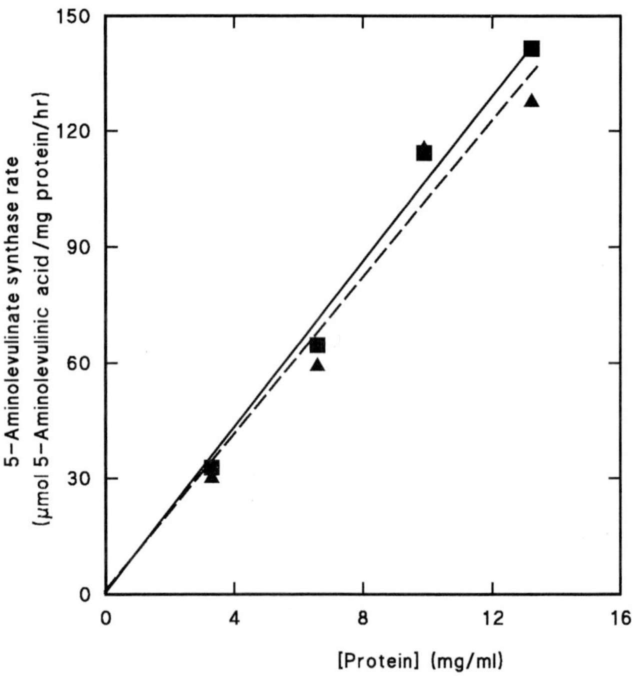

FIG. 2. Linear dependence of the pyruvate dehydrogenase and Mauzerall and Granick 5-aminolevulinate synthase assays on protein concentration. (■). Data obtained from the pyruvate dehydrogenase assay; (▲) data obtained from the Mauzerall and Granick assay.

with respect to both time and protein concentration (see Figs. 1 and 2). A Lineweaver–Burk plot of dependence of activity of the *R. spheroides* enzyme on succinyl-CoA gave values for K_m and V_{max} of 6 μM and 138 μmol/mg protein/hr, respectively. The K_m for glycine was 5 mM. These values agree well with those in the literature.[5] Data obtained using the coupled assay showed excellent agreement with the data obtained using the discontinuous assay[18] (Fig. 2).

Acyl-CoA Substrate Specificity of Rhodobacter spheroides and Mouse Erythroid 5-Aminolevulinate Synthases

The acyl-CoA specificities of 5-aminolevulinate synthases from *R. spheroides* and mouse erythrocytes, as representatives of prokaryote and eukaryote enzymes, have been investigated using the new coupled assay. The following acyl-CoA derivatives were employed: acetyl-CoA, propionyl-CoA, malonyl-CoA, methylmalonyl-CoA, butyryl-CoA, DL-β-hydroxybutyryl-CoA, *n*-valeryl-CoA, glutaryl-CoA, and acetoacetyl-CoA. As with succinyl-CoA, both enzymes demonstrated Michaelis–Menten kinetics with all the acyl-CoA derivatives that behaved as substrates. The kinetic constants of acyl-CoA subtrates for both 5-aminolevulinate synthases relative to those for succinyl-CoA are shown in Table I. Acetyl-CoA and *n*-valeryl-

TABLE I
KINETIC CONSTANTS FOR ACYL-COA THIOESTERS WITH *Rhodobacter spheroides* AND MOUSE ERYTHROID 5-AMINOLEVULINATE SYNTHASES[a]

	Kinetic constant[b,c]			
	R. spheroides		Mouse	
Acyl-CoA derivative	K_m	V_{max}/K_m	K_m	V_{max}/K_m
Succinyl	5	1.000	50	1.000
Propionyl	1700	0.003	1400	0.018
Malonyl	65	0.090	238	0.045
Methylmalonyl	550	0.020	250	0.073
Butyryl	1700	0.010	1900	0.025
β-Hydroxybutyryl	140	0.115	192	0.320
Glutaryl	1100	0.006	1700	0.023
Acetoacetyl	250	0.147	188	0.130

[a] Determined using pyruvate dehydrogenase-linked assay.

[b] Values for K_m are micromolar.

[c] The V_{max}/K_m value for succinyl-CoA has been normalized to 1.00. The actual values are 0.26 and 0.38 ml/hr/mg for the *R. spheroides* and mouse enzymes, respectively.

CoA were inactivate as substrates with both enzymes. Because 5-aminolevulinate synthases from other sources may accept acetyl-CoA as a substrate this point should be checked carefully before adoption of the assay.

Interestingly, both prokaryote and eukaryote 5-aminolevulinate synthases exhibited a similar preference for the acyl-CoA derivatives. Both enzymes showed an affinity for C_4 acids containing oxygen functions in the β-position, indicating that acyl chain length and hydrophilicity are important determinants in the catalytically productive binding of acyl-CoA substrates.

The new assay described in this chapter has the advantages of being both continuous and applicable to the use of other acyl-CoA derivatives as substrates. This contrasts with the related continuous assay that involves the regeneration of succinyl-CoA with α-ketoglutarate dehydrogenase.[19] It also overcomes the major problem of analyzing highly unstable enzymatic products using the Knorr pyrrole reaction with acetylacetone devised by Mauzerall and Granick.[18] This method, while satisfactory for 5-aminolevulinic acid analysis, is unreliable with some of the more unstable α-aminoketones.

The new assay has allowed, for the first time, a comprehensive analysis of the specificity of 5-aminolevulinate synthase for acyl-CoA substrates other than the natural substrate, succinyl-CoA. In principle the use of pyruvate dehydrogenase to monitor CoA formation is not limited to the 5-aminolevulinate synthase reaction but could be applied to the study of other acyl-CoA-dependent α-aminoketone synthases such as 7-ketopelargonic acid synthase, 2-amino-3-ketobutyrate synthase, and 3-ketodihydrosphingosine synthase.[17]

Acknowledgments

We are grateful to the BBSRC for funding this study and to Prof. G. Ferreira for providing the *E. coli* strain JM 109 harboring the overexpression vector pGF23 encoding mouse erythroid 5-aminolevulinate synthase.

[36] Dipyrromethane Cofactor Assembly of Porphobilinogen Deaminase: Formation of Apoenzyme and Preparation of Holoenzyme

By PETER M. SHOOLINGIN-JORDAN, MARTIN J. WARREN, and SARAH J. AWAN

Porphobilinogen deaminase catalyzes the formation of preuroporphyrinogen from four molecules of porphobilinogen (Scheme I). Preuroporphyrinogen is a highly unstable 1-hydroxymethylbilane that acts as the substrate for uroporphyrinogen III synthase to yield uroporphyrinogen III, the common tetrapyrrole precursor for other tetrapyrroles. Porphobilinogen deaminases have been isolated from a number of sources and their properties have been well established (for a review see Ref. 1). All deaminases exist as monomeric species with molecular weight values between 33,000 and 45,000. The nucleotide sequences of genes/cDNAs specifying the deaminases from bacterial, plant, and animal sources show considerable conservation in the deduced protein sequences,[2] suggesting that all of the enzymes are likely to be structurally related to one another. Investigations with the deaminase from *Escherichia coli* have identified a novel prosthetic group,[3,4] named the dipyrromethane cofactor,[3] made up of two porphobilinogen-derived units linked together and covalently attached to the enzyme (Scheme I). The cofactor acts as a primer for the synthesis of the linear tetrapyrrole (bilane) chain that is built onto the free α-position of the cofactor. This occurs by the sequential condensation of four porphobilinogen molecules with the holoenzyme through enzyme intermediate complexes, termed ES, ES_2, ES_3, and ES_4.[5-8] The product, preuroporphyrinogen, is liberated from ES_4 by hydrolysis, regenerating the holoenzyme with the cofactor still covalently attached (Scheme I).

[1] P. M. Jordan, *N. Comprehens. Biochem.* **19,** 1 (1991).
[2] P. M. Jordan, *Ciba Found. Symp.* **180,** 70 (1994).
[3] P. M. Jordan and M. J. Warren, *FEBS Lett.* **225,** 87 (1987).
[4] G. J. Hart, A. D. Miller, F. J. Leeper, and A. R. Battersby, *J. Chem. Soc. Chem. Commun.* 1762 (1987).
[5] P. M. Anderson and R. J. Desnick, *J. Biol. Chem.* **255,** 1993 (1980).
[6] M. J. Warren and P. M. Jordan, *Biochemistry* **27,** 9020 (1988).
[7] G. J. Hart, A. D. Miller, and A. R. Battersby, *Biochem. J.* **252,** 909 (1988).
[8] R. T. Aplin, J. R. Baldwin, C. Pichon, C. A. Roessner, A. I. Scott, C. J. Schofield, N. J. Stolowich, and M. J. Warren, *Bioorg. Med. Chem. Lett.* **1,** 503 (1991).

0076-6879/97 $25

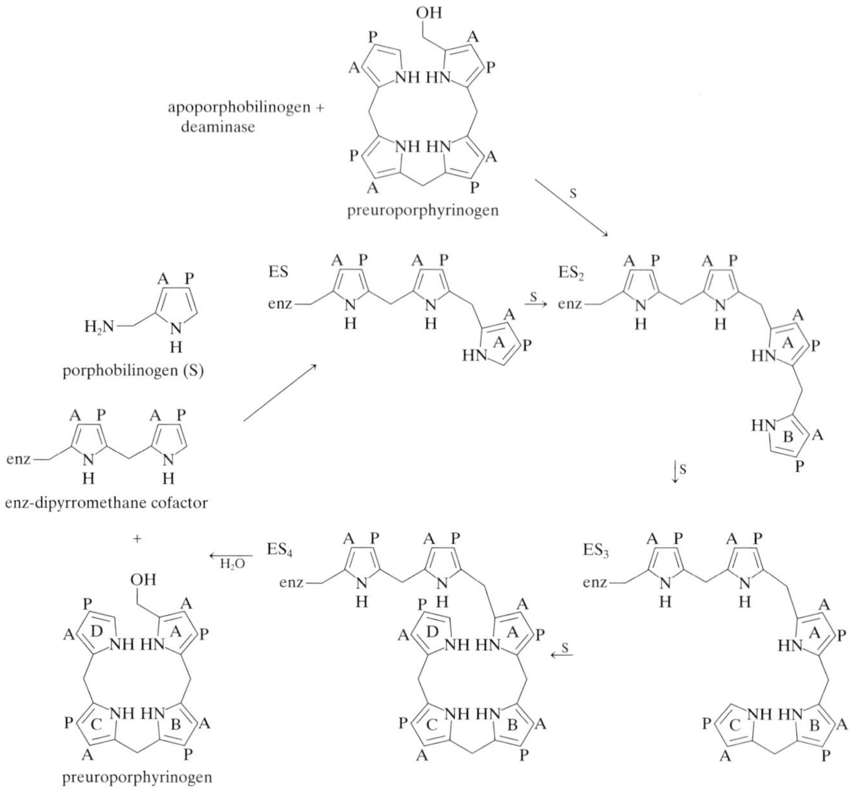

SCHEME I. The stepwise assembly of preuroporphyrinogen attached to the dipyrromethane cofactor of porphobilinogen deaminase and the assembly of the "nascent" holoenzyme (ES_2) from apoporphobilinogen deaminase by reaction with preuroporphyrinogen. The nascent holoenzyme (ES_2) is converted into the holoenzyme by completion of the catalytic cycle through the sequence $ES_2 \rightarrow ES_3 \rightarrow ES_4 \rightarrow E$. The nascent holoenzyme may also be hydrolyzed via ES by heat treatment, with the loss of rings A and B, to yield the holoenzyme (E) containing the dipyrromethane cofactor. PBG, Porphobilinogen; A, CH_2CO_2H; P, $CH_2CH_2CO_2H$.

In the holoenzyme, the four carboxylic acid groups of the dipyrromethane cofactor interact with highly conserved arginine residues at positions 131, 132, 149, and 155 in the *E. coli* enzyme. Alteration of either Arg-131 or -132 by site-directed mutagenesis gives rise to inactive mutant proteins unable to assemble the cofactor.[9,10] Porphobilinogen deaminase lacking the

[9] P. M. Jordan and S. C. Woodcock, *Biochem. J.* **280,** 445 (1991).

[10] M. Lander, A. R. Pitt, P. R. Alefounder, D. Bardy, C. Abell, and A. R. Battersby, *Biochem. J.* **275,** 447 (1991).

cofactor, termed the *apoenzyme,* may be isolated from genetically engineered bacterial strains[6,11] in which the ability to synthesize the early precursors, 5-aminolevulinic acid and porphobilinogen, has been disrupted. Alternatively, the apoenzyme may be generated from the holoenzyme by removal of the cofactor by acid treatment.[7] An apoenzyme has also been isolated from a deaminase mutant in which Cys-242 was substituted by serine.[12]

The most efficient regeneration of the holoenzyme is achieved by reacting the apoenzyme directly with preuroporphyrinogen, the natural precursor,[13] but because this is technically quite difficult to achieve, it is easier to incubate the apoenzyme in the presence of porphobilinogen and a catalytic amount of holodeaminase to generate the preuroporphyrinogen *in situ.* Regeneration of the holoenzyme from the apoenzyme by incubation with porphobilinogen for 2–4 hr has also proved possible,[6,7,11] although the yields of holoenzyme are low with variable recoveries of between 10 and 40% over a period of 2–4 hr. It is likely that regeneration in this case also involves the participation of preuroporphyrinogen, formed by trace amounts of holoenzyme present in the apoenzyme preparation.

The X-ray structure of the *E. coli* deaminase[14,15] has revealed a protein with three domains linked to one another by flexible hinge regions. The dipyrromethane cofactor resides in a large cleft between domains 1 and 2, attached covalently by a thioether linkage to Cys-242 on a loop from domain 3. Aspartate at position 84 has been identified from the X-ray structure as a key catalytic group.[16] Site-directed mutagenesis of Asp-84 to glutamate (D84E) generates an enzyme with less than 0.5% of the wild-type catalytic activity, whereas mutations to alanine (D84A) or asparagine (D84N) yield completely inactive enzymes. Both the D84A and D84N mutants exist as inactive holoenzyme ES_2 intermediate complexes, indicating that Asp-84 is not essential for the binding and reaction of preuroporphyrinogen with the apoenzyme.

This chapter outlines two methods for generating the apodeaminase and two methods for regenerating the holoenzyme from the apoenzyme. The methods may be adapted for labeling the dipyrromethane cofactor with either radioactive or stable isotopes.

[11] A. I. Scott, K. R. Clemens, N. J. Stolowich, P. J. Santander, M. D. Gonzalez, and C. A. Roessner, *FEBS Lett.* **242,** 319 (1989).

[12] A. I. Scott, C. A. Roessner, N. J. Stolowich, P. Karuso, H. J. Williams, S. K. Grant, M. D. Gonzalez, and T. Hoshino, *Biochemistry* **27,** 7984 (1988).

[13] P. M. Shoolingin-Jordan, M. J. Warren, and S. J. Awan, *Biochem. J.* **316,** 373 (1996).

[14] G. V. Louie, P. D. Brownlie, R. Lambert, J. B. Cooper, S. P. Wood, T. L. Blundell, M. J. Warren, S. C. Woodcock, and P. M. Jordan, *Nature (London)* **359,** 33 (1992).

[15] G. V. Louie, P. D. Brownlie, R. Lambert, J. B. Cooper, T. L. Blundell, S. P. Wood, V. N. Malashkevich, A. Hädener, M. J. Warren, and P. M. Shoolingin-Jordan, *Proteins Struct. Funct. Genet.* **25,** 48 (1996).

[16] S. C. Woodcock and P. M. Jordan, *Biochemistry* **33,** 2688 (1994).

Materials and Methods

All routine chemicals are obtained from Sigma (St. Louis, MO) or Merck (Poole, Dorset, UK) or are prepared as indicated below. Porphobilinogen is synthesized enzymatically from 5-aminolevulinic acid using purified 5-aminolevulinate dehydratase.[17] Alternatively, porphobilinogen is purchased from Sigma.

Reagents for Assay of Porphobilinogen Deaminase

Tris-HCl buffer (pH 8.2), 20 m*M*
Porphobilinogen (Sigma), 2 m*M* in 20 m*M* Tris-HCl buffer (pH 8.2): Prepare by dissolving 0.5 mg of the monohydrate per milliliter of buffer
Ehrlich's reagent[18]: Dissolve 1 g of 4-dimethylaminobenzaldehyde in 42 ml of glacial acetic acid, followed by the addition of 8 ml of 70% (w/v) perchloric acid. The reagent is prepared fresh
Benzoquinone (0.1%, w/v) in methanol (prepare fresh)
Sodium metabisulfite (saturated solution)
HCl, 1 and 5 *M*

Porphobilinogen Deaminase Assay Method

An aliquot (25 μl) of porphobilinogen deaminase (20 U/ml; 0.5 mg/ml)[19] is preincubated with 75 μl of 20 m*M* Tris-HCl buffer, pH 8.2, for 2 min at 37°. The reaction is initiated by the addition of 50 μl of a prewarmed 2 m*M* porphobilinogen solution. After 5 min at 37°, 65 μl of 5 *M* HCl is added to terminate the reaction, followed by 25 μl of benzoquinone [0.1% (w/v) in methanol] to oxidize the porphyrinogens to porphyrins. Samples are then incubated in the dark, on ice, for 20 min and any remaining benzoquinone is decolorized by the addition of 50 μl of saturated sodium metabisulfite. The solution is then diluted 10-fold with 1 *M* HCl, centrifuged to remove any precipitated protein, and the absorbance is determined at 405.5 nm (E_{mM} = 548 M^{-1} cm^{-1}). One unit (1 U) is defined as the amount of porphobilinogen deaminase enzyme needed to consume 1 μmol of porphobilinogen per hour.

Preparation of Preuroporphyrinogen

Preuroporphyrinogen (0.1 μmol) is generated from 0.5 μmol of porphobilinogen in a final volume of 1 ml of degassed Tris-HCl buffer, pH 9.1,

[17] P. M. Jordan and J. S. Seehra, *Methods Enzymol.* **123,** 427 (1986).
[18] D. Mauzerall and S. Granick, *J. Biol. Chem.* **219,** 435 (1956).
[19] P. M. Jordan, S. D. Thomas, and M. J. Warren, *Biochem. J.* **254,** 427 (1988).

using 100 μg of purified porphobilinogen deaminase over a period of 1 min at 37°. The sample is rapidly cooled to 0° in liquid nitrogen and the preuroporphyrinogen is separated from the holodeaminase by ultrafiltration through an Amicon (Danvers, MA) PM10 membrane fitted to a 5-ml concentration cell under nitrogen at 4° in a cold room. The preuroporphyrinogen is used at once or frozen in liquid nitrogen for up to 1 hr, under nitrogen, until required.

Isolation of Porphobilinogen Deaminase from Escherichia coli

The deaminase is conveniently isolated from a variety of sources (for reviews see Refs. 1, 2, and 20). In this case, porphobilinogen deaminase is expressed from strain BM3, constructed from *E. coli* strain TB1 harboring a plasmid (pBM3) constructed by cloning a 1.68-kb *Bam*HI–*Sal*I DNA fragment containing the *E. coli hemC* gene from pST48[21] into pUC18 by standard methods.[22]

The isolation procedure is as follows.

Growth of Bacteria, Sonication, and Heat Treatment. Sterilized bacterial medium (4 liters) containing ampicillin (100 μg/ml) is prewarmed to 37° in five baffled flasks (2 liter) before the addition of 1% (v/v) inocula. After growth with rotary shaking (160–180 rpm) for 18 hr at 37°, the cells are centrifuged at 5500 rpm for 30 min at 4° and resuspended (3–4 ml/g of cells) in 0.1 *M* potassium phosphate buffer, pH 8.0, containing 14 m*M* 2-mercaptoethanol. Phenylmethylsulfonyl fluoride (PMSF) solution is added to give a final concentration of 0.1 m*M* and, immediately, four 45-sec sonications are carried out (disintegrator amplitude, 10 μm, peak to peak). The sample is then heat treated for 10 min at 60° followed by cooling to 0° in an ice–salt bath. The precipitated protein is removed by centrifugation at 8000 rpm for 20 min at 4°.

Ammonium Sulfate Fractionation. Solid ammonium sulfate is added slowly to the preceding extract (protein, 30 mg/ml) to give 30% saturation (176 g/liter). The solution is allowed to equilibrate with stirring for 40 min at 4° and the supernatant is removed by centrifugation at 8000 rpm for 20 min at 4°. Further ammonium sulfate is added to give 60% saturation (an additional 198 g/liter) and the solution is equilibrated as before. The pellet containing the enzyme is collected by centrifugation and resuspended in 30 ml of 0.1 *M* potassium phosphate buffer, pH 8.0, containing 14 m*M*

[20] P. M. Shoolingin-Jordan, *J. Bioenerget. Biomembr.* **27,** 181 (1995).

[21] S. D. Thomas and P. M. Jordan, *Nucleic Acids Res.* **14,** 6215 (1986).

[22] J. Sambrook, E. F. Fritsch, and T. Maniatis, "Molecular Cloning Laboratory Manual," 2nd Ed. Cold Spring Harbor Laboratory Press, Cold Spring Harbor, New York, 1989.

2-mercaptoethanol. The sample is then dialyzed against 2 liters of the same buffer at 4° for at least 4 hr, with stirring.

Ion-Exchange Chromatography. Ion-exchange chromatography is performed using a DEAE-Sephacel column (2.5 × 20 cm) equilibrated and eluted with 0.1 *M* potassium phosphate buffer, pH 8, containing 14 m*M* 2-mercaptoethanol. The column fractions containing the deaminase enzyme are located by assay and by sodium dodecyl sulfate-polyacrylamide gel electrophoresis (SDS–PAGE).[23] Active fractions found to be free from any major contaminating proteins are concentrated to 10–15 ml by ultrafiltration, using a 100-ml Amicon ultrafiltration cell fitted with a PM10 membrane, and the deaminase is desalted into AnalaR water (Merck, Poole, Dorset, UK) using a PD10 column. The purified enzyme (specific activity, 30–40 U/mg) is then lyophilized to yield a white solid that is stable for several months when stored at −20° under nitrogen.

Porphobilinogen Deaminase Apoenzyme Preparation

Preparation from a hemB⁻ Strain of Escherichia coli. Wild-type porphobilinogen deaminase apoenzyme is prepared from an *E. coli hemB⁻* strain RP523[24] grown on Luria Bertani (LB) medium supplemented with 7.8 μ*M* hemin and harboring the plasmid pBM3. The apoenzyme is prepared as follows: The cell paste from 2.0 liters of culture is sonicated for four 1-min periods in 30 ml of 20 m*M* Tris-HCl buffer, pH 8, containing 5 m*M* 2-mercaptoethanol, using an ultrasonic disintegrator at an amplitude of 10 μm. The extract is cooled by immersing the sonication vessel in an ice–water bath and by allowing 3 min for cooling between the periods of sonication. The resulting lysate is centrifuged at 10,000 *g* for 10 min at 4° and the supernatant is then applied directly to a Mimetic Orange 1 (Affiniti Chromatography, Ltd., Isle of Man, UK) column (2.5 × 14 cm) and eluted with steps of NaCl in the buffer described above. The apoenzyme elutes at 500 m*M* and 1 *M* NaCl. The apoenzyme is desalted into 1 m*M* Tris-HCl buffer, pH 7.5, containing 2 m*M* dithiothreitol, using a PD10 column (Pharmacia, Piscataway, NJ). The desalted solution is freeze dried for storage at −20°. The apoenzyme is judged to be in excess of 90% pure by SDS–PAGE[23] and contains no measurable porphobilinogen deaminase holoenzyme. The yield is approximately 50%.

Preparation of Porphobilinogen Deaminase Apoenzyme by HCl Treatment of Holoenzyme.[7] Freeze-dried holodeaminase (10 mg) is dissolved in 6 ml of 20 m*M* Tris-HCl buffer, pH 6, containing dithiothreitol (2 mg/ml)

[23] U. K. Laemmli and M. Favre, *J. Mol. Biol.* **80,** 573 (1973).
[24] J. M. Li, C. S. Russell, and S. D. Cosloy, *Gene* **75,** 177 (1988).

and concentrated HCl is added slowly to a final concentration of 1 *M*. The solution is allowed to stand overnight in the dark at room temperature. Precipitated protein is collected by centrifugation and resuspended in 35 ml of 50 m*M* potassium phosphate buffer, pH 7.2, containing 6 m*M* EDTA, 1 m*M* dithiothreitol, 0.1 *M* NaCl, and 0.6 *M* urea. The solution is dialyzed against the same buffer overnight at 4° and then against 50 m*M* potassium phosphate buffer, pH 7.4, containing 6 m*M* EDTA and 13 m*M* 2-mercaptoethanol for a further 48 hr, with one change of buffer. The solution is concentrated by ultrafiltration using a PM10 Amicon membrane and desalted using a PD10 column into 1 m*M* Tris-HCl buffer, pH 7.5, containing 2 m*M* dithiothreitol. The solution containing apoenzyme is freeze-dried and stored at −20°. If further purification is required, the preceding procedure using a Mimetic Orange 1 affinity column, is employed. The yield is 30%.

Nondenaturing Polyacrylamide Gel Electrophoresis

A method is employed[23] in which separations using nondenaturing conditions are carried out in the absence of SDS and 2-mercaptoethanol. Electrophoresis using nondenaturing gels is conducted at 4° in a cold room.

Discussion and Comments

Reconstitution of Porphobilinogen Deaminase Holoenzyme from Apoenzyme with Preuroporphyrinogen

Preuroporphyrinogen is not only the product of porphobilinogen deaminase and the substrate for uroporphyrinogen III synthase, but is also the precursor of the dipyrromethane cofactor. Preuroporphyrinogen is generated separately from porphobilinogen by incubation with holoporphobilinogen deaminase as described above, under conditions that consume all of the porphobilinogen and generate preuroporphyrinogen in a yield of approximately 80%. The preuroporphyrinogen is separated from the holodeaminase by rapid ultrafiltration and incubated immediately with apodeaminase. The results shown in Fig. 1a indicate that an active porphobilinogen deaminase is formed rapidly from a 10-fold molar excess of preuroporphyrinogen over apodeaminase.

The formation of the holoenzyme may also be followed by nondenaturing gel electrophoresis. Porphobilinogen deaminase holoenzyme migrates as a doublet (Fig. 2a) whereas the apoenzyme appears as a broad smudge migrating at lower mobility (Fig. 2b). The regenerated holoenzyme formed initially (Fig. 2d) appears as a protein band with a mobility coincident with

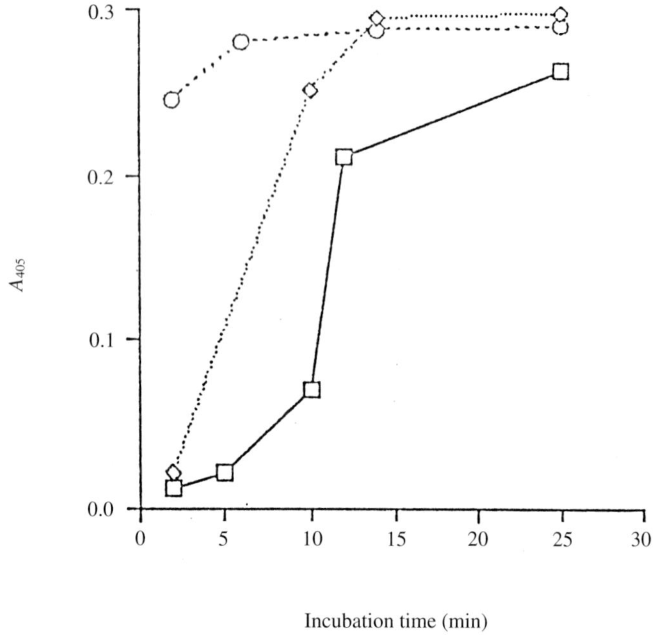

FIG. 1. Formation of porphobilinogen deaminase holoenzyme from porphobilinogen apoenzyme under various conditions. Apoenzyme (1 mg/ml) was incubated in 20 m*M* Tris-HCl buffer, pH 8, containing 2 m*M* dithiothreitol with (a) preuroporphyrinogen filtered free of deaminase (○); (b) 10 mole equivalents of porphobilinogen with added active holoenzyme (holoenzyme : apoenzyme ratio, 1 : 50) (◇); (c) 10 mole equivalents of porphobilinogen without holoenzyme (□). The procedures are described in Materials and Methods. Aliquots were removed and analyzed for porphobilinogen deaminase activity. Values are corrected for the nonenzymatic formation of uroporphyrinogen (measured as uroporphyrin) in (a) and the presence of holodeaminase in (b).

that of an ES_2 complex (Fig. 2f) and is termed the "nascent" holoenzyme[13] (apoenzyme reacted with preuroporphyrinogen).

Because *E. coli* porphobilinogen deaminase intermediate complexes, such as ES_2, undergo enzyme-catalyzed hydrolysis to yield free enzyme (E) on heat treatment at 60°,[6] the holoenzyme (E) may be obtained by heat treatment of the initially formed nascent holoenzyme (ES_2). During this heat treatment, the two terminal pyrrole units are lost (rings A and B of ES_2 in Scheme I), resulting in the conversion of most of the ES_2 complex into free holoenzyme (E) (Fig. 2e). The heat treatment also has the desirable effect of denaturing any remaining apoenzyme, because it is clear that the broad smear of the residual apoenzyme seen in Fig. 2b is removed in Fig. 2e.

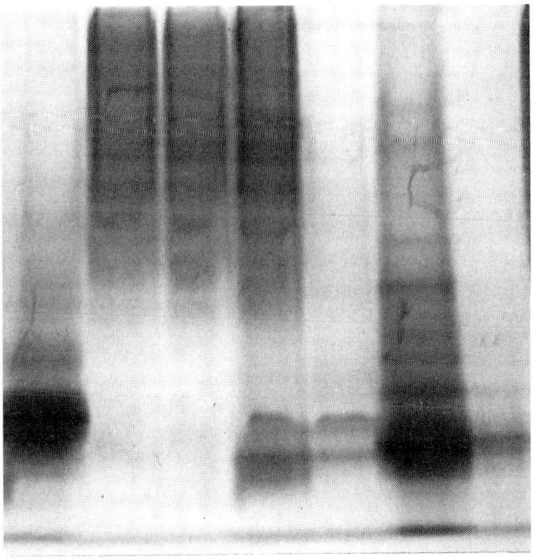

FIG. 2. Nondenaturing polyacrylamide gel electrophoresis of apoporphobilinogen deaminase and its conversion to holoporphobilinogen deaminase. (a) Holoporphobilinogen deaminase; (b) apoporphobilinogen deaminase; (c) apoenzyme incubated with porphobilinogen (as in Fig. 1c); (d) apoenzyme incubated with preuroporphyrinogen; (e) heat-treated sample from (d); (f) purified ES_2 complex. Details of experiments are described in Materials and Methods.

Reconstitution of Wild-Type Holoporphobilinogen Deaminase from Apoenzyme with Porphobilinogen in Presence of Catalytic Amounts of Holodeaminase

Because of the technical difficulties of generating and handling the unstable natural cofactor precursor, preuroporphyrinogen, the holoenzyme may be reconstituted from preuroporphyrinogen that has been generated *in situ* by incubation of porphobilinogen with a "catalytic" quantity of holoenzyme. A suitable apoenzyme : holoenzyme ratio for this purpose is 50 : 1. Porphobilinogen deaminase apoenzyme (1 mg/ml) and holoenzyme (0.02 mg) are incubated with porphobilinogen (10 mole equivalents) at 4° for various times and aliquots are analyzed for active holoenzyme by assay with porphobilinogen. The formation of the holoenzyme (Fig. 1b) is appreciably slower when compared to the use of preuroporphyrinogen (Fig. 1a). The apoenzyme → holoenzyme transformation may also be followed by electrophoresis using 10% (w/v) polyacrylamide nondenaturing gels as described in the previous section. The maximum yield of holoenzyme generated in this way is about 50% of that of the holoenzyme generated from preuroporphyrinogen (data not shown).

Incubation of porphobilinogen with apoenzyme generates the holoenzyme slowly, with a pronounced lag phase (Fig. 1c). It is likely that the lag phase is necessary to generate preuroporphyrinogen from trace amounts of holodeaminase. Polyacrylamide gel electrophoresis indicates that the holoenzyme generated in this way is barely detectable after 10 min (Fig. 2c).

Formation of Holoporphobilinogen Deaminase Containing Isotopically Labeled Dipyrromethane Cofactor

The preceding methodology may be readily applied to the preparation of isotopically labeled holoenzyme. Two broad approaches may be used.

1. Labeled 5-amino[4-^{14}C]levulinic acid or 5-amino[5-^{14}C]levulinic acid (10 μCi) (Amersham International, Amersham, UK) may be introduced into the bacterial culture described above and used for generating the holoenzyme. Alternatively, a mutant deficient in 5-aminolevulinic acid synthesis, such as the *E. coli hemA*$^-$, may be cultured in the presence of labeled 5-aminolevulinic acid as described previously.[6] The latter method yields a far more highly labeled dipyrromethane cofactor because there is no dilution of the label from endogenously biosynthesized 5-aminolevulinic acid. The disadvantage of the preceding method is that it is inconvenient for handling large volumes of radiolabeled bacterial media and for isolating a radioactive enzyme.
2. The alternative method to form labeled holoenzyme is to adopt the method described above for the *in vitro* regeneration of holoenzyme from apoenzyme and labeled porphobilinogen. Labeled porphobilinogen is readily prepared from labeled 5-aminolevulinic acid using 5-aminolevulinate dehydratase (porphobilinogen synthase).[17]

The methods described above provide two alternatives for generating the *E. coli* porphobilinogen deaminase apoenzyme. One method relies on the growth of either a *hemA*$^-$ or a *hemB*$^-$ heme-requiring mutant harboring a plasmid encoding the *hemC* gene. The resulting apoenzyme may then be isolated by an affinity column purification method using Mimetic Orange 1. The other method generates the holoenzyme from a normal *E. coli* strain overexpressing the *hemC* gene and uses acid treatment to convert the holoenzyme into the apoenzyme. The holoenzyme may be generated from the apoenzyme by reaction with preuroporphyrinogen, the product of the deaminase reaction itself. Preuroporphyrinogen is the natural precursor for the pyrrole rings of the cofactor. However, incubation of the apoenzyme with porphobilinogen in the presence of catalytic amounts of holoenzyme also leads to a significant regeneration of holoenzyme. These methods may be adapted to prepare the holodeaminase in which the cofactor is labeled isotopically.

Although much is known about the cofactor assembly mechanism with *E. coli* porphobilinogen deaminase, little is known about other deaminases. For instance, the cofactor assembly is believed to be defective in some human deaminase mutants that are responsible for the genetically inherited disease, acute intermittent porphyria. The molecular basis of this disease has been reviewed[25] and the structures of related *E. coli* mutants have been determined.[26] The methods presented thus provide a platform for the further study of this unique cofactor in *E. coli* and permit additional work in human and other deaminases to be pursued.

Acknowledgments

Grants to P.M.S.J. and M.J.W. from the BBSRC are gratefully acknowledged. We are grateful to Dr. Woodcock (Institute of Ophthalmology) for helpful advice. Thanks are due to L. A. McNeill for help with the manuscript.

[25] P. D. Brownlie, R. Lambert, G. V. Louie, P. M. Jordan, T. L. Blundell, M. J. Warren, J. B. Cooper, and S. P. Wood, *Protein Sci.* **3,** 1644 (1995).

[26] R. Lambert, P. D. Brownlie, S. C. Woodcock, G. V. Louie, J. C. Cooper, M. J. Warren, P. M. Jordan, T. L. Blundell, and S. P. Wood, *Ciba Found. Symp.* **180,** 97 (1994).

[37] Coupled Assay for Uroporphyrinogen III Synthase

By PETER M. SHOOLINGIN-JORDAN and ROBERT LEADBEATER

The enzyme uroporphyrinogen III synthase (EC 4.2.1.75; also known as uroporphyrinogen III cosynthase) catalyzes a remarkable reaction in which preuroporphyrinogen is rearranged and cyclized to yield uroporphyrinogen III (Scheme I). Uroporphyrinogen III is the common precursor for hemes, chlorophylls, vitamin B_{12}, and all other tetrapyrroles (for a review see Ref. 1). The uroporphyrinogen III synthase substrate, preuroporphyrinogen, is generated by the preceding enzyme of the tetrapyrrole pathway, porphobilinogen deaminase (also known incorrectly as uroporphyrinogen I synthase and more recently as hydroxymethylbilane synthase, HMBS) by a reaction that involves the polymerization of four molecules of the monopyrrole precursor porphobilinogen. Preuroporphyrinogen has

[1] P. M. Jordan (ed.), *N. Comprehens. Biochem.* **19,** 1 (1991).

0076-6879/97 $25

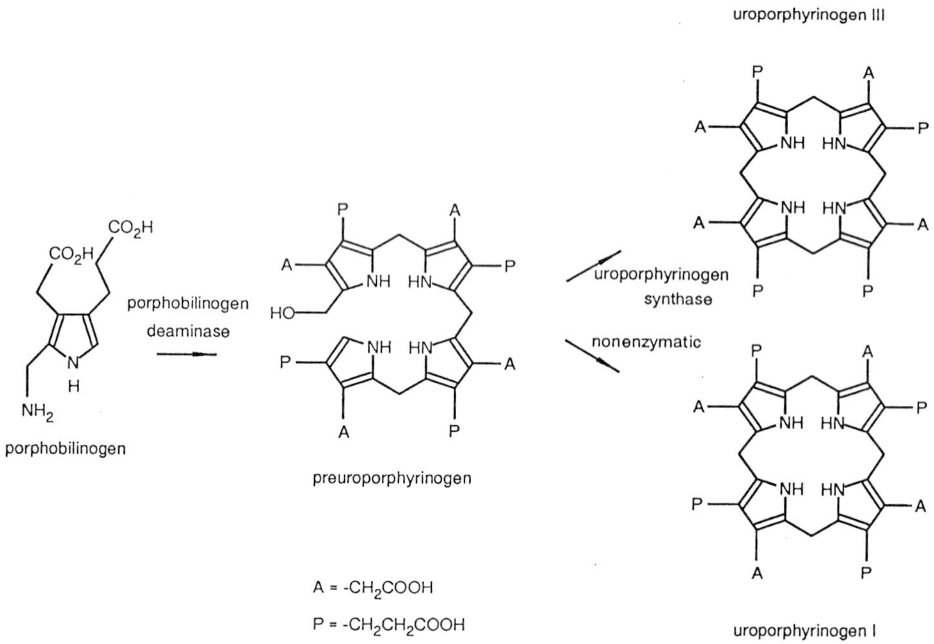

SCHEME I. The transformation of porphobilinogen into preuroporphyrinogen and uroporphyrinogens I and III.

a half-life of less than 5 min at neutral pH values,[2] cyclizing spontaneously to uroporphyrinogen I, a physiologically unimportant isomer (Scheme I). For reviews on both enzymes see Refs. 1, 3, and 4.

Before the discovery of preuroporphyrinogen as the substrate for uroporphyrinogen III synthase,[2,5] the assay of the enzyme was carried out indirectly by measuring the ability of increasing amounts of the enzyme (then called cosynthetase) to increase the uroporphyrinogen III : uroporphyrinogen I ratio in the presence of a fixed amount of porphobilinogen deaminase. The proportion of the two isomers was then determined as their uroporphyrin methyl esters by paper chromatography,[6,7] although, owing to molecular stacking, separations were not always reproducible.

[2] P. M. Jordan, G. Burton, H. Nordlöv, M. M. Schneider, L. Pryde, and A. I. Scott, *J. Chem. Soc. Chem. Commun.* 204 (1979).

[3] F. J. Leeper, *Ciba Found. Symp.* **180,** 111 (1994).

[4] P. M. Shoolingin-Jordan, *J. Bioenerg. Biomembr.* **27,** 181 (1995).

[5] P. M. Jordan and A. Berry, *FEBS Lett.* **112,** 86 (1980).

[6] P. A. D. Cornford and A. Benson, *J. Chromatogr.* **10,** 141 (1963).

[7] J. E. Falk and A. Benson, *Biochem. J.* **55,** 101 (1953).

A more reliable assessment of the III : I isomer ratio was achieved by decarboxylation of the uroporphyrins to coproporphyrins and chromatographic analysis of the methyl esters.[8] All these methods were time consuming and prone to inaccuracy. The application of high-performance liquid chromatography (HPLC) to the latter method using coproporphyrin esters greatly improved the accuracy of the analysis and also allowed isomers II and IV to be identified.[9] The further development of HPLC methods[10,11] for the analysis of the esters of uroporphyrin isomers without resorting to decarboxylation greatly assisted the quantitative analysis. Later refinements to HPLC methods permitted the separation of the free uroporphyrin acids.[12–14]

The discovery that the porphobilinogen deaminase and uroporphyrinogen III synthase reactions occurred sequentially with preuroporphyrinogen as a free intermediate[2] allowed, for the first time, the design of a rapid coupled assay method for the synthase.[15] This method relies on the fact that preuroporphyrinogen is converted into uroporphyrinogen III far more rapidly by uroporphyrinogen III synthase than it is converted to uroporphyrinogen I by chemical cyclization. Thus the difference between the uroporphyrinogens formed (after oxidation to uroporphyrin) in the presence and absence of the synthase provides a measure of the uroporphyrinogen III formed and hence the activity of synthase enzyme. By carrying out the assay over a short period, the amount of uroporphyrinogen I formed nonenzymatically is minimized, thus permitting low levels of the synthase to be determined. This approach has since been adapted by others,[16,17] although the principle is the same.

In the method presented here, preuroporphyrinogen is generated *in situ* from porphobilinogen by the action of purified porphobilinogen deaminase and the amount of uroporphyrinogen III formed by the synthase from the preuroporphyrinogen is then quantified, against the unavoidable formation of uroporphyrinogen I. This is achieved by spectrophotometric or fluorometric analysis of uroporphyrins I and III, the formation of which

[8] T. C. Chu and E. J.-H. Chu, *J. Biol. Chem.* **227,** 505 (1957).

[9] A. R. Battersby, D. G. Buckley, G. L. Hodgson, R. E. Markwell, and E. McDonald, *in* "High Pressure Chromatography in Clinical Chemistry," p. 63. Academic Press, London, 1976.

[10] J. Bommer, B. Burnham, R. Carlson, and D. Dolphin, *Anal. Biochem.* **95,** 444 (1979).

[11] H. Nordlöv, P. M. Jordan, G. Burton, and A. I. Scott, *J. Chromatogr.* **190,** 221 (1980).

[12] A. W. Wayne, R. C. Straight, E. E. Wales, and E. Englert, Jr., *J. High Resolut. Chromatogr. Chromatogr. Commun.* **2,** 621 (1979).

[13] D. Wright and C. K. Lim, *Biochem. J.* **213,** 85 (1983).

[14] C. K. Lim and T. J. Peters, *Methods Enzymol.* **123,** 383 (1986).

[15] P. M. Jordan, *Enzyme* **28,** 158 (1982).

[16] S. Tsai, D. Bishop, and R. J. Desnick, *Anal. Biochem.* **166,** 120 (1987).

[17] G. J. Hart and A. R. Battersby, *Biochem. J.* **232,** 151 (1985).

is validated by HPLC separation of the isomers. The method is particularly useful during enzyme purifications in which large numbers of samples need to be assayed rapidly. The assay method has also been adapted to measure enzyme levels in blood and other tissues.[15]

Materials and Methods

Synthesis and Purification of Porphobilinogen

Porphobilinogen is synthesized enzymatically from 5-aminolevulinic acid using purified 5-aminolevulinate dehydratase (porphobilinogen synthase).[18] Alternatively, porphobilinogen is purchased from Sigma (St. Louis, MO).

Materials for Purification of Porphobilinogen Deaminase

Bacterial medium: Combine Bacto-tryptone (10 g), Bacto-yeast extract (5 g), and NaCl (5 g), make up to 1 liter with distilled water, and autoclave immediately
Ampicillin (50 mg/ml)
Potassium phosphate buffer (pH 8.0), 0.1 *M*, containing 14 m*M* 2-mercaptoethanol
Phenylmethylsulfonyl fluoride (PMSF), 0.1 m*M* (in ethanol)
Ammonium sulfate (low in heavy metals)
DEAE-Sephacel anion exchanger (wet bead size, 40–160 μm) (Pharmacia, Piscataway, NJ)
AnalaR water (Merck, Rahway, NJ)
PD10 gel-filtration columns (Pharmacia)

Isolation of Porphobilinogen Deaminase from Escherichia coli

Porphobilinogen deaminase is required to generate preuroporphyrinogen, the substrate for uroporphyrinogen III synthase. The deaminase is conveniently isolated from a variety of sources (for reviews see Refs. 1 and 4). In the following method, porphobilinogen deaminase is isolated from a recombinant strain of *Escherichia coli* harboring the cloned *hemC* gene[19] that overexpresses porphobilinogen deaminase.[19,20] The procedure is described in [36] in this volume.[21]

[18] J. S. Seehra and P. M. Jordan, *Methods Enzymol.* **123,** 427 (1986).
[19] S. D. Thomas and P. M. Jordan, *Nucleic Acids Res.* **14,** 6215 (1986).
[20] P. M. Jordan, S. D. Thomas, and M. J. Warren, *Biochem. J.* **254,** 427 (1988).
[21] P. M. Shoolingin-Jordan, M. J. Warren, and S. J. Awan, *Methods Enzymol.* **281,** [36], 1997 (this volume).

Reagents for Assay of Porphobilinogen Deaminase

Tris-HCl buffer (pH 8.2), 20 m*M*
Porphobilinogen (Sigma), 2 m*M* in 20 m*M* Tris-HCl buffer, pH 8.2: Prepare by dissolving 0.5 mg of the monohydrate per milliliter of buffer
Ehrlich's reagent: Dissolve 1 g of 4-dimethylaminobenzaldehyde in 42 ml of glacial acetic acid, followed by the addition of 8 ml of 70% (w/v) perchloric acid. Prepared fresh
Benzoquinone (0.1%, w/v) in methanol
Sodium metabisulfite (saturated solution)
HCl, 1 and 5 *M*

Porphobilinogen Deaminase Assay Method. The assay of uroporphyrinogen III synthase requires a known quantity of porphobilinogen deaminase to generate the required amount of preuroporphyrinogen. An aliquot (25 μl) of porphobilinogen deaminase (20 U/ml; 0.5 mg/ml) is preincubated with 75 μl of 20 m*M* Tris-HCl buffer, pH 8.2, for 2 min at 37°. The reaction is initiated by the addition of 50 μl of a prewarmed 2 m*M* porphobilinogen solution. After 5 min at 37°, 65 μl of 5 *M* HCl is added to terminate the reaction, followed by 25 μl of benzoquinone [0.1% (w/v) in methanol] to oxidize the porphyrinogens to porphyrins. Samples are then incubated in the dark, on ice, for 20 min and any remaining benzoquinone is decolorized by the addition of 50 μl of saturated sodium metabisulfite. The solution is then diluted 10-fold with 1 *M* HCl, centrifuged to remove any precipitated protein, and the absorbance is determined at 405 nm (E_{mM} = 548 M^{-1} cm^{-1}). A standard curve is prepared using authentic uroporphyrin I (Fig. 1).

One unit (1 U) is defined as the amount of porphobilinogen deaminase enzyme needed to consume 1 μmol of porphobilinogen per hour.

Purification of Escherichia coli Uroporphyrinogen III Synthase. If required this enzyme may be prepared from recombinant strains of *E. coli* in a procedure involving Sephacryl S-200 gel filtration, hydroxylapatite chromatography, and FPLC (fast protein liquid chromatography).[22]

Reagents for Assay of Uroporphyrinogen III Synthase

Potassium phosphate buffer (pH 8.0), 0.1 *M*
Tris-HCl buffer (pH 8.2), 20 m*M*
Porphobilinogen deaminase: 20 U/ml (0.5–1 mg/ml) in 20 m*M* Tris-HCl buffer, pH 8.2
Porphobilinogen, 2 m*M* in 20 m*M* Tris-HCl buffer, pH 8.2

[22] A. F. Alwan, B. I. A. Mgbeje, and P. M. Jordan, *Biochem. J.* **264,** 397 (1989).

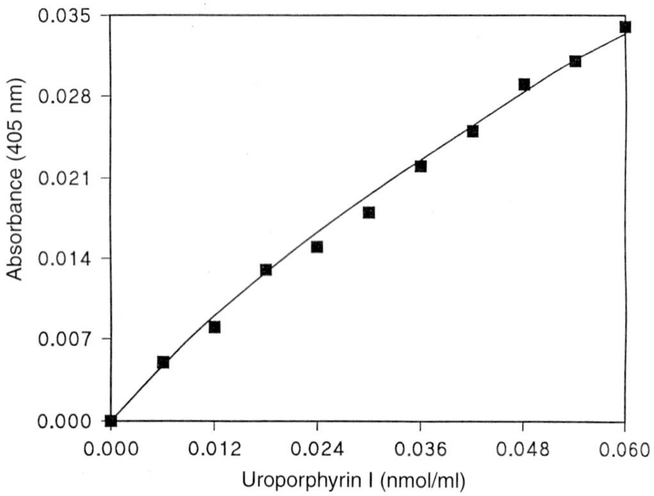

FIG. 1. Relationship between absorbance and molar concentration of uroporphyrin.

Alcohol–dry ice bath: Ethanol/solid CO_2 (1:1, approximately)
HCl, 0.1 *M*
Iodine solution: 0.1% (w/v) in 1% aqueous KI
Sodium metabisulfite (saturated solution)
Trichloroacetic acid (10%, w/v)

Rapid Coupled Assay for Determination of Uroporphyrinogen III Synthase Activity. The assay of uroporphyrinogen synthase is based on the rapid method reported by Jordan[15] and modified by Tsai *et al.*[16] and Hart and Battersby.[17]

Samples (50 μl; 1 U) of porphobilinogen deaminase and 65 μl of 20 m*M* Tris-HCl buffer, pH 8.2, are preincubated at 37° for 2 min before the addition of 50 μl of a 2 m*M* porphobilinogen solution that has also been brought to 37° in a 2-min preincubation. After rapid vortex mixing incubation is carried out for 1 min to generate preuroporphyrinogen. Samples of uroporphyrinogen III synthase (10 μl) appropriately diluted within the range 0.5–5 U/ml in 0.1 *M* potassium phosphate buffer, pH 8, are then added to each tube; the content of each tube is mixed by rapid vortex mixing and further incubation is carried out for 1 min to transform the preuroporphyrinogen into uroporphyrinogen III. If 2-mercaptoethanol has been incorporated in the buffers during the extraction or isolation of the uroporphyrinogen III synthase it should be removed using a PD10 column (Pharmacia) prior to assaying. A blank, containing assay buffer in place of the enzymes, is used to zero spectrometers during measurements. A control, containing deaminase, porphobilinogen, buffer and additional 10 μl of buffer instead of the synthase, is used to determine the background uroporphyrinogen I level.

All reactions are terminated after the 1-min incubation by plunging the tubes into ethanol:solid CO_2 and the uroporphyrinogens formed are oxidized to uroporphyrins by the addition of 50 μl of I_2/KI solution to the frozen solution. The samples are placed on ice in the dark for 10–20 min to allow oxidation, after which excess iodine is decolorized by the addition of 50 μl of sodium metabisulfite solution. Protein is precipitated by the addition of 500 μl trichloroacetic acid and, after centrifugation, 100 μl of the supernatant is removed and diluted to 1 ml with 0.1 *M* HCl. The change in absorbance with the addition of the synthase enzyme is detected at 405 nm (E_{mM} = 548 M^{-1} cm^{-1}).

Alternatively, the uroporphyrinogen formed may be quantitized by fluorescence, using a standard curve (Fig. 2) and an excitation wavelength of 399 nm and emission wavelength of 615 nm. This method is particularly useful for low concentrations of porphyrins.

The relative proportions of the type I and III isomers are also determined from analysis of the reaction mixtures by reversed-phase HPLC (Fig. 3). The experimental conditions for the separation of uroporphyrin isomers by HPLC are similar to those previously reported.[12] Typical retention times for uroporphyrin I and III isomer standards are 16 and 20 min, respectively, for ammonium acetate (pH 5.16)–acetonitrile (86:14, v/v).

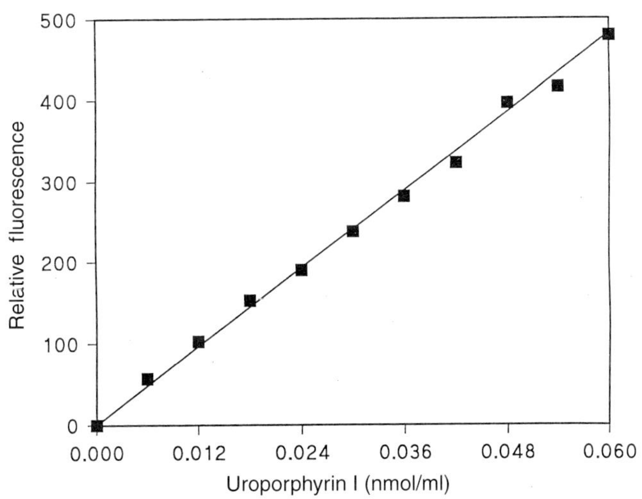

FIG. 2. Relationship between relative fluorescence and molar concentration of uroporphyrins (see text for details).

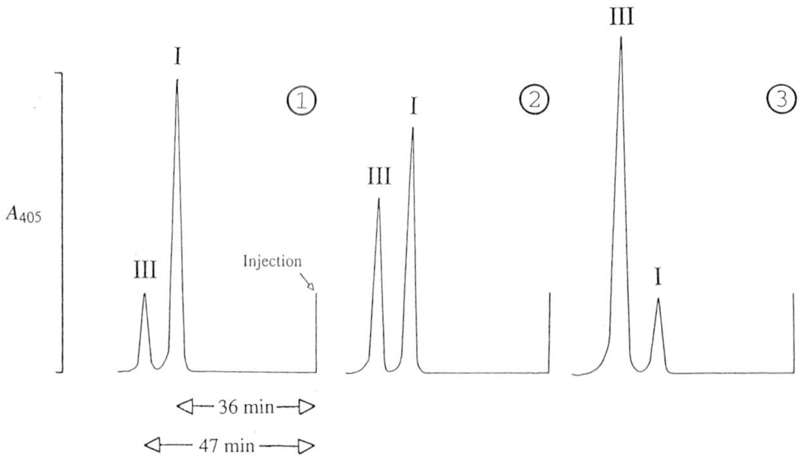

FIG. 3. HPLC trace showing the separation of uroporphyrin I and III isomers, with increasing uroporphyrinogen III synthase units ranging from 1 to 3. It can be seen that with increasing units of uroporphyrinogen III synthase the III isomer peak increase is accompanied by a decrease in the I isomer. The separation was carried out using ammonium acetate buffer (pH 5.16)–acetonitrile (87 : 13, v/v) as described in Materials and Methods.

The elution times increased to 36 and 47 min using ammonium acetate (pH 5.16)–acetonitrile (87 : 13, v/v).

Slight changes in the method may also be necessary depending on the source of the uroporphyrinogen III synthase enzyme under investigation.

Reagents and Materials for HPLC Analysis of Uroporphyrins

Ammonium acetate buffer (pH 5.16) 1 *M*: Adjust with glacial acetic acid (28 ml/liter)
Acetonitrile (Rathburn Chemicals, Walkerburn, Scotland)
AnalaR water (Merck, Poole, Dorset, UK)

All solutions and solvents are passed through a 0.2-μm (pore size) membrane filter (Whatman, Clifton, NJ) filter and thoroughly degassed before use.

Uroporphyrin I and III standards (Porphyrin Products, Logan, UT)
Shandon 5-μm Hypersil BDS C_{18} HPLC column (250 × 4.6 mm)
HPLC equipment with the capacity to analyze at 405 nm

Calculations

The units of uroporphyrinogen III synthase are calculated from the following formula:

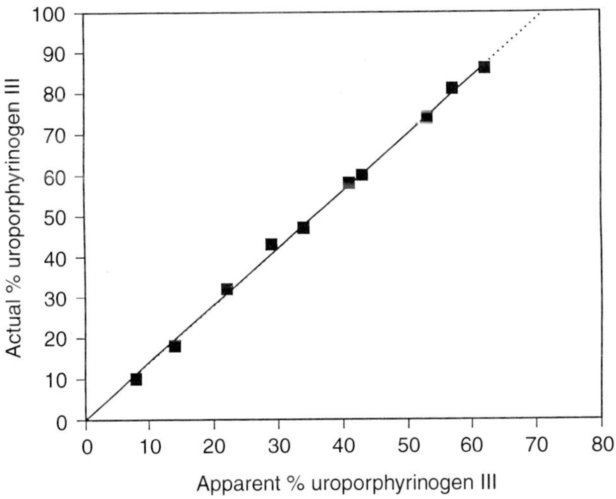

FIG. 4. Relationship between apparent percentage of uroporphyrinogen III formed (from absorbance data) and actual corrected percentage of uroporphyrinogen III as determined by HPLC. This correction is due to the lower amounts of uroporphyrogen I formed in the presence of the uroporphyrinogen III synthase enzyme.

$$\text{Activity } (\mu\text{mol/hr}) = A_{405} \times 10^{(a)} \times 775^{(b)} \times 60^{(c)} \times 10^{6(d)}\, 5.48 \times 10^{5(e)} \times 10^{(f)} \times 10^{3(g)}$$

where (*a*) is the dilution in the cuvette; (*b*) is the total volume (in microliters); (*c*) is the conversion of minutes to hours; (*d*) is the conversion of moles to micromoles; (*e*) is the molar extinction coefficient of uroporphyrin; (*f*) is the enzyme sample size (in microliters); and (*g*) is the conversion of liters to milliliters.

One unit (1 U) of uroporphyrinogen III synthase is defined as the amount of enzyme required to produce 1 μmol of uroporphyrinogen III per hour.

Discussion and Comments

The difficult synthetic organic chemistry[23] needed to prepare and handle 1-hydroxymethylbilanes such as preuroporphyrinogen makes the assay of uroporphyrinogen III synthase with enzymatically generated preuroporphyrinogen the only feasible method for most laboratories. The availability

[23] A. R. Battersby, C. J. R. Fookes, G. W. J. Matcham, E. McDonald, and R. Hollenstein, *J. Chem. Soc. Perkin Trans.* **1,** 2427 (1982).

of porphobilinogen deaminase from human or animal erythrocytes or from animal liver and the stability of the enzyme, together with the commercial availability of porphobilinogen, mean that preuroporphyrinogen may be prepared in any biochemical or medical laboratory. More recently, the expression of the *hemC* gene specifying the deaminase from recombinant bacterial strains has aided further the isolation of the enzyme.

The ability to assay the activity of uroporphyrinogen III synthase relies on the fact that the enzyme-catalyzed transformation of the unstable 1-hydroxymethylbilane, prueroporphyrinogen, into uroporphyrinogen III is much faster than the spontaneous nonenzymatic cyclization to form uroporphyrinogen I. However, the involvement of a substrate with a half-life of less than 5 min under the assay conditions means that there will always be a background formation of uroporphyrinogen I not only during the preuroporphyrinogen generation stage but also during the uroporphyrinogen III synthesis stage. The unavoidable formation of uroporphyrinogen I is the main difficulty in the assay of the synthase, because at low levels of synthase there will be a proportional increase in the preuroporphyrinogen available for chemical cyclization to give uroporphyrinogen I. In contrast, at high levels of the synthase there will be less uroporphyrinogen I formed because preuroporphyrinogen will be utilized by the synthase. This will lead to an underestimate of the synthase activity at high levels of the synthase enzyme. The comparison of the direct assay readings and predicted ratios of isomer III to isomer I with the ratios determined by HPLC provides an important validation of the rapid assay (Fig. 4).

Acknowledgment

We are grateful to the BBSRC for financial support.

[38] Expression and Purification of Mammalian 5-Aminolevulinate Synthase

By HARRY A. DAILEY and TAMARA A. DAILEY

The first step in the biosynthesis of any tetrapyrrole is the synthesis of the compound 5-aminolevulinate (ALA). In nature two ways have evolved to produce this compound.[1] In one the starting material is glutamate in

[1] P. M. Jordan, *in* "Biosynthesis of Heme and Chlorophylls" (H. A. Dailey, ed.), p. 55. McGraw-Hill, New York, 1990.

0076-6879/97 $25

the form of glutamyl-tRNA. Two enzymes, glutamyl-tRNA reductase and glutamate-1-semialdehyde aminotransferase, catalyze the formation of ALA from glutamyl-tRNA. This pathway is referred to as the C_5 pathway and is found in plants, archaebacteria, and many eubacteria.[2] In contrast, nonplant eukaryotes and a few eubacteria utilize a single enzymatic step to form ALA from succinyl-CoA and glycine.[1] This enzyme is named ALA synthase (ALAS) (EC 2.3.1.37) and synthesis of ALA by ALAS is referred to as the C_4 pathway.

In chickens and mammals two ALAS enzymes are found.[3,4] These two enzymes, which are highly homologous in structure and function, are encoded by separate genes and are subject to different forms of regulation. ALAS-1, or the housekeeping form of ALAS, is present in all nonerythroid cell types. It has a short half-life and is subject to regulation by the end product of the pathway, protoheme. Details of the regulation of ALAS can be found in reviews by May *et al.*[5] and Dierks.[3]

A second form of the enzyme is ALAS-2, or the erythroid-specific form. This enzyme, which is encoded on the human and mouse X chromosome,[6] is induced only during the period of increased heme synthesis in developing erythroid cells. Unlike ALAS-1, it does not appear to be regulated by heme, but is subject to regulation by iron, due to the presence of a 5′ iron-responsive element (IRE), as well as a variety of erythroid-specific factors.[7] It does not appear, however, that ALAS-2 alone is responsible for regulation of heme synthesis in these cells.[8] Deficiency of activity of ALAS-2 has been demonstrated to be the cause of X-linked sideroblastic anemia (see Ref. 9). A number of genetic mutations have been identified in individuals with pyridoxine-responsive X-linked sideroblastic anemia, and some have been demonstrated to result in decreased enzyme activity.[10–12]

[2] Y. J. Avissar, J. G. Ormerod, and S. I. Beale, *Arch. Microbiol.* **151,** 513 (1989).

[3] P. Dierks, *in* "Biosynthesis of Heme and Chlorophylls" (H. A. Dailey, ed.), p. 201. McGraw-Hill, New York, 1990.

[4] R. D. Ridable, M. Yamamoto, and J. D. Engel, *Proc. Natl. Acad. Sci. U.S.A.* **86,** 792 (1989).

[5] B. K. May, C. R. Bhasker, M. J. Bawden, and T. C. Cox, *Mol. Biol. Med.* **7,** 405 (1990).

[6] P. D. Cotter, H. F. Willard, J. L. Gorski, and D. F. Bishop, *Genomics* **13,** 211 (1992).

[7] T. C. Cox, M. J. Bawden, M. Alison, A. Martin, and B. K. May, *EMBO J.* **10,** 1891 (1991).

[8] H. Lake-Bullock and H. A. Dailey, *Mol. Cell. Biol.* **13,** 7122 (1993).

[9] S. S. Bottomly, *in* "Wintrobe's Clinical Hematology" (G. R. Lee, T. C. Bithell, J. Forester, J. W. Athens, and J. N. Lakens, eds.), 9th Ed., Vol. 1. Lea & Febiger, Philadelphia, 1993.

[10] E. Prades, C. Chambon, T. A. Dailey, H. A. Dailey, J. Briére, and B. Grandchamp, *Hum. Genet.* **95,** 424 (1995).

[11] P. D. Cotter, M. Baumann, and D. F. Bishop, *Proc. Natl. Acad. Sci. U.S.A.* **89,** 4028 (1992).

[12] T. C. Cox, S. S. Bottomley, J. S. Wiley, M. J. Bacoden, C. S. Mathews, and B. K. May, *N. Engl. J. Med.* **330,** 675 (1994).

All forms of ALAS appear to be homodimers with pyridoxal 5′-phosphate as a cofactor. In eukaryotes the enzyme is nuclear encoded, cytoplasmically synthesized as a precursor with a mitochondrial targeting sequence, and then translocated into the mitochondrial matrix where it is processed into its mature form (see Ref. 3 for review).

Expression and Purification of Murine 5-Aminolevulinate Synthase 2

The cDNA for mouse ALAS-2 is supplied by P. Curtis (Wistar Institute, Philadelphia, PA) and is in two fragments (pMS-6 and pMS-12[13]). These fragments have been combined to give a full-length ALAS-2 cDNA in plasmid pHD2.[14] Using this plasmid as starting material two different expression vectors have been constructed in our laboratory. One, pGF23, encodes the production of a mature-length ALAS-2 under the control of an alkaline phosphatase promoter[14] and a second, pTAD-ALAS2, encodes the production of a mature-length ALAS-2 under the control of a Tac promoter.[10] Both plasmids when expressed in *Escherichia coli* JM109 and grown under appropriate conditions give high yields of protein. The deciding factor in selection of vector has only to do with choice or ease of induction technique.

Following growth and induction of protein synthesis, cells are harvested by centrifugation at 6000 *g* for 15 min at 4°. The cell pellet is suspended in buffer A [20 m*M* potassium phosphate (pH 7.2), 1 m*M* EDTA, 5 m*M* mercaptoethanol, 10% (v/v) glycerol, and 20 μM pyridoxal phosphate] that contains 100 μM phenylmethylsulfonyl fluoride (PMSF) and the cells disrupted by sonication at maximum power (twice for 30 sec). The disrupted cells are centrifuged at 15,000 *g* for 20 min at 4° to remove cellular debris and the supernatant is carefully decanted. Saturated ammonium sulfate solution, pH 8.0, is slowly added to yield 40% (w/v) saturation and after 30 min at 4° this solution is centrifuged (15,000 *g* for 20 min at 4°). The resulting pellet, which contains ALAS-2, is resuspended in a minimal volume of buffer A with PMSF and applied to an Ultrogel AcA 44 column (1.5 × 30 cm) equilibrated with buffer A. Fractions from this column that contain ALAS activity are pooled and applied to a DEAE-Sephacel column (2.5 × 12 cm) that is equilibrated with buffer A. The column is successively washed with 100 ml of buffer A and 100 ml of buffer A with 40 m*M* KCl before the protein is eluted with buffer A with 100 m*M* KCl. Fractions containing ALAS activity are pooled and stored at −20°. From 1 liter of culture one obtains approximately 50 mg of enzyme.

[13] D. S. Schoenhaut and P. J. Curtis, *Gene* (*Amsterdam*) **48,** 55 (1986).
[14] G. C. Ferreira and H. A. Dailey, *J. Biol. Chem.* **268,** 584 (1993).

5-Aminolevulinate Synthase Assay

A variety of ALAS assays are available and choice of assay will be dependent on enzyme activity and level of purity of the sample. A radiochemical assay described by Brooker *et al.*[15] is effective and sensitive, although costly. We have used this procedure only on crude tissue or cell extracts. The most common assay procedure is the one based on use of modified Ehrlich's reagent.[16] When proper precautions are taken this procedure is accurate and inexpensive. There is also a coupled continuous spectrophotometric assay that should prove useful in assaying purified ALAS.[17]

Properties of Recombinant Murine 5-Aminolevulinate Synthase 2

ALAS-2 as prepared herein has a molecular weight of approximately 56,000 as determined by sodium dodecyl sulfate (SDS)–polyacrylamide gel electrophoresis. This along with the observation that gel filtration of the purified protein on Sephadex G-200 yields a molecular weight of just over 100,000 supports previous reports that the enzyme is a homodimer.[1] The enzyme contains a noncovalently bound pyridoxal 5′-phosphate, which may be removed by dialysis. Functional holoenzyme is restored by incubation with pyridoxal 5′-phosphate. Although it has been shown that Lys-313 of the enzyme will form a covalent adduct with the cofactor,[18] site-directed mutagenesis studies indicate that this residue is not required for binding of cofactor, but may be involved in catalysis.[19] Determination of the actual residues involved awaits X-ray crystallographic structure determination. The kinetic constants for recombinant murine ALAS-2 are similar to those determined for the enzyme from mammalian tissues. The K_m for glycine is 51 mM and for succinyl-CoA is 55 μM. The k_{cat} is approximately 0.5 min^{-1}.[14]

Summary

We have described a procedure for production and purification of recombinant, mature-length mouse ALAS-2. The fact that *E. coli* utilizes the C_5 path for ALA production means that there is no problem with contamination of the recombinant ALAS-2 by host cell enzyme, such as one may have with a yeast expression system. While the detailed procedure

[15] J. D. Brooker, G. Srivastava, B. K. May, and W. H. Elliott, *Enzyme* **28,** 109 (1982).

[16] L.-F. Lien and D. S. Beattie, *Enzyme* **28,** 120 (1982).

[17] G. A. Hunter and G. C. Ferreira, *Anal. Biochem.* **226,** 221 (1995).

[18] G. C. Ferreira, P. J. Neame, and H. A. Dailey, *Protein Sci.* **2,** 1959 (1993).

[19] G. C. Ferreira, U. Vajapey, O. Hafey, G. A. Hunter, and M. J. Barber, *Protein Sci.* **4,** 1001 (1995).

produces enzyme in good yield with relatively common protein purification techniques, future expression systems may be developed to take advantage of the rapid purification achieved by the use of a 6-histidine (His_6) amino-terminal tag and metal chelate chromatography. Such approaches in this laboratory with protoporphyrinogen oxidase,[20,21] coproporphyrinogen oxidase,[22] and uroporphyrinogen decarboxylase have resulted in the production and purification of enzymes whose kinetic and physical parameters are essentially identical to those of proteins lacking the His_6 tag.

Acknowledgment

This work was supported by NIH Grants DK32303 and DK35898 (to H.A.D.).

[20] H. A. Dailey and T. A. Dailey, *J. Biol. Chem.* **271,** 8714 (1996).
[21] T. A. Dailey and H. A. Dailey, *Protein Sci.* **5,** 98 (1996).
[22] A. Medlock and H. A. Dailey, *J. Biol. Chem.* **271,** 32507 (1996).

[39] Expression, Purification, and Characteristics of Mammalian Protoporphyrinogen Oxidase

By TAMARA A. DAILEY and HARRY A. DAILEY

The penultimate enzyme of the heme biosynthetic pathway, protoporphyrinogen oxidase (PPO) (EC 1.3.3.4), catalyzes the six-electron oxidation of protoporphyrinogen IX to form protoporphyrin IX (Scheme I).[1-3] The enzyme has been purified from mouse[4] and bovine[5] liver, and from yeast.[6] More recently, the enzyme has been cloned from *Bacillus subtilis,*[7-9] *Myxo-*

[1] R. J. Porra and J. E. Falk, *Biochem. J.* **90,** 69 (1964).
[2] R. Poulson and W. J. Polglase, *J. Biol. Chem.* **250,** 1269 (1975).
[3] H. A. Dailey, *in* "Biosynthesis of Heme and Chlorophylls" (H. A. Dailey, ed.), p. 123. McGraw-Hill, New York, 1990.
[4] H. A. Dailey and S. W. Karr, *Biochemistry* **26,** 2697 (1987).
[5] L. J. Siepker, M. Ford, R. de Kock, and S. Kramer, *Biochim. Biophys. Acta* **913,** 349 (1987).
[6] J. M. Camadro, F. Thome, N. Brouillet, and P. Labbe, *J. Biol. Chem.* **269,** 32085 (1994).
[7] M. Hansson and L. Hederstedt, *J. Bacteriol.* **174,** 8081 (1992).
[8] M. Hansson and L. Hederstedt, *J. Bacteriol.* **176,** 5962 (1994).
[9] T. A. Dailey, P. N. Meissner, and H. A. Dailey, *J. Biol. Chem.* **269,** 813 (1994).

0076-6879/97 $25

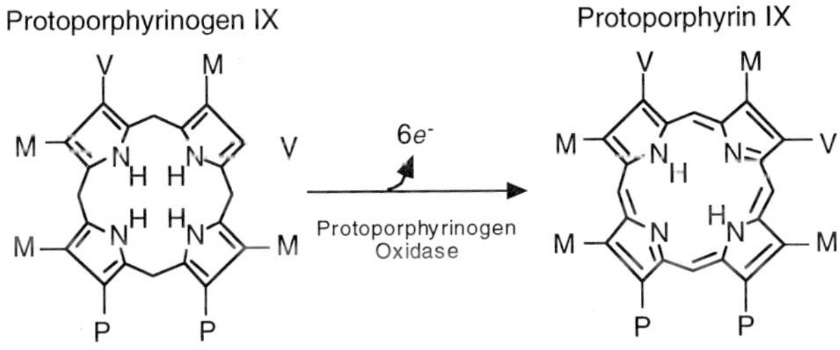

SCHEME I. Oxidation of protoporphyrinogen IX catalyzed by protoporphyrinogen oxidase.

coccus xanthus,[10] human placenta,[11,12] mouse,[13,14] and yeast.[15] The coding regions of the cDNAs for these enzymes are similar in length, and with the possible exception of the yeast enzyme, do not contain a typical membrane targeting sequence. The gene for human PPO has been mapped to chromosome 1q23 by fluorescence *in situ* hybridization.[16] It is approximately 4.5 kb, and consists of 13 exons.[16] The mouse enzyme is located on chromosome 1H2 and is also approximately 4.3 kb with 13 exons.[17]

Protoporphyrinogen oxidase from the aerobic prokaryotic and eukaryotic organisms studied to date is structurally and biochemically distinct from the enzyme found in *Escherichia coli*[18,19] and *Rhodopseudomonas sphaeroides*[20] (both facultative organisms) and from that of *Desulfovibrio gigas*,[21] an anerobe. These anaerobic or facultatively anaerobic prokaryotes all appear to have a protoporphyrinogen oxidase activity that exists in a multienzyme complex coupled to the respiratory chain of the cell. In these organisms, a variety of compounds can serve as the terminal electron acceptor for the oxidase. In contrast, the "aerobic" protoporphyrinogen oxidases are not coupled to a respiratory chain and require O_2 directly as the electron acceptor.

[10] H. A. Dailey and T. A. Dailey, *J. Biol. Chem.* **271,** 8714 (1996).
[11] T. A. Dailey and H. A. Dailey, *Protein Sci.* **5,** 98 (1996).
[12] K. Nishimura, S. Taketani and H. Inokuchi, *J. Biol. Chem.* **270,** 8076 (1995).
[13] T. A. Dailey, H. A. Dailey, P. N. Meissner, and A. R. K. Prasad, *Arch. Biochem. Biophys.* **324,** 379 (1995).
[14] S. Taketani, T. Yoshinaga, T. Furukawa, H. Kohno, R. Tokunaya, K. Nishimura, and H. Inokuchi, *Eur. J. Biochem.* **230,** 760 (1995).
[15] J. M. Camadro and P. Labbe, *J. Biol. Chem.* **271,** 9120 (1996).
[16] A. G. Roberts, S. D. Whatley, J. Daniels, P. Holmans, Z. Fenton, M. J. Owen, P. Thompson, C. Long, and G. H. Elder, *Hum. Mol. Genet.* **4,** 2387 (1995).
[17] J. F. McManus and H. A. Dailey, manuscript in preparation (1997).
[18] N. J. Jacobs and J. M. Jacobs, *Biochem. Biophys. Res. Commun.* **65,** 435 (1975).
[19] N. J. Jacobs and J. M. Jacobs, *Biochim. Biophys. Acta* **459,** 141 (1977).
[20] N. J. Jacobs and J. M. Jacobs, *Arch. Biochem. Biophys.* **211,** 305 (1981).
[21] D. J. Klemm and L. L. Barton, *J. Bacteriol.* **169,** 5209 (1987).

TABLE I
MUTATIONS IN HUMAN PROTOPORPHYRINOGEN OXIDASE GENE

Mutation	Ref.
R59W	*a*
R168C	*a*
A433P	*b*
G358R	*c*
G169E	*c*
Insertion of 12 bp (AAGGCCAACGCC) following bp 656 of cDNA	*c*
Point insertion of a G at bp 1022 of cDNA	*d*
G232R	*d*

[a] P. N. Meissner, T. A. Dailey, R. J. Hift, M. Ziman, A. V. Corrigall, A. G. Roberts, D. M. Meissner, R. E. Kirsch, and H. A. Dailey, *Nature Genet.* **13,** 95 (1996).
[b] A. G. Roberts, S. D. Whatley, T. A. Dailey, H. A. Dailey, and G. H. Elder, *Hepatology* **23,** 309 (1996).
[c] G. H. Elder, personal communication (1996).
[d] J. C. Deybach, H. Puy, A. M. Robreau, J. Lamoril, V. Da Silva, B. Grandchamp, and Y. Nordmann, *Hum. Mol. Genet.* **5,** 407 (1996).

In humans, a deficiency in protoporphyrinogen oxidase leads to a genetic disorder known as variegate porphyria.[22,23] This is an autosomally dominant disorder that manifests with both photosensitivity and neurovisceral abnormalities.[24] To date, there have been several mutations identified in the human enzyme that are associated with clinical symptoms (Table I). Of these the R59W mutation is of particular interest because it has been shown to be the cause of the well-characterized variegate porphyria founder effect in South Africa,[25] where it is estimated that as many as 20,000 individuals may possess this same mutation.[24]

In our laboratory, we have engineered the *B. subtilis,*[9] *M. xanthus,*[10] and human[11] PPO cDNAs to contain an NH_3-terminal 6-histidine (His_6) tag, which has greatly facilitated their purification and characterization. By using a metal-chelating matrix, it is possible to purify to apparent homogeneity 10–20 mg of recombinant protein from 1 liter of *E. coli* culture. In this chapter we describe the techniques employed to obtain a homogeneous enzyme preparation and give some characteristics of the purified enzyme.

[22] D. A. Brenner and J. R. Bloomer, *N. Engl. J. Med.* **302,** 765 (1980).
[23] J. C. Deybach, H. DeVerneuil, and Y. Nordmann, *Hum. Genet.* **58,** 425 (1981).
[24] R. S. Day, *Semin. Dermatol.* **5,** 138 (1986).
[25] P. N. Meissner, T. A. Dailey, R. J. Hift, M. Ziman, A. V. Corrigall, A. G. Roberts, D. M. Meissner, R. E. Kirsch, and H. A. Dailey, *Nature Genet.* **13,** 95 (1996).

Purification of Recombinant Human Enzyme

The 6-histidine-tagged recombinant human enzyme is easily purified from 1-liter culture of *E. coli* JM109 containing the previously described plasmid pHPPO-X,[11] grown at 37° with shaking. Maximal yields of protein are obtained when the culture is grown using a "superbroth" such as Circlegrow (Bio 101, La Jolla, CA) and by incubating the culture for 18–22 hr to allow maximal induction of the plasmid promoter. The cells are harvested by centrifugation at 7000 *g* at 4° for 15 min, then resuspended in 60 ml of 50 m*M* sodium phosphate (pH 8.0), 0.2% (w/v) *n*-octyl-β-D-glucopyranoside (BOG), and phenylmethylsulfonyl fluoride (PMSF) (10 μg/ml). After the cells are resuspended, an additional 4.8 ml of 10% (v/v) BOG is added to bring the final BOG concentration to 1.0% (v/v). The cells are sonicated at maximum power (twice for 30 sec) on ice, and centrifuged at 100,000 *g* for 1 hr at 4°. The resulting supernatant is decanted and stirred for 30 min at room temperature with 3.5 ml of a 50% slurry of TALON matrix (Clontech, Palo Alto, CA). This suspension is then poured into a small chromatography column such as an Econo-Column (Bio-Rad, Richmond, CA), and the flow-through allowed to drip through at approximately 2 drops/sec. The column is then washed with 5 ml of the cell pellet resuspension buffer, followed by a wash with 10 bed volumes of 50 m*M* sodium phosphate (pH 8.0), 0.2% (w/v) BOG, 20% (v/v) ethanol, PMSF (10 μg/ml). The purified protein is eluted from the column with 6 ml of 50 m*M* sodium phosphate (pH 8.0), 0.2% (w/v) BG, 300 m*M* imidazole, PMSF (10 μg/ml). Fractions of 0.5 ml are collected, and the protein typically elutes off the column in the third to seventh fractions. The PPO-containing fractions are easily identified by their bright yellow color, indicating the presence of the flavin cofactor. The purified enzyme is stable for several days at 4°.

Notes: (1) The yield of purified PPO enzyme is dependent on the age of the *E. coli* culture. We obtained maximal yields with cultures grown 18–22 hr. Induction of the plasmid with isopropylthiogalactoside (IPTG) does not increase the yields obtained. (2) Other metal-chelate resins, for example Ni-NTA matrix (Qiagen, Chatsworth, CA), are also effective in purifying recombinant PPO, but despite more stringent washing conditions, in our hands the purity of the final product is decreased compared to that obtained with the TALON matrix.

Protoporphyrinogen Oxidase Assay Procedure

The assay procedure for PPO is essentially that described by Brenner and Bloomer.[26]

[26] D. A. Brenner and J. R. Bloomer, *Clin. Chim. Acta* **100,** 259 (1980).

Preparation of Substrate

Protoporphyrinogen IX is freshly prepared from protoporphyrin IX by reduction with 3% (w/v) sodium amalgam in low light. Briefly, an approximately 1 m*M* solution of protoporphyrin IX (Porphyrin Products, Logan, UT) is prepared by weighing out 4.2 mg of the solid, wetting with concentrated ammonium hydroxide, then adding 8.0 ml of 10 m*M* potassium hydroxide (KOH) and 2 ml of 100% ethanol to give a final volume of 10 ml. Five milliliters of this solution is diluted with 5 ml of 10 m*M* KOH in a foil-covered flask. This solution of protoporphyrin is then shaken in the stoppered flask with approximately 35 g of finely crushed sodium amalgam. The reaction vessel is kept dark and occasionally the stopper is loosened to allow pressure to escape. The reaction of the protoporphyrin solution with the sodium amalgam generates heat as well as gas, so proper protective eye guards, gloves, and clothing should be worn and all manipulations should be done in a fume hood.

When the solution of protoporphyin changes from red to colorless in approximately 3–5 min, the protoporphyrinogen solution is poured over glass wool packed about half full in a 10-ml syringe and collected in a foil-wrapped test tube. This column serves a dual function of separating the solution from residual amalgam and also removes some residual protoporphyrin that sticks to the glass wool. The recovered mixture is flushed with N_2, then titrated with approximately 2 ml of 2 *M* 3-(*N*-morpholino)propanesulfonic acid (MOPS) solution to obtain a final pH of pH 8. The final concentration of protoporphyrinogen obtained can be calculated either by allowing the porphyrinogen to autoxidize or by adding aliquots to some purified PPO enzyme to convert the protoporphyrinogen into protoporphyrin. The protoporphyrin absorbance in 2.7 *N* hydrochloric acid (HCl) is measured at 408 nm, using a E_m of 262.[27]

Assay Procedure

The assay procedure for protoporphyrinogen oxidase is based on the fact that the substrate, protoporphyrinogen IX, is colorless and does not fluoresce, while the product, protoporphyrin IX, is red and fluoresces strongly with an emission around 636 nm. The reaction can be followed using a fluorimeter with excitation at 405 nm and measuring emission at 635 nm. The assay procedure must be carried out in low light to avoid photooxidation of the substrate. For more purified preparations of enzyme where there is no heme or porphyrin present in the sample, one may also

[27] J. H. Fuhrhop and K. M. Smith, *in* "Porphyrins and Metalloporphyrins" (K. M. Smith, ed.), p. 757. Elsevier, New York, 1975.

follow the reaction by spectrophotometrically following the increase in protoporphyrin being produced by monitoring the porphyrin Soret band absorbance at 410 nm.

The assay buffer contains 0.1 *M* sodium phosphate monobasic (NaH_2PO_4), pH 8.0, and 0.1% (v/v) Tween 20. The cuvette buffer is a 1 : 10 dilution in distilled, deionized water of the assay buffer. The buffers should be equilibrated to 37° before use. To set up for the assay, pipette 0.8 ml of assay buffer into an empty tube for each sample to be assayed, as well as one tube for a blank. Add 50 μl of 50 m*M* glutathione to each tube, then add the enzyme sample to be assayed (except for the blank). Maintain these tubes at 37° in a darkened environment throughout the assay procedure.

To begin the assay, pipette 200 μl of substrate as prepared above into a tube containing assay buffer, glutathione, and enzyme sample. Mix, take out a 100-μl aliquot, and put it into a cuvette. Add 0.9 ml of cuvette buffer and read immediately in the fluorimeter. This is the zero time point for each enzyme sample to be assayed. Repeat the procedure at 10, 20, and 30 min or until the reaction rate is no longer linear. Background autoxidation is determined from the reaction mixture to which no enzyme is added. To quantitate the amount of product formed, a standard curve is established in the same buffer system with known amounts of protoporphyrin present.

The diluted aliquot method is employed, rather than following the reaction mixture directly, because the rate of nonenzymatic photooxidation of porphyrinogen catalyzed by the fluorimeter light source is appreciable and the amounts of porphyrin that are produced in some reactions can lead to inner filter effects. Rates of photooxidation should be lower in the spectroscopic assays assuming that the lamp intensity is lower than would be found in a spectrofluorimeter. Under any circumstances, however, the fluorometric detection of product will be more sensitive, but care must be exercised to maintain a constant cuvette temperature because the fluorescence yield is temperature dependent.

Discussion

A molecular weight of 51,000 is calculated for the human PPO by both sodium dodecyl sulfate (SDS) gel electrophoresis and from its amino acid sequence. Similar sizes are found for all other PPOs. The ultraviolet–visible spectrum of the human enzyme is shown in Fig. 1. On the basis of the amino acid composition, an extinction coefficient of 48,000 at 275 nm was calculated for the human enzyme. From the spectrum it was determined that there is 0.5 FAD per PPO monomer.[11] Gel-exclusion chromatography demonstrates that PPO is a homodimer, so there exists one FAD per

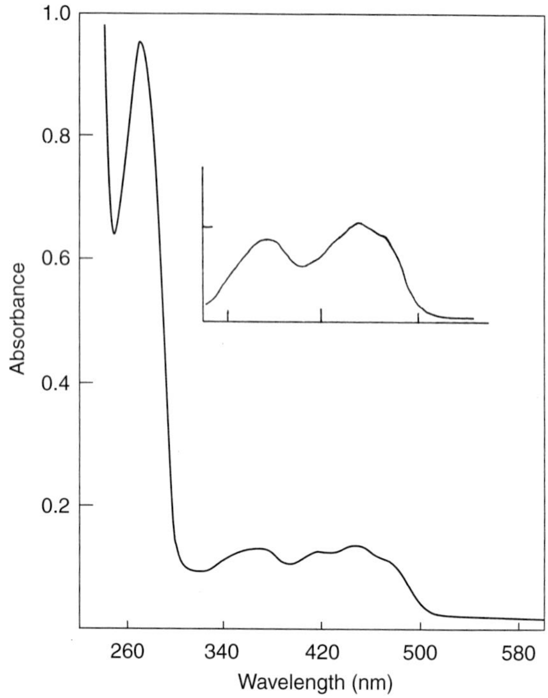

FIG. 1. Visible–ultraviolet spectrum of human protoporphyrinogen oxidase. The enzyme sample (18 μM) was in 50 mM sodium phosphate (pH 8.0), 0.2% (w/v) *n*-octyl-β-D-glucopyranoside, 300 mM imidazole. Because of the presence of imidazole the spectrum shown was obtained against a reference that contained this buffer but without enzyme. *Inset*: Spectrum of the acid-extracted FAD obtained from the enzyme preparation.

dimer.[11] This stoichiometry has been found for all preparations of human and *Myxococcus* PPOs with no preparation ever being obtained with more than one FAD per dimer. The FAD is noncovalently bound.

Figure 2 shows a comparison of the currently verified PPO sequences of *B. subtilis, M. xanthus,* yeast, human, and mouse. Although there is little overall identity, there are a few regions of homology. The largest of these is near the amino terminus, where one finds the putative dinucleotide-binding motif for the FAD.[28] There exist no obvious transmembrane sequences that correlate with the observation that the protein is relatively easily solubilized.

The six-electron oxidation of protoporphyrinogen IX to protoporphyrin IX occurs in three steps, consuming 3 mol of O_2[29] and producing 3 mol of

[28] R. K. Wierenga, P. Terpstra, and W. G. J. Hol, *J. Mol. Biol.* **187,** 101 (1986).
[29] G. C. Ferreira and H. A. Dailey, *Biochem. J.* **250,** 597 (1988).

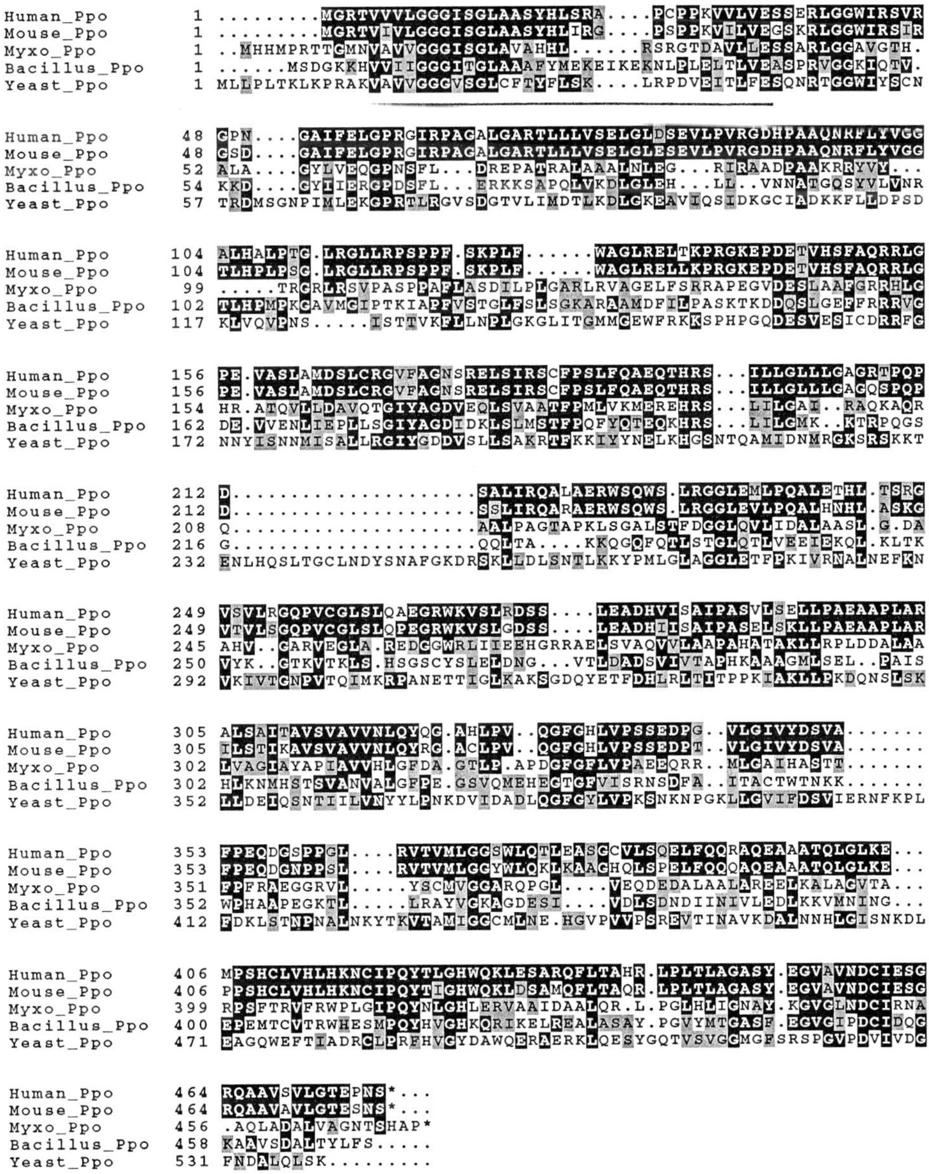

FIG. 2. Sequence comparison of the currently verified protoporphyrinogen oxidases. The dinucleotide-binding motif is underlined.

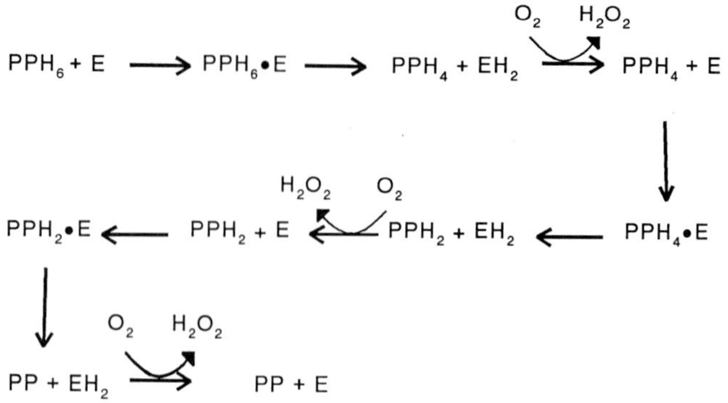

FIG. 3. Proposed model for the mechanism of action of protoporphyrinogen oxidase (E). PPH_6, Protoporphyrinogen IX; PPH_4, tetrahydroprotoporphyrin; PPH_2, dihydroprotoporphyrin; PP, protoporphyrin.

H_2O_2.[11] Data acquired using stopped-flow methodology suggest that the dihydroporphyrin and tetrahydroporphyrin intermediates of the reaction are released and subsequently rebound by the enzyme for further oxidation to the end product, protoporphyrin (Fig. 3). The mechanism of the reaction has been proposed to occur by three desaturations of the cofacial meso-

TABLE II
KINETIC CONSTANTS FOR PROTOPORPHYRIN OF PROTOPORPHYRINOGEN OXIDASE

Source	K_m (μM)	k_{cat} (min^{-1})	Ref.
Bacillus	10.4		*a*
Myxococcus	1.6	5.2	*b*
Human	1.7	10.5	*c*
Yeast	0.02		*d*

[a] T. A. Dailey, P. Meissner, and H. A. Dailey, *J. Biol. Chem.* **269,** 813 (1994).
[b] H. A. Dailey and T. A. Dailey, *J. Biol. Chem.* **271,** 8714 (1996).
[c] T. A. Dailey and H. A. Dailey, *Protein Sci.* **5,** 98 (1996).
[d] J. M. Camadro and P. Labbe, *J. Biol. Chem.* **271,** 9120 (1996).

hydrogens followed by a rearrangement at the fourth position on the opposite side of the porphyrin ring.[30]

The four PPO enzymes expressed to date appear to be highly similar with respect to their physical and enzymological characteristics (Table II). The *B. subtilis* enzyme, however, differs from the others in that it is not inhibited by micromolar concentrations of the diphenyl ether herbicide, acifluorfen, and it can oxidize not only protoporphyrin IX, but also coproporphyrinogen III.[9]

Acknowledgment

This work was supported by NIH Grants DK32303 and DK35898 (to H.A.D.).

[30] M. Aktar, *in* "Biosynthesis of Tetrapyrroles" (P. M. Jordan, ed.), p. 67. Elsevier, New York, 1991.

[40] Purification and Properties of Uroporphyrinogen Decarboxylase from Human Erythrocytes

By ANDREW G. ROBERTS and GEORGE H. ELDER

Introduction

Uroporphyrinogen decarboxylase (UROD) (EC 4.1.1.37) is a cytosolic enzyme of the heme biosynthetic pathway. It catalyzes the sequential removal of the four acetic acid substituents of uroporphyrinogen III to form coproporphyrinogen III with the consequent formation of hepta-, hexa-, and pentacarboxylic porphyrinogen as intermediates.[1] At high substrate concentrations, the sequence of decarboxylation is random but, under physiological conditions, the preferred route is clockwise starting at the acetic acid substituent on ring D.[1,2] Uroporphyrinogen decarboxylase also converts uroporphyrinogen I to coproporphyrinogen I and decarboxylates acetic acid-substituted porphyrinogens of the II- and IV-isomer series that do not occur naturally.[3] Current evidence suggests that these reactions all take place at a single catalytic site.[4]

[1] A. Jackson, H. A. Sancovich, A. M. Ferramola, N. Evans, D. E. Games, S. A. Maitlin, G. H. Elder, and S. G. Smith, *Philos. Trans. R. Soc. Lond. Biol.* **273,** 191 (1976).
[2] J. Luo and C. K. Lim, *Biochem. J.* **289,** 529 (1993).
[3] D. Mauzerall and S. Granick, *J. Biol. Chem.* **232,** 1141.
[4] A. G. Roberts and G. H. Elder, *J. Bioenerg. Biomembr.* **27,** 207 (1995).

0076-6879/97 $25

The structure and properties of URODs from different species have been reviewed.[4] Human UROD is encoded by a single gene on the short arm of chromosome 1 (1p34).[5] The gene contains 10 exons spread over 3 kb and has a 5′ promoter region whose organization is consistent with the ubiquitous expression of a single enzyme.[6,7] However, the rate of transcription increases during erythroid differentiation to a level four-fold higher than in other tissues[6]; the mechanism of this tissue-specific effect has not been determined. Erythrocytes are, therefore, a convenient source from which homogeneous preparations of UROD can be obtained in relatively good yield.[8–10] The enzyme has also been purified to homogeneity from chicken erythrocytes,[11] bovine liver,[12] yeast,[13] and *Rhodobacter sphe-roides.*[14] The procedure described here is a development of our previously published method.[9] It has been used on several occasions and has proved to be a reliable and robust method that results in a highly purified product.

Purification Procedure

General

All stages of the purification procedure are carried out at 4°. Dithiothreitol (DTT) is added to all buffers immediately before use to a final concentration of 0.5 m*M*.

Step 1. Preparation of Hemolysate and Removal of Hemoglobin

Packed human erythrocytes (approximately 2 liters) are washed twice with an equal volume of 0.9% (w/v) NaCl and hemolyzed by resuspending the packed cells in an equal volume of ice-cold water, with gentle stirring, for 2 hr. DEAE-cellulose (DE52; Whatman, Maidstone, Kent, UK), fined

[5] M. G. Mattei, A. Dubart, D. Beaupain, M. Goosens, and J. F. Mattei, *Cytogenet. Cell. Genet.* **40,** 892 (1985).

[6] M. Romana, A. Dubart, D. Beaupain, C. Chabret, M. Goosens, and P. H. Romeo, *Nucleic Acids Res.* **15,** 7343 (1987).

[7] M. J. Moran-Jimenez, C. Ged, M. Romana, R. Enriquez de Salamanca, A. Taieb, G. Topi, L. D'Allessandro, and H. de Verneuil, *Am. J. Hum. Genet.* **58,** 712 (1996).

[8] H. de Verneuil, S. Sassa, and A. Kappas, *J. Biol. Chem.* **258,** 2454 (1983).

[9] G. H. Elder, J. Tovey, and D. Sheppard, *Biochem. J.* **215,** 45 (1983).

[10] P.-H. Romeo, N. Raich, A. Dubart, D. Beaupain, M. Pryor, J. Kushner, M. Cohen-Solal, and M. Goossens, *J. Biol. Chem.* **261,** 9825 (1986).

[11] S. Kawanishi, Y. Seki, and S. Sano, *J. Biol. Chem.* **258,** 4285 (1983).

[12] J. G. Straka and J. P. Kushner, *Biochemistry* **22,** 4664 (1983).

[13] F. Felix and N. Brouillet, *Eur. J. Biochem.* **188,** 393 (1990).

[14] R. M. Jones and P. M. Jordan, *Biochem. J.* **293,** 703 (1993).

by repeated washing with water, is equilibrated with 0.1 *M* potassium phosphate buffer (pH 7.0) and then extensively washed with water. The hemolysate (4.1 liters) is added to the resin (500 g dry weight), gently stirred for 2 hr, and allowed to settle for 16–20 hr, after which a further 200 g of dry unequilibrated DE52 is added. The resin is stirred for another 2 hr. The resin is then transferred to a large glass column (120 × 300 mm), extensively washed with 2 m*M* potassium phosphate buffer (pH 7.0) until most of the hemoglobin had been eluted, and then washed with 2 m*M* potassium phosphate buffer (pH 7.0)–0.1 *M* KCl until the absorbance of the supernatant at 280 nm is less than 0.29 cm^{-1}. Finally, the resin is washed with 2 m*M* potassium phosphate buffer (pH 7.0)–0.5 *M* KCl and the eluted fractions containing UROD are pooled.

Step 2. Ammonium Sulfate Precipitation

Solid $(NH_4)_2SO_4$ (25 g/100 ml) is added slowly with stirring to the pooled eluate and the resulting solution is stirred for 2 hr. After centrifugation for 20 min at 25,000 *g* the pellet is discarded and additional $(NH_4)_2SO_4$ (16 g/100 ml) added. Precipitated protein is collected by centrifugation as before and resuspended in a minimum volume of 50 m*M* potassium phosphate buffer (pH 7.0).

Step 3. Phenyl-Sepharose Chromatography

The redissolved $(NH_4)_2SO_4$ precipitate is mixed with an equal volume of 2 m*M* potassium buffer (pH 7.0) containing 2 *M* $(NH_4)_2SO_4$ and applied, at a flow rate of 2 ml/min, to a phenyl-Sepharose CL-48 (Pharmacia Fine Chemicals, Uppsala, Sweden) column (35 × 120 mm) equilibrated with 2 m*M* potassium phosphate buffer (pH 7.0)–1 *M* $(NH_4)_2SO_4$. The column is washed with 200 ml of this buffer followed by 200 ml of 2 m*M* potassium phosphate buffer containing 0.5 *M* $(NH_4)_2SO_4$. Uroporphyrinogen decarboxylase is eluted using a linear gradient from 2 m*M* buffer–0.5 *M* $(NH_4)_2SO_4$ to 2 m*M* buffer (gradient volume, 300 ml) followed by 40 ml of 2 m*M* buffer and 300 ml of 2 m*M* buffer containing ethanediol (60%, v/v). All fractions containing UROD are pooled, concentrated by precipitation with $(NH_4)_2SO_4$ (62 g/100 ml), and redissolved in 50 m*M* potassium phosphate buffer (pH 7.0).

Gel Filtration

The redissolved $(NH_4)_2SO_4$ precipitate (18 ml) is applied to an Ultragel AcA 54 (LKB Instruments, Ltd., Croydon, Surrey UK) column (44 × 1500 mm) previously equilibrated with 50 m*M* potassium phosphate buffer (pH

7.0)–0.2 *M* KCl. Five-minute fractions are collected at a flow rate of 100 ml/hr. Only those fractions containing a high ratio of UROD to total protein are pooled.

Ion-Exchange High-Performance Liquid Chromatography

The pooled fractions are concentrated by dialysis against 1 liter of 20 m*M* Tris-HCl buffer (pH 8.0) containing 30% (w/v) polyethylene glycol 6000 (PEG 6000) to a volume of 10 ml and further dialyzed overnight using two changes of Tris-HCl buffer.

Concentrated enzyme preparations (2-ml aliquots) are centrifuged at 15,000 *g* for 2 min and injected onto a TSK DEAE-5PW high-performance liquid chromatography (HPLC) column (7.5 × 75 mm) (Anachem, Ltd., Luton, UK) equilibrated with 20 m*M* Tris-HCl buffer (pH 8.0) at a flow rate of 1 ml/min. Protein is eluted from the column using an NaCl gradient from 0.005 to 0.45 *M* over 60 min. The NaCl concentration is increased from 0.005 to 0.025 *M* over 10 min, then to 0.15 *M* over 5 min, then to 0.30 *M* over 35 min, and finally to 0.45 *M* over 5 min. Fractions (1 ml) are collected into polypropylene tubes containing DTT (0.05 ml) to give a final concentration of 0.5 m*M* DTT. Fractions containing the highest UROD activity are pooled and stored in aliquots at −70°. The enzyme is stable under these conditions for at least 3 months.

Assay Methods

Uroporphyrinogen decarboxylase activity is monitored at each stage of the purification procedure by measuring the conversion of pentacarboxylate porphyrinogen III to coproporphyrinogen III, as described previously,[9] or by a modification in which porphyrins are not converted to their methyl esters but separated directly by reversed-phase high-performance liquid chromatography.[15] A convenient alternative procedure using pentacarboxylate porphyrinogen I, which is available commercially (Porphyrin Products, Logan, UT), has been described by Smith *et al.*[16] For some preparations, UROD concentration is measured by electroimmunoassay using a rabbit polyclonal antiserum to purified human erythrocyte UROD.[9]

Comments on Purification Procedure

More than 80% of UROD elutes from phenyl-Sepharose at low salt concentration; the rest is eluted by 60% (v/v) ethanediol. These two frac-

[15] C. K. Lim and T. J. Peters, *Clin. Chim. Acta* **193,** 55 (1984).

[16] A. G. Smith, J. E. Francis, S. J. E. Kay, J. B. Greig, and F. P. Stewart, *Biochem. J.* **238,** 871 (1986).

tions can be pooled without compromising the purity of the final enzyme. Only peak fractions should be pooled after gel filtration and following ion-exchange HPLC.

Inclusion of DTT in all buffers, especially during the later stages of purification, is essential for preservation of enzyme activity. We have also repeatedly observed loss of enzyme activity after concentration using collodion membranes. A similar observation was made by Straka and Kushner,[12] who reported a high affinity of bovine liver UROD for glass or plastics, and may reflect a relatively high (10% in humans) aromatic amino acid content.

The results of a representative purification of UROD from 2 liters of packed human erythrocytes are summarized in Table I. A total of 2.2 mg of enzyme was obtained, representing a yield of 11%. The 24,360-fold purification is similar to that obtained by de Verneuil *et al.*[8] Sodium dodecyl sulfate–polyacrylamide gel electrophoresis (SDS–PAGE) and reversed-phase high-performance liquid chromatography of the purified enzyme on a Synchropak RP-P (Synchron, Inc., Linden, Indiana) column eluted with a 1% (v/v) trifluoroacetic acid–acetronitrile gradient [0–45% (v/v) acetonitrile over 60 min] confirmed that UROD represented more than 85% of the total recovered protein. Measurements of enzyme activity and immunoreactive enzyme concentration showed that the specific activity of the purified enzyme using either pentacarboxylate porphyrinogen III or uroporphyrinogen III as substrate was unchanged from that of the crude hemolysate.

Properties

Physical Properties

Human UROD contains 367 amino acid residues and has a molecular mass of 40.831 kDa.[10] Sodium dodecyl sulfate-polyacrylamide gel chroma-

TABLE I

PURIFICATION OF UROPORPHYRINOGEN DECARBOXYLASE FROM HUMAN ERYTHROCYTES[a]

Step	Volume (ml)	Protein (mg/ml)	Activity (U/mg)	Purification (-fold)	Yield (%)
Hemolysate	2,125	230.0	0.06	0	100
DEAE-cellulose	730	6.6	3.11	54	53
$(NH_4)_2SO_4$ fractionation	43	24.0	16.65	240	51
Phenyl-Sepharose	18	13.0	48.61	854	41
Ultrogel AcA 54	10	4.3	66.37	1,167	10
TSK DEAE-5 PW HPLC	11	0.2	1,384.80	24,360	11

[a] Enzyme activity was measured using pentacarboxylate porphyrinogen III as substrate; 1 unit of enzyme activity is equivalent to the formation of 1 μmol of coproporphyrinogen per minute.

tography of the enzyme, purified as described above, shows a single protein band at 39.5–41 kDa but determination of molecular mass by gel filtration gives a value of 58 kDa.[8,9] The isoelectric point of human UROD is 4.60.[8,9]

Kinetic Properties

The kinetics of purified human UROD have been investigated by de Verneuil *et al.*[8] At pH 6.8, K_m values for uroporphyrinogen I, uroporphyrinogen III, pentacarboxylate porphyrinogen I, and pentacarboxylate porphyrinogen III were 0.8, 0.35, 0.07, and 0.05 μM, respectively. Relative activities for these substrates at this pH were as follows: pentacarboxylate porphyrinogen I (1.0) and III (6.7); uroporphyrinogen I (2.2) and III (3.4). The optimum pH for decarboxylation of both series III isomers was pH 6.8 but was lower for the series I-isomers: pH 6.2 for uroporphyrinogen and pH 5.4 for pentacarboxylate porphyrinogen. At their respective pH optima, decarboxylation rates for series I- and III-isomers were similar.

Cofactors

No cofactor has been identified in human UROD or in enzymes from other species.[4,8]

Inhibitors

In addition to competitive inhibition between alternative substrates, human UROD is inhibited by uroporphyrin I and III and by other porphyrins produced by oxidation of its substrates. Detailed studies of the kinetics of inhibition of purified human UROD by porphyrins have not been reported. The enzyme is also rapidly inactivated by uroporphyrin-catalyzed photooxidation.

Human UROD contains six cysteine residues and requires reducing agents for full activity *in vitro*. It is significantly inhibited by reaction with 5,5′-dithiobis(2-nitrobenzoate), *p*-hydroxymercuribenzoate, Hg^{2+}, or *N*-ethylmaleimide (NEM). Inactivation by NEM is prevented by prior incubation with pentacarboxylate porphyrinogen III and requires reaction with at least three of the six cysteine residues; two of the remaining residues do not react with this reagent unless the enzyme is denatured with 6 *M* guanidine hydrochloride. Other SH-reactive reagents, such as iodoacetamide and iodoacetic acid, have little or no effect on activity. Human UROD is also inhibited by other group-specific reagents. Inhibition by phenylglyoxal or butanedione is prevented by prior incubation with substrate or oxidized substrate, suggesting that arginine residues may be required for binding of substrate carboxyl groups as for other enzymes of heme biosynthesis. The enzyme is also inhibited by diethyl pyrocarbonate (DEPC). This inhibition is reversed by hydroxylamine, indicating reaction of DEPC with at least one histidine residue, but is not prevented by prior incubation with substrate.

Inhibition by metals has been studied because iron is involved in the pathogenesis of the human UROD deficiency disorder, porphyria cutanea tarda. In addition to Hg^{2+}, human UROD is inhibited by Cu^{2+} and Pt^{2+} but not by Co^{2+}, Mg^{2+}, Mn^{2+}, Pb^{2+}, or Sb^{3+}. Although inhibition of partially purified human enzyme by iron has been reported, neither Fe^{2+} nor Fe^{3+} appears to inhibit purified human UROD.[8]

[41] Purification and Properties of Coproporphyrinogen III Oxidase from Bovine Liver

By TAKEO YOSHINAGA

Introduction

Coproporphyrinogen oxidase (CPO, EC 1.3.3.3), the sixth enzyme of the heme biosynthetic pathway, catalyzes the conversion of coproporphyrinogen III to protoporphyrinogen IX by oxidative decarboxylation of two propionate groups at positions 2 and 4 of the tetrapyrrole to yield vinyl groups.[1-3] The mammalian enzyme is located in the intermembrane space of mitochondria and is soluble or very loosely bound to the membrane[4,5]; in contrast, CPO in yeast is cytosolic.[6] Since 1961[1] there have been many attempts to purify the enzyme. Yoshinaga and Sano succeeded in obtaining a preparation from bovine liver that was almost pure[7] and CPOs from yeast[6] and mouse liver[8] were also purified. However, some discrepancies in molecular properties of these enzymes were seen; for example, the molecular mass of bovine liver CPO was 71.6 kDa in contrast with CPO of mouse or yeast, which was 35 kDa. A modified purification procedure of bovine

[1] S. Sano and S. Granick, *J. Biol. Chem.* **236,** 1173 (1961).
[2] A. M. del C. Batlle, A. Benson, and C. Rimington, *Biochem. J.* **97,** 731 (1965).
[3] H. G. Elder, J. O. Evans, J. R. Jackson, and A. H. Jackson, *Biochem. J.* **169,** 215 (1978).
[4] H. G. Elder and J. O. Evans, *Biochem. J.* **172,** 345 (1978).
[5] B. Grandchamp, N. Phung, and Y. Nordmann, *Biochem. J.* **176,** 97 (1978).
[6] J. M. Camadro, H. Chambon, J. Jolles, and P. Labbe, *Eur. J. Biochem.* **156,** 579 (1986).
[7] T. Yoshinaga and S. Sano, *J. Biol. Chem.* **255,** 4722 (1980).
[8] M. Bogard, J. M. Camadro, Y. Nordmann, and P. Labbe, *Eur. J. Biochem.* **181,** 417 (1989).

0076-6879/97 $25

liver has permitted us to clone the enzymes from mouse[9] and from human.[10] The analysis of these cDNAs provided a molecular mass of 37 kDa from amino acid compositions. Coproporphyrinogen oxidase genes from *Saccharomyces cerevisiae*,[11] *Salmonella typhimurium*,[12] *Escherichia coli*,[13] and soybean[14] have been also isolated and sequenced.

Recombinant mouse CPO expressed in *E. coli* has a specific activity that is almost equivalent to that of the bovine liver enzyme, and contains copper responsible for the enzyme activity.[15] Site-directed mutagenesis of highly conserved His-158 caused the complete loss of the activity with concomitant release of bound copper.[15]

This chapter describes an improved method of purification of CPO from bovine liver, purification of recombinant mouse CPO from *E. coli*, properties of the enzymes, and results of point mutation of histidine residues.

Purification of Coproporphyrinogen Oxidase

Enzyme Assay

Methods to assay CPO are separated into two groups. The first utilizes the absorbance of protoporphyrin after extraction from the organic phase using different concentrations of HCl[1] and the second method uses radioactivity in the determination of the separated protoporphyrin.[16] Separation of protoporphyrin from coproporphyrin and determination by high-performance liquid chromatography (HPLC) equipped with a reversed-phase column and fluorometric monitor can be used.[17] The assay method described here can be used to treat many samples at the same time.

Coproporphyrinogen III as the substrate is prepared by hydrolysis of coproporphyrin tetramethyl ester with 6 *N* HCl for 6 hr at room temperature and pooled in 0.01 *N* NaOH in the dark at a concentration of 10 m*M*. Just

[9] H. Kohno, T. Furikawa, Y. Yoshinaga, R. Tokunaga, and S. Taketani, *J. Biol. Chem.* **268,** 21359 (1993).

[10] S. Takenani, H. Kohno, T. Furukawa, T. Yoshinaga, and R. Tokunaga, *Biochim. Biophys. Acta* **1183,** 547 (1994).

[11] M. Zagorec, L.-M. Buhler, I. Treich, T. Keng, L. Guarente, and R. Labbe-Bios, *J. Biol. Chem.* **263,** 9718 (1988).

[12] K. Xu and T. Elliott, *J. Bacteriol.* **175,** 4990 (1993).

[13] B. Troup, M. Jahn, C. Hungerer, and D. Jahn, *J. Bacteriol.* **176,** 673 (1994).

[14] O. Madsen, L. Sandal, N. N. Sandal, and K. A. Macker, *Plant Mol. Biol.* **23,** 35 (1993).

[15] H. Kohno, T. Furukawa, R. Tokunaga, S. Taketani, and T. Yoshinaga, *Biochim. Biophys. Acta* **1292,** 156 (1996).

[16] B. Grandchamp and Y. Nordmann, *Enzyme* **28,** 196 (1982).

[17] R. Guo, C. K. Lim, and T. J. Peter, *Clin. Chim. Acta* **177,** 245 (1988).

before use, coproporphyrin is quickly reduced by granular 3% (w/v) sodium amalgam (1 g/ml) in the presence of 0.1 *M* 2-mercaptoethanol with vigorous shaking. When the solution becomes colorless or pale yellow, it is drawn through a small cotton ball with a pipette and transferred to a new test tube. The reduced solution is neutralized with 6 *N* H_3PO_4, and then diluted with 4 vol of ice-cooled 0.1 *M* Tris-HCl buffer, pH 7.6.

Reaction mixtures contain 5–100 μl of enzyme solution, 100 μM coproporphyrinogen, and 20 m*M* 2-mercaptoethanol in 80–100 m*M* Tris-HCl buffer, pH 7.6, at a final volume of 0.5–0.1 ml in 10-ml glass tubes. The mixtures are incubated at 37° for 15–60 min in the dark. After incubation, 4 ml of acetic acid–ethyl acetate (1 : 3, v/v) is added followed by exposure to strong light (500-W tungsten light) for 5 to 30 min; the color of the mixture becomes pinkish orange and then the color diminishes. Two milliliters of H_2O is added followed by stirring with a Vortex mixer for 5 sec. After the mixture separates into organic and aqueous phases, the aqueous phase is removed with a long 18-gauge needle connected to an aspirator through a stop cock or to a cylinder. One milliliter of 2.5 *N* HCl is added to the organic phase, which is then vortexed for 5 sec; porphyrins are extracted into the aqueous phase and then the organic phase is removed. The pH of the aqueous phase is brought to pH 3.5 by addition of 1.6 ml of saturated sodium acetate. Four milliliters of ethyl ether is added to each tube and porphyrins are extracted into the organic phase. The aqueous phase is then aspirated off. Unreacted coproporphyrin is removed by washing the organic phase with 2 ml of 0.12 *N* HCl three or four times. The amount of remaining substrate can be determined by measuring the absorbance of the washing solution at 402 nm (E_{mM} = 470 M^{-1} cm^{-1}). Protoporphyrin is extracted with 1.0 or 2.0 ml of 3 *N* HCl from the organic phase and the amount is determined by measuring absorbance at 408 nm (E_{mM} = 262 M^{-1} cm^{-1}).

This method is convenient as only one tube is necessary for each sample throughout the procedure from incubation to product extraction. Ten-milliliter conical centrifuge tubes with screw caps are recommended to prevent overflow of contents on vortexing.

One unit of CPO is defined as the amount that catalyzes the conversion of 1 nmol of protoporphyrinogen per hour at 37°.

Preparation of Crude Enzyme Source

In mammalian cells CPO is located in the intermembrane space of mitochondria,[4,5] and is almost soluble or very loosely bound to the membrane. When the mitochondrial fraction is prepared carefully from fresh tissue to obtain an intact preparation, the specific activity of CPO is higher among the other fractions; however, the total amount of activity in these other fractions is not large. If a large column (62 × 900 mm) is available, it is not important to start with pure mitochondria.

All the subsequent operations are carried out at temperatures between 0 and 4°. One kilogram of fresh bovine liver (loss of the enzyme activity is less than 50% over 1 month when fresh liver is stored at −80°) is sliced and homogenized in 4 liters of 20 m*M* Tris-HCl buffer, pH 7.6, containing 20 m*M* 2-mercaptoethanol and 0.2 m*M* phenylmethylsulfonyl fluoride (TMP buffer), with a Waring blender under cooling. The homogenate is centrifuged at 15,000 *g* for 20 min and the supernatant is collected.

Ammonium Sulfate Fractionation. Powdered ammonium sulfate is added to the supernatant to 35% saturation, the pH of the mixture is adjusted to pH 7.8, and the solution is stirred for 30 min. After removal of the precipitate by centrifugation at 15,000 *g* for 30 min, ammonium sulfate is added to the solution to 45% saturation. The mixture is stirred for 30 min, then the enzyme is precipitated by centrifugation at 15,000 *g* for 30 min. The activity of CPO in the precipitated state is not stable, so the precipitate obtained is dissolved with a minimum volume of TMP buffer and dialyzed twice against 20–50 vol of TMP buffer for 6 hr.

Purification of Enzyme

DEAE-Cellulofine Column Chromatography. Dialyzed crude enzyme is centrifuged at 30,000 *g* for 20 min to remove insoluble materials, followed by loading onto a column of DEAE-Cellulofine (62 × 900 cm) (Seikagaku-Kogyo, Tokyo, Japan) preequilibrated with TMP buffer at a flow rate of 100 ml/hr. The procedure for preparation of the column is important to obtain high resolution, especially when using a large column. No air bubbles should be trapped in the column. The column is washed with 0.9 liter of TMP buffer followed by elution of the enzyme with a linear gradient of NaCl from 0 to 0.5 *M* in 5 liters of TMP buffer at a flow rate of 200 ml/hr. Thirty-milliliter fractions are collected. Typical elution patterns of protein and CPO activity are shown in Fig. 1. Coproporphyrinogen oxidase is eluted at 0.1 *M* NaCl. Fractions with CPO activity over 50% of peak activity are pooled and the volume ratio of this pooled fraction to total eluate is 6–7%.

Hydroxylapatite Column Chromatography. The pH of the combined fractions is adjusted to pH 6.5 with H_3PO_4. Coproporphyrinogen oxidase is not stable under acidic conditions, and thus care must be taken to avoid dropping below pH 6.5. The fractions are loaded on a column of hydroxylapatite (26 × 300 mm) pretreated with 20 m*M* potassium phosphate buffer, pH 6.5, containing 10 m*M* 2-mercaptoethanol and 0.2 m*M* phenylmethylsulfonyl fluoride (PMP buffer) at a flow rate of 50 ml/hr. Hydroxylapatite gel is usually handled with phosphate buffer, because the gel is protected from

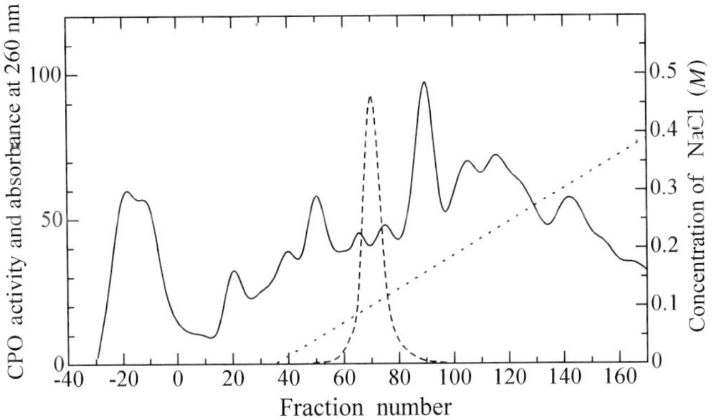

FIG. 1. Elution profile of DEAE-Cellulofine column chromatography for purification of CPO from bovine liver. Details are described in text. Fraction 1 indicates the starting position of the gradient for elution. The CPO activities in the fractions are shown with a dashed line and absorbance of eluate monitored at 260 nm is shown with a solid line. The scales of these parameters are expressed in arbitrary units on the *y* axis. The concentrations of NaCl in the fractions are shown with a dotted line.

dissolving by phosphate ions. In this case, however, the fractions are loaded onto the column in Tris-HCl buffer from the previous step, which exposes the fresh surface of the gel and promotes binding of CPO. The column is washed with 200 ml of PMP buffer at a flow rate of 70 ml/hr. The enzyme is eluted with a linear gradient of 500 ml of 20 m*M* PMP buffer to 500 ml of 300 m*M* potassium phosphate buffer, pH 7.6, containing 10 m*M* 2-mercaptoethanol and 0.2 m*M* phenylmethylsulfonyl fluoride at a flow rate of 70 ml/hr. The eluate is collected in 10-ml fractions. CPO was eluted at a phosphate concentration of 110 m*M*. The elution position of CPO is variable because the performance of hydroxylapatite differs between preparations, manufacturers, and even within the same batch. Hydroxylapatite for column chromatography from Nacalai Tesque (Kyoto, Japan) is sold as a dried powder, is easy to handle, and has good performance with regard to both capacity and flow rate. Fractions with CPO activity over 50% of peak activity are combined and 13 m*M* 2-mercaptoethanol and solid ammonium sulfate up to 1.0 *M* are added. The volume of combined fractions is about 15% of that of total eluate.

Hydrophobic Column Chromatography by Butyl-Toyopearl. Fractions with increased ionic strength are loaded onto a column (26 × 250 mm) of butyl-Toyopearl (TSK gel butyl-Toyopearl 650S; Tosoh Corporation, Tokyo, Japan) preequilibrated with TMP buffer containing 1.0 *M* ammonium sulfate at a flow rate of 30 ml/hr. The column is washed with TMP buffer containing 0.8 *M* ammonium sulfate at a flow rate of 50 ml/hr until the absorbance monitored at 260 nm returns to the initial level. The enzyme is eluted with a linear gradient of ammonium sulfate from 0.8 to 0 *M* in 600 ml of TMP buffer at a flow rate of 40 ml/hr and the eluate is collected in 6-ml fractions. Fractions with CPO activity over 60% of peak activity are combined. The peak of activity is eluted at a position corresponding to 0.12 *M* ammonium sulfate and the volume of fractions combined is about 13% of that of the total eluate. Coproporphyrinogen oxidase in the fractions is collected by precipitation using 60% saturation of ammonium sulfate. The enzyme pellet is dissolved in 2.0 ml of 50 m*M* Tris-HCl buffer containing 10 m*M* 2-mercaptoethanol.

Cellulofine GCL-2000m Column Chromatography. After removal of insoluble materials by centrifugation at 30,000 *g* for 20 min, the concentrated CPO is loaded on a column for gel filtration with Cellulofine GCL-2000m (21 × 1100 mm) (Seikagaku-Kogyo, Tokyo, Japan) preequilibrated with 50 m*M* Tris-HCl buffer containing 10 m*M* 2-mercaptoethanol. The enzyme is eluted with the same buffer used for preequilibration of the column at a flow rate of 10 ml/hr. After 5 hr, the flow rate is increased to 20 ml/hr. The initial slow flow rate is to prevent irregularities in migration pattern caused by the high viscosity of the sample. Fractions of 3.3 ml are collected. Proteins are separated primarily in two peaks detected at 260 nm, and CPO activity is eluted with the later peak. Fractions with CPO activity over 40% of peak are combined and ammonium sulfate is added to a concentration of 0.8 *M*.

Phenyl-Sepharose CL-6B Column Chromatography. The fraction obtained in the previous step is loaded on a column of phenyl-Sepharose CL-6B (16 × 200 mm) preequilibrated with 0.8 *M* ammonium sulfate in 20 m*M* Tris-HCl, pH 7.4, containing 10 m*M* 2-mercaptoethanol at a flow rate of 15 ml/hr. The column is washed with 30 ml of the same buffer used for preequilibration. The enzyme is eluted with a linear gradient from 0.2 to 0 *M* ammonium sulfate in 200 ml of the buffer, followed by further washing of the column with 50 ml of the buffer containing 5% (v/v) 2-propanol. The eluate is collected in 2-ml fractions. Coproporphyrinogen oxidase activity is eluted at the end of the gradient or at the boundary to the buffer containing 2-propanol as a single peak that coincides with absorbance at 260 nm. Fractions with the same specific activity as the peak are combined and concentrated by ultrafiltration [Amicon (Danvers, MA) filter membrane YM10; molecular weight cutoff 10,000].

The results of typical purification are summarized in Table I.

Purification of Recombinant Mouse Coproporphyrinogen Oxidase from *Escherichia coli*

In the same manner as purified bovine liver CPO, mouse cDNA is isolated[9] and the recombinant cDNA expressed in mammalian cells[9] or in

TABLE I
PURIFICATION OF COPROPORPHYRINOGEN OXIDASE FROM BOVINE LIVER

Fraction	Protein (mg)	Total activity ($\times 10^{-3}$ units)	Specific activity (units/mg protein)	Recovery (%)
Homogenate	92,000	213	2.31	100
Ammonium sulfate fractionation (35–45%)	8,200	98.1	11.9	46.1
DEAE-Cellulofine	630	51.2	81.3	24.0
Hydroxylapatite	68	28.3	416	13.3
Butyl-Toyopearl	8.7	19.8	2,280	9.3
Cellulofine GCL-2000m	4.3	15.8	3,670	7.4
Phenyl-Sepharose CL-6B	0.18	7.6	42,200	3.6

E. coli[15] to produce a protein with coproporphyrinogen oxidase activity. To elucidate the properties and reaction mechanism of the enzyme, recombinant mouse CPO is purified from *E. coli* and yields more protein than when purified from mammalian cells.

Construction of Expression Vector of Mouse Coproporphyrinogen Oxidase in pUC18

Mouse full-length CPO is 3.0 kbp and CPO is encoded by 1062 bp, corresponding to 354 amino acid residues. Amino acid terminal analysis of the bovine enzyme suggests that the N terminus of the mature protein starts at residue 32 (serine), corresponding to nucleotide 94.[9] The mouse CPO cDNA has restriction sites for *Sac*I at nucleotide 13 and for *Kpn*I at nucleotide 1385. The *Sac*I–*Kpn*I fragment of CPO cDNA is excised from the full-length cDNA and ligated into the multiple cloning site of pUC18 to generate the expression vector pMCO-1. This vector is used to transform *E. coli* strain JM109. The enzyme expressed in *E. coli* is composed of 355 residues, 32 amino acid residues longer than the mature enzyme, and the sequence of the N terminus is **MITNS**SSGAR, in contrast with that of the original cDNA encoding **MVPK**SSGAR-.

Cultivation of Escherichia coli Carrying Expression Vector

Escherichia coli strain JM109 transformed with pMCO-1 is precultured in 10 ml of Luria–Bertani (LB) medium in the presence of sodium ampicillin (100 μg/liter) overnight with shaking at 37°. The medium is inoculated into 1 liter of LB medium supplemented with sodium ampicillin (100 μg/liter) and 2 m*M* isopropyl-β-D-thiogalactopyranoside (IPTG) for 10 hr with shaking at 37°. The cells are collected by centrifugation with a yield of 3–6 g of cells (wet weight) per liter.

Purification of Recombinant Mouse Enzyme from Escherichia coli

The procedures for large-scale purification of CPO (50–100 g) have been described previously.[15] In this section, a simple small-scale purification procedure using 5–10 g cells is described.

Escherichia coli cells (5–10 g wet weight) immediately after culture or from freezer storage are suspended in 30 ml of 0.1 *M* Tris-HCl buffer, pH 7.8, containing 50 m*M* 2-mercaptoethanol. The suspension is sonicated until no intact cells remain, taking care to avoid heating, and then diluted with 50 ml of cold distilled water containing 10 m*M* 2-mercaptoethanol. The pH value of the solution is adjusted to pH 7.5 and cell debris is removed by centrifugaton at 30,000 *g* for 30 min.

The supernatant is loaded on a column of DEAE-Cellulofine (26 × 400 mm) equilibrated with 20 m*M* Tris-HCl buffer, pH 7.6, containing 10 m*M* 2-mercaptoethanol at a flow rate of 20 ml/hr. The column is washed, at a flow rate of 40 ml/hr, with the buffer used for equilibration of the column until the absorbance monitored at 260 nm returns to the initial level (about 200 ml of buffer). The enzyme is eluted with a linear gradient of NaCl from 0 to 0.5 *M* in 800 ml of the buffer at a flow rate of 40 ml/hr and the eluate is collected in 7-ml fractions. The CPO activity is eluted at a position corresponding to 0.1 *M* NaCl. Fractions with CPO activity over 60% of the peak are combined and the volume of this pooled fraction is less than 5% of the total eluate volume. The pH value of the pooled fraction is adjusted to pH 6.5 with 3 *N* H_3PO_4, taking care not to go below pH 6.5.

The combined fractions are loaded on a column of hydroxylapatite (16 × 200 mm) preequilibrated with 20 m*M* potassium phosphate buffer, pH 6.5, containing 10 m*M* 2-mercaptoethanol at a flow rate of 12 ml/hr. The column is washed with 40 ml of the same buffer used for equilibration of the column. The enzyme is eluted with a linear gradient of 100 ml of the washing buffer to 100 ml of 0.3 *M* potassium phosphate buffer, pH 7.6, containing 10 m*M* 2-mercaptoethanol. The activity of CPO is eluted at a position corresponding to 110 m*M* phosphate, almost coinciding with the peak of absorbance at 260 nm (Fig. 2). Fractions with CPO activity over 60% of the peak value are combined.

The enzyme in the pooled fractions is concentrated by ultrafiltration with an Amicon membrane (YM10) or by precipitation with 60% saturation of ammonium sulfate. A typical purification is summarized in Table II. The enzyme purified by this method has a specific activity of more than 90% of that purified by the large-scale method.[15]

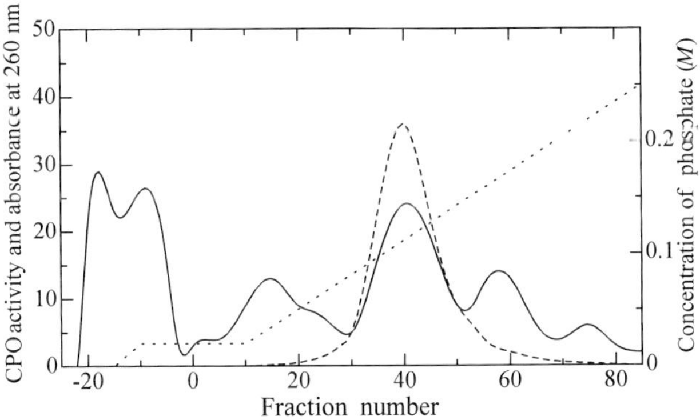

FIG. 2. Elution profile of hydroxylapatite column chromatography for purification of recombinant mouse CPO from *Escherichia coli*. Details are described in text. Fraction 1 indicates the starting position of the gradient for elution. The CPO activities in the fractions are shown with a dashed line and absorbance of eluate monitored at 260 nm is shown with a solid line. The scales of these parameters are expressed in arbitrary units of the *y* axis. The concentrations of phosphate in the fractions are shown with a dotted line.

Properties of Bovine and Mouse Coproporphyrinogen Oxidase

Molecular Properties of Coproporphyrinogen Oxidase. The molecular size of purified CPO from bovine liver was reported to be 71.6 kDa[7]; however, that of CPO from yeast was reported to be 35 kDa,[6] as was that from mouse liver.[8] The accurate molecular sizes of CPOs from many sources have been elucidated by molecular cloning.[9–14] Molecular properties such as molecular weight, amino acid composition, and even conformation of the active sites of enzyme are similar in different mammalian species. The molecular size of mouse CPO is 37,255 Da,[9] that of human is 36,842 Da,[10] and even that of yeast is 37,673 Da.[11] The molecular size of bovine CPO is estimated to be 37 kDa. The molecular size of the bovine enzyme has been determined again by sodium dodecyl sulfate

TABLE II
PURIFICATION OF RECOMBINANT COPROPORPHYRINOGEN OXIDASE FROM *Escherichia coli*

Fraction	Protein (mg)	Total activity ($\times 10^{-3}$ units)	Specific activity (units/mg protein)	Recovery (%)
Crude extract	375	393	1,050	100
DEAE-Cellulofine	39.3	221	5,620	56.2
Hydroxylapatite	8.3	148	17,800	37.7

(SDS) electrophoresis and gel filtration, in which purified recombinant mouse CPO and other proteins were used as standards and confirmed to be 37 kDa.[18] The discrepancy of the reported molecular size of 71.6 and 35 kDa can probably be explained as follows. Purified CPOs from bovine liver and recombinant mouse enzyme are eluted by gel-filtration column chromatography in a sharp peak corresponding to positions of 37 and 41 kDa, respectively. However, both enzymes elute at an earlier position and in a broad peak corresponding to 60 to 70 kDa in the presence of 0.2% (v/v) Tween 20. Both enzymes after purification have a tendency to aggregate gradually to form a polymer if left on ice for 1 week.

The purified CPO from bovine liver has no absorption band in the visible region, suggesting that the enzyme has no prosthetic groups.[7] Purified recombinant mouse CPO also shows no characteristic absorption bands of prosthetic groups when highly concentrated.[15]

Metal bound to purified CPO from bovine liver was analyzed with inductivity-coupled plasma atomic emission spectrometry with the following results: 0.07 to 0.14 atom of iron/enzyme and 0.06 atom of copper/enzyme. However, no other metals (Sn, As, Pb, Ni, Hg, Cd, Be, Cr, Co, Se) were detected.[7] In contrast, recombinant mouse CPO contained 1.05 atoms of copper/enzyme and trace amounts of Fe, Cr, and Ni were detected by inductivity coupled plasma atomic emission spectrophotometry (Shimadze ICPS-8000).[15] As described above, it is hard to recognize that the metal bound to CPO is different between bovine and mouse CPOs. Copper is the metal bound to mammalian CPO.

Kinetic Properties of Coproporphyrinogen Oxidases. The K_m value of the purified CPO from bovine liver is 48 μM[7] and the value of purified mouse recombinant CPO is 47 μM.[15] These K_m values are different by an order of magnitude from the values of 0.3 and 0.05 μM obtained from mouse liver CPO[8] and yeast CPO,[6] respectively. The K_m values reported by the other groups are 32 μM for yeast,[19] 36 μM for tobacco,[20] 0.15 μM for rat liver,[16] and 35 μM for *Chromatium D.*[21] These K_m values can be separated into two groups: over 10 μM and less than 1 μM. This may be caused by the two types of assay method used (i.e., extraction of protoporphyrin from an organic phase with various concentrations of HCl and the use of radioactivity). In the determination of the K_m value it is important to note that CPO activity decreases at higher ionic strength (apparent K_i

[18] T. Yoshinaga and S. Sano, *J. Biol. Chem.* **255,** 4727 (1980).
[19] R. Poulson and J. W. Polglase, *J. Biol. Chem.* **249,** 6367 (1974).
[20] W. P. Hsu and G. W. Miller, *Biochem. J.* **117,** 215 (1970).
[21] M. Mori and S. Sano, *Biochim. Biophys. Acta* **264,** 252 (1972).

is 0.2 *M*), which may be brought about by the substrate. The K_m value for the crude enzyme is 25 to 30 μM.[7]

Purified CPO from bovine liver is activated by various phospholipid or nonionic detergents two to fourfold, despite inhibitory effects of ionic detergents such as sodium deoxycholate or sodium dodecyl sulfate.[7] Purified mouse CPO is also activated by lecithin or neutral detergents.[8] Activation by these compounds is caused by an approximately twofold increase in V_{max} and decrease in K_m by 40–50%, and this is the reason for the lower K_m value of the crude enzyme.[7]

Coproporphyrinogen oxidase from bovine liver is slightly activated by reducing agents such as dithiothreitol or 2-mercaptoethanol, presumably by protecting the substrate from autoxidation during incubation, because such reducing agents are less effective with shorter incubation periods. Various agents that react with SH groups such as *p*-chloromercuribenzoic acid, *N*-ethylmaleimide (NEM), or iodoacetate do not inhibit the activity, significantly suggesting no participation of thiol groups in the enzyme reaction.[7]

Inhibitory effects of metal chelators on CPO activity were reported[1,20] using crude enzyme. However, no such effects were found with purified bovine liver enzyme.[7] In addition, no effects of metal chelators such as D-penicillamine were seen even with recombinant mouse CPO containing one atom of copper per peptide. Addition of metal ions showed no effect on CPO activity of purified enzyme.[7]

As described above, purified CPO has no color at the wavelength where many prosthetic groups show absorption. In contrast, addition of many compounds regarded as prosthetic groups to the assay system of purified bovine CPO resulted had no effect.[7] Coproporphyrinogen oxidase from mammalian cells grown under aerobic conditions seems to have no need for prosthetic groups, in contrast to anaerobic cells.

Molecular oxygen is believed to be necessary for the catalysis of CPO from aerobic cells. The activity of CPO was not significantly decreased when the reaction mixture was maintained under vacuum with a rotary oil pump without substitution of nitrogen in the reaction tube, suggesting that CPO from aerobic cells has higher affinity for oxygen.[22]

General Properties of Coproporphyrinogen Oxidase. No prosthetic groups or factors other than copper have been found. The effects of limiting uptake of copper during enzyme synthesis were examined using minimum medium or LB medium in the presence of a copper-chelating agent, and recombinant mouse CPO was synthesized although its activity was reduced. This result strongly supported participation of copper in the enzyme reaction.[15]

[22] T. Yoshinaga, unpublished data (1996).

In many enzymes, copper coordinates with histidine residues. Four histidine residues and one cysteine residue (H148, H158, H197, H227, and C219), conserved in many CPOs, were replaced with alanine residues by site-directed mutagenesis. Strain H158A (H158 to A158) showed complete loss of CPO activity, whereas the amount of protein produced was slightly increased. The enzyme of strain H148A showed decreased stability and low specific activity.[15] Coproporphyrinogen oxidase from strain H158A was purified to homogeneity according to the procedure described for normal recombinant mouse CPO, using immunoactivity as an indicator of CPO. The protein of H158A behaved in the same way as the normal enzyme during purification, suggesting that the conformation of the enzyme is conserved after replacement of histidine residue with alanine. No copper atoms were detected in the preparation of H158A. From these results, copper atoms in CPO seem to coordinate with a histidine residue and participate in enzyme catalysis.

Chemical modification of bovine CPO also suggested that tyrosine residues play a role in catalysis[18] and this should be confirmed by site-directed mutagenesis. Genetic analyses of hereditary coproporphyria indicated that substitution of some amino residues caused decreased CPO activity.[23,24]

Studies to elucidate the reaction mechanism of the enzyme began more than 20 years ago, using substrate analogs because of the difficulty in obtaining the enzyme.[25,26] β-Hydroxypropionate at positions 2 and 4 of porphyrin is thought to be one of the intermediates of the reaction.[18,27,28] The question has been raised as to why CPO selectively catalyzes only two propionyl groups of four at positions 2 and 4. Elder *et al.* proposed a mechanism in which porphyrins containing a sequence of substituents of methyl-methyl-**propionyl**-methyl or **vinyl**-methyl-**propionyl**-methyl have higher affinity for the enzyme and the **propionyl** group is converted to a **vinyl** group.[3] According to this proposal, the substrate will rotate 90° after conversion of the first propionyl group at position 2 to a vinyl group.[3] This was also confirmed using porphyrins with a β-hydroxypropionyl group at position 2, position 4, or both.[18]

[23] P. Martasek, Y. Nordmann, and B. Grandchamp, *Hum. Mol. Genet.* **3,** 477 (1994).

[24] H. Fujita, H. Kohno, S. Taketani, N. Nomura, K. Furuyama, R. Akagi, T. Terajima, R. A. Galbrain, and S. Sassa, *Hum. Mol. Genet.* **3,** 1807 (1994).

[25] A. R. Battersby and E. McDonald, *in* "Porphyrins and Metalloporphyrins" (K. M. Smith, ed.), p. 61. Elsevier, New York, 1975.

[26] A. R. Battersby, E. Hunt, E. Edword, and J. Moron, *J. Chem. Soc. Perkin* **I,** 2917 (1972).

[27] S. Sano, *J. Biol. Chem.* **241,** 5276 (1966).

[28] M. Akhtar, *in* "Biosynthesis of Tetrapyrroles" (P. M. Jordan, ed.), p. 67. Elsevier, New York, 1991.

Some points remain unclear concerning the subunit structure of CPO (i.e., (1) whether the molecular size of CPO is 71 or 35 kDa, and (2) whether CPO has a monomeric or a homodimeric structure). The molecular size of the peptide was determined to be 37 kDa by genetic analysis. As described above, purified bovine liver enzyme and the recombinant enzyme behave as monomeric structures during gel-filtration column chromatography. It is possible that CPO is a dimeric enzyme that forms an active site between the subunits, but which in the absence of substrate dissociates to monomers. It is difficult to determine the molecular size in the presence of substrate by gel-filtration column chromatography, because the substrate is not sufficiently stable during chromatography. To examine this hypothesis, recombinant mouse CPO activity was measured in the presence of various molar ratios of inactive point-mutated recombinant mouse CPO (H158A). Because H158A-CPO conserves the conformation of the native enzyme, active CPO and inactive H158A-COP should form heterodimers with an incomplete active site between the subunits and the total enzyme activity would consequently decrease. However, the enzyme activity was not decreased by the addition of the inactive CPO.[22] The hypothesis is also unlikely with reference to the results concerning the rotation mechanism of substrate described above.[3,18] Coproporphyrinogen oxidase is therefore likely to be a monomeric enzyme in mammalian cells.

[42] Purification and Properties of Coproporphyrinogen III Oxidase from Yeast

By PIERRE LABBE

Introduction

Coproporphyrinogen III oxidase (CPO) is an enzyme of the heme and chlorophyll biosynthetic pathways that catalyzes the oxidative decarboxylation of the 2- and 4-propionyl groups of coproporphyrinogen III (coprogen) into vinyl groups to form protoporphyrinogen IX (protogen).[1,2]

[1] S. Sano and S. Granick, *J. Biol. Chem.* **236,** 1173 (1961).
[2] R. J. Porra and J. E. Falk, *Biochem. J.* **90,** 69 (1964).

0076-6879/97 $25

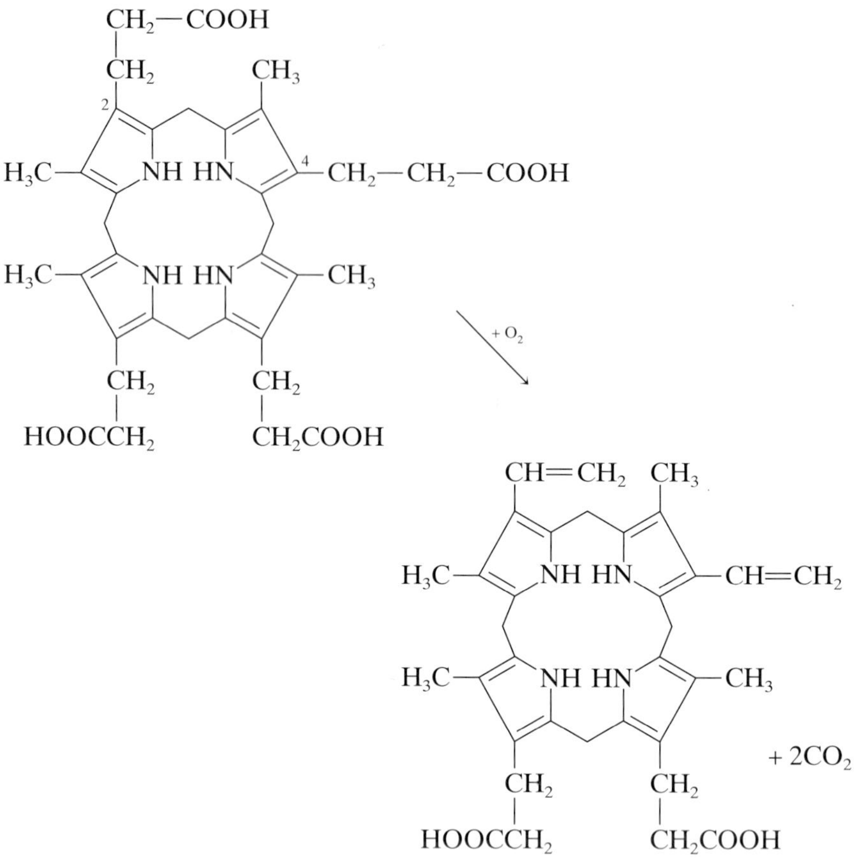

In mammalian liver cells, CPO is located in the mitochondrial intermembrane space and probably interacts with the inner mitochondrial membrane, because 20% of the enzymatic activity is consistently found associated with that membrane fraction.[3,4] Yoshinaga and Sano[5,6] were the first to purify the enzyme to homogeneity from beef liver and found it is a monomer (M_r 71,600) devoid of any cofactor or metal susceptible to react with molecular oxygen, with tyrosine residue(s) possibly involved in the enzymatic activity. Later, CPO was purified from mouse liver and the native enzyme appeared to be a dimer of 70 kDa.[7]

[3] G. H. Elder and J. O. Evans, *Biochem. J.* **172,** 345 (1978).
[4] B. Grandchamp, N. Phung, and Y. Nordmann, *Biochem. J.* **176,** 97 (1978).
[5] T. Yoshinaga and S. Sano, *J. Biol. Chem.* **255,** 4722 (1980).
[6] T. Yoshinaga and S. Sano, *J. Biol. Chem.* **255,** 4727 (1980).
[7] M. Bogard, J. M. Camadro, Y. Nordmann, and P. Labbe, *Eur. J. Biochem.* **181,** 417 (1989).

The yeast *Saccharomyces cerevisiae* is a facultative aerobe that represents the simplest and best-known model of the eukaryotic cell. Yeast heme deficient mutants can grow with a fermentative carbon source, as does the wild-type strain grown in the absence of oxygen, provided that oleic acid and ergosterol are present in the growth medium. However, anaerobically grown wild-type cells do contain some hemoproteins. All the enzymes of the heme pathway are present at roughly similar levels both in aerobically and anaerobically grown cells,[8] except for CPO activity, which is much higher (40- to 50-fold) in anaerobically grown cells and in heme-deficient mutants.[9] Miyake and Sugimura[10] found 70–100% of the total CPO activity in the soluble fraction of cell-free extracts prepared from either the wild strain or a heme-deficient mutant. Poulson and Polglase[11] partially purified the yeast enzyme from mitochondria on the basis of a specific activity slightly higher than the activity in the cytoplasmic fraction. The molecular mass of CPO was around 75 kDa and the enzyme displayed the unusual property of equal activity in the presence or absence of oxygen provided that ATP, methionine (or *S*-adenosylmethionine), and $NADP^+$ as electron acceptor were present in the anaerobic enzymatic assay, as previously shown by Tait[12] for the enzyme of the photosynthetic anaerobic bacteria *Rhodopseudomonas spheroides*. These results suggested that the facultative aerobe *S. cerevisiae* had either one bifunctional CPO or two CPOs functioning differently. This confusing situation was thereafter clarified biochemically and genetically: (1) yeast CPO purified to homogeneity is a homodimer of two 35-kDa subunits displaying only an oxygen-dependent activity,[13] and (2) the enzyme is a cytosolic protein encoded by the *HEM13* gene, whose deletion yields a respiratory mutant totally heme deficient and devoid of any CPO activity. The *HEM13* gene encodes a 328-amino acid polypeptide[14]; its expression is controlled negatively by heme and oxygen at the transcriptional level.[15]

Nine complete sequences of genes or cDNAs encoding CPO are currently available in databanks that include, in addition to the *HEM13* gene, bacterial, plant, and mammalian genes. The derived amino acid sequences show 30% identity extending over the entire proteins. These sequences

[8] P. Labbe, *Biochimie* **53,** 1001 (1971).
[9] D. Urban-Grimal and R. Labbe-Bois, *Mol. Gen. Genet.* **183,** 85 (1981).
[10] S. Miyake and T. Sugimura, *J. Bacteriol.* **96,** 1997 (1968).
[11] R. Poulson and W. J. Polglase, *J. Biol. Chem.* **249,** 6367 (1974).
[12] G. H. Tait, *Biochem. J.* **128,** 1159 (1972).
[13] J. M. Camadro, H. Chambon, J. Jollès, and P. Labbe, *Eur. J. Biochem.* **156,** 579 (1986).
[14] M. Zagorec, J. M. Buhler, I. Treich, T. Keng, L. Guarente, and R. Labbe-Bois, *J. Biol. Chem.* **263,** 9718 (1988).
[15] M. Zagorec and R. Labbe-Bois, *J. Biol. Chem.* **261,** 2506 (1986).

show no amino acid sequence similarity with the bacterial putative anaerobic CPO′ from *Escherichia coli*,[16] *Salmonella typhimurium*,[17] and *R. spheroides*.[18]

Little is known, however, about the structural and catalytic characteristics of the protein, and the mechanism of the enzymatic reaction remains unknown. This is in part because of the low abundance of CPO, which hampers the obtaining of sufficient quantities of enzyme. That point was overcome by overexpression in yeast of the yeast CPO, which now represents up to 10–12% of the total proteins and can be easily purified in hundreds of milligrams.

Coproporphyrinogen III Oxidase Enzymatic Assay

Principle

Porphyrinogens are colorless molecules that on oxidation yield the corresponding highly conjugated porphyrins, which absorb visible light and fluoresce strongly under certain conditions. The enzymatic assay[19] is a coupled assay in which the protogen arising from the oxidative decarboxylation of coprogen by CPO is immediately oxidized enzymatically to the fluorescent protoporphyrin IX by an excess of protoporphyrinogen oxidase (PPO). The added PPO is present in yeast mitochondrial membranes (devoid of endogenous CPO) prepared from commercial baker's yeast. A neutral detergent (Tween 80) is added to the assay to minimize a possible dimerization of protoporphyrin that would not fluoresce. A chelator is also added to the assay medium to inhibit the ferrochelatase activity bound to the mitochondrial membranes to inhibit heme and/or Zn-protoporphyrin formation that would interfere with the assay. Dithiothreitol (DTT) minimizes the nonenzymatic formation of coproporphyrin III from coprogen, which is a competitive inhibitor of CPO activity.

Experiments in which the presence of PPO was not desirable are performed by using the radiochemical assay of Grandchamp and Nordmann.[20]

Special Reagents

Coprogen and yeast mitochondrial membranes are prepared and stored as described in Ref. 19.

[16] B. Troup, C. Hungerer, and D. Jahn, *J. Bacteriol.* **177,** 3326 (1995).
[17] K. Xu and T. Elliott, *J. Bacteriol.* **176,** 3196 (1994).
[18] S. A. Coomber, R. M. Jones, P. M. Jordan, and C. N. Hunter, *Mol. Microbiol.* **6,** 3159 (1992).
[19] P. Labbe, J. M. Camadro, and H. Chambon, *Anal. Biochem.* **149,** 248 (1985).
[20] B. Grandchamp and Y. Nordmann, *Biochem. Biophys. Res. Commun.* **74,** 1089 (1977).

Procedure

Measurements are made with a Jobin-Yvon JY3D spectrofluorimeter (Jobin-Yvon, Longjumeau, France) equipped with a Hamamatsu R928H4 photomultiplier to enhance the sensitivity in the red region of the emission spectra of porphyrins. To avoid excessive light on the cuvette leading to a possible photooxidation of porphyrinogens and to ensure good selectivity, the bandwidths of the four slits are set as follows: 2 nm each for entry and exit for the excitation monochromator; 2 nm for entry and 20 nm for exit for the emission monochromator. The spectrofluorimeter is equipped with a thermostatted cell holder. Mixing of the cuvette content (3 ml) is carried out by magnetic stirring. The kinetics of fluorescence emission or the excitation/emission spectra are recorded with a potentiometric recorder. Under our experimental conditions, the excitation and emission wavelengths are 410 and 632 nm, respectively. The standard reaction medium (3 ml in glass or disposable cuvettes) contains air-saturated Tris buffer (100 m*M*, pH 7.6 at 30°, 1 m*M* EDTA, 1 m*M* DTT, 1 mg of Tween 80,[21] ~2–3 mg of yeast mitochondrial membrane protein,[22] and CPO to be measured. The reaction is started by adding 2–3 μM coprogen (final concentration) and the initial velocity is recorded. A standard calibration curve for protoporphyrin fluorescence is used to quantify CPO activity (see Ref. 19).

One unit of CPO is the amount of enzyme that catalyzes the formation of 1 nmol of protogen/hr at 30° in the standard assay system.

Overproduction and Purification of Yeast Coproporphyrinogen III Oxidase

Principle

The fact that yeast CPO biosynthesis is negatively regulated by heme and oxygen (see Ref. 15) prompted us to develop a homologous system of overexpression. A yeast CPO-overproducing strain has been constructed by introducing the *HEM13* gene with its promoter into a high-copy plasmid in a yeast strain disrupted for the *HEM1* gene [encoding 5-aminolevulinate (ALA) synthase] and therefore heme deficient. The cells are first pregrown in a small volume of synthetic medium containing ALA to maintain the plasmid at a high copy level. That preculture is used to inoculate heavily a large volume (several liters) of rich medium lacking ALA. Overnight growth at 30° until the onset of stationary growth allows maximal induction

[21] Tween 80, Polyoxyethylene sorbitan monooleate, a very low critical micelle concentration (CMC) neutral detergent.

[22] Corresponding to 20–30 PPO units (1 unit equals 1nmol of protogen oxidized per hour).

of CPO (low heme, low O_2). The purification scheme used for overproduced yeast CPO is similar to that previously described (see Ref. 13).

Strain, Growth Conditions, and Harvesting

Yeast Strain. The heme-deficient strain S150-2B/H1Δ (*Mata leu2-3,112 ura3-52 his3-Δ1 trp1-289 hem1-Δ1*) is obtained by replacing the *HEM1* gene with a deleted null allele *hem1-Δ1* (R. Labbe-Bois, unpublished, 1996). It is transformed with the episomic plasmid YEp351[23] (2μ, *LEU2*) carrying the 2,68-kb *Eco*RV–*Sph*I *HEM13* gene encompassing 1079 nucleotides of the 5′ noncoding region, and 660 nucleotides of the 3′ noncoding region (see Ref. 14).

Preculture. Yeast is inoculated in a 500-ml Erlenmeyer flask containing 100 ml of synthetic medium [0.67% (w/v) Difco (Detroit, MI) Difco yeast nitrogen base without amino acids, 2% (w/v) glucose, tryptophan, histidine, uracil (each at 20 mg/liter), and ALA (50 mg/liter)]. Cells are grown overnight at 30° with gentle agitation until the onset of stationary phase (A_{600} 3.5). At that stage, CPO activity is measured in a 5- to 10-ml aliquot of the preculture. Yeast cells are pelleted by 1 min of centrifugation, transferred to a 2-ml Eppendorf tube, and washed by centrifugation (1 min at 10,000 *g*, 4°) with ice-cold 100 m*M* Tris buffer, pH 7.6. Cells are suspended in 0.7 ml of ice-cold Tris buffer. One milliliter of 0.5-mm-diameter glass beads is added and the tube is vortexed vigorously for 3 min. CPO activity and proteins are measured in a 50- to 100-fold dilution in Tris buffer of the supernatant obtained after 1 min of centrifugation at 10,000 *g* for 1 min at 4°.

Culture. The rest of the preculture is used to inoculate 9 liters of rich medium consisting of 1% (w/v) Difco yeast extract, 1% (w/v) Difco Bacto-peptone, 2% (w/v) glucose, 0.1% (v/v) Tween 80, and ergosterol (30 mg/ml). Tween 80 provides the oleic acid (as a soluble form) that the heme mutant is unable to make. Cells are grown overnight at 30° with gentle magnetic stirring.

Harvesting. Cells are harvested at the onset of stationary growth (A_{600} 5–6) by 10-min centrifugations at 4° and 5000 *g* (RC5-B, rotor GS3; Sorvall, Newtown, CT) followed by two washings of the pellets by centrifugation (10 min at 5000 *g*, 6°), one with cold water, and one with cold breaking buffer [100 m*M* Tris buffer (pH 7.6) containing two antiprotease agents: 1 m*M* EDTA and phenylmethylsulfonyl fluoride (PMSF, 60 μg/ml)].

In these growth conditions, the maximum yield of yeast cells was 9 grams wet weight/liter. The pellet of yeast cells (80 g wet weight) was immediately processed.

[23] J. E. Hill, A. M. Myers, T. J. Koerner, and A. Tzagoloff, *Yeast* **2,** 163 (1986).

Protein Assays

Protein concentrations are determined by the method of Lowry *et al.*[24] during the first steps of purification and by the method of Bradford[25] during the final steps. Bovine serum albumin (BSA) is used as standard.

Sodium dodecyl sulfate (SDS)-10% (w/v) polyacrylamide gel electrophoresis is performed according to Laemmli.[26]

Purification of Coproporphyrinogen III Oxidase from Yeast Cells

All of the following operations are carried out at 0–4° from 80 g (wet weight) of cells as starting material.

Step 1: Preparation of Cell-Free Extracts. Yeast cells are broken with a Braun homogenizer MSK (B. Braun, Melsungen, Germany). A 50% (w/v) suspension of yeast cells in breaking buffer is made and 40-ml aliquotes of that suspension are introduced into glass flasks (75-ml capacity), each containing 32 ml of 0.5-mm-diameter precooled glass beads. Breaking for 1.5 min is carried out at high speed (4000 oscillations/min) with the homogenizer cooled with liquid CO_2. The resulting viscous homogenate is decanted, and the beads are rinsed five or six times with 10 ml each of breaking buffer. Combined washings are centrifuged for 10 min at 3000 *g* to remove unbroken cells and debris. The supernatant is saved and the pellets washed twice by centrifugation (10 min at 3000 *g*, 6°) in the breaking buffer. Supernatant and washings are pooled and referred to as cell homogenate.

Step 2: Preparation of Soluble Fraction. The cell homogenate is centrifuged at 45,000 *g* for 1 hr (rotor SS34, Sorvall RCB). The supernatant is saved and the pellet washed by centrifugation (1 hr at 45,000 *g*, 6°). The combined supernatants represent the intracellular soluble fraction and contain all CPO activity.

Step 3: Ammonium Sulfate Fractionation. Solid ammonium sulfate is slowly added to the soluble fraction to produce a 40% (w/v) saturated solution (22.1 g/100 ml). The mixture is gently stirred for 30 min and centrifuged at 30,000 *g* for 20 min. The pellet is discarded and more ammonium sulfate is added to bring the supernatant to 70% (w/v) saturation (18.7 g/100 ml). The precipitated proteins are collected by centrifugation and dissolved in the minimum amount of breaking buffer.

The protein fraction (~5 g) is divided in two subfractions. At this point the fractions can be kept frozen at −20° for 1 year without any loss of activity.

[24] O. H. Lowry, N. Rosebrough, A. L. Farr, and R. J. Randall, *J. Biol. Chem.* **193,** 265 (1951).
[25] M. M. Bradford, *Anal. Biochem.* **72,** 248 (1976).
[26] U. K. Laemmli, *Nature (London)* **227,** 680 (1970).

Step 4: Chromatofocusing. One subfraction is thawed and dialyzed three times against 2 liters of 25 m*M* imidazole, pH 7.4, containing 1 m*M* EDTA and PMSF (60 μg/ml). The dialysate is passed through a column (1.6 × 60 cm, 110 ml) of Polybuffer Exchanger 96 (Pharmacia, Piscataway, NJ) equilibrated in 25 m*M* imidazole, pH 7.4 (containing EDTA and PMSF), until pH and conductivity equilibration. Protein elution is achieved with a linear gradient of from pH 7.4 to pH 5, produced by 1400 ml of an eightfold dilution of Polybuffer 74 (Pharmacia) adjusted to pH 5 with 1 *M* HCl. The flow rate is 70 ml/hr and 4.5-ml fractions are collected. CPO activity is recovered mainly between pH 6.6 and 5.9 as a huge peak, and is concentrated by ammonium sulfate precipitation (100% saturation; 69.8 g/100 ml) followed by centrifugation (30,000 *g*, 20 min) and dissolution of the precipitate in the minimum amount of 100 m*M* Tris buffer, pH 7.6 (without EDTA and PMSF).

The second subfraction is processed in exactly the same way.

Step 5: DEAE-Sepharose Chromatography. Both concentrated chromatofocusing subfractions are pooled and dialyzed two times against 2 liters of 25 m*M* Tris, pH 8.4, until pH and conductivity equilibration. The dialysate is loaded onto a column (1.6 × 30 cm; 60 ml) of DEAE-Sepharose CL-6B equilibrated with 25 m*M* Tris buffer, pH 8.4. The column is washed with 100 ml of Tris buffer and CPO is eluted by a 500-ml linear gradient of KCl (0–0.2 *M*) in the same buffer. The flow rate is 50 ml/hr and 4.5-ml fractions are collected. CPO activity is recovered around 0.04–0.05 *M* KCl as a narrow symmetrical peak. Active fractions are pooled and concentrated by ultrafiltration using an Amicon (Danvers, MA) ultrafilter (series 8000) and Diaflo Amicon membranes (YM30). At that step CPO is pure, as judged from protein-overloaded SDS-polyacrylamide gels. A summary of the purification procedure is given in Table I.

The relatively low yield (~50%) in CPO recovery between steps 4 and 5 is the consequence of the strategy used for purifying the enzyme by using

TABLE I
PURIFICATION OF COPROPORPHYRINOGEN III OXIDASE

Fraction	Volume (ml)	Protein (mg)	Total units	Specific activity (nmol/hr/mg)	Yield (%)
Homogenate	447	7,830	4,100,000	555	100
Soluble fraction	430	5,375	4,100,000	732	100
Ammonium sulfate fraction	74.5	2,150	3,570,000	804	87
Chromatofocusing eluate	114	585	3,503,000	4,830	85
DEAE-Sepharose eluate	80	336	1,710,000	5,090	42

the minimum of steps (e.g., only a few fractions of the CPO protein peak from step 4 are pooled). However, the yield in CPO could be increased by pooling more fractions around the CPO peak at step 4, and by adding a gel-filtration step between steps 4 and 5

Step 6: Gel Filtration. Each concentrated chromatofocusing subfraction from step 4 is applied separately to a column (2.6 × 100 cm, 500 ml) of Ultrogel Ac44 equilibrated with 100 m*M* Tris buffer, pH 7.6. Proteins are eluted with the same buffer at a flow rate of 30 ml/hr and 4.5-ml fractions are collected. In a separate experiment, a mixture of pure proteins is used to calibrate the column. CPO elutes as a single peak corresponding to a 70-kDa protein. Both gel-filtration subfractions containing CPO are pooled, concentrated, and dialyzed as described in step 4 before starting step 5.

Comments. One gram (wet weight) of cells contains consistently 10–12 mg of active CPO, which represents 10–12% of the total proteins. This is the result of a tremendous overexpression (>2000-fold) of that protein when compared to the parental strain containing the chromosomal *HEM13* gene. The fact that yeast CPO is a cytoplasmic enzyme—and not a mitochondrial one as for the mammalian cells—probably led to that high level of overexpression.

Hundreds of milligrams of pure CPO can be obtained within 2 weeks starting from the preculture of the mutant.

Properties

Physicochemical Properties

The enzyme[27] is colorless, even at concentrations up to 20 mg/ml, and the absorption spectrum shows only the classic absorption maximum at λ 283 nm with a shoulder at 290 nm. Fluorescence excitation and emission maxima were 290 and 350 nm, respectively.

Although the protein contains three cysteine residues per monomer, no SH group was titratable by 5,5′-dithiobis(2-nitrobenzoic acid) in the native protein.

CPO was stable for years when kept frozen at −20°, and for months at room temperature in the presence of 1% (w/v) NaN_3 (100 m*M* Tris, pH 7.6). The enzyme was heat sensitive, because a 2-min incubation of the protein at 56° led to a total loss of enzymatic activity. However, heat denaturation was completely prevented by adding coprogen or coproporphyrin III at a concentration equimolar to that of the dimeric enzyme.

[27] Some of these properties have already appeared in Ref. 13.

Uroporphyrin III had no protective effect, but protogen and protoporphyrin showed a small protective effect.

When incubated with 3 *M* urea, CPO was slowly denatured at pH 6.3 and much faster at pH 7.6 (with a concomitant total loss of enzymatic activity). However, on dilution, some activity (30–40%) was recovered. At pH 7.6, some protection against 3 *M* urea denaturation at pH 7.6 was provided by coprogen.

Yeast CPO does not contain any known redox cofactor. In particular, all attempts to detect a "built-in" cofactor[28,29] [topaquinone (TQ) and tryptophan-tryptophyl quinone (TTQ)] were unsuccessful (see Inhibitors, below).

Yeast CPO does not contain a transition metal because insignificant amounts of copper and/or iron were repeatedly found (<0.1 atom/enzyme subunit). Metal analyses were performed by inductively coupled plasma atomic emission spectroscopy (ICP-AES),[30] or by spectrophotometric determination by chelation with bathocuproin sulfonate (BCS), bathophenanthroline sulfonate (BPS), or diethyl dithiocarbamate (DETC), after mineralization of the protein samples (~2 mg) with concentrated hot nitric acid. Detection of copper and iron by chelation with BCS, BPS, or DETC of denatured samples [30 min of boiling in 2% (w/v) SDS] was also unsuccessful. The enzymatic activity, measured by the coupled fluorimetric assay, was insensitive to the presence of the metal-specific chelators (mainly copper or iron specific[31]). The fact that we previously found iron in purified CPO (Ref. 13) was probably due to contamination during purification, but others have claimed that mouse CPO overexpressed in *E. coli* needed copper for maximal activity and that one histidine residue was essential for CPO activity and probably for copper binding.[32]

Catalytic Properties

The pH dependence of CPO activity showed a broad peak centered near pH 7.6; the p*I* was 6.2.

CPO activity measured *in vitro* without PPO was increased 2.5-fold by neutral detergents (1 mg of Triton X-100 or Tween 80 per milliliter), 1.7-fold by phospholipids (1 mg of soybean lecithin or yeast mitochondrial

[28] J. P. Klinman and D. Mu, *Annu. Rev. Biochem.* **63,** 299 (1994).

[29] W. S. McIntire, *FASEB J.* **8,** 513 (1994).

[30] Service Central d'Analyses CNRS, 69 390 Vernaison, France.

[31] Fe: BPS, 8-hydroxyquinoline 5-sulfonate, and desferri-ferrioxamine B; Cu: BCS, BPS, and DETC.

[32] H. Kohno, T. Furukawa, R. Tokunaga, S. Taketani, and T. Yoshinaga, *Biochim. Biophys. Acta* **1292,** 156 (1996).

crude lipid fraction per milliliter), and 2-fold by the added PPO (yeast mitochondrial membranes used for the fluorimetric assay of CPO activity; see above).

Free SH groups did not seem to be involved in the catalytic activity of CPO, which was resistant to SH reagents. Kohno *et al.* (see Ref. 32) mutated the unique conserved cysteine residue (over nine sequences) of the mouse CPO expressed in *E. coli* without any loss of enzymatic activity.

The enzyme was active only when molecular oxygen was used as electron acceptor; no anaerobic activity could be detected (using $NADP^+$ and *S*-adenosylmethionine) as claimed by others (see Ref. 11). The K_m value for coprogen was low ($\sim 0.05\ \mu M$) and that for oxygen was estimated to be of the same order of magnitude or less.

During the enzymatic reaction, hydrogen peroxide was evolved. This was shown by incubating concentrated coprogen with concentrated CPO—without PPO—in a closed O_2–electrode vessel and by adding catalase at different times during the incubation. The exact stoichiometry of the reaction could not be determined accurately because the protogen, which accumulated rapidly, was autoxidized (slowly and nonenzymatically) to protoporphyrin, a process that generates H_2O_2.

Inhibitors

As shown previously by Elder and Evans for the mammalian CPO,[33] coproporphyrin III is a competitive inhibitor of the enzyme.

N-Bromosuccinimide (NBS), a specific modifier of tryptophan residues, was the only molecule among the chemicals tested that inhibited CPO activity. The putative inhibitors, including iron and copper chelators (see above), or derivatizing chemicals were chosen to assess the possible involvement of either a metal or a "built-in" coenzyme or both in the catalytic activity.[30,34] *N*-Bromosuccinimide reacted rapidly with CPO, inactivating the enzyme at an NBS:CPO concentration ratio of about 15–20 (1.5–2 NBS per tryptophan residue of CPO homodimer, each monomer having eight tryptophan residues). This suggests the possible involvement of tryptophan residue(s) in the activity.

The readily available CPO from a *S. cerevisiae* overproducing strain now presents opportunities to obtain structural information on this enzyme and to determine the mechanism of CPO activity.

[33] G. Elder and J. O. Evans, *Biochem. J.* **169,** 205 (1978).

[34] Fe and/or Cu: carbon monoxide, BCS, BPS, *cis*- and *trans*-1,2-diaminocyclohexane, ferricyanide, hydroxyurea, imidazole, KCN, NaN_3; "built-in" coenzymes: (dinitro)phenylhydrazine, NH_2OH, Na-$CNBH_3$, Li-BH_4, 3-methyl-2-benzothiazolinone; glycinate radical: sodium hypophosphite.

Acknowledgments

This work was supported by grants from the CNRS and Paris-7 University. The author thanks R. Labbe-Bois for constructing the CPO-overproducing mutant strain and for advice, and J. M. Camadro and the late H. Chambon for so efficiently helping to initiate these studies.

[43] Expression, Purification, and Characterization of Recombinant Mammalian Ferrochelatase

By VERA M. SELLERS and HARRY A. DAILEY

The terminal step in heme biosynthesis is the insertion of ferrous iron into protoporphyrin IX, a reaction that is catalyzed by the enzyme ferrochelatase (EC 4.99.1.1). In mammals, this step represents the convergence of two critical pathways, namely cellular iron metabolism and tetrapyrrole biosynthesis. With the expression of recombinant mammalian ferrochelatases in *Escherichia coli,*[1,2] the current view of the possible role of ferrochelatase in both heme biosynthesis and in regulation of cellular iron metabolism is rapidly expanding.

Ferrochelatase has been isolated and cloned from several bacterial,[3–10] yeast,[11] plant,[12,13] and mammalian[14–18] sources. Human ferrochelatase is

[1] D. A. Brenner, J. M. Didier, F. Frasier, S. R. Chrisrensen, G. A. Evans, and H. A. Dailey, *Am. J. Hum. Genet.* **50,** 1203 (1992).
[2] H. A. Dailey, V. M. Sellers, and T. A. Dailey, *J. Biol. Chem.* **269,** 390 (1994).
[3] K. Miyamoto, K. Nakahigashi, K. Nishimura, and H. Inokuchi, *J. Mol. Biol.* **219,** 393 (1991).
[4] A. C. Kessler, A. Haase, and P. R. Reeves, *J. Bacteriol.* **175,** 1412 (1993).
[5] M. Skurnick, GenBank Accession No. Z47767 X63827 (1995).
[6] O. White, GenBank Accession No. U32742 L42023 (1996).
[7] J. Frustaci and M. R. O'Brian, *J. Bacteriol.* **174,** 4223 (1992).
[8] A. J. Biel, *J. Bacteriol.* **177,** 6693 (1996).
[9] M. Hansson and L. Hederstedt, *J. Bacteriol.* **174,** 8081 (1992).
[10] J. A. Gutierrez and L. N. Csonka, *J. Bacteriol.* **177,** 390 (1995).
[11] R. Labbe-Bois, *J. Biol. Chem.* **265,** 7278 (1990).
[12] K. Miyamoto, R. Tanaka, H. Teramoto, T. Masuda, H. Tsuji, and H. Inokuchi, *Plant Physiol.* **105,** 769 (1994).
[13] A. G. Smith, M. A. Santana, A. D. M. Wallace-Cook, J. M. Roper, and R. Labbe-Bois, *J. Biol. Chem.* **269,** 13405 (1994).
[14] Y. Nakahashi, S. Taketani, M. Okuda, K. Inoue, and R. Tonuga, *Biochem. Biophys. Res. Commun.* **173,** 748 (1990).
[15] S. Taketani, Y. Nakahashi, T. Osumi, and R. Tokunaga, *J. Biol. Chem.* **265,** 19377 (1990).
[16] D. A. Brenner and F. Frasier, *Proc. Natl. Acad. Sci. U.S.A.* **88,** 849 (1991).

0076-6879/97 $25

encoded by a single nuclear gene located at chromosome 18q21.3.[1] The various ferrochelatase enzymes range in size from 35 to 42 kDa. Excluding the *Bacillus subtilis* ferrochelatase, all are membrane proteins that require detergents to remain in solution. Hydropathy plots predict no membrane-spanning segments. In prokaryotes, ferrochelatase is associated with the cytosolic side of the cytoplasmic membrane.[19] Eukaryotic ferrochelatase is located on the inner mitochondrial membrane with the active site facing the mitochondrial matrix,[20] or in the chloroplast.[21]

The various ferrochelatase protein sequences have little overall homology, although there are several highly conserved regions (Fig. 1). The highest homology, 88%, is between the human and mouse ferrochelatases, yet there is less than 30% homology between the human and *B. subtilis* enzymes. The *B. subtilis* enzyme is the only ferrochelatase that differs significantly at any of the consensus regions. Mammalian and yeast ferrochelatases possess an amino-terminal mitochondrial targeting sequence that is proteolytically cleaved on translocation into the mitochondrion.[22,23] Likewise, plant ferrochelatases contain a chloroplast-targeting sequence at the amino terminus[13] and within this region is also located a proposed consensus sequence motif for phosphorylation.[23] All of the eukaryotic ferrochelatases contain a carboxyl-terminal extension of at least 30 additional residues. Only in the animal ferrochelatases, however, do these additional residues serve to coordinate a labile, redox-active iron–sulfur cluster.[24,25]

Many previous attempts at purification of mammalian ferrochelatase have failed to isolate large amounts of the holoenzyme, complete with the iron–sulfur cluster. Therefore, most previous experiments were conducted with small amounts of enzyme that was contaminated with significant amounts of apoprotein. This chapter provides several modifications of the previously published purification procedure, which now allow consistent production of large quantities of recombinant ferrochelatase.

[17] H. Shibuya, D. Nannemann, M. Tamassia, O. L. Allphin, and G. S. Johnson, *Biochim. Biophys. Acta* **1231,** 117 (1995).

[18] B. H. Parsons and H. A. Dailey, manuscript in preparation (1997).

[19] H. A. Dailey, *Methods Enzymol.* **123,** 408 (1986).

[20] B. M. Harbin and H. A. Dailey, *Biochemistry* **24,** 366 (1985).

[21] M. Matringe, J. M. Camadro, P. Labbe, and R. Scalla, *Biochem. J.* **269,** 231 (1989).

[22] J. M. Camadro and P. Labbe, *J. Biol. Chem.* **263,** 11675 (1988).

[23] S. R. Karr and H. A. Dailey, *Biochem. J.* **254,** 799 (1988).

[24] H. A. Dailey, M. G. Finnegan, and M. K. Johnson, *Biochemistry* **33,** 403 (1994).

[25] G. C. Ferreira, R. Franco, S. G. Lloyd, A. S. Pereira, I. Moura, J. J. G. Moura, and B. H. Huynh, *J. Biol. Chem.* **269,** 7062 (1994).

```
Yerps_Fc        1 ............................................................
Yeren_Fc        1 ............................................................
Ecoli_Fc        1 ............................................................
Hinf_Fc         1 ............................................................
Braja_Fc        1 ............................................................
Rcap_Fc         1 ..............................................MTIANRILPHAPAD
Cucumber_Fc     1 ............................................................
Barley_Fc       1 ............................................................
Arab_Fc         1 ............................................................
Bovine_Fc       1 ................VLLRDRLLYGGSRACQPRRCQSGAATAAAATETAQRARSPKPQA
Human_Fc        1 MRSLGANMAAALRAAGVLLRDPLASSSWRVCQPWRWKSGAAAAAVTTETAQHAQGAKPQV
Mouse_Fc        1 MLSASANMAAALRAAGALLREPLVHGSSRACQPWRCQSGAAVAA.TTEKVHHAKTTKPQA
Chicken_Fc      1 .......MAAAGRAA.....RPLVAGGRQLRVPLRWRGQVAAAAPST.........KPQA
Yeast_Fc        1 ......MLSRTIRTQGSFLRRSQLTITRSFSVTF........................NM
Bsub_Fc         1 ............................................................

Yerps_Fc        1 ............................................................
Yeren_Fc        1 ...MKQSKLGVLMVNLEHRMPPTPQAVKRYLAEFLSVRRVVDTSPWLWWPLLRGVILPIR
Ecoli_Fc        1 ...MRQTKTGILLANLGTPDAPTPEAVKRYLKQFLSDRRVVDTSRLLWWPLLRGVILPLR
Hinf_Fc         1 ..MTKPAKIGVLLANLGTPDSPTPKSISRYLWQFLTDPRVVDLPRCKWYPLLKAIILPLR
Braja_Fc        1 ...SGQKRVGVLLVNLGTPDTADAPGVRVYLKEFLSDARVIEDQGLVWKVVLNGIILRQR
Rcap_Fc        15 HPPVPVPRVGVLLANLGTPDATDYWSMRRYLNEFLSDRRVIDYPIWKWQPLLQLIILSKR
Cucumber_Fc     1 ...AVDDKVGVLLLNLGGPETLD..DVQPFLYNLFADPDIIRLPRLFR..FLQEPLAKLI
Barley_Fc       1 ...AVEEKVGVLLLNLGGPETLN..DVQPFLFNLFADPDIIRLPRLFR..FLQRPLAKLI
Arab_Fc         1 ...VAEDKIGVLLLNLGGPETLN..DVQPFLYNLFADPDIIRLPRPFQ..FLQGTIAKFI
Bovine_Fc      45 QPGNRKPRTGILMLNMGGPETVE..EVQDFLQRLFLDQDLMTLPV.......QDKLGPFI
Human_Fc       61 QPQKRKPKTGILMLNMGGPETLG..DVHDFLLRLFLDQDLMTLPI.......QNKLAPFI
Mouse_Fc       60 QPERRKPKTGILMLNMGGPETLG..EVQDFLQRLFLDRDLMTLPI.......QNKLAPFI
Chicken_Fc     40 EPETRKPKTGILMLNMGGPERLD..DVHDFLLRLFLDRDLMTLPA.......QNKLAPFI
Yeast_Fc       31 QNAQKRSPTGIVLMNMGGPSKVE..ETYDFLYQLFADNDLIPISAKY.....QKTIAKYI
Bsub_Fc         1 ...MSRKKMGLLVMAYGTPYKEE..DIERYYTHI....................RRGRKP

Yerps_Fc        1 ............................................................
Yeren_Fc       58 SPRVAKLYQSVWMDE..GSPLLVYSRRQQKALAERMP......EI.PVELGMSYGSPNLP
Ecoli_Fc       58 SPRVAKLYASVWMEG..GSPLMVYSRQQQQALAQRLP......EM.PVALGMSYGSPSLE
Hinf_Fc        59 SKRIAKNYQAIWTEQ..GSPLLAISRQQKDALQAYLDNQ..NIDT.QVEIAMTYGNPSMQ
Braja_Fc       58 PRSKALDYQKIWNNEKNESPLKTITRSQSAKLAAALSDR...DHV.VVDWAMRYGNPSIK
Rcap_Fc        75 PFTSGNNYRSIWNEERDESPLMTITRDQVRKLRAAVETRYGAGNV.VVDFCMRYGNPSTR
Cucumber_Fc    54 STYRAPKSKEGYASIGGGSPLRKITDEQAQALKMALAEKNMST...NVYVGMRYWYPFTE
Barley_Fc      54 STFRAPKSNEGYASIGGGSPLRKITDEQANALKVALKSKNLEA...DIYVGMRYWYPFTE
Arab_Fc        54 SVVRAPKSKEGYAAIGGGSPLRKITDEQADAIKMSLQAKNIAA...NVYVGMRYWYPFTE
Bovine_Fc      96 AKRRTPKIQEQYRRIGGGSPIKMWTSKQGEGMVKLLDELSPHTAPHKYYIGFRYVHPLTE
Human_Fc      112 AKRRTPKIQEQYRRIGGGSPIKIWTSKQGEGMVKLLDELSPNTAPHKYYIGFRYVHPLTE
Mouse_Fc      111 AKRRTPKIQE..RRIGGGSPIKMWTSKQGEGMVKLLDELSPATAPHKYYIGFRYVHPLTE
Chicken_Fc     91 AKRRTPRIQEQYSRIGGGSPIKKWTAVQGEGMVKLLDSMSPQTAPHKYYIGFRYVHPLTE
Yeast_Fc       84 AKFRTPKIEKQYREIGGGSPIRKWSEYQATEVCKILDKTCPETAPHKPYVAFRYAKPLTA
Bsub_Fc        36 EPEMLQDLKDRYEAIGGISPLAQITEQQAHNLEQHLNEI.QDEITFKAYIGLKHIEPFIE

Yerps_Fc        1 ............................................................
Yeren_Fc      109 DAIDKLLAQGVTKLVVLPLYPQYSWSTSAAVWDAVARILASYRRLPSI..SFIRDYAEHP
Ecoli_Fc      109 SAVDELLAEHVDHIVVLPLYPQFSCSTVGAVWDELARILARKRSIPGI..SFIRDYADNH
Hinf_Fc       114 SAVKNLLKNQVERIIVLPLYPQYSSSTTGAVFDAFANALKEERGLLPF..DFIHSYHIDE
Braja_Fc      114 SGIDALIG.GMRPHLAVPLYPQYSASTSATVCDEVFRVLARLRAQPTL..RVTPPYYEDE
Rcap_Fc       134 DVLDDMLAQGCERILFLPLYPQYAGATSATANDQFFRALMQVKRQPAA..RTVPEYFARP
Cucumber_Fc   111 EAIQQIKRDGITRLVVLPLYPQYSISTTGSSIRVLQKMFREDAYLSSLPVSIIKSWYQRE
Barley_Fc     111 EAIDQIKKDKITKLVVLPLYPQYSISTSGSSIRVLQNIVKEDPYFAGLPISIIESWYQRE
Arab_Fc       111 EAVQQIKKDKITRLVVLPLYPQYSISTTGSSIRVLQDLFRKDPYLAGVPVAIIKSWYQRR
Bovine_Fc     156 EAIEEMERDGLERAVAFTQYPQYSCSTTGSSLNAIYRYYNEVGRKPTMKWSTIDRWPTHP
Human_Fc      172 EAIEEMERDGLERAIAFTQYPQYSCSTTGSSLNAIYRYYNQVGRKPTMKWSTIDRWPTHH
Mouse_Fc      169 EAIEEMERDGLERAIAFTQYPQYSCSTTGSSLNAIYRYYNEVGQKPTMKWSTIDRWPTHP
Chicken_Fc    151 EAIEEMEDDGIERAIAFTQYPQYSCSTTGSSLNAIYRYYNKKGKKPKMKWSIIDRWPTHP
Yeast_Fc      144 ETYKQMLKDGVKKAVAFSQYPHFSYSTTGSSINELWRQIKALDSERSISWSVIDRWPTNE
Bsub_Fc        95 DAVAEMHKDGITEAVSIVLAPHFSTFSVQSYNK...RAKEEAEKLGGLTITSVESWYDEP
```

FIG. 1. Comparison of ferrochelatase amino acid sequences. In order, the sequences shown are for *Yersinia pseudotuberculosis*,[4] Yersinia enterocolitica,[5] *Escherichia coli*,[3] *Haemophilus influenzae*,[6] *Bradyrhizobium japonicum*,[7] *Rhodobacter capsulatis*,[8] cucumber,[12] barley,[12] *Arabidopsis thaliana*,[13] bovine,[17] human,[14] mouse,[15,16] chicken,[18] yeast,[11] and *Bacillus subtilis*.[9] Highly conserved residues are heavily shaded. Moderately conserved residues are lightly shaded. Cysteine residues that coordinate the $[2Fe\text{-}2S]^{2+}$ cluster are marked with an asterisk.

```
Yerps_Fc        1  ............................................................
Yeren_Fc      167  AYISALKQSVENSFVQ.HGKPD...RLVLSFHGIPKPYA.QLGDDYPQRCEDTSRALRAE
Ecoli_Fc      167  DYINALANSVRASFAK.HGEPD...LLLLSYHGIPQRYA.DEGDDYPQRCRTTTRELASA
Hinf_Fc       172  NYINALADSIKVRLKS.D...E...FLLFSYHGIPLRYE.KMGDYYREHCKQTTIAVVNK
Braja_Fc      171  AYIEALAVSIETHLATLPFKPE...LIVASFHGMPKSYV.DKGDPYQEHCIATTEALRAA
Rcap_Fc       192  SYIEALASSVERVYATLDTRPD...VLVASYHGMPKRYH.REGDPYHCQCQKTSRLLRER
Cucumber_Fc   171  GYIKSMADLMQAELKNFANPQ..EVMIFFSAHGVPVGYVENAGDPYKDQMEECICLIMQE
Barley_Fc     171  GYVKSMADLIEKELSVFSNPE..EVMIFFSAHGVPLTYVKDAGDPYRDQMEDCIALIMEE
Arab_Fc       171  GYVNSMADLIEKELQTFSDPK..EVMIFFSAHGVPVSYVENAGDPYQKQMEECIDLIMEE
Bovine_Fc     216  LLIQCFADHILKELDHFPPEKRREVVILFSAHSLPMSVV.NRGDPYPQEVGATVQRVMDK
Human_Fc      232  LLIQCFADHILKELDHFPLEKRSEVVILFSAHSLPMSVV.NRGDPYPQEVSATVQKVMER
Mouse_Fc      229  LLIQCFADHILKELNHFPEEKRSEVVILFSAHSLPMSVV.NRGDPYPQEVGATVHKVMEK
Chicken_Fc    211  LLIQCFADHIQKELDLFPPDKRKDVVILFSAHSLPMSVV.NRGDPYPQEVGATVQRVMEK
Yeast_Fc      204  GLIKAFSENITKKLQEFPQPVRDKVVLLFSAHSLPMDVV.NTGDAYPAEVAATVYNIMQK
Bsub_Fc       152  KFVTYWVDRVKETYASMPEDERENAMLIVSAHSLPEK.IKEFGDPYPDQLHESAKLIAEG

Yerps_Fc        1  .........................TPYTDETLKSLPSQ.GVKHIQLICPGFSADCLETL
Yeren_Fc      222  I.ALPAETKSRVHIQSRF.GREPWLTPYTDETLKSLPSQ.GVKHIQLICPGFSADCLETL
Ecoli_Fc      222  L.GMAPE.KVMMTFQSRF.GREPWLMPYTDETLKMLGEK.GVGHIQVMCPGFAADCLETL
Hinf_Fc       224  L.GL.TENQWRMTFQSRF.GREEWLQPYTDKFLESAAAQ.NIQKIAVICPGFSVDCLETI
Braja_Fc      227  R.RLDAS.KLLLTFQSRF.GNDEWLQPYTDKTMERLAKE.GVRRIAVVTPGFAADCLETL
Rcap_Fc       248  L.GWGPD.SIDTTFQSVF.GTEEWLRPYTVEHVVQLAEA.GKKNIAVISPAFSADCIETL
Cucumber_Fc   229  LKARGIGNEHTLAYQSRV.GPVQWLKPYTDEVLVELG.QKGIKSLLAVPVSFVSEHIETL
Barley_Fc     229  LKSRGTLNDHTLAYQSRV.GPVQWLKPYTDEVLVELG.QKGVKSLLAVPVSFVSEHIETL
Arab_Fc       229  LKARGVLNDHKLAYQSRV.GPVQWLKPYTDEVLVDLG.KSGVKSLLAVPVSFVSEHIETL
Bovine_Fc     275  L...GYSNPYRLVWQSKV.GPMPWLGPQTDEAIKGLC.KRGRKNILLVPIAFTSDHIETL
Human_Fc      291  L...EYCNPYRLVWQSKV.GPMPWLGPQTDESIKGLC.ERGRKNILLVPIAFTSDHIETL
Mouse_Fc      288  L...GYPNPYRLVWQSKV.GPVPWLGPQTDEAIKGLC.ERGRKNILLVPIAFTSDHIETL
Chicken_Fc    270  L...NHSNPYRLVWQSKV.GPMPWLGPQTDETIKGLC.QRGKKNMLLVPIAFTSDHIETL
Yeast_Fc      263  L...KFKNPYRLVWQSQV.GPKPWLGAQTAE.IAEFL.GPKVDGLMFIPIAFTSDHIETL
Bsub_Fc       211  ....AGVSEYAVGWQSEGNTPDPWLGPDVQDLTRDLFEQKGYQAFVYVPVGFVADHLEVL

Yerps_Fc       35  EEIKEQNREFF.LHAGGEKFEYIPALNDDEGHIALLEQLIRHNI................
Yeren_Fc      279  EEIKEQNREIF.LHAGGEKFEYIPALNDDKGHIDLLEQLVRDHLSC..............
Ecoli_Fc      278  EEIAEQNREVF.LGAGGKKYEYIPALNATPEHIEMMANLVAAYR................
Hinf_Fc       280  EEIDEENRENF.LNNGGQSYQYIPALNVEHAHIEMMGKLILEKLT*..............
Braja_Fc      283  EEIAQENAEIF.KHNGGETFSAIPCLNDSEPGMDVIRTLVLRELQG..............
Rcap_Fc       304  EEINGEIREAF.EHAGGESFTYVPCLNDDDLHIAALLEVVEENLAGWID...........
Cucumber_Fc   287  EEIDMEYKH.LALESGIQNWGRVPALNCNSSFISDLADAVIEALPSATALAPHTS...ST
Barley_Fc     287  EEIDMEYRE.LALESGIENWGRVPALGCTSSFISDLADAVVEALPSASAMATRKV...KD
Arab_Fc       287  EEIDMEYRE.LALESGVENWGRVPALGLTPSFITDLADAVIESLPSAEAMSNPNAVVDSE
Bovine_Fc     330  YELDIEYSQVLASECGLENIRRAESLNGNPLFSKALADLVHSHLQS............KE
Human_Fc      346  YELDIEYSQVLAKECGVENIRRAESLNGNPLFSKALADLVHSHIQS............NE
Mouse_Fc      343  YELDIEYSQVLAQKCGAENIRRAESLNGNPLFSKALADLVHSHIQS............NK
Chicken_Fc    325  YELDIEYAQVLANECGVENIRRAESLNGNPLFSKALADLVCSHIQS............NE
Yeast_Fc      317  HEIDLG...VIGESEYKDKFKRCESLNGNQTFIEGMADLVKSHLQS............NQ
Bsub_Fc       267  YDNDYECK.VVTDDIGASYY.RPEMPNAKPEFIDALATVVLKKLGR..............

Yerps_Fc           ...............................................................
Yeren_Fc           ...............................................................
Ecoli_Fc           ...............................................................
Hinf_Fc            ...............................................................
Braja_Fc           ...............................................................
Rcap_Fc            ...............................................................
Cucumber_Fc   343  DADDHDPFLYAIKLLFGSVLAFILLLSPKAFMVFRNNFLLNYTRIYGYRGERSEFFWVRLIFT
Barley_Fc     343  TDSDMDMMHYLTKMFLGSVLAFFLLLSPRLVSAFRNTLQ........................
Arab_Fc       346  DSESSDAFSYIVKMFFGSILAFVLLLSPKMFHAFRNL..........................
Bovine_Fc     378  RCSTQLTLSCPLCVNPTCRETKSFFTSQQL.................................
Human_Fc      394  LCSKQLTLSCPLCVNPVCRETKSFFTSQQL.................................
Mouse_Fc      391  LCSTQLSLNCPLCVNPVCRKTKSFFTSQQL.................................
Chicken_Fc    373  ICSKQLTLCCPLCVNPVCRETKAFFTNQQL.................................
Yeast_Fc      362  LYSNQLPLDFALGKSNDPVKDLSLVFGNHEST...............................
Bsub_Fc            ...............................................................
                            *  *    *
```

FIG. 1. (*continued*)

Purification

Plasmids encoding mouse or human ferrochelatase, minus the amino-terminal mitochondrial targeting sequence, are engineered with an additional upstream ribosomal binding site as described elsewhere.[2] Expression in *E. coli* strain JM109 allows maximal protein production as compared to other strains tested. *Escherichia coli* strains JM109 and DH5 have both been found to be acceptable for production of holoenzyme. A variety of other available strains, including those that employ the inducible T7 promoter system, have been tested and found to be less acceptable. Some strains will produce large amounts of apoprotein, as evidenced by sodium dodecyl sulfate (SDS) gel electrophoresis, but produce little active enzyme. Such strains should be acceptable for expression of ferrochelatases that do not contain the metal center.

One-liter cultures of JM109 carrying one of the previously described plasmids[2] are grown in Circlegrow (Bio 101, La Jolla, CA) medium containing ampicillin (100 μg/ml) at 37° for 20 hr. Cells are harvested by centrifugation (5000 g, 15 min, 4°), and suspended to a volume of 120 ml in cold solution A [20 m*M* Tris–morpholinepropanesulfonic acid (MOPS) (pH 8.1), 20% (v/v) glycerol, 1 m*M* sodium dithionite, and phenylmethylsulfonyl fluoride (PMSF, 10 μg/ml)]. All procedures are performed at 4°. After sonicating four times (20–30 sec each) to disrupt the cells, the membrane fraction is isolated from the cytoplasmic fraction by centrifugation at 100,000 *g* for 60 min. The membrane fraction is then suspended in 60 ml of solubilization buffer [20 m*M* Tris–MOPS (pH 8.1), 1% (w/v) sodium cholate, 1.0 *M* KCl, 1 m*M* sodium dithionite, and PMSF (10 μg/ml)], sonicated four times (20–30 sec each), and then centrifuged at 100,000 *g* for 60 min to pellet the membrane fraction. The volume of the supernatant, containing solubilized ferrochelatase, is measured and added to that amount of saturated ammonium sulfate solution, pH 7.5, calculated to bring the resulting solution to 45% (w/v) ammonium sulfate solution. Following centrifugation at 10,000 *g* for 15 min the clear supernatant is discarded, and the amber-colored pellet is suspended in 3 to 5 ml of equilibration buffer [20 m*M* Tris–MOPS (pH 8.1), 1% (v/v) Triton X-100, 0.5 *M* KCl, 1 m*M* sodium dithionite, and PMSF (10 μg/ml)]. A 1.0-ml Pharmacia (Piscataway, NJ) Hi Trap Blue affinity column is washed with 10 ml of equilibration buffer prior to loading the protein. The protein fraction is applied to the column, and washed with 3 to 5 ml of equilibration buffer, followed by 3 to 5 ml of wash I [20 m*M* Tris–MOPS (pH 8.1), 1% (v/v) Triton X-100, 1.0 *M* KCl, 1.0 m*M* sodium dithionite, PMSF (1 μg/ml)], and last by 3 to 5 ml of wash II [20 m*M* Tris–MOPS (pH 8.1), 0.5% (w/v) sodium cholate, 1.0 *M* KCl, 1.0 m*M* sodium dithionite, PMSF (1 μg/ml)]. Ferrochelatase is eluted from the column with 5 ml of elution buffer [20 m*M* Tris–MOPS (pH 8.1), 1% (w/v) sodium cholate, 1.5 *M* KCl, 1.0 m*M* sodium dithionite, PMSF (1 μg/ml)] into 1-ml fractions.

To produce a porphyrin-free preparation of ferrochelatase, an additional step is taken prior to ammonium sulfate precipitation. The detergent-solubilized fraction is applied to a 10-ml bed volume DEAE-Sepharose column, which effectively removes any residual porphyrin from the preparation. The DEAE-Sepharose column is washed first with distilled water, then with 100 m*M* Tris–MOPS, pH 8.1, before being equilibrated with solubilization buffer. The 60-ml supernatant fraction containing solubilized ferrochelatase is then passed through the column.

After the purification scheme detailed above, ferrochelatase may be concentrated by salt precipitation. The eluted fractions from the Hi Trap Blue column are pooled, brought to 150% of the original volume by addition of distilled water, and then brought to 50% (w/v) saturation with ammonium sulfate by the addition of an appropriate volume of saturated ammonium sulfate solution, pH 7.5. Following centrifugation at 10,000 *g* for 20 min, the resulting dark red "pellet" may be floating or solidly pelleted. The floating pellet can be easily removed with a glass rod, and the precipitated enzyme is suspended in the desired amount of buffer (0.1 *M* Na-HEPES or 0.1 *M* Tris–MOPS, pH 7.5 to 8.1) containing 0.6 to 1% (w/v) β-octylglucopyranoside.

In addition to the procedure outlined above there has been a report on the expression and purification of a soluble mouse ferrochelatase.[26] The particular preparation also requires the presence of detergents during the purification procedure, but makes use of ferrochelatase that is in the cytoplasmic fraction of *E. coli*. In our preparations we have noted that a fraction of ferrochelatase is found in the cytoplasm, but we have found that this enzyme does not contain a stoichiometric amount of the $[2Fe\text{-}2S]^{2+}$ cluster and so must contain a significant amount of apoferrochelatase. As a result of this, we utilize only the membrane-associated enzyme to purify protein for kinetic or biophysical studies. The enzyme derived from the cytoplasmic fraction is adequate for antibody production.

Enzyme Assay

Recombinant mammalian ferrochelatase can be assayed at any point from crude cell extract to purified enzyme. The standard assay used in this laboratory employs iron and porphyrin as substrates and quantitates the

[26] G. C. Ferreira, *J. Biol. Chem.* **269,** 4396 (1994).

product as the pyridine hemochromogen.[27] Assay components include 0.5 ml of 0.1 *M* Tris–MOPS (pH 8.1), 100 μl of 50 m*M* 2-mercaptoethanol, 100 μl of 1.0 *M* sodium bicarbonate, 50 μl of 4 m*M* ferrous citrate (or ferrous ammonium sulfate), and 100 μl of 1 m*M* protoporphyrin IX. The physiological substrate, protoporphyrin IX, is used most frequently, yet mesoporphyrin, deuteroporphyrin, and hematoporphyrin are all suitable substrates. Protoporphyrin stock solution is made by weighing 5 mg of protoporphyrin IX in a glass tube, adding 50 μl of 2 *N* NH_4OH, vortexing to form a paste, then adding 0.5 ml of 20% (v/v) Triton X-100, and vortexing once more. Distilled water (4.5 ml) is added. The protoporphyrin concentration is quantitated using visible absorbance at 408 nm in 2.7 *N* HCl, and using an E_{mM} of 262.[28] Distilled water is added to bring the solution to a 1 m*M* concentration. The stock solution must be made fresh prior to use and covered from exposure to light.

Ferrous citrate or ferrous ammonium sulfate stock solutions can be made using distilled water. Ferrous iron concentration is determined by adding 50 μl of the approximately 4 m*M* iron stock solution to 100 μl of 0.1 *M* ferrozine (Sigma, St. Louis, MO). This is brought to 5 ml with distilled water. The visible absorbance of this solution at 562 nm is used with an E_{mM} of 27.8 to determine the concentration of ferrous iron.[29]

To start the assay, ferrochelatase is routinely added last. Assays are incubated in open air at 37°, in the dark. The reaction is stopped after 30 min by adding 0.5 ml of 50 m*M* iodoacetamide. NaOH (1 *N*, 0.25 ml) is added to create alkaline conditions, followed by addition of 0.25 ml of pyridine to form the bispyridine heme complex. After mixing, the sample is then divided into two quartz cuvettes. Approximately 1 mg of dithionite is used to reduce the heme in one of the two cuvettes. The difference spectrum between 557 and 541 nm is used to quantitate the pyridine hemochrome of protoheme (ΔE_{mM} 20.7).[28]

Properties

Spectroscopy

All spectra are obtained with ferrochelatase in elution buffer, or in buffer containing β-octylglucopyranoside. Ultraviolet–visible (UV–Vis) absorption is measured from 620 to 250 nm. Protein concentration can be

[27] R. S. Porra and O. T. G. Jones, *Biochem. J.* **87,** 186 (1963).

[28] J.-H. Furhop and K. M. Smith, *in* "Porphyrins and Metalloporphyrins" (K. M. Smith, ed.), pp. 784 and 806. Elsevier, Amsterdam, 1976.

[29] H. A. Dailey, *J. Bacteriol.* **132,** 302 (1977).

determined from the UV absorption band at 278 nm, using a millimolar extinction coefficient of 47. The presence of a visible absorption band at 415 nm results from enzyme-associated high-spin heme. Characteristic features at 330, 460, and 550 nm are indicative of the $[2Fe\text{-}2S]^{2+}$ cluster.[30] Electron paramagnetic resonance (EPR), variable temperature magnetic circular dichroism (MCD),[30] resonance Raman,[31] and Mössbauer[32] spectroscopies provide more detailed information concerning this metal center. The EPR and variable-temperature MCD spectra of dithionite-reduced enzyme indicate that the metal center is a paramagnetic $[2Fe\text{-}2S]^{+}$ cluster with spin 1/2. The rhombic EPR signal has $g = 2.00$, 1.94, and 1.94. This signal saturates below 10 K and is observable up to 100 K with no broadening.

At this time, three of the four cysteine residues in the carboxyl terminus (C403, C406, and C411) are believed to serve as ligands to the metal center.[31] A fourth ligand remains to be identified. Although the metal center does not appear to be involved in the actual catalytic process, loss of the iron–sulfur cluster corresponds to loss of enzyme activity.[30,31] While the exact role of the iron–sulfur cluster remains to be determined, a structural or regulatory function is supported. By acting as a target for nitric oxide, the iron–sulfur cluster has been shown to serve as a trigger for the inactivation of mammalian ferrochelatases.[33,34] As part of an immune response, inactivation of ferrochelatase by nitric oxide may limit heme availability to pathogens, and shunt iron to ferritin for storage.[34]

Kinetics

Recombinant human and mouse ferrochelatases exhibit similar kinetic properties. The human enzyme has a K_m of 8.19 μM for protoporphyrin, and a K_m of 9.35 μM for iron. Mouse ferrochelatase has K_m values of 15.3 μM for protoporphyrin and 11.1 μM for iron.[2] All of these values are consistent with published rates for ferrochelatase purified from native sources, although no data exist for human liver ferrochelatase assayed with protoporphyrin (see Ref. 35 for review). The addition of bicarbonate to ferrochelatase assays serves to improve enzyme activity at low iron substrate

[30] K. Waegemann and J. Soll, *J. Biol. Chem.* **271,** 6545 (1996).

[31] B. R. Crouse, V. M. Sellers, M. G. Finnegan, H. A. Dailey, and M. K. Johnson, *Biochemistry* **35,** 6222 (1996).

[32] R. Franco, J. J. G. Moura, I. Moura, S. G. Lloyd, B. H. Huynh, W. S. Forbes, and G. C. Ferreira, *J. Biol. Chem.* **270,** 26352 (1995).

[33] T. Furukawa, H. Kohno, R. Tokunaga, and S. Taketani, *Biochem. J.* **310,** 533 (1995).

[34] V. M. Sellers, M. K. Johnson, and H. A. Dailey, *Biochemistry* **35,** 2699 (1996).

[35] H. A. Dailey, *in* "Mechanisms of Metallocenter Assembly" (R. P. Hausinger, G. L. Eichhorn, and L. G. Marzilli, eds.), pp. 77–95. VCH, New York, 1996.

concentrations and lowers the observed K_m for iron. The role of bicarbonate in the reaction remains to be determined.

Substrates

The preference of ferrochelatase for divalent cation substrates varies among species. For example, the utilization of copper(II) as a substrate is unique to the *B. subtilis*-type ferrochelatase, which does not use cobalt. Mammalian ferrochelatases can use divalent iron, cobalt, and zinc as substrates. Manganese, cadmium, lead, and mercury are competitive inhibitors.[36]

Acceptable porphyrin substrates include only the dicarboxylated porphyrin IX isomers. N-Alkylated porphyrins are strong inhibitors, particularly porphyrins with A or B ring alkylations. The N-alkylated porphyrin, in which the alkylated ring is tilted approximately 30° out of the plane of the macrocycle, has been proposed to be a transition state analog for the porphyrin substrate of ferrochelatase.[37] Metallation of bent porphyrins is known to proceed at a much faster rate than that of planar porphyrins.[38] In fact, an antibody made against N-alkylated mesoporphyrin exhibits ferrochelatase-type activity.[37] Ferrochelatase may serve to bring the substrates into close proximity, distort the porphyrin plane, and thereby increase the rate of metallation.[35] Specific residues that could induce such a ring distortion include conserved tryptophan residues.

Stability

The stability of purified recombinant mammalian ferrochelatase is enhanced by the presence of detergents, glycerol, and dithiothreitol. Human ferrochelatase is stable with maximal levels of activity at 37°, yet the enzyme becomes inactive and precipitates from solution at temperatures just over 37°. Mouse ferrochelatase is stable at least to 42°. Concentrated human or mouse ferrochelatase can be stored at 4° for several days following purification. Although there are no detailed studies currently available, the mouse enzyme appears to be stable at 4° for a greater length of time than human ferrochelatase and its $[2Fe\text{-}2S]^{2+}$ cluster may be less sensitive to inactivation by nitric oxide.

In mammals, decreased ferrochelatase activity results in a clinical disorder known as erythropoietic protoporphyria (EPP). EPP is generally consid-

[36] H. A. Dailey, *Ann. N.Y. Acad. Sci.* **514,** 81 (1987).

[37] A. G. Cochran and P. G. Schultz, *Science* **249,** 781 (1990).

[38] D. K. Lavallee, *in* "Mechanistic Principles of Enzyme Activity" (J. F. Liedman and A. Greenberg, eds.) VCH Publishers, New York, 1988.

ered to be an autosomal dominant hereditary disease yet bovine protoporphyria is transmitted as an autosomal recessive tract. Biochemically, EPP is characterized by an accumulation of protoporphyrin in erythrocytes, serum, and the liver. Elevated levels of protoporphyrin IX cause cutaneous photosensitivity as the major symptom.[39] In rare cases protoporphyrin IX deposition in hepatobiliary structures results in severe liver disease.[40] Naturally occurring mutations that cause human EPP have been identified and characterized. A variety of single amino acid changes caused by missense mutations, several exon deletions resulting from posttranscriptional splicing abnormalities, and a total coding region deletion demonstrate the heterogeneous character of the genetic anomalies that cause EPP (see Ref. 35).

Interestingly, ferrochelatase activity in affected individuals ranges from less than 25 to 50% of normal enzyme activity. One possible theory advanced to explain this has been that ferrochelatase may function as a homodimer *in vivo,* where two normal subunits are required for enzyme activity. Although there has been speculation that ferrochelatase may function as a dimer or other multimer *in vivo,* at present the best support in data comes from radiation inactivation target size studies.[41] As isolated here, ferrochelatase runs as a single band on an SDS gel with a molecular weight of 42,000, and elutes from a gel-filtration column equilibrated with detergent at a position consistent with a molecular weight of 42,000, implying that ferrochelatase, as prepared, is a monomer.

Acknowledgment

This work was supported by NIH Grants DK32303 and DK35898 to H. A. D.

[39] Y. Nordman and J.-C. Deybach, *in* "Biosynthesis of Heme and Chlorophylls" (H. A. Dailey, ed.), p. 491. McGraw-Hill, New York, 1990.

[40] J. R. Bloomer and R. Enriquez, *Gastroenterology* **82,** 569 (1982).

[41] J. G. Straka, J. R. Bloomer, and E. S. Kempner, *J. Biol. Chem.* **266,** 24637 (1991).

Section IV

Miscellaneous Vitamins and Coenzymes

[44] Structural Characterization of Modified Folates in Archaea

By ROBERT H. WHITE

I. Introduction

At some time during the evolution of living systems, a fundamental divergence occurred in the biochemistry of the coenzymes responsible for the metabolism of C_1 units. This fact is confirmed by the presence of a host of different C_1 carriers, which are referred to here collectively as the modified folates, that have been characterized in members of the domain Archaea.[1] Thus far these modified folates have been found only in members of the Archaea and have the structures shown in Fig. 1. The modified folates, found in the methanogenic members of the Archaea, include methanopterin (MPT; the first modified folate to be characterized), sarcinapterin, taliopterin, taliopterin-0, taliopterin-1, and thermopterin. A series of modified folates containing poly-$\beta(1\rightarrow4)$-linked *N*-acetylglucosamine side chains have also been identified in the thermophilic sulfur-dependent members of the Archaea, *Pyrococcus furiosus* and *Thermococcus litoralis*. A nonmethylated modified folate, sulfopterin, having structural features common to both folates and methanopterin, has also been partially characterized in *Sulfolobus solfataricus*. Biosynthetic[2] and stereochemical data[3] indicate that all of these modified folates have their roots in folate biochemistry, and thus their description here as modified folates.

Establishing the chemical structures of these modified folates, considering the wide range of concentrations at which they occur among members of the Archaea,[4] has required the use of several different analytical methods. In the case of methanopterin, a sufficient amount of the cofactor was obtainable from kilogram amounts of the archaebacterium, *Methanobacterium thermoautotropicium* (2000 nmol/g dry weight), so that standard organic structural methods [^{1}H nuclear magnetic resonance (NMR) and ^{13}C NMR spectroscopy] were used as the major tool in establishing its chemical structure. In general, however, neither the level of the modified folates among members of the Archaea nor the amount of cell material available is large enough to allow this type of chemical characterization to be carried out on a routine basis.

[1] R. H. White, *Biochemistry* **32,** 745 (1993).
[2] R. H. White, *Biochemistry* **35,** 3447 (1996).
[3] R. H. White, *Chirality*, **8,** 332 (1996).
[4] R. H. White, *J. Bacteriol.* **173,** 1978 (1991).

0076-6879/97 $25

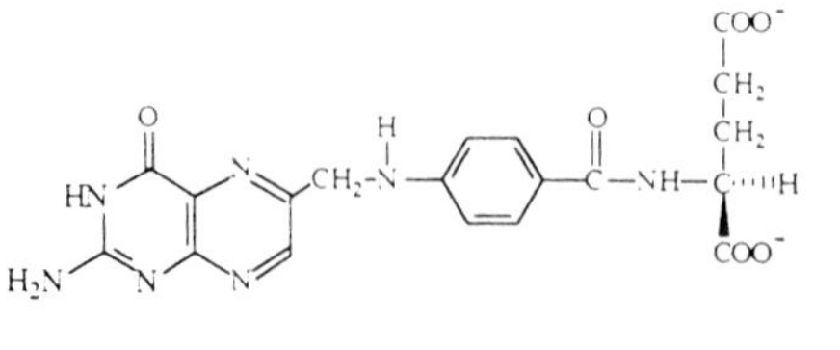

Folate

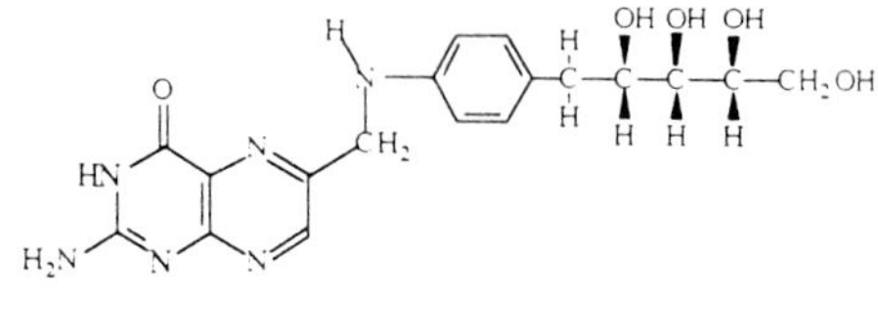

Fragment of Sulfopterin

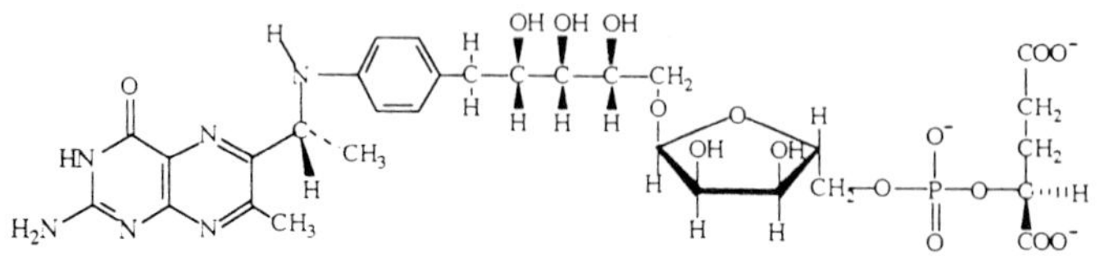

Methanopterin (MPT)

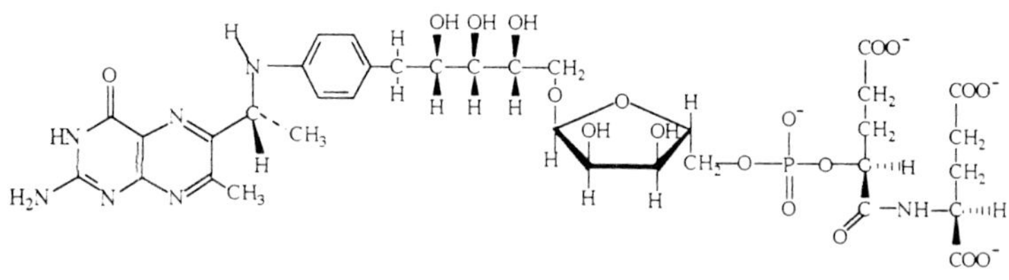

α-Glutamylmethanopterin (Sarcinapterin)

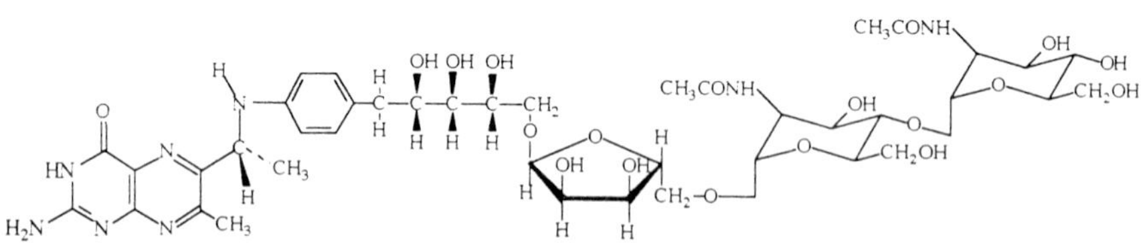

One of the modified folates found in *Pyrococcus furiosus*

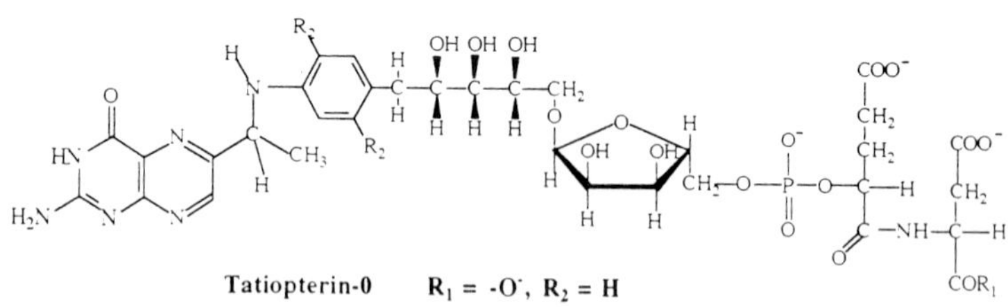

Tatiopterin-0	R_1 = -O⁻, R_2 = H
Tatiopterin-I	R_1 = -glutamic acid, R_2 = H
Thermopterin	R_1 = -O⁻, R_2 = OH

FIG. 1. Structures of modified folates.

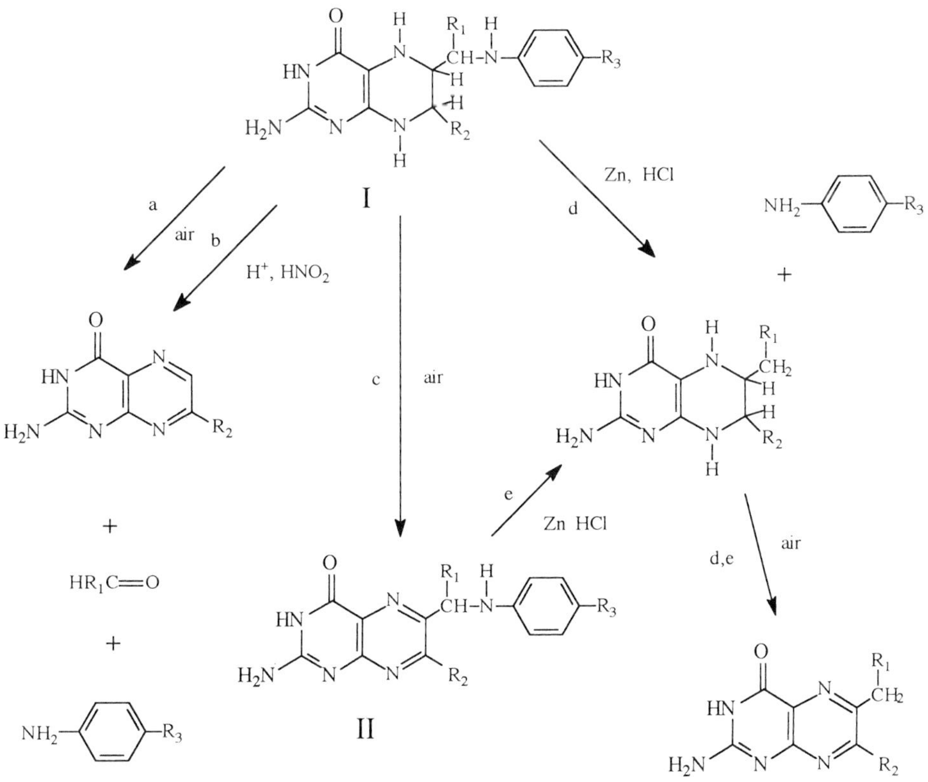

FIG. 2. Reactions of modified folates.

Thus, more sensitive analytical methods had to be developed to allow for the quantitative and structural characterization of these modified folates at the levels at which they occur in these bacteria. Like the folates, these modified folates occur in the cells as the 5,6,7,8-tetrahydro (H_4) derivatives, which can undergo cleavage reactions[5] to produce characteristic fragments. These fragments, because of their specific chemical and physical properties, allow for the structural characterization of the cofactors, using much smaller amounts of material than would be possible by working with the intact cofactor. These procedures are outlined in Fig. 2, and include (1) the oxidative cleavage of the reduced modified folates to arylamines and either pterins or 7-substituted pterins (Fig. 2, reactions a and b),[6] and (2) the

[5] E. A. Cossins, *in* "Folates and Pterins" (R. L. Blakely and S. J. Benkovic, eds.), Vol. 1, pp. 1–59. John Wiley & Sons, New York, 1984.
[6] R. H. White, *J. Bacteriol.* **162,** 516 (1985).

reductive cleavage of the modified folates to the same substituted arylamines and either 6-substituted or 6,7-disubstituted pterins as outlined in reaction d (Fig. 2). The substituted pattern of the pterins generated by these procedures can be established by comparison to known pterins, either by thin-layer chromatography (TLC) or high-performance liquid chromatography (HPLC), using the inherent fluorescence of the pterins for their detection. The substituted arylamines can be assayed directly as their Bratton–Marshall azo dye derivatives[7] by chromatographic separation of the azo dye derivatives on BioGel P-4 columns (Bio-Rad, Richmond, CA) under acidic conditions, followed by TLC analysis and/or C_{18} reversed-phase HPLC of the separated compounds. The substituted arylamines can also be chromatographically separated on the basis of their charge by DEAE-Sephadex column chromatography and assayed by fluorescence after reaction with fluorescamine.[8] In either case, the separated arylamines can be subjected to different chemical and/or enzymatic treatments that are helpful in establishing their structures.

In cases in which the amount of reduced, modified folate(s) present in a given cell is relatively high, as in the methanogens, a small amount of the intact, air-oxidized, modified folate can be recovered during cell extraction. This material can then be chromatographically purified, and then assayed and characterized by the preceding reductive cleavage methods.

II. Procedures

Two methods for the extraction of modified folates from bacterial cells have been used. In the first method, the cells are extracted aerobically with warm 50% (v/v) ethanol, leading to the air-oxidative cleavage of a large percentage of the reduced modified folates present in the cells (Fig. 2, route a). The resulting extract is then assayed for the presence of 7-substituted pterins and substituted arylamines.

The second method consists of analysis of the modified folates present in anaerobic cell extracts obtained either by sonication[9] or French press methods,[10] and employs nitrous acid oxidative cleavage of the modified H_4folates under the conditions of the Bratton–Marshall reaction (Fig. 2, route b). The Bratton–Marshall azo dye derivatives prepared from these extracts can be directly assayed as purple bands on BioGel P-4 columns.

[7] D. Zhou and R. H. White, *J. Bacteriol.* **174,** 4576 (1992).
[8] R. A. H. Furness and P. C. Loewen, *Anal. Biochem.* **117,** 126 (1981).
[9] R. H. White, *Biochemistry* **27,** 7458 (1988).
[10] M. J. K. Nelson and J. G. Ferry, *J. Bacteriol.* **160,** 526 (1984).

The Zn/HCl reduction of the modified H_4folates in these extracts can also be used to produce 6,7-disubstituted pterins and substituted arylamines (Fig. 2, route d) In some cases a small portion of the modified H_4folates can be oxidized, on exposure to air, to the intact, oxidized, modified folate (Fig. 2, structure II), which can then be separated (DEAE-Sephadex column chromatography), purified, and assayed. Aerobic reduction of these intact, oxidized, modified folate(s) with Zn/HCl then generates 6,7-disubstituted pterins and substituted arylamines (Fig. 2, route e).

A. *Cell Extraction and Analysis of Modified Folates Using Air-Oxidative Cleavage*

Bacterial cell pellets (1–5 g), isolated by centrifugation of growing bacterial cultures, are extracted aerobically with 10 ml of 50% (v/v) ethanol for 15 min at 80° in a capped tube. After centrifugation the resulting pellet is reextracted by the same procedure; the combined, clear supernatants are exposed to air in an open dish, in the absence of light, for at least 12 hr to complete the air-oxidative cleavage of the modified H_4folate(s). The sample is reduced in volume with a stream of N_2 gas and then dissolved in 5 ml of water. After centrifugation to remove any precipitate, the sample is applied to a DEAE-Sephadex column (1 × 25 cm) and eluted with a linear gradient of either sodium chloride (0–2 *M*) in 40 m*M* potassium phosphate buffer, pH 5.8, or ammonium bicarbonate (0–1.5 *M*). Pterins are detected in the column fractions by measuring the fluorescence intensity at 450 nm with excitation at 350 nm. No separation of pterin from 7-methylpterins is observed in either the sodium chloride or the ammonium bicarbonate elutions of the DEAE-Sephadex columns. Substituted arylamines in the fractions eluted with sodium chloride are detected by reacting 1 ml of the fraction with 20 μl of fluorescamine (15 mg/ml in acetone) for 5 min, and measuring the fluorescence intensity of the resulting fluorescamine adduct (excitation, 407 nm; emission, 519 nm). Analysis for arylamines in the fractions eluted with ammonium bicarbonate is conducted in the same manner, after first mixing 0.5 ml of each fraction with an equal volume of 1 *M* potassium phosphate buffer, pH 5.8, before reaction with fluorescamine. The fluorescamine-coupling reactions are performed at pH 5.8 because of the much greater specificity of the reaction for aromatic amines at this pH.[8] If the cell extracts contain *p*-aminobenzoyl glutamates [*p*-AB-$(Glu)_x$] as the oxidative cleavage product of pteroylglutamates, then the amount of these folates present in the cells can be readily determined from the fluorescamine assay of the separated fractions as previously described.[11]

[11] P. C. Loewen, *Methods Enzymol.* **122,** 330 (1986).

1. Analysis of Pterins. The pterin-containing fractions from the sodium chloride-eluted DEAE-Sephadex column are combined and concentrated by evaporation with a stream of N_2 gas, until the appearance of solid sodium chloride. The resulting solution is applied to a C_{18} Sep-Pak cartridge (Waters Associates, Milford, MA) to trap the pterin and/or 7-methylpterin; the cartridge is then washed with water, and the pterins eluted with methanol. The pterins in the pterin-containing fractions from the ammonium bicarbonate-eluted column are isolated by heating the sample to 100° while evaporating the volatile buffer with a stream of N_2 gas. In most cases, identification of the pterins in the column fractions is readily accomplished by TLC analysis on silica gel 60 F-254 TLC plates (E. Marck AG, Darmstadt, Germany) using acetonitrile–water–formic acid (88%) (40 : 10 : 5, v/v/v) as the developing solvent. In this solvent system, each pterin has the indicated R_f value: 7-methylpterin, 0.524; 6-methylpterin, 0.554; 6,7-dimethylpterin, 0.600; 6-ethylpterin, 0.644; and 6-ethyl-7-methylpterin, 0.658. The pterins are readily identified on the TLC plate by exposure of the plates to ultraviolet light.

Sometimes the presence of large amounts of impurities still present in the samples after column separation precludes their direct identification by TLC. When this occurs, the column-purified pterins are further purified from an acidic solution by retention on a small Dowex 50W-8X H^+ column (4 × 6 mm); the column is washed with water (3 ml), the pterins eluted with aqueous 3 *M* ammonia (3 ml), and then identified using TLC analysis.

The pterins could also be identified and quantitated by HPLC with fluorescence detection using known pterin samples for comparison. A Perkin-Elmer (Norwalk, CT) HPLC system, equipped with two Picosphere C_{18} cartridge columns in series, and interfaced with an LS-3 fluorescence detector, is used. The pterin samples are eluted from the columns with a linear gradient (0–100%) of methanol. This solvent gradient system readily resolves pterin, 7-methylpterin, 6-methylpterin, 6,7-dimethylpterin, 6-ethylpterin, and 6-ethyl-7-methylpterin. Because the yield of pterin or 7-methylpterin from the oxidative cleavage of a modified H_4folate is expected to be only 20–50%, the true amount of the modified folate could be two to five times larger than the amount of pterin isolated.

2. Analysis of p-Aminobenzoylpolyglutamates. Fractions from the sodium chloride gradient-eluted DEAE-Sephadex column believed to contain *p*-AB-$(Glu)_x$ (fluorescamine reaction), on the basis of their elution position, are adjusted to pH 7.6 with 1 *M* NaOH, and applied to a second DEAE-Sephadex column to resolve the individual *p*-AB-$(Glu)_x$ as described by Loewen.[11] The *p*-AB-$(Glu)_x$ eluted from the column with a sodium chloride gradient are measured and quantitated by reaction with fluorescamine. The ammonium bicarbonate gradient used does not elute the *p*-AB-$(Glu)_x$ from the DEAE-Sephadex column.

B. Cell Extraction and Analysis of Modified Folates Using Nitrous Acid Oxidative Cleavage and Analysis of Azo Dye Derivatives of Arylamines on BioGel P-4 Columns

1. Preparation of Anaerobic Cell Extracts. Procedures for the preparation of anaerobic cell extracts either by sonication[9] or French press methods[10] have been described.

2. Formation of Azo Dyes of Substituted Arylamines and Their Separation on BioGel P-4 Columns. Anaerobically prepared cell-free extracts (0.5 ml) are mixed with 0.1 ml of 6 *M* HCl and centrifuged to remove the precipitated proteins. The modified folates contained in the resulting supernatants are then cleaved and converted to their azo dye derivatives under the conditions of the Bratton–Marshall assay. This assay procedure is carried out at room temperature, with thorough mixing after each reagent addition. First, add 0.1 ml of 1.5% (w/v) aqueous $NaNO_2$ to the sample, followed, after 2 min, with 0.1 ml of 7.5% (w/v) aqueous ammonium sulfamate to decompose the excess nitrous acid. If excessive foaming of the sample occurs, a small amount (~0.1 μl) of antifoam A (Sigma, St. Louis, MO) is added. After an additional 2 min, 0.3 ml of 1% (v/v) naphthylethylenediamine dihydrochloride is added, and the solution is left for 1 hr at room temperature to complete the formation of the purple-colored azo dye. Each sample is then separated and purified by absorption chromatography using a BioGel P-4 column (1 × 48 cm) previously equilibrated with 50 m*M* HCl. The azo dye derivatives are eluted with 50 m*M* HCl and quantitated in the fractions (5.5 ml) by their absorbance at 562 nm, ε 48.0 mM^{-1} cm^{-1}. On BioGel P-4 columns, the relative retention volumes for a series of the azo dye derivatives of known arylamines, which are related to the known modified folates, are shown in Table I. From the data presented, it is clear that the degree of retention of these azo dyes on this column is correlated with the polarity of the side chains attached to the arylamine. The more polar the side chain, the lower the retention volume. The azo dyes are recovered from the fractions, either by lyophilization or by evaporation with a stream of N_2 gas, and identified, when known standards are available, by TLC (visual inspection of the purple spots) and/or by HPLC (A_{556}). The known standards can be obtained by chemical synthesis or specific degradation of the natural products.[3] The structures of the azo dyes can be further identified by the chemical and/or enzymatic treatments discussed below.

TABLE I
CHROMATOGRAPHIC CHARACTERISTICS OF AZO DYES OF ARYLAMINES DERIVED FROM MODIFIED FOLATES

	Chromatographic parameters		
Azo dye from:	BioGel P-4[a]	TLC (R_f)[b]	HPLC (min)
APDR-ribose-(GluNAc)$_5$	0.35	0.017	25.32
APDR-ribose-(GluNAc)$_4$	0.40	0.033	26.02
APDR-ribose-(GluNAc)$_3$	0.46	0.38	26.74
APDR-ribose-(GluNAc)$_2$	0.60	0.10	27.47
α-Glutamylmethaniline	0.71	0.13	27.57
Methaniline	0.74	0.11	27.57
4-(β-D-Ribofuranosyl)aminobenzene-5′-phosphate	0.79	0.10	25.75
APDR-ribose-GluNAc	0.73	0.17	28.08
1-(4-Aminophenyl)-5-deoxy(1-α-D-ribofuranosyl)-D-ribitol (APDR-ribose)	0.89	0.25	29.44
1-(4-Aminophenyl)-1-deoxy-D-ribitol-1-phosphate (APDR-P)	0.93	0.11	28.30
4-(α-D-Ribofuranosyl)aminobenzene	0.97	0.63	—
1′-(4-Aminophenyl)-1′-deoxy-D-ribitol (APDR)	1.00	0.61	30.60
4-(β-D-Ribofuranosyl)aminobenzene (β-RFA)	1.1	0.63	—
Aniline	1.15	0.76	—
4-Aminobenzoic acid	1.7	0.65	—

[a] The values are relative to an elution volume of 1.00 for azo dye derivative 1′-(4-aminophenyl)-1′-deoxy-D-ribitol.

[b] The solvent system used was acetonitrile–water–formic acid (88%, w/w) (40 : 10 : 5, v/v/v).

3. Thin-Layer Chromatography Analysis of Azo Dye Derivatives of Arylamines. The azo dye derivatives of the arylamines are identified by TLC analysis using the preceding system described for the TLC analysis of pterins. With this TLC system, the azo dye derivatives of some known arylamines have R_f values as listed in Table I. The purple color of the azo dye derivatives is most intense immediately after evaporation of the solvent from the TLC plate, and completely disappears after a few hours of exposure of the plate to light and air. This fading of the TLC spots is prevented by placing the TLC plate in a closed chamber containing HCl vapors.

4. High-Performance Liquid Chromatography Analysis of Azo Dye Derivatives. The azo dye derivatives of the arylamines are assayed by reversed-phase chromatography using two different HPLC systems and columns. One system consists of a Du Pont (Wilmington, DE) instrument, 8800 series, gradient liquid chromatographic system fitted with a Whatman (Clifton, NJ) Partisil PX5 10/25 ODS column. The azo dyes are eluted with a linear gradient [methanol–water–trifluoroacetic acid (50 : 50 : 1, v/v) to methanol–trifluoroacetic acid (100 : 1, v/v)] over a 20-min period at a flow rate of 1 ml/min. Under these conditions, the azo dye derivatives of methaniline, 1′-(4-aminophenyl)-1′-deoxy-D-ribitol (APDR ribose, and APDR elute at 5.16, 5.78, and 6.01 min, respectively.

The other system consists of a Shimadzu liquid chromatography system (model LC-6A) mounted with an Axxiom octadecylsilane column (5 μm, 4.6 × 250 mm) (Cole Scientific, Calabasas, CA) and equipped with a Shimadzu SPD 6AV UV–Vis detector. The azo dye derivatives are eluted using a gradient generated with 0.1% (v/v) aqueous trifluoroacetic acid as solvent A and 0.075% (v/v) trifluoroacetic acid in acetonitrile as solvent B. The gradient is linear from 3.5% (v/v) solvent B to 45% (v/v) solvent B over a 40-min period, and the flow rate is 1 ml/min. Elution is monitored by the absorbance at 556 nm. The retention times of the azo dye derivatives of a series of known arylamines are shown in Table I.

5. *Chemical and Enzymatic Analysis of Azo Dye Products*

a. ACID HYDROLYSIS. Aliquots of the BioGel P-4 column-isolated azo dyes are subjected to acid hydrolysis or phosphatase cleavage to obtain further information about their chemical structure. Because of the sensitivity of the α-ribosidic bond to acid hydrolysis, in MPT and all of the currently characterized modified folates, the presence of the α-ribosidic bond in any given molecule is readily established by a brief acid hydrolysis of the sample. This is accomplished by heating the azo dye derivative at 100° for 2 min in 1 *M* HCl, evaporating the acid, and determining the production of the azo dye derivative of APDR by TLC and/or HPLC.

b. PHOSPHOMONOESTERASE HYDROLYSIS. Azo dye products are tested for the presence of phosphate monoesters by dissolving 1–100 nmol in 50 μl of 0.1 *M* glycine buffer, pH 10.4, containing 1 m*M* $ZnCl_2$ and 1 m*M* $MgCl_2$, and treating with 0.3 unit of *Escherichia coli* phosphatase for 2 hr at 37°. The sample is then acidified and the azo dye isolated by chromatography on a small BioGel P-4 column (0.4 × 0.8 cm), eluted with 50 m*M* HCl. The azo dye fractions are concentrated by evaporation and examined by TLC and/or HPLC. Changes in the R_f or retention time of the azo dye confirm the presence of one or more monophosphate esters in the molecule.

c. PHOSPHODIESTERASE I HYDROLYSIS. The selected azo dye (10–20 nmol) is dissolved in 200 μl of 0.1 *M* glycine buffer, pH 8.8. To this solution is added ~1 unit of phosphodiesterase I from *Crotalus atrox*. This solution is incubated for 12 hr at 37°, and the reaction mixture is then separated on a small column of BioGel P-4, eluted with 50 m*M* HCl. The column fractions containing the azo dye derivative are then assayed by TLC and/or HPLC.

d. β-*N*-ACETYLGLUCOSAMINIDASE HYDROLYSIS. A lyophilized sample (10–20 nmol) of the individual azo dyes purified on BioGel P-4 columns is dissolved in 100–200 μl of 100 m*M* citrate buffer, pH 4.0. To this sample is added 0.04 to 0.4 unit of β-*N*-acetylglucosaminidase, prepared by dissolving a small portion of the ammonium sulfate-precipitated enzyme in the same citrate buffer. At timed intervals, aliquots of the enzymatically digested sample are assayed by TLC and HPLC to identify the products. In this manner, the successive loss of individual *N*-acetylglucosamine residues can be measured.

e. CHITINASE HYDROLYSIS. A lyophilized sample (10–20 nmol) of the individual azo dyes purified on BioGel P-4 columns is dissolved in 1–2 ml of 50 m*M* phosphate buffer, pH 6.0. To 30-μl portions of this sample is added 2 μl of a chitinase solution containing ~0.02 unit of activity. These reaction mixtures are incubated at room temperature, and at timed intervals (5, 15, 30, and 60 min) a reaction is terminated by the addition of 1 *M* HCl. The products are assayed by TLC.

f. GAS CHROMATOGRAPHY–MASS SPECTROMETRIC IDENTIFICATION OF PRODUCTS. The azo dye derivatives of APDR and 4-(β-D-ribofuranosyl) aminobenzene (β-RFA) can be identified as their tifluoroacetate derivatives by using gas chromatography–mass spectrometry (GC–MS). This is accomplished by the reductive cleavage of the azo dye (>1 nmol) dissolved in 0.5 ml of 1 *M* HCl with zinc dust (1 mg) for 2 min at room temperature. After removal of the zinc by filtration and evaporation of the HCl with a stream of nitrogen, the arylamines are isolated by absorption on Dowex 50W-X8 H^+ and eluted with 3 *M* aqueous ammonia after evaporation of the solvents. The sample is reacted for 12 hr at room temperature with 0.2 ml of an equal mixture of methylene chloride and trifluoroacetic anhydride. After evaporation of the excess solvent, the trifluoroacetate derivatives are separated by GC–MS on a DB-1 column (30 m × 0.32 mm, J&W Scientific Co., Folsom, CA) programmed from 75° at 10°/min. 1′-(4-Aminophenyl)-1′-deoxy-D-ribitol has a molecular ion at *m/z* 707 and 4-(β-D-ribofuranosyl)aminobenzene has a molecular ion at *m/z* 609.

C. Analysis of Intact Oxidized Cofactor Using Reductive Cleavage

Reductive Cleavage. The position of elution of the intact cofactor present in the DEAE-Sephadex-purified fractions obtained (Section II,A,2), can be determined by the Zn/HCl reduction of each fraction and the measurement of the amount of 6- and/or 6,7-disubstituted pterin(s) produced. Thus, 0.5 ml of each fraction is mixed with 0.5 ml of 1 *M* HCl and 0.1 ml of a suspension of zinc dust [0.5 g in 1 ml of 0.5% (w/v) gelatin], shaken for 8 min, and then centrifuged to remove the insoluble material. The resulting pellet is washed with 0.5 ml of 0.5 *M* HCl, and 1 *M* NaOH is then slowly added to the combined clear aqueous layers until the first appearance of a $Zn(OH)_2$ precipitate. After adjustment of the solution to pH 4.5 by the addition of acetic acid, the precipitate redissolves and the resulting clear solution is applied to a C_{18} Sep-Pak cartridge column. After the column is washed with water, the pterins are eluted with methanol. The residue resulting from evaporation of the methanol is assayed by TLC and HPLC as described above (Section II,A,1). The arylamines produced in these Zn/HCl-cleaved fractions can also be measured after BioGel P-4 chromatography of the azo dyes as discussed in Section II,B,2.

III. Concluding Remarks

Methods are now in place for the chemical characterization of the modified folates that are present among members of the Archaea. Because of the expected similarities in the chemistry of all these different modified folates (i.e., they all must function as C_1 carriers), these methods can be applied to the characterization of the modified folates in any organism. Because the presence of folate has been confirmed only in very few bacteria, it is possible that bacteria, as with archaebacteria, will be found to contain a wide range of modified folates.

Section V

Methodologies Broadly Applicable to Vitamins and Coenzymes

[45] Biokinetic Analysis of Vitamin Absorption and Disposition in Humans

By JANOS ZEMPLENI

Introduction

Biokinetics (classically, "pharmacokinetics") describes the metabolic fate of administered compounds in organisms by applying mathematical procedures. Such kinetic approaches require the administration of a vitamin test dose to subjects and the subsequent sampling of blood or urine. The time courses of the vitamin and its metabolites in these body fluids provide *in vivo* information about absorption, distribution, biotransformation, and excretion of the compounds considered.

Timing of Sampling

Following the determination of vitamin baseline concentrations in body fluids (see the next section) and the administration of the vitamin test dose, blood and urine samples are taken repeatedly to allow kinetic analysis of the vitamin concentration-versus-time curves. The duration of sampling (time when last sample is taken) is determined by the compliance of the subjects to obey the study protocol, and the requirement of a sufficient number of data points during the terminal phase of vitamin disposition. Favorably, vitamin baseline concentrations as observed before administration of the vitamin test dose are reached again when the trial is terminated. The achievement of baseline levels is mainly determined by the disposition characteristics of the vitamin: Intravenously (i.v.) administered pyridoxine hydrochloride has a plasma half-life of ~0.11 hr (i.e., it is eliminated from plasma by 97% within 1 hr).[1] The terminal half-life of its metabolite pyridoxal 5′-phosphate is 31.5 hr (i.e., ~158 hr is necessary to reach baseline levels again). However, methods exist for extrapolation of the terminal slopes of concentration-versus-time curves; these approaches are discussed later.

No general recommendations exist concerning the frequency of sampling, but the interval between two samples should not exceed the half-life of the vitamin. Samples should be drawn more frequently during the phases of vitamin absorption and when absorption ceases or immediately after i.v.

[1] J. Zempleni and W. Kübler, *Clin. Chim. Acta* **229,** 27 (1994).

0076-6879/97 $25

administration. During the terminal phase of disposition a less frequent sampling is usually sufficient. Absorption was 95% complete within 1.5 hr for pyridoxine hydrochloride,[1] within 5 hr for riboflavin,[2] and within 12 hr for ascorbic acid.[3] The lag time (i.e., the difference between oral vitamin administration and onset of absorption) was small for both pyridoxine hydrochloride and riboflavin (<20 min). The initial phase of disposition is usually fast with half-lives of 0.10–0.15 hr found for i.v. thiamin hydrochloride,[4] pyridoxine hydrochoride,[1] or riboflavin.[2] Such phases are missed if sampling is not done appropriately. However, sampling is limited by the volume of blood drawn or by the ability of the subjects to give urine voids as required.

Baseline Correction: Endogenous Levels

The vitamin baseline concentrations in body fluids as observed before administration of the test doses result from regular dietary intake. Because these endogenous levels are not caused by the administered test doses, they are subtracted from the postdose values. Plasma concentrations of thiamin,[5] flavocoenzymes,[2] and pyridoxal 5′-phosphate[6,7] were found to be constant throughout the day, or their variations were negligible as described for riboflavin.[2] Under such circumstances it is sufficient to subtract a constant baseline value from all postdose concentrations [Eq. (1)]. Urinary vitamin excretion is transformed into amount excreted per unit of time (moles per hour) for this purpose.

$$C_t = C_t^* - C_b \tag{1}$$

where C_t is the concentration at time t, corrected by the baseline concentration (mol/liter); C_t^* is the concentration at time t, before subtraction of the baseline value (mol/liter); and C_b is the baseline concentration as determined before administration of the test dose (mol/liter).

If a pronounced circadian variation in baseline values is found, as was the case for amino acids in serum,[8] the subtraction of a constant value would no longer be sufficient. This problem can be solved by obtaining a day profile of the considered vitamin without administering a test dose

[2] J. Zempleni, J. R. Galloway, and D. B. McCormick, *Am. J. Clin. Nutr.* **63,** 54 (1996).
[3] W. Kübler and J. Gehler, *Int. Z. Vitamin. Forsch.* **40,** 442 (1970).
[4] J. Zempleni, M. Hagen, S. Vogel, U. Hadem, and W. Kübler, *Nutr. Res.* **16,** 1479 (1996).
[5] H. Mascher and C. Kikuta, *J. Pharm. Sci.* **82,** 56 (1993).
[6] J. Leinert, I. Simon, and D. Hötzel, *Int. J. Vitam. Nutr. Res.* **53,** 166 (1983).
[7] H. Mascher, *J. Pharm. Sci.* **82,** 972 (1993).
[8] K. J. Moch and W. Kübler, *Z. Ernahrungswiss* **32,** 2 (1993).

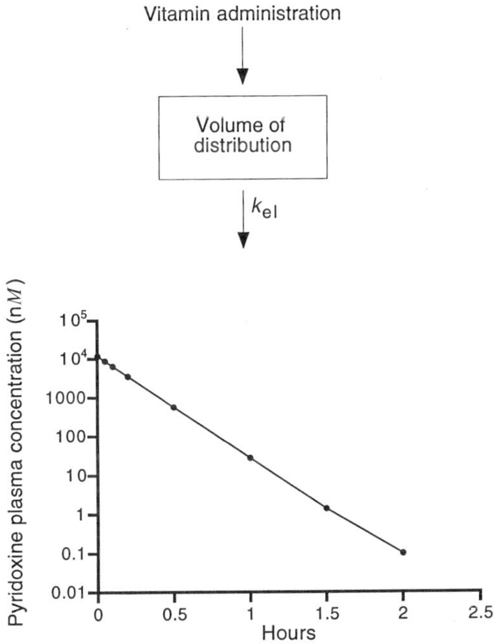

FIG. 1. One-compartment open model of vitamin metabolism and log-transformed plasma concentration-versus-time curves as calculated for an i.v. bolus injection of pyridoxine hydrochloride. [Data from J. Zempleni and W. Kübler, *Clin. Chim. Acta* **229,** 27 (1994).]

(first trial), and subsequent fitting of the baseline by applying polynomials of an appropriate degree [Eq. (2)].[9] After obtaining the best fit of the curve by least-squares nonlinear regression, the corresponding baseline value matching the time of each postdose concentration (second trial) is calculated. The further procedure is as described in Eq. (1).

$$f(x) = a_0 + a_1x + a_2x^2 + a_3x^3 + \cdots \qquad (2)$$

Models for Kinetic Analyses

The kinetics of water-soluble vitamins were described by applying one-, two-, or three-compartment open models.[1-4,10] Figures 1 to 3 illustrate these three alternatives, in which vitamin elimination from the body and transfer between compartments are assumed to occur in a first-order fash-

[9] L. Saunders and R. Fleming, "Mathematics and Statistics for Use in the Biological and Pharmaceutical Sciences." The Pharmaceutical Press, London, 1971.

[10] J. Gehler and W. Kübler, *Int. Z. Vitam. Forsch.* **40,** 454 (1970).

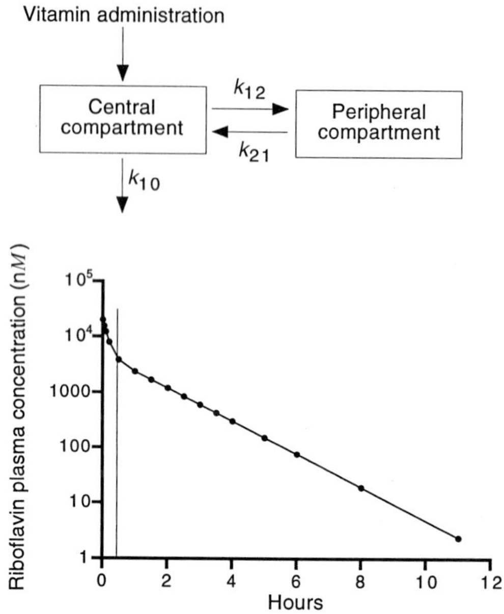

FIG. 2. Two-compartment open model of vitamin metabolism and log-transformed plasma concentration-versus-time curves as calculated for an i.v. bolus injection of riboflavin. The vertical line in the plasma curve indicates the time when pseudoequilibrium between compartments has been established. [Data from J. Zempleni, J. R. Galloway, and D. B. McCormick, *Am. J. Clin. Nutr.* **63,** 54 (1996).]

ion. All subsequent equations are based on the assumption of exclusive elimination from the central compartment and not from the peripheral compartment or a combination of both. The one-compartment model (Fig. 1) describes the body as a single, kinetically homogeneous unit. It is useful for the pharmacokinetic analysis of vitamins that distribute rapidly throughout the body. However, most vitamins entering the systemic circulation require a finite time to distribute fully throughout the available space (i.e., until a pseudoequilibrium has been established) (multicompartment models; Figs. 2 and 3). During this distribution phase, vitamin concentrations in plasma will decrease more rapidly than during the postdistributive phase. To choose the appropriate model for further kinetic analysis, the postdose vitamin plasma concentrations are log-transformed as shown in Figs. 1 to 3 for the rapid i.v. injection of pyridoxine hydrochloride (one compartment), riboflavin (two compartments), and thiamin hydrochloride (three compartments). If a one-compartment model applies, the plot of concentration versus time will show one straight line. When a vitamin is administered

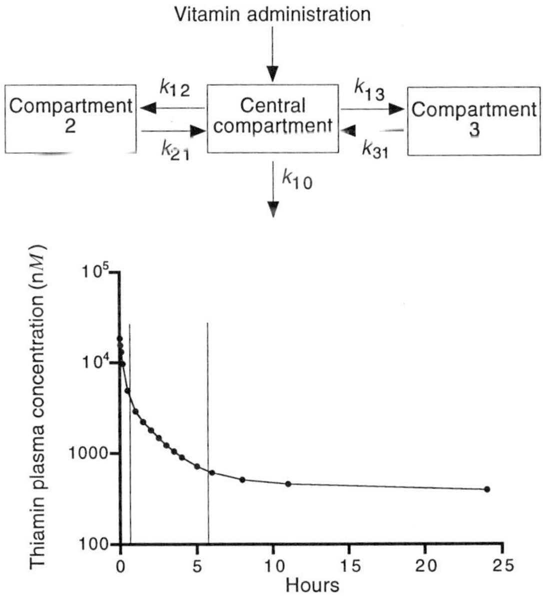

FIG. 3. Three-compartment open model of vitamin metabolism and log-transformed plasma concentration-versus-time curves as calculated for an i.v. bolus injection of thiamin hydrochloride. The vertical lines in the plasma curve indicate the times when pseudoequilibrium between compartments has been established. No volumes of distribution were available for thiamin; therefore, fictive values were assumed in this case. [Data from J. Zempleni, M. Hagen, S. Vogel, U. Hadem, and W. Kübler, *Nutr. Res.* **16,** 1479 (1996).]

orally, a straight line will be obtained when absorption ceases. In those cases in which two- or three-compartment models apply, a plot of concentrations versus time will show two or three straight lines, respectively. Kinetic homogeneity does not necessarily mean that the vitamin concentrations in the tissues of the central compartment (one-compartment model: volume of distribution) at any given time are the same as in blood plasma (the anatomical reference compartment). However, it does assume that any change that occurs in the plasma level of a vitamin quantitatively reflects a change that occurs in central compartment tissue levels. Compartments are usually not assigned to anatomic entities, but they reflect units of equilibration of vitamin concentrations.

Even though a great deal of kinetic information can be gained by the use of a single route of administration (usually i.v. or oral), the use of both i.v. and oral routes of administration on different occasions is desirable.

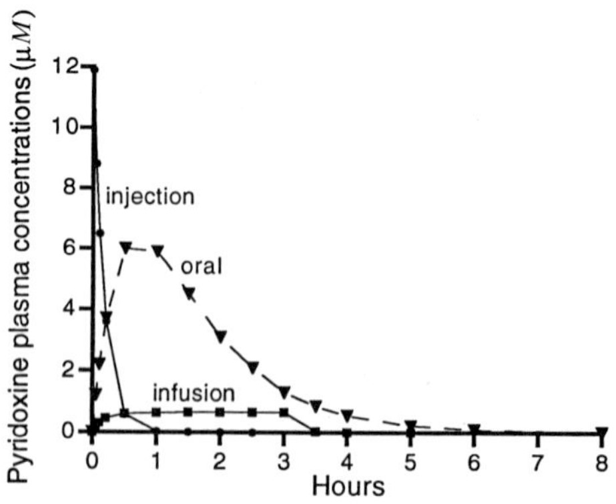

FIG. 4. Comparison of the kinetics of identical doses (0.5 mmol) of pyridoxine hydrochloride given orally, as rapid i.v. injection, or as continuous constant rate i.v. infusion (3 hr) to a 70-kg male subject. Curves were calculated using the kinetic data provided for pyridoxine hydrochloride. [Data from J. Zempleni, *J. Am. Coll. Nutr.* **14,** 579 (1995).]

The disposition kinetics of thiamin,[11] pyridoxine,[12] biotin,[13] and riboflavin[2] differ depending on the route of administration. The choice of the appropriate model may even differ for the same vitamin when different routes of administration are used.[3,10,11] Also, the i.v. administration (as rapid injection or continuous infusion) provides a reference for oral bioavailability studies.

One-Compartment Model

Several pathways of vitamin metabolism involve enzymatic processes that are limited in terms of capacity. However, at low concentrations of vitamins the rates of these enzymatic processes can be approximated well by first-order kinetics. In the case of a rapid i.v. injection of a vitamin bolus (Fig. 4), plasma concentration-versus-time curves are fitted by nonlinear regression according to Eq. (3). Regression analysis is done from the point of highest plasma vitamin concentration until baseline values are achieved or until the time when the last sample is drawn. It is emphasized that the kinetic parameters must be calculated from the individual curves of the subjects and not from the mean curves of several subjects.

$$C_t = C_0 e^{-k_{el} t} \quad (3)$$

[11] C. M. E. Tallaksen, A. Sande, T. Bøhmer, H. Bell, and J. Karlsen, *Eur. J. Clin. Nutr. Pharmacol.* **44,** 73 (1993).

[12] J. Zempleni, *J. Am. Coll. Nutr.* **14,** 579 (1995).

[13] M. Frigg, D. Hartmann, and O. C. Straub, *Int. J. Vitam. Nutr. Res.* **64,** 36 (1994).

where C_t is the vitamin concentration at time t (mol/liter); C_0 is the vitamin concentration at time 0 (zero-time intercept; mol/liter); e equals 2.718; k_{el} is the apparent first-order elimination rate constant (hr^{-1}); and t is the time after vitamin administration (hr).

A fit through the measured vitamin concentrations in plasma provides two parameters: the zero-time intercept (C_0) and the apparent first-order elimination rate constant (k_{el}). All first-order rate constants described in this chapter can be transformed into half-lives according to Eq. (4). The zero-time intercept is a fictive variable, assuming an instant distribution of the vitamin at the time of injection but without any part of the compound being eliminated at that time. Such a concentration cannot be measured, because vitamin distribution needs a brief period of time, but glomerular filtration and metabolism start immediately after injection. However, the importance of C_0 for further calculations will be demonstrated.

$$t_{1/2} = \ln 2/k \tag{4}$$

where $t_{1/2}$ is the apparent first-order half-life (hr); ln 2 is the natural logarithm of 2; and k is the first-order rate constant of disposition, transfer, or elimination (hr^{-1}).

In the case of continuous constant-rate intravenous infusion, vitamin concentrations in plasma increase according to Eq. (5) during infusion (Fig. 4). If infusion is done over a sufficient period of time, a plasma steady state vitamin concentration will be reached. This steady state concentration is calculated by applying Eq. (6). Infusion of a vitamin for a period of time equal to seven biological half-lives results in concentrations within 1% of steady state. After termination of infusion, plasma vitamin concentration follows Eq. (7).

$$C_t = \frac{D}{Vk_{el}T}(1 - e^{-k_{el}t}) \tag{5}$$

where C_t, k_{el}, and t are as defined for Eq. (3); D is the amount of vitamin administered (mol); e equals 2.718; V is the apparent volume of distribution [liters; see Eq. (30)]; and T is the duration of continuous infusion (hr).

$$C_t = \frac{D}{Vk_{el}T} \tag{6}$$

where C_t and k_{el} are as defined for Eq. (3); D and T are as defined for Eq. (5); and V is the apparent volume of distribution [liters; see Eq. (30)].

$$C_t = [D/(Vk_{el}T)](e^{k_{el}T} - 1)e^{-k_{el}t} \tag{7}$$

where C_t and k_{el} are as defined for Eq. (3); D and T are as defined for Eq. (5); and V is the apparent volume of distribution [liters; see Eq. (30)].

Following the oral administration of vitamins (Fig. 4), their concentration-versus-time curve in plasma is analyzed by applying Eq. (8).[14]

$$C_t = \frac{Fk_a D}{V(k_a - k_{el})}(e^{-k_{el}t} - e^{-k_a t}) \tag{8}$$

where C_t and k_{el} are as defined for Eq. (3); F is the fraction of the vitamin dose that is absorbed following extravascular administration; k_a is the apparent first-order absorption rate constant (hr^{-1}); D is as defined for Eq. (5); and V is the apparent volume of distribution [liters; see Eq. (30)].

Apparent Zero-Order Absorption. Under certain conditions the absorption of vitamins may be better described by assuming zero-order (constant rate) rather than first-order kinetics. Equation (9) describes the vitamin concentrations in plasma under these conditions.[14] It simplifies to Eq. (10) for the time during which absorption takes place.

$$C_t = \frac{k_0(e^{-k_{el}T} - 1)e^{-k_{el}t}}{Vk_{el}} \tag{9}$$

where C_t, k_{el}, and t are as defined for Eq. (3); k_0 is the rate of vitamin absorption (mol/hr); T equals t during absorption (hr); after absorption apparently ceases, T is a constant (equaling the time at the end of absorption) corresponding to the absorption time; and V is the apparent volume of distribution [liters; see Eq. (30)].

$$C_t = \frac{k_0(1 - e^{-k_{el}t})}{Vk_{el}} \tag{10}$$

where C_t, k_{el}, and t are as defined for Eq. (3); k_0 is as defined for Eq. (9); and V is the apparent volume of distribution [liters; see Eq. (30)].

The maximal vitamin concentration (C_{max}) in plasma after oral administration and the time necessary to achieve this concentration (t_{max}) can be recorded as measured, but it is also possible to provide extrapolated data, which do not depend on an appropriate sampling time (close to maximum). The time of peak concentration is calculated by Eq. (11) and C_{max} is calculated by Eq. (12).[14]

$$t_{max} = \frac{2.303}{k_a - k_{el}} \log \frac{k_a}{k_{el}} \tag{11}$$

where t_{max} is the time of maximal plasma concentration (hr); k_a is as defined for Eq. (8); and k_{el} is as defined for Eq. (3).

[14] M. Gibaldi and D. Perrier, "Pharmacokinetics." Marcel Dekker, New York, 1982.

$$C_{max} = \frac{FD}{V} e^{-k_{el}t_{max}} \quad (12)$$

where C_{max} is the maximal vitamin plasma concentration (mol/liter); F is as defined for Eq. (8); D is as defined for Eq. (5); e equals 2.718; k_{el} is as defined for Eq. (3); t_{max} is as defined for Eq. (11); and V is the apparent volume of distribution [liters; see Eq. (30)].

The time course of metabolites formed in the body from the administered vitamin is described by Eq. (13), if a one-compartment model applies. The calculation is done analogously to that applied for oral vitamin administration [see Eq. (8)]. By analyzing the metabolite curve alone, one cannot distinguish if the terminal slope of the curve provides the elimination rate constant of the administered vitamin or its metabolite.[14] Therefore, the kinetics of the administered parent compound should always be considered, even if the metabolite is of primary interest.

$$C_{t(m)} = \frac{k_f D}{V_{met}[k_f - k_{el(met)}]} [e^{-k_{el(met)}t} - e^{-k_f t}] \quad (13)$$

where $C_{t(m)}$ is the metabolite concentration at time t (mol/liter); k_f is the apparent first-order rate constant of metabolite formation (hr^{-1}); D is the absorbed vitamin dose (mol); $k_{el(met)}$ is the apparent first-order elimination rate constant of the metabolite (hr^{-1}); and V_{met} is the apparent volume of distribution of the metabolite [liters; see Eq. (30)].

Multicompartment Model

The plasma vitamin concentration-versus-time curve after rapid i.v. injection is described by Eqs. (14) and (15) for two- and three-compartment models, respectively.

$$C_t = C_{01}e^{-k_\alpha t} + C_{02}e^{-k_\beta t} \quad (14)$$

where C_t and t are as defined for Eq. (3); C_{01} and C_{02} are vitamin concentrations at time 0, zero-time intercepts of the fast and slow phase (mol/liter); e equals 2.718; and k_α and k_β are apparent first-order disposition rate constants of the fast and slow phase (hr^{-1}).

$$C_t = C_{01}e^{-k_\alpha t} + C_{02}e^{-k_\beta t} + C_{03}e^{-k_\gamma t} \quad (15)$$

where C_t and t are as defined for Eq. (3); C_{01}, C_{02}, and C_{03} are vitamin concentrations at time 0, zero-time intercepts of the fast and slow phases (mol/liter); e equals 2.718; and k_α, k_β, and k_γ are apparent first-order disposition rate constants of the fast and slow phases (hr^{-1}).

The plasma vitamin concentration-versus-time curve for continuous i.v. infusion is described by Eq. (16) during the time infusion takes place and by Eq. (17) after termination of infusion when a two-compartment model applies.[15] The plasma concentration at the termination of vitamin administration may be calculated from Eq. (16) by setting t equal to T (time of termination of infusion).

$$C_t = \frac{C_{01}}{k_\alpha t}(1 - e^{-k_\alpha t}) + \frac{C_{02}}{k_\beta t}(1 - e^{-k_\beta t}) \tag{16}$$

where C_t is as defined for Eq. (3); C_{01}, C_{02}, k_α, and k_β are as defined for Eq. (14), obtained by rapid i.v. injection of the vitamin; e equals 2.718; and t is the time after onset of vitamin infusion (hr).

$$C_t = \frac{C_{01}(1 - e^{-k_\alpha T})\, e^{-k_\alpha (t-T)}}{k_\alpha T} + \frac{C_{02}(1 - e^{-k_\beta T}) e^{-k_\beta (t-T)}}{k_\beta T} \tag{17}$$

where C_t is as defined for Eq. (3); C_{01}, C_{02}, k_α, and k_β are as defined for Eq. (14), obtained by rapid i.v. injection of the vitamin; e equals 2.718; T is as defined for Eq. (5); and t is as defined for Eq. (16).

Disposition kinetics in a two-compartment model following oral vitamin administration are calculated according to Eq. (18) by applying the method of residuals.[14] Space does not allow a detailed description of this process. The same procedure may be used for the calculation of biphasically eliminated metabolites formed from the administered vitamin. The first-order absorption rate constant is calculated according to Loo and Riegelman if both oral and i.v. data are available.[16]

$$C_t = C_{01} e^{-k_a t} + C_{02} e^{-k_\alpha t} + C_{03} e^{-k_\beta t} \tag{18}$$

where C_t and t are as defined for Eq. (3); C_{01}, C_{02}, and C_{03} are vitamin concentrations at time 0, zero-time intercepts (mol/liter); e equals 2.718; k_a is as defined for Eq. (8); and k_α and k_β are as defined for Eq. (14).

The first-order intercompartmental transfer rate constants and the first-order elimination rate constant for a multicompartment model (see Figs. 2 and 3) are calculated from the first-order disposition rate constants and the zero-time intercepts. For two-compartment models the transfer and elimination rate constants are calculated from Eqs. (19)–(21).[14]

$$k_{21} = \frac{C_{01} k_\beta + C_{02} k_\alpha}{C_{01} + C_{02}} \tag{19}$$

[15] R. E. Notari, "Biopharmaceutics and Clinical Pharmacokinetics." Marcel Dekker, New York, 1987.

[16] J. C. K. Loo and S. Riegelman, *J. Pharm. Sci.* **57,** 918 (1968).

where k_{21} is the apparent first-order intercompartmental transfer rate constant, assigned for transfer from the peripheral to the central compartment (hr^{-1}); and C_{01}, C_{02}, k_α, and k_β are as defined for Eq. (14).

$$k_{10} = \frac{k_\alpha k_\beta}{k_{21}} \tag{20}$$

where k_{10} is the apparent first-order elimination rate constant (hr^{-1}); k_α and k_β are as defined for Eq. (14); and k_{21} is as defined for Eq. (19).

$$k_{12} = k_\alpha + k_\beta - k_{21} - k_{10} \tag{21}$$

where k_{12} is the apparent intercompartmental transfer rate constant, assigned for transfer from the central to the peripheral compartment (hr^{-1}); k_α and k_β are as defined for Eq. (14); k_{21} is as defined for Eq. (19); and k_{10} is as defined for Eq. (20).

For a three-compartment model, the calculations of the transfer rate constants and the elimination rate constant are done according to Eqs. (22)–(26).[14]

$$k_{21} = 0.5\left\{-\frac{k_\alpha C_{03} + k_\alpha C_{02} + k_\gamma C_{01} + k_\gamma C_{02} + k_\beta C_{01} + k_\beta C_{03}}{C_{01} + C_{02} + C_{03}} + \left[\left(\frac{k_\alpha C_{03} + k_\alpha C_{02} + k_\gamma C_{01} + k_\gamma C_{02} + k_\beta C_{01} + k_\beta C_{03}}{C_{01} + C_{02} + C_{03}}\right)^2 - 4\left(\frac{k_\alpha k_\beta C_{03} + k_\alpha k_\gamma C_{02} + k_\beta k_\gamma C_{01}}{C_{01} + C_{02} + C_{03}}\right)\right]^{1/2}\right\} \tag{22}$$

where k_{21} is the intercompartmental transfer rate constant, assigned to transfer from compartment 2 to the central compartment (hr^{-1}); and k_α, k_β, k_γ, C_{01}, C_{02}, C_{03} are as defined for Eq. (15).

$$k_{31} = 0.5\left\{-\frac{k_\alpha C_{03} + k_\alpha C_{02} + k_\gamma C_{01} + k_\gamma C_{02} + k_\beta C_{01} + k_\beta C_{03}}{C_{01} + C_{02} + C_{03}} - \left[\left(\frac{k_\alpha C_{03} + k_\alpha C_{02} + k_\gamma C_{01} + k_\gamma C_{02} + k_\beta C_{01} + k_\beta C_{03}}{C_{01} + C_{02} + C_{03}}\right)^2 - 4\left(\frac{k_\alpha k_\beta C_{03} + k_\alpha k_\gamma C_{02} + k_\beta k_\gamma C_{01}}{C_{01} + C_{02} + C_{03}}\right)\right]^{1/2}\right\} \tag{23}$$

where k_{31} is the intercompartmental transfer rate constant, assigned to transfer from compartment 3 to the central compartment (hr^{-1}); and k_α, k_β, k_γ, C_{01}, C_{02}, and C_{03} are as defined for Eq. (15).

$$k_{10} = \frac{k_\alpha k_\beta k_\gamma}{k_{21} k_{31}} \tag{24}$$

where k_{10} is the apparent first-order elimination rate constant (hr^{-1}); k_α, k_β, and k_γ are as defined for Eq. (15); k_{21} is as defined for Eq. (22); and k_{31} is as defined for Eq. (23).

$$k_{12} = \frac{(k_\beta k_\gamma + k_\alpha k_\beta + k_\alpha k_\gamma) - k_{21}(k_\alpha + k_\beta + k_\gamma) - k_{10}k_{31} + k_{21}^2}{k_{31} - k_{21}} \tag{25}$$

where k_{12} is the intercompartmental transfer rate constant, assigned to transfer from the central compartment to compartment 2 (hr^{-1}); k_α, k_β, and k_γ are as defined for Eq. (15); k_{21} is as defined for Eq. (22); k_{10} is as defined for Eq. (24); and k_{31} is as defined for Eq. (23).

$$k_{13} = k_\alpha + k_\beta + k_\gamma - (k_{10} + k_{12} + k_{21} + k_{31}) \tag{26}$$

where k_{13} is the intercompartmental transfer rate constant, assigned to transfer from the central compartment to compartment 3 (hr^{-1}); k_α, k_β, and k_γ are as defined for Eq. (15); k_{10} is as defined for Eq. (24); k_{12} is as defined for Eq. (25); k_{21} is as defined for Eq. (22); and k_{31} is as defined for Eq. (23).

Bioavailability of Vitamins

Both extent and rate of absorption can become characterized by kinetic approaches. Most commonly, the rate of absorption is characterized by the time necessary to achieve peak plasma concentrations (t_{max}) and by calculation of the apparent first-order absorption rate constant (k_a). The time of peak concentration is reported as the highest concentration measured or by calculation of a theoretical time as was shown in Eq. (11). However, t_{max} is influenced by the lag time between vitamin administration and onset of absorption, and the rate constant of disposition. An increase in the disposition rate constant results in earlier achievement of t_{max}.

The extent of vitamin absorption is usually judged by C_{max} in plasma, by the total area under the plasma vitamin concentration-versus-time curve (AUC), or by the amount of vitamin or metabolites excreted in urine. By using C_{max} for evaluation of the extent of absorption, two different values (measured or extrapolated value) can be used as was described for t_{max}. The AUC is usually calculated by the linear trapezoidal rule, which is the sum of the areas of trapezoids under the plasma curve [Eq. (27)].[14] It covers the area from the time of vitamin administration until the last observable or measured data point. If samples were not taken for a sufficient period of time to reach baseline levels again, several methods exist for extrapolation of the AUC for the terminal part of the curves. These methods include fitting of the remaining curve by the model chosen for curve analysis or fitting by simple exponential passing through the last observable data point.[17] However, frequently no extrapolation of the AUC over the last quantifiable concentration is necessary or such extrapolation might even introduce substantial error, if an inappropriate model was chosen. In case calculation of the AUC is needed for the time after the last sample was taken, it is most easily done by dividing the concentration value of the last data point by the terminal first-order disposition rate constant [Eq. (28)]. The extrapolated part of the AUC is then added to that calculated for the time during which samples were taken [Eq. (27)]. When estimating the bioavailability of vitamins by their plasma AUC, a correction for different doses or elimination half-lives might be necessary [Eq. (29)].[14] This is obvious from Fig. 4, where different disposition rates as determined by the route of administration influenced the AUC to a large extent.

$$\mathrm{AUC} = \sum_{i=0}^{n=1} \frac{t_{i+1} - t_i}{2} [C_{\mathrm{t}(i)} + C_{\mathrm{t}(i+1)}] \tag{27}$$

where AUC is the total area under the plasma vitamin concentration-versus-time curve (mol · hr · liter^{-1}); and t and C_{t} are as defined for Eq. (3).

$$\mathrm{AUC}_{\mathrm{extr}} = C_{\mathrm{t}(n)}/k \tag{28}$$

where $\mathrm{AUC}_{\mathrm{extr}}$ is the extrapolated part of the AUC, calculated from the time of the last plasma sample drawn until the time when the plasma vitamin concentration would have reached the baseline value (mol · hr · liter^{-1}); $C_{\mathrm{t}(n)}$ is the plasma vitamin concentration of the last sample drawn (mol/liter); and k is the apparent first-order disposition rate constant of the terminal phase of disposition (hr^{-1}).

$$\mathrm{BV} = \frac{D_{\mathrm{i.v.}}\, t_{1/2(\mathrm{i.v.})}\, \mathrm{AUC}_{\mathrm{oral}}}{D_{\mathrm{oral}} t_{1/2(\mathrm{oral})} \mathrm{AUC}_{\mathrm{i.v.}}} \cdot 100 \tag{29}$$

where BV is the bioavailability (%); $D_{\mathrm{i.v.}}$ and D_{oral} are i.v. and oral vitamin doses (mol); $t_{1/2(\mathrm{i.v.})}$ and $t_{1/2(\mathrm{oral})}$ are half-lives after i.v. and oral vitamin administration (hr); and $\mathrm{AUC}_{\mathrm{i.v.}}$ and $\mathrm{AUC}_{\mathrm{oral}}$ are the total area under plasma vitamin concentration-versus-time curve after i.v. and oral vitamin administration (mol · hr · liter^{-1}).

[17] F. Y. Bois, T. N. Tozer, W. W. Hauck, M.-L. Chen, R. Patnaik, and R. L. Williams, *Pharm. Res.* **11,** 715 (1994).

Urinary excretion data are useful when the vitamin or its metabolites are excreted in significant amounts via the kidney. The urinary excretion of riboflavin was also used to describe its zero-order kinetics of absorption.[18] The reciprocals of the urinary riboflavin excretion after different oral doses were plotted against the reciprocals of the different riboflavin doses given. Linear regression provided information about the maximum amount of riboflavin that can be absorbed from one single dose and the riboflavin dose that is half-saturating the absorption site.

Apparent Volumes of Distribution

The proportionality constant relating the vitamin concentration in blood or plasma to the amount of drug in the body has been termed the *apparent volume of distribution.* It must not be confused with the real volume of distribution, which cannot exceed total body water. The apparent volume of distribution approximates the real volume of distribution when binding to plasma proteins and tissue proteins is negligible. For most vitamins this is not the case. Vitamins that are predominantly bound to plasma proteins have apparent volumes of distribution that are smaller than their real volumes of distribution, whereas vitamins that are predominantly bound to tissue proteins have apparent volumes of distribution that are larger than their real distribution space and might exceed total body water.

The total body water accounts for approximately 60% of the body weight in normal-weight adults.[19] It comprises two major compartments; the intracellular fluid (approximately two-thirds of total body water) and the extracellular fluid (approximately one-third). The extracellular fluid is further subdivided into interstitial fluid and blood plasma. The interstitial fluid comprises three-fourths of the extracellular fluid volume.

All methods for estimating the apparent volumes of distribution require that the vitamin be given i.v. so that the amount reaching the systemic circulation is equivalent to the administered dose and known. In the case of monoexponential decline of vitamin plasma concentration (one-compartment model), the apparent volume of distribution (V) after i.v. bolus injection is calculated as shown in Eq. (30).[14] If the vitamin was administered by continuous i.v. infusion, V may be calculated from Eq. (31). It is necessary to infuse for a period long enough to attain steady state plasma concentrations.

[18] M. Mayersohn, *Eur. J. Pharmacol.* **19,** 140 (1972).
[19] R. M. Berne and M. L. Levy, "Physiology." Mosby–Year Book, St. Louis, 1993.

$$V = \frac{D}{C_0} \tag{30}$$

where V is the apparent volume of distribution (liters); D is as defined for Eq. (5); and C_0 is as defined for Eq. (3).

$$V = \frac{k_0}{k_{el} C_{ss}} \tag{31}$$

where V is as defined for Eq. (30); k_0 is the rate of vitamin infusion (mol/hr); k_{el} is as defined for Eq. (3); and C_{SS} is the vitamin plasma concentration at steady state (mol/liter).

However, most plots of vitamin concentration versus time after i.v. bolus injection must be described by multiexponential equations (multicompartment models). The apparent volume of the central compartment of distribution after i.v. bolus injection is calculated by Eq. (32).

$$V_c = \frac{D}{\Sigma C_{0n}} \tag{32}$$

where V_c is the apparent central volume of distribution (liters); D is as defined for Eq. (5); and values of C_{0n} are zero-time intercepts as defined for Eq. (14).

The volume of distribution during the terminal exponential phase of elimination (V_β) in plasma for any multicompartment model where elimination occurs from the central compartment is calculated after i.v. vitamin bolus injection from Eq. (33). V_β may also be obtained after continuous i.v. infusion [Eq. (34)].

$$V_\beta = \frac{D}{k_\beta \mathrm{AUC}} \tag{33}$$

where V_β is the apparent volume of distribution during the terminal phase of eliminaton (liters); D is as defined for Eq. (5); k_β is as defined for Eq. (14) (may be replaced by the rate constant of the terminal phase of elimination); and AUC is as defined for Eq. (27).

$$V_\beta = \frac{k_0}{k_\beta C_{ss}} \tag{34}$$

where V_β is as defined for Eq. (33); k_0 and C_{SS} are as defined for Eq. (31); and k_β is as defined for Eq. (14).

The most useful volume term to describe the apparent distribution space in a multicompartment system is the apparent volume of distribution at steady state. It is independent of vitamin elimination and can be calculated from Eq. (35) for any linear multicompartment model in which elimination occurs from the central compartment.

$$V_{ss} = \frac{D \sum_{i=1}^{n} C_{0n}/k_n^2}{(\text{AUC})^2} \tag{35}$$

where V_{SS} is the apparent volume of distribution at steady state (liters); D is as defined for Eq. (5); C_{0n} is as defined for Eq. (32); values of k_n are disposition rate constants, as defined for Eq. (14); and AUC is as defined for Eq. (27).

Clearance

The renal clearance (Cl_r) describes the amount of plasma cleared by urinary vitamin excretion per unit of time. Because binding of vitamins to macromolecules in plasma prevents them from glomerular filtration, binding must be assessed and vitamin concentrations in plasma are expressed in terms of non-protein-bound concentrations. Glomerular filterability is 98% for inulin (M_r 5500), 75% for myoglobin (M_r 17,000), and <1% for serum albumin (M_r 69,000).[19] Renal clearance is calculated by either Eq. (36) or Eq. (37).

$$\text{Cl}_r = \left(\frac{dX_u}{dt}\right)/C_m \tag{36}$$

where Cl_r is the renal clearance (ml/min); dX_u/dt is the urinary vitamin excretion rate (mol/hr); and C_m is the plasma concentration of unbound vitamin at the midpoint of the urine collection period (mol/liter).

$$\text{Cl}_r = \frac{X_u}{\text{AUC}} \tag{37}$$

where Cl_r is as defined for Eq. (36); X_u is the urinary vitamin excretion of the total observation period (mol); and AUC is the total area under the plasma unbound vitamin concentration-versus-time curve ($\text{mol} \cdot \text{hr} \cdot \text{liter}^{-1}$).

The use of Eq. (36) requires information about the plasma concentration at the midpoint of the urine collection period. Usually, plasma samples are not obtained exactly at the midpoints of the urine-sampling intervals. In this case, the required plasma concentrations are extrapolated by using the models described above for plasma curve analysis. This procedure for calculation of Cl_r is more laborious than using Eq. (37), but it provides clearance data over a wide range of plasma concentrations. This was of interest for riboflavin,[20] which undergoes both tubular secretion and reabsorption. The reabsorption process was saturated at lower plasma concentrations than was the secretion, resulting in increased clearances at higher concentrations.

Usually, Cl_r is expressed in units of milliliters per minute (i.e., it is necessary to transform the calculated values) (units of liters/hr are obtained by use of the above equations). Because Cl_r refers to units of volume, it is necessary to correct it for the different body sizes of subjects or species. Frequently, Cl_r is expressed referring to a body surface area of 1.73 m^2 or per kilogram body weight.

The systemic clearance describes the sum of all processes (e.g., renal and biliary excretion, metabolism) that contribute to the removal of vitamin from blood plasma. Its calculation requires data obtained from i.v. administration [Eq. (38)]. It should be expressed in units of milliliters per minute, corrected for body surface area or body weight. Vitamin plasma concentration is not corrected for protein binding. The difference between renal and systemic clearance comprises all elimination processes not related to urinary excretion and is termed the *nonrenal clearance*.

$$Cl_{sys} = \frac{D}{AUC} \tag{38}$$

where Cl_{sys} is the systemic clearance (ml/min); D is the i.v. administered vitamin dose (mol); and AUC is as defined for Eq. (27).

Rate Constants for Renal Vitamin Excretion

Urinary vitamin excretion rates are estimated by collecting single urine voids after vitamin administration, calculating the amount of compound excreted with each void (volume times concentration), and dividing the amount excreted by the collection duration of the corresponding void. An individual collection period should not exceed one biologic half-life of the vitamin. The values obtained (amount of vitamin excreted per hour) are plotted against the midpoints of the urine collection periods (Fig. 5) and the rate constants for renal excretion are calculated. For this purpose, it is sufficient to perform a least-squares nonlinear regression according to Eq. (3), (14), or (15) for the time after onset of the pure elimination process (straight line after log transformation, indicated by the arrow in Fig. 5).

[20] W. J. Jusko and G. Levy, *J. Pharm. Sci.* **59,** 765 (1970).

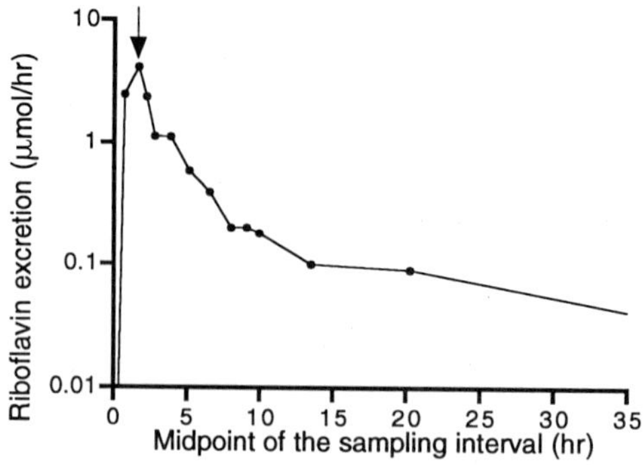

FIG. 5. Urinary excretion of riboflavin in a healthy male subject after the oral administration of 20 mg of riboflavin. The arrow indicates the onset of pure elimination processes. [Data from J. Zempleni, J. R. Galloway, and D. B. McCormick, *Am. J. Clin. Nutr.* **63,** 54 (1996).]

Enterohepatic Recycling of Vitamins

Frequently, the existence of enterohepatic recycling can be detected by using pharmacokinetic methods. If such a cycle exists, the vitamin plasma curve might show two peak concentrations (Fig. 6). The first peak results from the absorption of the administered test dose, whereas the second peak is caused by biliary excretion of a fraction of this test dose and reabsorption following emptying of the gall bladder. Such a concentration profile may

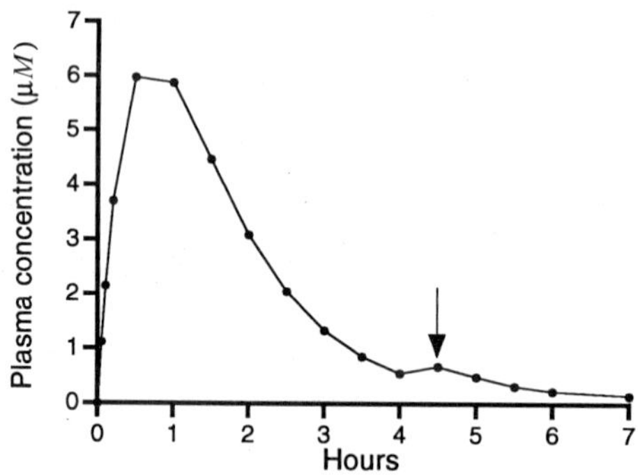

FIG. 6. Fictive plasma concentration-versus-time curve, calculated for a substance that is subject to enterohepatic recycling. The following assumption were made: one-compartment open model, dose administered was 0.5 mmol, no lag time between vitamin administration and onset of absorption, first-order absorption rate constant was 2.1 hr^{-1}, absorption was 100%, volume of distribution was 42 liters, first-order elimination rate constant was 0.9 hr^{-1}, biliary excretion contributes 5% to overall elimination. Emptying of the gall bladder was assumed to take place 4 hr after vitamin administration.

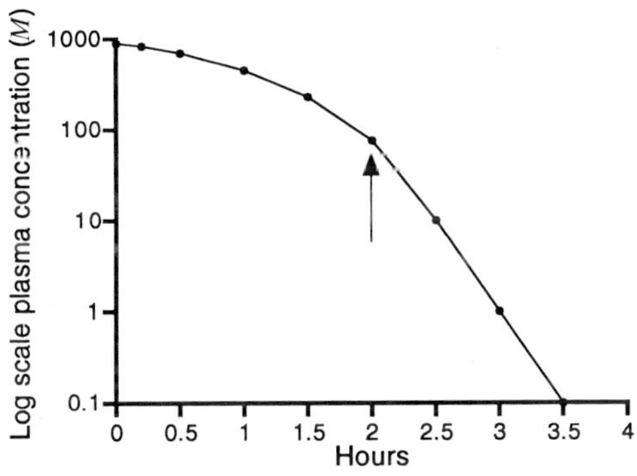

FIG. 7. Log-transformed plasma concentration-versus-time curve of a fictive compound that is subject to capacity-limited elimination processes. The arrow indicates the onset of first-order kinetics ($C_t \leq K_m$).

even be reproduced from urinary data, as was reported for riboflavin.[21] Similar findings were made for 7α-hydroxyriboflavin in plasma.[22]

Nonlinear Kinetics

The biokinetics of several vitamins were adequately described by first-order processes. However, some vitamins may have nonlinear absorption or distribution characteristics[3,18] or they are eliminated from the body in a nonlinear fashion.[22] Frequently, the kinetic of these compounds can be described by the Michaelis–Menten equation [Eq. (39)].[14]

$$-\frac{dC}{dt} = \frac{V_m C}{K_m + C} \tag{39}$$

where $-dC/ct$ is the rate of decline of vitamin concentration at time t ($\text{mol} \cdot \text{liter}^{-1} \cdot \text{h}^{-1}$); V_m is the theoretical maximum rate of the process ($\text{mol} \cdot \text{liter}^{-1} \cdot \text{h}^{-1}$); t is as defined for Eq. (3); and K_m is the Michaelis constant (mol/liter).

To assess whether a vitamin possesses nonlinear kinetic properties, a series of single doses of varying size should be administered. The extent of absorption, disposition rate constants, etc., might be altered as determined by the dose. However, even when nonlinear kinetics apply, the terminal slope of the plasma curve ($C_t \ll K_m$) follows first-order kinetics (Fig. 7). This terminal slope might be under the detection limit of the

[21] W. J. Jusko and G. Levy, *J. Pharm. Sci.* **56,** 58 (1967).

[22] J. Zempleni, J. R. Galloway, and D. B. McCormick, *Int. J. Vitam. Nutr. Res.* **66,** 151 (1996).

method used for vitamin assay. For a vitamin that is eliminated by a single capacity-limited process, K_m and V_m can be estimated from plasma concentration data in the postabsorptive–postdistributive phase. Such estimates require the determination of the rate of change of the plasma concentration from one sampling time to the next, dC/dt, as a function of the plasma concentration C_m at the midpoint of each sampling interval. The data are usually plotted according to one of the linearized forms of the Michaelis–Menten equation, such as the Lineweaver–Burk expression [Eq. (40)]. A plot of the reciprocal of dC/dt versus the reciprocal of C_m yields a straight line with the intercept $1/V_m$ and slope K_m/V_m. Alternative approaches are available.[14]

$$\frac{1}{dC/dt} = \frac{K_m}{V_m C_m} + \frac{1}{V_m} \tag{40}$$

TABLE I
BIOKINETIC PARAMETERS OF WATER-SOLUBLE VITAMINS[a,b]

Vitamin administered	k_a	k_α	k_β	k_γ	V_c	V_{ss}
Ascorbic acid						
i.v.	—	1.313	0.265	—	0.21	0.62
Orally	0.30	0.240	—	—	—	—
Pyridoxine hydrochloride						
i.v.	—	6.049	—	—	0.60	—
Orally	2.10	0.913	—	—	—	—
Riboflavin						
i.v.	—	6.892	0.686	—	0.37	0.61
Orally	0.64	1.236	0.174	—	—	—
Thiamin hydrochloride						
i.v.	—	4.557	0.561	0.012	—	—
Orally	6.60					

[a] From J. Gehler and W. Kübler, *Int. Z. Vitam. Forsch.* **40,** 454 (1970); W. Kübler and J. Gehler, *Int. Z. Vitam. Forsch.* **40,** 442 (1970); J. Zempleni, *J. Am. Coll. Nutr.* **14,** 579 (1995); J. Zempleni, J. R. Galloway, and D. B. McCormick, *Am. J. Clin. Nutr.* **63,** 54 (1996); J. Zempleni, M. Hagen, S. Vogel, U. Hadem, and W. Kübler, *Nutr. Res.* **16,** 1479 (1996); and C. M. E. Tallaksen, A. Sande, T. Bøhmer, H. Bell, and J. Karlsen, *Eur. J. Clin. Nutr. Pharmacol.* **44,** 73 (1993).

[b] For rate constants, data given in hr^{-1}. k_a, Apparent first-order absorption rate constant; k_α, k_β, and k_γ, apparent first-order disposition rate constants (fast and slow phases); in case a one-compartment model applies, only one rate constant is calculated and should be termed apparent first-order elimination rate constant (k_{el}). Data given in liters/kg body weight for V_c, central volume of distribution, and V_{ss}, distribution space under steady-state conditions. i.v., Intravenously.

where dC/dt is the rate of change of the plasma concentration from one sampling time to the next ($mol \cdot liter^{-1} \cdot hr^{-1}$); K_m and V_m are as defined for Eq. (39); and C_m is the plasma concentration at the midpoint of each sampling interval (mol/liter).

For certain vitamins that exhibit parallel capacity-limited and first-order elimination, it may be possible to administer sufficiently high i.v. doses so that initial vitamin concentrations are substantially larger than the K_m. Under these conditions methods are available for calculation of K_m, V_m, and the apparent first-order disposition rate constant.[14] They are too complex to be discussed here. However, if the initial plasma concentrations are not high enough, the apparent first-order disposition rate constant will be overestimated, which produces errors in the estimates of V_m and K_m. Moreover, care must be taken when using the plasma AUC for estimation of bioavailability when zero-order elimination kinetics apply. The AUC increases more than proportionally with an increase in dose.

Biokinetic Parameters Obtained for Water-Soluble Vitamins

Table I shows biokinetic data that have been reported for water-soluble vitamins. Although no complete sets of kinetic parameters are provided, Table I gives an idea of the magnitude of the processes described.

[46] *In Situ* Kinetics: An Approach to Recommended Intake of Vitamin C

By MARK LEVINE, STEVEN C. RUMSEY, YAOHUI WANG, JAE PARK, ORAN KWON, and NOBUYUKI AMANO

Introduction

Recommendations for nutrient intakes promise to be one of the most powerful tools in medicine for promoting health and preventing disease. To fulfill this potential, recommendations should be for optimal or ideal intake.[1,2] But what does optimal intake really mean?[3] We believe that key

[1] M. Levine, K. R. Dhariwal, P. W. Washko, J. D. Butler, R. W. Welch, Y. H. Wang, and P. Bergsten, *Am. J. Clin. Nutr.* **54,** 1157S (1991).

[2] M. Levine, K. R. Dhariwal, R. W. Welch, Y. Wang, and J. B. Park, *Am. J. Clin. Nutr.* **62**(Suppl.), 1347S (1995).

[3] M. Levine, *N. Engl. J. Med.* **314,** 892 (1986).

challenges for nutrition scientists are to define optimal nutrient intakes, to develop methods to determine them, and to test these methods.

Given the challenges, perhaps there is some comfort in remembering that the problems are not new. Nutrition scientists long ago recognized that vitamin amounts that prevented deficiency might differ from amounts for ideal health.

> Our conception of what constitutes an adequate intake of vitamin . . . is the crude determination of growth maintenance and absence of deficiency. We have no way of estimating the maximum need for normal physiologic activity. . . . The optimum requirement of vitamin may be greatly in excess of the accepted physiological requirement . . . (Perla and Marmortston[4,4a])

Although written about vitamins A and thiamin, the concepts apply to many vitamins. To address the problem of optimal vitamin intake, we use vitamin C as a model vitamin. Like vitamin A and thiamin, concepts for vitamin C are applicable to other nutrients.

One early concept of the ideal amount of a nutrient was that amount that prevented infection or promoted survival in the face of altered homeostasis.[4–6] Vitamin amounts that prevented deficiency symptoms in control animals were not sufficient for infected animals, using survival or absence of infection as an end point. Years later, gram doses of vitamin C were proposed to prevent and treat the common cold.[7] The rationale was based on the earlier original concept that nutrient intake could influence infection outcome. Indeed, experiments to test this concept about vitamin C in humans were first published in 1942, and were the basis for claims about the vitamin 30 years later.[7,8] The inconclusiveness of these experiments is an example of the difficulty in using disease outcome by itself as an end point without additional knowledge of the molecular pathophysiology.

Besides infection, other perturbations of homeostasis were used to test nutrient recommendations. The concept was that "stress" could influence nutrient needs, and that stress could reveal otherwise occult needs.[4,4a,9]

[4] D. Perla and J. Marmortston, The effect of vitamin A on resistance. *In* "Natural Resistance and Clinical Medicine," p. 926. Little, Brown, Boston, 1941.

[4a] D. Perla and J. Marmortston, The effect of vitamin B and resistance. *In* "Natural Resistance and Clinical Medicine," p. 1011. Little, Brown, Boston, 1941.

[5] L. I. Kryzhanovskaya, E. S. London, and F. I. Rivosh, *J. Physiol.* **24,** 212 (1938).

[6] C. C. Torrance, *J. Biol. Chem.* **132,** 575 (1940).

[7] L. Pauling, "Vitamin C, the Common Cold, and the Flu." W. H. Freeman, San Francisco, 1976.

[8] R. W. Cowan, H. S. Diehl, and A. B. Baker, *J. Am. Med. Assoc.* **120,** 1268 (1942).

[9] C. L. Pirani, *Metabolism* **1,** 197 (1952).

Stress events used to test ascorbate needs included physical exertion, incidence of bleeding gums, heat tolerance, cold tolerance, wound healing, and altitude acclimatization.[10] Investigations were inconclusive, in part because the mechanism of ascorbate effect was unclear, because the end-point measurements may not have directly related to the effect of ascorbate, or because the measurements were subjective. Survival is another end point used to test ideal nutrient requirements.[4,4a,11,12] Although survival can be measured simply in animal experiments, survival in humans is a vastly more complex outcome. To test human recommendations, what are necessary are end points or outcomes that are related mechanistically to nutrient amount.

The ability to make nutrient recommendations is further complicated by interactions among different nutrients. Nutrients act in concert such that the effect of one nutrient alone might be difficult to detect in animals or clinically, except when frank deficiency occurs.[13] In retrospect this is not surprising, because many different nutrients may be needed in a single biosynthetic pathway. Nevertheless, recommendations for optimal nutrient intake have been hampered because effects of single nutrient changes may be difficult to measure unless direct action of the nutrient is tested.

In Situ Kinetics

The central problem remains: How do we measure what is ideal? What is an indication of "optimal"?[3]

We proposed that ideal vitamin intake should be based on how different vitamin concentrations affect vitamin function.[2,3] Detection of nutrient action should be targeted to specific biochemical and molecular action of the nutrient. Vitamin function should be studied *in situ* (i.e., in position). *In situ* refers to organelles, cells, animals, or ideally humans. Because studies on humans are difficult, experiments with normal human tissue may provide useful information. *In situ* studies are essential, because vitamin function in living tissue may differ from function in isolated reactions or with isolated proteins.[3,14,15] The goal is to determine vitamin function in relation to different nutrient concentrations, and therefore this approach to optimal vitamin ingestion is termed *in situ* kinetics.

[10] M. I. Irwin and B. K. Hutchins, *J. Nutr.* **106,** 821 (1976).

[11] D. Perla and J. Marmorston, *Arch. Pathol.* **23,** 543 (1937).

[12] J. E. Enstrom, L. E. Kanim, and M. A. Klein, *Epidemiology* **3,** 194 (1992).

[13] R. J. Williams, What are the basic principles which underlie nutritional science? *In* "Physician's Handbook of Nutritional Science," pp. 20–25. Charles C Thomas, Springfield, Illinois, 1975.

[14] K. R. Dhariwal, P. Washko, W. O. Hartzell, and M. Levine, *J. Biol. Chem.* **264,** 15404 (1989).

[15] K. R. Dhariwal, C. D. Black, and M. Levine, *J. Biol. Chem.* **266,** 12908 (1991).

TABLE I
In Situ KINETICS

Biochemical component: Vitamin biochemical and molecular function in relation to vitamin concentration
Assay for vitamin
Ability to have different vitamin concentrations *in situ:* Transport, depletion, repletion
Distribution of vitamin *in situ*
Assay for vitamin function
Determination of vitamin function *in vitro* and *in situ*
Localization of vitamin function *in situ* and relationship to vitamin distribution
Specificity of vitamin function *in vitro* and *in situ*
Vitamin function in relation to different vitamin concentrations *in situ*
Clinical component: Achieving effective vitamin concentrations in humans
Availability of vitamin in diet
Steady-state vitamin concentration in plasma as function of dose
Steady-state vitamin concentration in tissues as function of dose
Vitamin bioavailability
Vitamin excretion
Vitamin safety and adverse effects
Beneficial effects in relation to dose: Direct effects and epidemiologic observations

In situ kinetics has biochemical and clinical components (Table I).[1,2] The biochemical component determines how vitamin concentration affects vitamin biochemical or molecular function *in situ*. The clinical component determines how the effective vitamin concentrations are achieved in people.

In Situ Kinetics: Biochemical Component

The goal of the biochemical component of *in situ* kinetics is to determine the kinetics of many vitamin-dependent reactions *in situ*. With kinetics available, optimal vitamin ingestion would be that amount that produced maximal function, or V_{max}, of different reactions, but with no toxicity.[1] The reactions at V_{max} would have to confer benefit to the host with no adverse consequences. On the steep portion of a kinetic curve, small changes in substrate concentration produce large changes in reaction rate. For this reason, the steep portion of a kinetic curve should be avoided for nontoxic substrates or reaction products that are not harmful. This is of particular relevance to exogenously ingested substances such as vitamins, in contrast to endogenously synthesized compounds. Although theoretically it might be advantageous for a vitamin-dependent reaction not to proceed at V_{max} because of adverse consequences, this is probably not the case for vitamin C as discussed below.

Vitamin C acts as an electron donor in enzymatic and chemical reactions.[2] Vitamin C is required by at least eight isolated enzymes and is also

TABLE II
ASCORBIC ACID AND BIOCHEMICAL FUNCTION

Enzymatic reactions	Chemical reactions
Procollagen-proline dioxygenase (EC 1.14.11.2)	Intracellular
Procollagen-proline dioxygenase (EC 1.14.11.7)	Quenching of reactive oxidants
Procollagen-lysine 5-dioxygenase (EC 1.14.11.4)	Iron–ferritin interaction
γ-Butyrobetaine dioxygenase (EC 1.14.11.1)	Extracellular
Trimethyllysine dioxygenase (EC 1.14.11.8)	Quenching of reactive oxidants
Dopamine β-monooxygenase (EC 1.14.17.1)	Electron transfer to oxidized tocopherol
Peptidylglycine monooxygenase (EC 1.14.17.3)	Prevention of LDL oxidation
4-Hydroxylphenylpyruvate dioxygenase (EC 1.13.11.27)	Iron absorption in gastrointestinal tract

necessary for maximal enzyme activity in cells or tissues (Table II). Vitamin C is also utilized in nonenzymatic reactions, for example, to quench reactive oxidants or to reduce ferric iron[16–18] (Table II). Kinetics *in situ* were described for norepinephrine biosynthesis, ascorbate transmembrane electron transfer, and inhibition of low-density lipid (LDL) oxidation, but are unknown for most of the other reactions.[14,19,20] With kinetics available, it becomes possible to predict the concentration of vitamin that regulates a given function. Ideal vitamin intake would be that amount of vitamin that yields V_{max} values for different functions that confer benefit, and with no adverse consequences.[1] Vitamin C-dependent biochemical functions that would confer benefit at maximal activity include proline and lysine hydroxylation for wound healing, oxidant quenching intracellularly and extracellularly, carnitine biosynthesis for ATP generation, and perhaps biosynthesis of some hormones.

It is likely that there are no directly toxic vitamin C-dependent biochemical functions. Ascorbate was proposed to accelerate oxidant damage under some conditions in the presence of free cations, especially iron.[21] However, the cation concentrations used were excessive, and because cations are bound and not free *in vivo* these conditions could not occur. There are no reports of vitamin C-associated toxicity from retroperitoneal fibrosis or other conditions of excess collagen synthesis. There is no evidence that

[16] B. Frei, L. England, and B. N. Ames, *Proc. Natl. Acad. Sci. U.S.A.* **86,** 6377 (1989).
[17] I. Jialal and S. M. Grundy, *Circulation* **88,** 2780 (1993).
[18] K. E. Hoffman, K. Yanelli, and K. R. Bridges, *Am. J. Clin. Nutr.* **54,** 1188S (1991).
[19] K. R. Dhariwal, M. Shirvan, and M. Levine, *J. Biol. Chem.* **266,** 5384 (1991).
[20] I. Jialal, G. L. Vega, and S. M. Grundy, *Atherosclerosis* **82,** 185 (1990).
[21] V. Herbert, S. Shaw, E. Jayatilleke, and T. Stopler-Kasdan, *Stem Cells (Dayt.)* **12,** 289 (1994).

high vitamin C concentrations act as feedback inhibitors in the enzymatic reactions in which it is involved. There is no known harm in synthesizing norepinephrine, because hypertension is not caused by enhanced catecholamine synthesis unless the etiology of hypertension is pheochromocytoma.[22,23]

There are also few adverse effects of vitamin C clinically. Vitamin C is remarkably nontoxic, as reviewed elsewhere.[2,24] Briefly, vitamin C will enhance iron absorption by maintaining iron in its reduced form. Enhanced absorption can be inadvertently harmful in patients who are iron overloaded, such as those with thalassemia major, hemochromatosis, and sideroblastic anemia. However, there are no data indicating that iron overabsorption occurs from vitamin C in healthy people or in people who are heterozygous for hemochromatosis.[25] Indeed, available data show that ascorbate does not increase iron stores in healthy volunteers.[26]

As biochemical examples, vitamin C may regulate the enzymatic reactions of proline and lysine hydroxylation for collagen biosynthesis *in situ*. In amount of vitamin C providing V_{max} for proline hydroxylation would be beneficial *in situ*, because of maximally hydroxylated collagen for wound healing and for maintenance of connective tissue. V_{max} and its associated vitamin C concentration would appear to have no adverse consequences except perhaps in those rare patients who have iron overabsorption. A similar case exists for the nonenzymatic reactions in which oxidants are quenched by vitamin C. Quenching may occur in LDL oxidation, in oxidant-mediated DNA damage, and in neutrophils activated to kill bacteria.[17,20,27–29] Again, V_{max} would confer benefit without apparent harm.

To test whether the biochemical component of *in situ* kinetics could actually be determined experimentally, a vitamin C-dependent enzymatic reaction was selected. The reaction was norepinephrine biosynthesis from the substrate dopamine, mediated by the enzyme dopamine β-monooxygenase and vitamin C[14,19,30,31] An enzymatic rather than chemical reaction was

[22] M. D. Esler, *Bailleres Clin. Endo. Metab.* **7,** 415 (1993).

[23] S. S. Werbel and K. P. Ober, *Med. Clin. North Am.* **79,** 131 (1995).

[24] A. Bendich and L. Langseth, *J. Am. Coll. Nutr.* **14,** 124 (1995).

[25] M. Levine, S. C. Rumsey, Y. Wang, J. K. Park, O., W. Xu, and N. Amano, Vitamin C. *In* "Present Knowledge in Nutrition" (L. J. Filer and E. E. Ziegler, eds.). International Life Sciences Institute, Washington, D.C., 1996.

[26] J. D. Cook and E. R. Monsen, *Am. J. Clin. Nutr.* **30,** 235 (1977).

[27] B. Frei, *Am. J. Clin. Nutr.* **54,** 1113S (1991).

[28] C. G. Fraga, P. A. Motchnik, M. K. Shigenaga, H. J. Helbock, R. A. Jacob, and B. N. Ames, *Proc. Natl. Acad. Sci. U.S.A.* **88,** 11003 (1991).

[29] P. W. Washko, Y. Wang, and M. Levine, *J. Biol. Chem.* **268,** 15531 (1993).

[30] M. Levine, K. Morita, E. Heldman, and H. B. Pollard, *J. Biol. Chem.* **260,** 15598 (1985).

[31] M. Levine, *J. Biol. Chem.* **261,** 7347 (1986).

initially chosen because an enzymatic reaction was likely to confer greater specificity for a reducing substance than a chemical reaction. The effect of varying endogenous concentrations of vitamin C on norepinephrine biosynthesis was investigated in bovine chromaffin cells and chromaffin secretory vesicles from adrenal medulla. These data indicated that kinetics could be determined *in situ*. For *in situ* kinetics to be valid, at least one reaction should occur at V_{max} *in situ*, and vitamin C-mediated norepinephrine biosynthesis proceeded at V_{max} *in situ*.[14] The experiments also showed that the action of vitamin C was different *in situ* compared to the isolated enzyme dopamine β-monooxygenase. Dopamine β-monooxygenase as an isolated enzyme simply accepts single electrons from ascorbate in solution.[32,33] By contrast, transmembrane electron transfer occurs *in situ* from vitamin C on different sides of the chromaffin granule membrane, mediated by cytochrome *b*-561.[34,35] Kinetics for norepinephrine biosynthesis *in situ*, but not for isolated enzyme, involve both the K_m of ascorbate for dopamine β-monooxygenase and the K_m of ascorbate for transmembrane electron transfer.[14,15]

The principles of the biochemical component of *in situ* kinetics evolved from these experiments, and are straightforward (Table I). A tissue is selected with a known or possible function for the vitamin. The function can be enzymatic or nonenzymatic. There must be a sensitive and specific means to measure the vitamin, and to adjust its concentration in the tissue over a wide range. Thus, knowledge of substrate transport and tissue distribution is necessary. There must also be a means to measure the reaction mediated by the vitamin *in situ*, and the reaction specificity for the vitamin must be determined. Using this information, reaction activity *in situ* as a function of vitamin concentration *in situ* can then be determined.

The norepinephrine biosynthesis system was the first example of the principles of *in situ* kinetics. Although this system was an excellent model, it had two disadvantages. For *in situ* kinetics, it is beneficial if the reaction selected is important physiologically. In the case of norepinephrine biosynthesis, it is unclear whether there is benefit to be at V_{max}, even though the reaction probably proceeds at V_{max} in situ.[14,36,37] The norepinephrine biosynthesis system was studied in animal tissue. For human vitamin C requirements, data from human tissue are needed. Goals for biochemical *in situ* kinetics are to utilize human tissue to study reactions whose benefits

[32] S. Friedman and S. Kaufman, *J. Biol. Chem.* **240,** 4763 (1965).
[33] S. Dhawan, L. T. Duong, R. L. Ornberg, and P. J. Fleming, *J. Biol. Chem.* **262,** 1869 (1987).
[34] P. J. Fleming and U. M. Kent, *Am. J. Clin. Nutr.* **54,** 1173S (1991).
[35] D. Njus, V. Jalukar, J. A. Zu, and P. M. Kelley, *Am. J. Clin. Nutr.* **54,** 1179S (1991).
[36] D. Robertson, C. Beck, T. Gary, and M. Picklo, *Int. Angiol.* **12,** 93 (1993).
[37] D. Elmfeldt, L. Elvelin, and M. Nordlander, *Blood Press* **3,** 356 (1994).

are straightforward, and to determine vitamin C function in tissues where it is currently unknown.

In Situ Kinetics: Clinical Component

The clinical goals of *in situ* kinetics are to determine how the relevant vitamin concentrations are achieved in humans and what clinical effects occur from these vitamin concentrations. Clinical *in situ* kinetics has seven parts (Table I). Several aspects are described in more detail in Ref. 37a and elsewhere.[2,25,38]

The vitamin should be available for consumption in foods.[2] Vitamin C is available in fruits and vegetables.[39] If fruit and vegetable ingestion is five servings daily, vitamin C intake will be greater than 210 mg.[40] If intake approaches nine fruits and vegetables as recommnded by some, intake will be greater than 300 mg daily. Although vitamin C is available in the diet, less than 28% of most Americans consume five fruits and vegetables daily.[41] Twenty to thirty percent of Americans consume less than 60 mg of vitamin C daily, the equivalent of less than two fruits and vegetables.[42–44]

Steady-state vitamin concentrations in plasma in relationship to dose must be determined. These data were experimentally difficult to obtain. Until recently the data were incomplete because of limitations of doses, assays, and study designs.[2,3,45–62] A new NIH study described the relation-

[37a] M. Levine, S. Rumsey, and Y. Wang, *Methods Enzymol.* **279,** [6], 1997.

[38] M. Levine, C. Conry-Cantilena, Y. Wang, R. W. Welch, P. W. Washko, K. R. Dhariwal, J. B. Park, A. Lazarev, J. Graumlich, J. King, and L. R. Cantilena, *Proc. Natl. Acad. Sci. U.S.A.* **93,** 3704 (1996).

[39] D. Haytowitz, *J. Nutr.* **125,** 1952 (1995).

[40] P. Lachance and L. Langseth, *Nutr. Rev.* **52,** 266 (1994).

[41] I. B. Life Sciences Research Office, "Third Report on Nutrition Monitoring in the United States," ES-14. U.S. Government Printing Office, Washington, D.C., 1995.

[42] J. P. Koplan, J. L. Annest, P. M. Layde, and G. L. Rubin, *Am. J. Public Health* **76,** 287 (1986).

[43] B. H. Patterson, G. Block, W. F. Rosenberger, D. Pee, and L. L. Kahle, *Am. J. Public Health* **80,** 1443 (1990).

[44] S. P. Murphy, D. Rose, M. Hudes, and F. E. Viteri, *J. Am. Diet. Assoc.* **92,** 1352 (1992).

[45] E. M. Baker, J. C. Saari, and B. M. Tolbert, *Am. J. Clin. Nutr.* **19,** 371 (1966).

[46] E. M. Baker, R. E. Hodges, J. Hood, H. E. Sauberlich, and S. C. March, *Am. J. Clin. Nutr.* **22,** 549 (1969).

[47] R. E. Hodges, E. M. Baker, J. Hood, H. E. Sauberlich, and S. C. March, *Am. J. Clin. Nutr.* **22,** 535 (1969).

[48] E. M. Baker, R. E. Hodges, J. Hood, H. E. Sauberlich, S. C. March, and J. E. Canham, *Am. J. Clin. Nutr.* **24,** 444 (1971).

[49] R. E. Hodges, J. Hood, J. E. Canham, H. E. Sauberlich, and E. M. Baker, *Am. J. Clin. Nutr.* **24,** 432 (1971).

[50] A. Kallner, D. Hartmann, and D. Hornig, *Am. J. Clin. Nutr.* **32,** 530 (1979).

[51] M. Mayersohn, *Eur. J. Pharmacol.* **19,** 140 (1972).

[52] G. Zetler, G. Seidel, C. P. Siegers, and H. Ivens, *Eur. J. Clin. Pharmacol.* **10,** 273 (1976).

[53] P. J. Garry, J. S. Goodwin, W. C. Hunt, and B. A. Gilbert, *Am. J. Clin. Nutr.* **36,** 332 (1982).

ship between dose and steady-state plasma concentrations in seven healthy men, each of whom was hospitalized for 5–6 months.[38] These data indicated that there was a steep sigmoidal relationship between dose and steady-state plasma concentration. The first dose beyond the sigmoid portion of the curve was 200 mg daily, producing a fasting plasma concentration of approximately 65 μM. This concentration was more than 80% of the maximal plasma concentration of 80 μM achieved at the 1000-mg daily dose. At the highest dose of 2500 mg daily there was no further increase in fasting plasma values. At all doses there was a transient increase in plasma values immediately after ingestion, which rapidly returned to the fasting value.

Steady-state vitamin concentrations in tissue in relation to dose must also be determined. These data were also previously limited.[28,55,59] New data were obtained as part of the NIH study. Vitamin C in the neutrophils, monocytes, and lymphocytes of the patients was accumulated to millimolar concentration. Transport saturated at the 100-mg daily dose, with no further increases in intracellular concentration even at 2500 mg daily.[38]

Bioavailability and urinary excretion are two additional parts of clinical *in situ* kinetics. True bioavailability and urine excretion data have not been available for vitamin C because of the need for frequent sampling, prior assay limitations, and the necessity for subjects to be at steady state for the tested dose. True bioavailability data from the NIH study showed that vitamin C was completely absorbed at single doses of 200 mg, but bioavailability declined at 500 mg and 1250 mg. As bioavailability declined, urine excretion increased. Virtually all of the absorbed doses of 500 and 1250 mg were excreted unchanged in urine.[38]

Vitamin safety must be characterized for *in situ* kinetics. Prior studies indicated that vitamin C is quite safe, although some adverse effects were described for select patient subsets.[2,24,25] In addition to enhanced iron absorption discussed above, ascorbate doses of 500 mg and higher enhanced oxalate excretion in patients with known oxalate nephrolithiasis.[63,64] In the

[54] S. L. Melethil, W. E. Mason, and C.-J. Chiang, *Int. J. Pharm.* **31,** 83 (1986).
[55] R. A. Jacob, J. H. Skala, and S. T. Omaye, *Am. J. Clin. Nutr.* **46,** 818 (1987).
[56] D. J. VanderJagt, P. J. Garry, and H. N. Bhagavan, *Am. J. Clin. Nutr.* **46,** 290 (1987).
[57] J. Blanchard, K. A. Conrad, R. A. Mead, and P. J. Garry, *Am. J. Clin. Nutr.* **51,** 837 (1990).
[58] J. Blanchard, *J. Nutr.* **121,** 170 (1991).
[59] R. A. Jacob, F. S. Pianalto, and R. E. Agee, *J. Nutr.* **122,** 1111 (1992).
[60] P. W. Washko, R. W. Welch, K. R. Dhariwal, Y. Wang, and M. Levine, *Anal. Biochem.* **204,** 1 (1992).
[61] M. Levine, C. C. Cantilena, and K. R. Dhariwal, *World Rev. Nutr. Diet.* **72,** 114 (1993).
[62] H. Heseker and R. Schneider, *Eur. J. Clin. Nutr.* **48,** 118 (1994).
[63] M. Urivetzky, D. Kessaris, and A. D. Smith, *J. Urol.* **147,** 1215 (1992).
[64] T. R. Wandzilak, S. D. D'Andre, P. A. Davis, and H. E. Williams, *J. Urol.* **151,** 834 (1994).

NIH study, oxalate and urate excretion were increased at 1000 mg daily compared to lower doses. The increased excretion, while statistically significant, was still small. It is not known if there would be any impact from these small increases on formation of kidney stones. The issue would be of concern if 1000 mg produced large increases in vitamin C concentration in plasma and tissues. However, cells were saturated at 100 mg daily, and plasma was almost completely saturated by 400 mg daily. Because there was no advantage of higher doses with respect to saturation of body stores, the possibility of adverse effects of ascorbate at 1000 mg is probably of little consequence. It remains possible that there are other advantages of higher doses of ascorbate, either as unabsorbed vitamin C in the gastrointestinal tract or as excess vitamin C in the urinary excretion system.[65–67] However, these advantages are unproven and remain to be characterized.

In addition to harmful effects, beneficial clinical effects from the ingested vitamin should be described. One way to do so would be to characterize a clinical effect directly related to the biochemical or molecular action of the vitamin. What would be tested would be the effect of different plasma concentrations of vitamin C on a specific clinical outcome related to vitamin C biochemical action. Unfortunately, such data are not yet available. These studies are dependent on prior knowledge of the relationship between dose and steady-state concentrations in healthy people. Once these relationships are understood, we can proceed to study the effects of vitamin C on clinical outcomes that are directly related to biochemical or molecular action of the vitamin.

Another means to characterize potentially beneficial clinical effect of vitamins is by epidemiologic studies. These studies can be observational or interventional. Observational studies report associations between ingestion and outcome. The true cause of the observed effect is not certain. Interventional studies report outcomes between groups that are matched and whose only difference is whether the substance is administered.

For both study types, the effects of vitamin C are controversial regarding colorectal adenoma, cancer, coronary heart disease, stroke, cataract, and overall survival.[12,68–78] There are several explanations for the discrepancies.

[65] S. K. Hetey, M. L. Kleinberg, W. D. Parker, and E. W. Johnson, *Am. J. Hosp. Pharm.* **37,** 235 (1980).

[66] M. A. Helser, J. H. Hotchkiss, and D. A. Roe, *Carcinogenesis* **13,** 2277 (1992).

[67] J. Sobel and D. Kaye, Urinary tract infections. *In* "Principles and Practice of Infectious Diseases" (G. L. Mandell, R. G. J. Douglas and J. E. Bennett, eds.), p. 594. Churchill Livingstone, New York, 1990.

[68] R. Byers and N. Guerrieri, *Am. J. Clin. Nutr.* **62**(Suppl.), 1385S (1995).

[69] W. J. Blot, J.-Y. Li, P. R. Taylor, W. Guo, S. Dawsey, G. Q. Wang, C. S. Yang, S. F. Zheng, M. Gail, and G. Y. Li, *et al., J. Natl. Cancer Inst.* **85,** 1483 (1993).

[70] J. E. Enstrom, L. E. Kanim, and L. Breslow, *Am. J. Public Health* **76,** 1124 (1986).

The observed effect of vitamin C may not have been related closely enough to the expected biochemical action to test benefit. In many of these studies, plasma and tissue concentrations of vitamin C were unknown. In at least some of these studies, it is quite possible that vitamin C ingestion in control subjects was beyond the steep portion of the dose–concentration curve for vitamin C.[38] In these cases, vitamin C plasma and tissue concentrations in the control subjects would be close to saturation. Effects of additional vitamin C would not be expected. In many studies ingestion is calculated using food-recall surveys, which are often inaccurate.[79] Because there is a steep relationship between dose and concentration at low doses, food surveys alone may be a poor indicator of true plasma concentrations at ingestion <200 mg. In epidemiologic studies, especially observational ones, it is always possible that other factors account for the observed event. Many more observational epidemiologic studies are available for vitamin C compared to intervention studies.

The most consistent epidemiologic data for vitamin C are those regarding fruit and vegetable ingestion. Diets with high fruit and vegetable ingestion are associated with lower cancer risk, in particular for gastrointestinal cancers. These data are abundant and consonant.[68] Five servings of fruits and vegetables appears to be protective, and this would provide more than 210 mg of vitamin C daily. However, it is unknown why fruits and vegetables are protective. The benefit may be due to complex interactions between vitamin C and other components of these foods, or independent of vitamin C altogether.

Studies to test benefit of vitamin C directly on cancer outcome have shown no effect.[69,72] However, these studies suffer from the deficiencies

---

[71] K. F. Gey, U. K. Moser, P. Jordan, H. B. Stahelin, M. Eichholzer, and E. Ludin, *Am. J. Clin. Nutr.* **57,** 787S (1993).

[72] E. R. Greenberg, J. A. Baron, T. D. Tosteson, D. H. Freeman, Jr., G. J. Beck, J. H. Bond, T. A. Colacchio, J. A. Coller, H. D. Frankl, and R. W. Haile, *N. Engl. J. Med.* **331,** 141 (1994).

[73] P. F. Jacques, L. T. Chylack, Jr., R. B. McGandy, and S. C. Hartz, *Arch. Ophthalmol.* **106,** 337 (1988).

[74] R. A. Riemersma, D. A. Wood, C. C. Macintyre, R. A. Elton, K. F. Gey, and M. F. Oliver, *Lancet* **337,** 1 (1991).

[75] E. B. Rimm, M. J. Stampfer, A. Ascherio, E. Giovannucci, G. A. Colditz, and W. C. Willett, *N. Engl. J. Med.* **328,** 1450 (1993).

[76] J. M. Seddon, U. A. Ajani, R. D. Sperduto, R. Hiller, N. Blair, T. C. Burton, M. D. Farber, E. S. Gragoudas, J. Haller, and D. T. Miller, *JAMA* **272,** 1413 (1994).

[77] J. M. Robertson, A. P. Donner, and J. R. Trevithick, *Am. J. Clin. Nutr.* **53,** 346S (1991).

[78] S. Vitale, S. West, J. Hallfrisch, C. Alston, F. Wang, C. Moorman, D. Muller, V. Singh, and H. R. Taylor, *Epidemiology* **4,** 195 (1993).

[79] D. M. Hegsted, *J. Am. Coll. Nutr.* **11,** 241 (1992).

described above, especially regarding whether vitamin C concentrations were truly different in control vs test subjects, and in the distance between the observation and the direct biochemical action of the vitamin. Epidemiologic studies have a role in determining vitamin recommendations, but should not be used alone.[80] Limitations of epidemiologic studies should be recognized, and the steep dose–concentration curve for ascorbate should be accounted for.

In Situ Kinetics and Recommended Ingestion of Vitamin C

The combined principles of *in situ* kinetics have been used to recommend vitamin C ingestion of 200 mg daily.[38] In diets of developed countries five daily servings of fruits and vegetables are easily available, providing more than 210 mg of vitamin C daily.[40] Vitamin C in fruits and vegetables is clearly preferred to vitamin C in supplements because of the association of fruit and vegetable ingestion with decreased cancer risk.[68] Vitamin C ingestion of less than 200 mg daily is on the steep portion of the ingestion curve, where small changes in ingestion produce large changes in plasma concentration.[38] For a nontoxic substance such as vitamin C, the steep portion of the ingestion curve should be avoided. Vitamin C doses of 500 mg daily and higher are incompletely absorbed, and the fraction of the dose that is absorbed is completely excreted. Such doses could be beneficial if there were advantages to unabsorbed vitamin C in the gastrointestinal tract and/or vitamin C in the urinary system. Because such advantages have not been clarified, we see no reason to ingest doses of 500 mg and higher.[65–67] Doses less than 200 mg are probably completely absorbed, although pharmacokinetics of these doses are complicated by nonlinearity of volume of distribution and clearance. Bioavailability was determined in the fasting state with vitamin C administered alone. It is possible that bioavailability will decline when vitamin C is present in foodstuffs. For example, glucose inhibits vitamin C transport in some cells types.[81,82] It is also possible that the ascorbate oxidation product dehydroascorbic acid is either in foods or is formed in the gastrointestinal tract, and glucose inhibits dehydroascorbic acid transport.[83,84] If food causes ascorbate bioavailability to decline, then 200 mg would be a minimal estimate for recommended ingestion.

Only limited data are available in humans for biochemical and molecular function of vitamin C related to dose. Some data, although indirect, indicate

[80] G. Block, *Am. J. Clin. Nutr.* **62**(Suppl.), 1517 (1995).
[81] P. Washko and M. Levine, *J. Biol. Chem.* **267,** 23568 (1992).
[82] R. W. Welch, P. Bergsten, J. D. Butler, and M. Levine, *Biochem. J.* **294,** 505 (1993).
[83] R. Bigley, M. Wirth, D. Layman, M. Riddle, and L. Stankova, *Diabetes* **32,** 545 (1983).
[84] J. T. Vanderslice and D. J. Higgs, *Am. J. Clin. Nutr.* **54,** 1323S (1991).

that vitamin C transport reaches V_{max} at approximately 70 μM in several human cell types.[82,85,86] This is the same concentration achieved by approximately 200 mg daily of vitamin C.[38] But is there an advantage in saturation of vitamin C transporters? Stated another way, what functional advantage does a dose of 200 mg have compared to a lower dose? LDL oxidation is minimized *in vitro* at concentrations of ascorbate ≥ 40 μM.[20] It is quite unclear, however, whether this *in vitro* system, which used micromolar concentrations of free cations, is relevant to humans. Nitrosamines may be carcinogenic, and ascorbate can inhibit nitrosamine formation.[66] Unfortunately, *in vivo* data concerning ascorbate concentrations and nitrosamine formation are also limited.

In the best case, recommendations for vitamin ingestion should include direct functional data. However, these data are limited for many vitamins, not just vitamin C. We believe that major challenges and responsibilities of nutritional scientists are to obtain data for vitamin function *in situ* in direct relation to vitamin concentration.

Although direct functional data are lacking, recommendations for vitamin C ingestion are still necessary.[87] To make recommendations for optimal vitamin C consumption, both biological and clinical components of *in situ* kinetics should be considered. Data from epidemiologic studies should not serve as the primary basis for recommendations, nor should data from pharmacokinetic or functional studies.[2,80] However, these data are synergistic when considered as a whole. Taken together the data provide a unique perspective that none can provide when taken separately. All aspects of *in situ* kinetics should be considered when developing recommendations for vitamin intake.

Using these principles, we have made recommendations for vitamin C consumption. The biochemical and clinical data support a recommended intake of 200 mg daily. Related recommendations may be divided categorically. For example, 60 mg daily could be considered the required dose to prevent subtle symptoms of deficiency, and an upper safe dose would be <1000 mg daily.[38] *In situ* kinetics is a useful approach for determining recommended ingestion of vitamin C, and may also be useful for determining recommended ingestion of other vitamins.

[85] P. Washko, D. Rotrosen, and M. Levine, *J. Biol. Chem.* **264,** 18996 (1989).

[86] P. Bergsten, R. Yu, J. Kehrl, and M. Levine, *Arch. Biochem. Biophys.* **317,** 208 (1995).

[87] Food and Nutrition Board, Institute of Medicine, "How Should the Recommended Dietary Allowances Be Revised?" pp. 1–24. National Academy Press, Washington, D.C., 1994.

Author Index

Numbers in parentheses are footnote reference numbers and indicate that an author's work is referred to although the name is not cited in the text.

A

B

C

D

E

F

G

H

I

J

K

M

R

S

X

Y

Z

Subject Index

B

C

D

N

P

ISBN 0-12-182182-X
90038
9 780121 821821